LOGIC COLLOQUIUM '69

STUDIES IN LOGIC

AND

THE FOUNDATIONS OF MATHEMATICS

VOLUME 61

NORTH-HOLLAND PUBLISHING COMPANY
AMSTERDAM • LONDON

LOGIC COLLOQUIUM '69

PROCEEDINGS OF THE SUMMER SCHOOL AND COLLOQUIUM IN MATHEMATICAL LOGIC, MANCHESTER, AUGUST 1969

Edited by

R. O. GANDY
C. M. E. YATES

1971

NORTH-HOLLAND PUBLISHING COMPANY
AMSTERDAM · LONDON

Library of Congress Catalog Card Number 71-146188

International Standard Book Number 0 7204 2261 2

PUBLISHERS:

NORTH-HOLLAND PUBLISHING COMPANY – AMSTERDAM

NORTH-HOLLAND PUBLISHING COMPANY, LTD. – LONDON

PRINTED IN THE NETHERLANDS

IN MEMORIAM
ALAN MATHISON TURING
1912-1954

PREFACE

A summer school and colloquium in mathematical logic, generally known as 'the 69 Logic Colloquium', was held at the University of Manchester from 3rd August to 23rd August, 1969. This volume constitutes its proceedings. Information about the scientific programme and its relation to what is printed in this book may be gleaned from the table of contents and the 'List of Participants'. The conference was recognised as a meeting of the Association for Symbolic Logic, and abstracts of most of the contributed papers will be included in the report of the meeting which will appear in the Journal of Symbolic Logic.

The organising committee for the 69 Logic Colloquium consisted of *R.O. Gandy* (Manchester), *H. Hermes* (Freiburg), *M.H. Löb* (Leeds), *D. Scott* (Princeton) and *C.E.M. Yates* (Manchester). The conference was supported financially by the Logic, Methodology and Philosophy of Science division of the International Union for Philosophy and History of Science, and by a generous grant from NATO. It was recognised as a NATO Advanced Study Institute *.

On behalf of the organising committee we wish to thank the above mentioned organisations, the Vice-Chancellor and University of Manchester, Mrs E.C. Connery (Administrative Assistant to the Department of Mathematics, Manchester), and all the people who by their help made the conference a success.

As will be seen from the table of contents the influence and inspiration of *Turing* may be traced in many of the contributions. Because of this, and because of his association with Manchester University (where he was a Reader from 1948 until his death in 1954), it seemed appropriate to us to dedicate this volume to his memory.

Robin Gandy
Michael Yates

* We record here that 36 of the participants signed a declaration dissociating themselves from NATO's aims and expressing their conviction that scientific conferences 'should not be linked with organisations of this character'.

CONTENTS

(Due to illness, notes of Carol Karp's course on Generalised recursion theory could not be included. The substance of Michael Rabin's course on Automata and problems of decidability, and Jack Silver's course on Independence results in model theory will appear elsewhere.)

(Other invited talks, by way of being surveys, were given by Ronald Jensen, Tony Martin and Robert Solovay.)

Part III: Invited papers on generalisations of recursion theory

(The papers by Kreisel and Hinman and Moschovakis were not presented at the conference. Friedman's paper was read by R.O. Gandy.)

Part IV: Invited papers on other topics

(The work described in the paper by Leeds and Putnam was presented in a talk by Putnam. Other invited talks, on topics related to the Theory of automata, were given by Richard Büchi, Marvin Minsky, Michael Patterson and Michael Rabin.)

Part V: Contributed papers

(The papers by Friedman, Specker and Yates were not presented at the conference.)

LIST OF PARTICIPANTS

The symbols in parentheses after any name record contributions to the programme of the colloquium according to the following code:–

L = Course Lecturer, I = Invited speaker, C = Contributor of a paper read at the conference, T = Contributor of a paper by title, J = Joint contributor of a paper,

a = An abstract of the paper will appear in the report of the meeting in the Journal of Symbolic logic,

b = The paper is printed in this volume.

Ash, C.J.
Ash, J.F.
Aczel, P. (Ca)
Aloisio, P.
Bacsich, P.D. (Ca)
Barendregt, H.P.
Barwise, J. (Ib, Ja)
Bezboruah, A.
Bird, M.R.
Boffa, M. (Ca)
Boos, W.
Booth, D.V. (C)
Bridge, J.E.
Büchi, J.R. (I)
Bukovsky, L. (C)
Carpenter, A.J.
Chudacek, J.
Clapham, C.R.J.
Cleave, J.P.
Cooper, S.B. (Ta, Ta)
Cramp, G.A.
Cusin, R.
Cutland, N.J. (Ca)
Daguenet, M.
Daub, W.
Davies, E.
Derrick, J.
Devlin, K.J.
Dickmann, M.A.
Drabbe, J.
Drake, F.R.
Dubinsky, A.
Dymacek, M.
Eklof, P.C. (Ja)
Elliot, J.C.
Engelfriet, J.
Feiner, L. (Ca)
Felgner, U. (Ca)
Fenstad, J.E. (Jab)
Fischer, G.
Fitting, M.
Flum, J.
Fredriksson, E. (Ca)
Friedrichsdorf, U.
Fris, M.A.
Gabbay, D. (Cab, Ta)
Gandy, R.O.
Gavalu, M.
Gielen, J.
Gielen, W.
Gilmore, P.C. (C)
Gjone, G.
Gloede, K.
Glorian, S.
Goodman, J.G.B.
Gordon, C.
Gornemann, S.
Gostanian, R.
Grant, P.W.
Greif-Muhlrad, C.
Grigorieff, S.
Grivet, C.
Hajek, P. (Ib)
Hamilton, A.G. (Ca)
Hart, J.
Hay, L.
Hayes, A.
Henrard, P.
Herman, G.T. (Ca)
Hermes, H. (Ib)
Hinman, P.G. (Jb)
Hirscheimann, A.
Hodges, W.
Hopkin, D.R.
Howson, C.
de Iongh, D.R.
Isaacson, D.R.
Jackson, A.D.
Jahr, E.
Jech, T.
Jensen, R.B. (I)
Johnson, D.M.
de Jongh, D.H.J. (Ca)
Jugasz, I.
Jung, J.
Karp, C. (L)
Keune, F.J.
Koppelberg, B.J. (C)
Krcalova, M.
Kripke, S. (I)
Kunen, K. (Ib)
Lablanquie, J.C.
Lacombe, D. (L)
Lassaigne, R.
Lolli, G.
Lucas, T.

Lucian, M.L.
Lynge, O.
McBeth, C.B.R.
Macchi, P.
Machover, M.
Machtey, M. (Ca)
Macintyre, A. (Ca, Ta)
Magidor, M. (C)
Marek, W. (Ja)
Martin, D.A. (I)
Meyer, E.R. (Ja)
Minsky, M. (I)
Monro, G.P.
van der Moore, W.A.
Morley, M.
Moschovakis, Y.N. (Ib, Jb)
Moss, B.J.
Moss, M. (Cab)
Myhill, J.
Nyberg, A.M. (Jab)
Ore, T. Jr.
Pacholski, L. (Ja)
Paris, J.B. (Ta)
Paterson, M.S. (Ia)
Pelletier, D.H.
Perlis, D. (C)
Philp, B.J.
Pilling, D.L.
Pincus, D. (Ca)
Platek, R. (Ib)
Potthoff, K. (Ca)
Pour - El, M.
Prestel, A.
Putnam, H. (Ib)
Rabin, M. (L, I)
Renc, Z.
Richter, M.
Richter, W. (Ib)
Rodino, G.
de Roever, W.P.
Rogers, H.
Rogers, P.
Rose, H.E.
Rousseau, G.
Rowse, W.G.
Sabbagh, G.
Sacks, G. (Ib)
Sakarovitch, J.
Scholten, F.P.
Scott, A.B.
Shepherdson, J.C.
Shoenfield, J.R. (Lb)
Siefkes, D. (Ca)
Silver, J. (L)
Simmons, H.
Slomson, A.
Smith, J.P.
Sochor, A.
Solovay, R.M. (I)
Staples, J.
Stepanek, P. (Ja)
Stomps, H.J.
Strauss, A.
Suzuki, Y.
De Swart, J.
Tall, F.D.
Telgarsky, R.
Thomas, M.
Treherne, A.A.
Truss, J.
Ungar, A.M.
Villemin, F.Y.
Volger, H.
Wainer, S.S.
Wasilewska, A.
Waszkiewicz, J. (C)
Weglorz, B. (I)
Weiner, M.
van Westrhenen, S.C.
Whitehouse, R.M.
Wilmers, G.M.
Worboys, M.F.
Worrall, J.
Yates, C.E.M.

PART I

SUMMER SCHOOL LECTURES

RECURSION THEORETIC STRUCTURE FOR RELATIONAL SYSTEMS

Daniel LACOMBE [1)]

Faculté des Sciences de Paris, Département de Mathématiques, Paris

Given an algebraic structure E there are many ways in which we can introduce into the study of E notions of effectiveness. We look at some natural methods of doing this. Section 1 is a short preliminary discussion on types of algebraic object. The remainder falls into two main parts. In Section 2 we define various notions of recursiveness and relative recursiveness over an algebraic structure, and the more interesting notions are singled out and various equivalences given. Maps $\alpha : N \to E$ will play an important part. In Section 2 they give a means of importing recursion theory into the structure of E. In Section 3 such maps – or rather the equivalence classes of such maps with respect to recursive interreducibility – become interesting objects in themselves. We develop a complete structure for the set fo these equivalence classes and trace relationships between this structure and the nature of E itself.

1. N will denote the set of non-negative integers and E will be any set. By means of relations $\{R_i\}$ and functions $\{\varphi_j\}$ we may give E an algebraic structure. We look at some examples. It will be seen that little can be expressed using only monadic predicates.

(i) Let $A \subseteq \mathcal{P}E$ (the power set of E) and also:
  (a) A is closed under the Boolean operations $\cap$, C,
  (b) $(\forall x \in E)(\{x\} \in A)$,
  and (c) E is infinite (denumerable or not).

Suppose A' satisfies the same conditions for E. Then the elementary theories of inclusion on A, A' are elementarily equivalent.

1) These notes on Lacombe's lectures were taken and reshapen by Barry Cooper.

(ii) Suppose $A \subseteq \mathcal{P}E$, $\overline{\overline{E}} = \aleph_0$, $\overline{\overline{A}} = \aleph_0$.

If we put on A the condition

$$(\forall X \in A)(\forall Y)(X \equiv Y \text{(modulo the finite subsets of } E) \to Y \in A)\,,$$

then there exist continuously many maps $\tau : E \xrightarrow[\text{onto}]{1-1} E$ for which $\tau A = A$. This holds because we only consider *monadic* relations. We obtain a counter-example if we take $E = N$ and let A = the set of all two-place recursive relations. In this case the set of maps τ is just the set of recursive permutations and there are only $\aleph_0$ of these.

(iii) Take $A \subseteq \mathcal{P}E$ where E, A satisfy the following properties:
(a) $\overline{\overline{E}} = \overline{\overline{A}} = \aleph_0$,
(b) A is closed under the Boolean operations $\cap$, C (and so is a ring),
(c) $(\forall x \in E)(\{x\} \in A)$,
(d) $(\forall X \in A)(X \text{ infinite} \to (\exists Y, Z \in A)(Y, Z \text{ infinite} \wedge Y \cup Z = X \wedge Y \cap Z = \emptyset))$.

Let E', A' satisfy the same conditions. Then we can prove that they are isomorphic:

$$(\exists \tau : E \xrightarrow[\text{onto}]{1-1} E')(\tau A = A')\,.$$

We use the fact that members of A are only monadic predicates and that A is a ring. If we only impose closure under the lattice operations $\cap$, $\cup$ then isomorphism may fail.

There are interesting problems involving denumerable lattices of subsets of N. Consider the following lattices: (a) the set of recursively enumerable subsets of N, (b) the Π_1^1 subsets of N, (c) the Π_1^1 subsets of all hyperarithmetic functions. For a long time it was a problem whether these are isomorphic. Owings [5] shows that not only are they not isomorphic, they are not even elementarily equivalent.

2. E will now be a denumerable set except where otherwise stated.

Definition 2.1. An *enumeration* of E is a map $\alpha : N \to E$ which is onto. A *numbering* of E is a map $\alpha : N \to E$ which is onto and one-one.

Given a structure $S = \{R_i\} \cup \{\varphi_j\}$ on E, a numbering $\alpha : N \to E$ will enable us to compare this structure with various well-known ones on N. In particular we consider what it means for S to be 'recursive' with respect to α. If P is some property on relations, functions on N, we can ask: does there exist a numbering α such that $\alpha^{-1}S$ has the property P? – or do we have for all numberings α that $\alpha^{-1}S$ has this property?

Definition 2.2. (i) S is *∃-recursive* (*of recursive type, decidable*) iff $(\exists\alpha : N \xrightarrow[\text{onto}]{1-1} E)(\alpha^{-1}S$ is recursive).

(ii) S is *∀-recursive* iff $(\forall\alpha : N \xrightarrow[\text{onto}]{1-1} E)(\alpha^{-1}S$ is recursive).

Property (i) looks like a Σ_1^1 relation. If we limit S then it may become arithmetical. Say we take $S = \{\varphi\}$ where φ is a permutation of E. Then (i) is equivalent to the property: $\{(m, n) \mid \exists$ at least m cycles of φ with n elements$\}$ is r.e. Another such case is that in which S = the equivalence relations on E.

We may ask: is property (i) properly Σ_1^1 and nothing less? We could consider a standard numbering of two-place r.e. predicates on E (= N), and take all indices which correspond to r.e. sets of recursive type. A strong result would be that this set is properly Σ_1^1.

Example. Let $\theta_{\text{rec}}(N)$ = the recursive permutations of N (or we could have taken the primitive recursive inverses).

We have the usual group operation $\times$ on members of $\theta_{\text{rec}}(N)$. Then we can choose $\alpha, \beta \in \theta_{\text{rec}}$ where the sub-group generated by α, β is not only not ∃-recursive but not even recursively presentable. All automorphisms of θ_{rec} are inner. Another question is whether all permutations of recursive type are of primitive recursive type. We can show that they are not but the counter-example is not easy to construct since, for instance, such a φ must have no infinite cycles and not infinitely many cycles of any given number of elements. On the other hand, for a well-ordering on a denumerable E to be of recursive type is equivalent to it being of primitive recursive type, or of arithmetic type, or even of Σ_1^1 type: it is to be a recursive ordinal.

Fraissé [2] makes two interesting conjectures about necessary and sufficient conditions for ∃-recursiveness. While these have since turned out not to be true, the most interesting part of his paper is devoted to a notion of relative recursiveness which will be shown to be equivalent to our ∀-recursiveness in (Definition 2.3).

Since we may replace functions by graphs we will only consider relations $\mathbf{P} = (P_1, ..., P_m)$, $\mathbf{Q} = (Q_1, ..., Q_n)$ on E.

Definition 2.3. (i) **P** is *∃-recursive in* **Q** iff $(\exists\alpha : N \xrightarrow[\text{onto}]{1-1} E)(\alpha^{-1}\mathbf{P}$ is recursive in $\alpha^{-1}\mathbf{Q})$.

(ii) **P** is *∀-recursive in* **Q** iff $(\forall\alpha : N \xrightarrow[\text{onto}]{1-1} E)(\alpha^{-1}\mathbf{P}$ is recursive in $\alpha^{-1}\mathbf{Q})$.

We cannot immediately define a notion of ∃!-recursive (in) since if α is a numbering for which $\alpha^{-1}\mathbf{P}$ is recursive (in $\alpha^{-1}\mathbf{Q}$) then recursive permutations of N will yield new such α's. We must divide the numberings into equivalence classes $\bar{\alpha}$ under recursive permutation and write $(\exists !\, \bar{\alpha})$.

Consider functions $\varphi_1, ..., \varphi_m$ on E. Suppose $\exists\, a_1, ..., a_n \in E$ such that every element of E is generated by $\varphi_1, ..., \varphi_m$ from these elements (i.e., E is the Skolem hull of $a_1, ..., a_n$ under $\varphi_1, ..., \varphi_m$) and suppose the structure is ∃-recursive. Then it is ∃!-recursive. And a necessary and sufficient condition for the structure to be ∃-recursive is that the set of true equations between Skolem terms be recursive.

Examples. (a) Taking E to be Z^+ (the positive integers) we can arrange a 1–1 map of N onto E such that $\leqslant$ is recursive but successor is not.

(b) $\langle Q$ (the rationals), $+\rangle$ is ∃!-recursive (and the only such $\bar{\alpha}$ also makes $\times$ recursive).

(c) $\langle R_A$ (the real algebraic numbers), $+, \times\rangle$ is ∃!-recursive.

(d) $\langle C_A$ (complex algebraic numbers), $+, \times\rangle$ is not ∃!-recursive, since there exist continuously many numberings which make $+, \times$ recursive. But we may obtain uniqueness by adding conjugation to the structure.

We write

$$F^{(p)}E = E^{E^p} \text{ (i.e., the set of functions } E^p \to E) ,$$

$$FE = \bigcup_{p\in N} F^{(p)}E ,$$

$$R^{(p)}E = \mathcal{P}E^p \text{ (i.e., the set of } p\text{-place relations on } E),$$

$$RE = \bigcup_{p\in N} R^{(p)}E .$$

We will see later that ∀-recursiveness is equivalent to the simple property of being reckonable (Definition 2.4). Corresponding to ∀-recursive *in* there will be two equivalent notions: (1) reckonable in (Definition 2.11), and (2) the notions due to Fraissé of F_0-recursiveness and F-recursiveness (Definition 2.12).

Definition 2.4. Let $P \in R^{(p)}E$ where E may be denumerable or not. Then:

(i) P is *radically reckonable* if $P(x_1, ..., x_p) \equiv \mathcal{B}(x_i = x_j)$ on E, where

$\mathcal{B}(x_i = x_j)$ is a formula built up by Boolean operations from formulae of the kind $x_i = x_j$. (Remark: $\mathcal{B}(x_i = x_j)$ could be any first order formula in =, since we know such a formula is always equivalent to a propositional one.)

(ii) P is *reckonable* if $\exists a_1, \ldots, a_n \in E$ such that $P(x_1, \ldots, x_p) \equiv \mathcal{B}(x_i = x_j, x_i = a_k)$ on E (i.e., we now allow a finite number of constants of E in the definition).

Theorem 2.5. If E is denumerable and $P \in RE$, then

$$P \text{ is reckonable} \leftrightarrow P \text{ is } \forall\text{-recursive.}$$

The proof uses only two properties of the recursive relations: (a) they contain the reckonable relations, and (b) they are denumerable. So we could write '$\forall$-elementary', '$\forall$-primitive recursive', etc. instead of '$\forall$-recursive'. We now define $\Theta(P) \subseteq R^{(p)}F$ by

$$\Theta(P) = \{\tau(P) \mid \tau : E \xrightarrow{1-1} F\}.$$

Then P is radically reckonable iff $\overline{\overline{\Theta(P)}} = 1$ (since P is defined by just one formula so is $\tau(P)$). When E is denumerable, P is reckonable but not radically reckonable iff $\overline{\overline{\Theta(P)}} = \aleph_0$. And P is not reckonable iff $\overline{\overline{\Theta(P)}} = 2^{\aleph_0}$. This is proved using Baire-like methods. An appropriate τ can be built by steps, each step involving a choice which gives not only different τ but different $\tau(P)$. $\overline{\overline{\Theta(P)}} = 2^{\aleph_0}$ is one gauge of the richness of the non-reckonable relations P.

The next result removes any interest from relative $\exists$-recursiveness.

Theorem 2.6. (1) Let $R_1, \ldots, R_m \in RE$ be reckonable and $P \in RE$ where E is denumerable. Then P is $\exists$-recursive iff P is $\exists$-recursive in $(R_1, \ldots, R_m)$.

(2) If now there is an R_i which is not reckonable, then every $P \in RE$ is $\exists$-recursive in $(R_1, \ldots, R_m)$.

(2) will follow from Theorem 2.8. First we make a definition.

Definition 2.7. (i) $P \in R^{(p)}E$ is *reducible to* $Q \in R^{(q)}E$ by functions $\varphi_1, \ldots, \varphi_q \in F^{(p)}E$ iff

$$P(x_1, \ldots, x_p) \equiv Q(\varphi_1(x_1, \ldots, x_p), \ldots, \varphi_q(x_1, \ldots, x_p)),$$

(ii) Let $M \subseteq FN$. Then P is *M-reducible to* Q iff

$\exists \varphi_1, ..., \varphi_q \in M$ such that P is reducible to Q by $\varphi_1, ..., \varphi_q$.

Theorem 2.8. If $Q \in R^{(q)}E$ is not reckonable, then for every $P \in R^{(p)}E$ there is a numbering α of E for which $\alpha^{-1}P$ is M-reducible to $\alpha^{-1}Q$ for every class $M \subseteq FN$ satisfying the conditions:

(i) M contains all constant functions,
(ii) $\forall p, q, \exists \varphi_1, ..., \varphi_q \in M$ for which
(a) for each i, φ_i is one-one from N^p to N,
(b) $\varphi_i(x_1, ..., x_p) \geqslant \max(x_1, ..., x_p)$,
(c) $i \neq j \rightarrow \varphi_i(N^p) \cap \varphi_j(N^p) = \emptyset$,
(d) the complement of $(\varphi_1(N^p) \cup ... \cup \varphi_q(N^p)$ is infinite.

We note that the conditions on M are very weak, and that all the following classes satisfy them: the recursive functions, the primitive recursive functions, the elementary functions, and the polynomials with non-negative integral coefficients. Since very simple M satisfy (i) and (ii), $\exists -R$ (where R replaces 'recursive' in Definition 2.3) is not an interesting notion for most natural reducibilities R. This leaves the more important $\forall -R$ predicate.

We now consider the notions:

(i) P is $\forall$-recursive in $R_1, ..., R_m$ (cf. Definition 2.3),
(ii) P is *recursively reckonable* from $R_1, ..., R_m$ (cf. Definition 2.11),
(iii) P is F_0 (or F)-*recursive* in $R_1, ..., R_m$ (cf. Definition 2.12).

(i) only has meaning for denumerable E and in this case all three notions will turn out to be equivalent (where finitely many constants may be added to the list $R_1, ..., R_m$ in the third case).

For the most part we only need denumerability and continuity of the recursive functionals, but to prove (ii) $\equiv$ (iii) we will use a theorem specifically about recursiveness due to Kleene [3] and Craig and Vaught [1].

We consider what it means for P to be 'computable' from $R_1, ..., R_m$ ($\in R^{(r_1)}E, ..., R^{(r_m)}E$, respectively). $(U, X_1, ..., X_m, x_1, ..., x_p)$ will be said to be an *instruction* for a *computation* if U is a finite subset of E and $X_1, ..., X_m \in R^{(r_1)}U, ..., R^{(r_m)}U$ respectively, and $x_1, ..., x_p \in E$ (but not necessarily $\in U$).

T will be the set of all instructions (assuming the numbers $r_1, ..., r_m, p$ to be given).

Let $\mathcal{C}^+, \mathcal{C}^- \subseteq T$ and define $\mathcal{C} = (\mathcal{C}^+, \mathcal{C}^-)$. We interpret an instruction $(U, X_1, ..., X_m, x_1, ..., x_p) \in \mathcal{C}^+$ (or $\mathcal{C}^-$) as saying that if $X_i = R_i \upharpoonright U$ each

$i = 1, ..., m$, then $P(x_1, ..., x_p)$ (or $\neg P(x_1, ..., x_p)$ respectively). As the definition stands, $\mathcal{C}$ may give rise to not only well-defined computations P at points $(x_1, ..., x_p)$, but also contradictory computations or perhaps no computations at all. There is an 'effective' procedure by which we can eliminate contradictions, but we cannot simultaneously effectively eliminate indeterminacy of computation. $\mathcal{C}$ determines a partial functional $\Phi_\mathcal{C}$ which is a mapping into $R^{(p)}E$ of the subset of $R^{(r_1)}E \times ... \times R^{(r_m)}E$ made of those $(R_1, ..., R_m)$ for which there is neither contradiction nor indeterminacy. Thus $\Phi_\mathcal{C}$ is continuous with respect to the product topologies on the sets $R^{(k)}E$. In the particular case $E = N$, if we were to impose the extra condition that $\mathcal{C}^+, \mathcal{C}^-$ are r.e. (where T is provided with an obvious canonical numbering) then $\Phi_\mathcal{C}$ becomes a partial recursive functional and our notion of a computation just gives relative recursiveness. Conversely any p.r. functional can be obtained from two appropriate r.e. sets of instructions $\mathcal{C}^+, \mathcal{C}^-$. Also, it is well-known that every partial recursive functional $\Phi_\mathcal{C}$ (where $\mathcal{C} = (\mathcal{C}^+, \mathcal{C}^-)$) is equal to a functional $\Phi_{\mathcal{C}'}$ where $\mathcal{C}' = (\mathcal{C}'^+, \mathcal{C}'^-)$ and $\mathcal{C}'^+, \mathcal{C}'^-$ are recursive. To obtain this $\mathcal{C}'$ we replace each instruction $(U, X_1, ..., X_m, x_1, ..., x_p)$ in $\mathcal{C}$ by finitely many instructions $(U', X_1', ..., X_m', x_1, ..., x_p)$, taking as U' a 'large enough' superset of U and as the $(X_1', ..., X_m')$ all possible extensions of $(X_1, ..., X_m)$ to U'.

Assume now that the only structure on E is given by $R_1, ..., R_m$. We will require $\mathcal{C}$ to be 'symmetric' and could ask $\mathcal{C}$ to satisfy:

Suppose there exists an isomorphism (in an obvious sense) of $(U', X_1', ..., X_m', x_1', ..., x_p')$ onto $(U, X_1, ..., X_m, x_1, ..., x_p)$. Then if $(U, ..., x_p) \in \mathcal{C}^+$ (or $\mathcal{C}^+$) then $(U', ..., x_p') \in \mathcal{C}^+$ (or $\mathcal{C}^-$ respectively). However, this simple symmetry condition turns out to be too strong and we only need symmetry with respect to a finite set of constants $a_1, ..., a_n \in E$.

Definition 2.9. The computation $(\mathcal{C}^+, \mathcal{C}^-)$ is said to be $(a_1, ..., a_n)$*-symmetric* if the following condition is satisfied:

If $I, I' \in T$ and there is an isomorphism of I onto I' which leaves each a_i invariant, then $I \in \mathcal{C}^+$ (or $\mathcal{C}^-$) implies $I' \in \mathcal{C}^+$ ($\mathcal{C}^-$ respectively). The $(a_1, ..., a_n)$*-symmetrization* of $\mathcal{C}$ is the smallest $(a_1, ..., a_n)$-symmetric $\mathcal{C}'$ such that $\mathcal{C} \subseteq \mathcal{C}'$. Of cource, this symmetrization may introduce new contradictions. We notice immediately that if $\Phi_\mathcal{C}$ (as a functional) is invariant under each permutation of E which leaves each a_i invariant then $\Phi_\mathcal{C} = \Phi_{\mathcal{C}'}$ where $\mathcal{C}'$ is the $(a_1, ..., a_n)$-symmetrization of $\mathcal{C}$.

We will need the following uniformity result:

Lemma 2.10. Assume E, F are denumerable and $R_1, ..., R_m, P \in RE$. Let us state on F a denumerable set M of (not necessarily symmetric) computations. Then the condition

$$(\forall \tau : E \xrightarrow[\text{onto}]{1-1} F)(\exists \mathcal{C} \in M)(\tau P = \Phi_{\mathcal{C}} (\tau R_1, ..., \tau R_m))$$

implies the condition

$$(\exists n)(\exists e_1, ..., e_n \in E)(\exists f_1, ..., f_n \in F)(\exists \mathcal{C} \in M)(\forall \tau : E \xrightarrow[\text{onto}]{1-1} F)$$

$$(\tau e_1 = f_1 \wedge ... \wedge \tau e_n = f_n \rightarrow \tau P = \Phi_{\mathcal{C}} (\tau R_1, ..., \tau R_m)) .$$

To prove this we assume that the latter is false and build up a τ step by step (thus using the denumerability of M) which contradicts the former statement.

If $\mathcal{C}$ is $(a_1, ..., a_n)$-symmetric we can abstract from all transforms of a given instruction $(U, X_1, ..., X_m, x_1, ..., x_p) \in \mathcal{C}$ to a *configuration* $(V, Y_1, ..., Y_m, y_1, ..., y_p, b_1, ..., b_n)$ independent of the set E. We can define the set of all such configurations (say J) knowing the numbers $r_1, ..., r_m, p, n$. J is denumerable. Given a pair $\mathcal{D} = (\mathcal{D}^+, \mathcal{D}^-)$ where $\mathcal{D}^+, \mathcal{D}^- \subseteq J$, we may define a computation $\mathcal{C}$ on E which is $(a_1, ..., a_n)$-symmetric by the set of all mappings of members of $\mathcal{D}^+, \mathcal{D}^-$ into a structure on E where each b_i is mapped onto a_i. If for this $\mathcal{C}$ we have $P = \Phi_{\mathcal{C}} (R_1, ..., R_m)$ then P is said to be $(a_1, ..., a_n)$-reckonable from $R_1, ..., R_m$ by means of $\mathcal{D}$. More formally:

Definition 2.11. (i) P is *$(a_1, ..., a_n)$-reckonable from $R_1, ..., R_m$* by means of $\mathcal{D}$ iff

$$P(x_1, ..., x_p) \leftrightarrow (\exists (V, Y_1, ..., Y_m, y_1, ..., y_p, b_1, ..., b_n) \in \mathcal{D}^+)$$

$$(\exists U(\text{finite}) \subset E) \, [(U, R_1 \upharpoonright U, ..., R_m \upharpoonright U, x_1, ..., x_p, a_1, ..., a_n)$$

$$\text{is isomorphic to } (V, ..., b_n)]$$

and

$$\neg P(x_1, ..., x_p) \leftrightarrow \text{the same formula with } \mathcal{D}^+ \text{ replaced by } \mathcal{D}^- .$$

(ii) If we can choose $n = 0$ in definition (i) then P is said to be *radically reckonable from $R_1, ..., R_m$ by means of $\mathcal{D}$.*

(iii) Assume that we are given a canonical numbering of J. Then P is *recur-*

sively $(a_1, \ldots, a_n)$*-reckonable from* $R_1, \ldots, R_m$ if it is $(a_1, \ldots, a_n)$-reckonable from $R_1, \ldots, R_m$ by means of some $\mathcal{D} = (\mathcal{D}^+, \mathcal{D}^-)$, where $\mathcal{D}^+, \mathcal{D}^-$ are r.e.

(iv) P is *recursively radically reckonable from* $R_1, \ldots, R_m$ if it is radically reckonable from $R_1, \ldots, R_m$ by means of some r.e. $\mathcal{D}$.

(v) P is *recursively reckonable from* $R_1, \ldots, R_m$ if it is recursively $(a_1, \ldots, a_n)$-reckonable from $R_1, \ldots, R_m$ for some $a_1, \ldots, a_n \in E$.

There is a close link between these definitions and Definitions 2.4. In Definition 2.11 (i) we can write $\mathcal{D}^{\pm} = \{I_n^{\pm}\}_{n \in N}$ and replace the conditions given there by the denumerable set $\{(C_n^{\pm})\}_{n \in N}$ where $(C_n^{\pm})$ is a condition corresponding to the configuration $I_n^{\pm}$. The condition $(C_n^{\pm})$ could be written in the form:

$$(\exists u_1, \ldots, u_k)\mathcal{B}_n^+(R_1, \ldots, R_m, u_1, \ldots, u_k, x_1, \ldots, x_p, a_1, \ldots, a_n)$$

$$\leftrightarrow P(x_1, \ldots, x_p)\,,$$

where k depends on n and is the cardinal of the finite set V in I_n^+, and where $\mathcal{B}^+$ is a Boolean combination of atomic formulas in $R_1, \ldots, R_m$ and $=$, with $u_1, \ldots, u_k, x_1, \ldots, x_p, a_1, \ldots, a_n$ as individuals. There is a similar condition for (C_n^-). Recursive reckonability and recursive radical reckonability are transitive relations.

We can show, as we did above for the p.r. functionals, that the condition that $\mathcal{D}$ be r.e. in the above definitions could be replaced by the condition of being recursive or primitive recursive with the same results.

We also note that reckonability by means of finite $\mathcal{D}$ is not a trivial notion.

We now move on to Fraissé's notions of F, F_0-recursiveness. We need to introduce a formal language. Our predicate symbols will be $\rho_1, \ldots, \rho_m, \pi$, $\sigma_1, \ldots, \sigma_n, =$. The ρ's will be interpreted as R's, π as P, and the σ's will be additional relational symbols.

Let Φ be a closed first-order formula in the language. $\langle E', R_1', \ldots, R_m', P', S_1', \ldots, S_n' \rangle$ will denote a realisation of this language. We write $C_\Phi^{E, R_1, \ldots, R_m}$ for the class of all models of Φ which are extensions (enriched by new relations) of the system $\langle E, R_1, \ldots, R_m \rangle$. ${}^0C_\Phi^{E, R_1, \ldots, R_m}$ is those realisations of $C_\Phi^{E, R_1, \ldots, R_m}$ for which $E = E'$ (we may omit the superscripts for convenience).

Definition 2.12. Assume that there exists a Φ in our language satisfying:

(1) $(\forall (E', R_1', \ldots, R_m', P', S', \ldots, S_n') \in C_\Phi)(P' \upharpoonright E = P)$,

(2) $C_\Phi \neq \emptyset$.

Then we say that P is *F-recursive in* $R_1, ..., R_m$. We notice that (2) excludes the case, for example, in which Φ is inconsistent.

If there is a Φ satisfying (1) and the (original Fraissé) condition:

$$(2_0) \qquad {}^0C_\Phi \neq \emptyset,$$

then we say that P is *F_0-recursive in* $R_1, ..., R_m$.

For each $x \in E$ let us introduce a formal constant $\vec{x}$. The *diagram* of R_i is defined by

$$\Delta(R_i) = \{\rho_i(\vec{x}_1, ..., \vec{x}_{r_i}) \mid R_i(x_1, ..., x_{r_i})\} \cup$$
$$\{\sim\rho_i(\vec{x}_1, ..., \vec{x}_{r_i}) \mid \neg R_i(x_1, ..., x_{r_i})\},$$

and $\Delta(R_1, ..., R_m) = \Delta(R_1) \cup ... \cup \Delta(R_m)$. Then (2) implies that $\Delta(R_1, ..., R_m) \cup \{\Phi\}$ is consistent. So if (2) or (2_0) is satisfied, (1) means that

$$P(x_1, ..., x_p) \leftrightarrow \Delta(R_1, ..., R_m) \cup \{\Phi\} \vdash \pi(\vec{x}_1, ..., \vec{x}_p)$$

and

$$\neg P(x_1, ..., x_p) \leftrightarrow \Delta(R_1, ..., R_m) \cup \{\Phi\} \vdash \sim\pi(\vec{x}_1, ..., \vec{x}_p) .$$

We obtain the notions of F_0, *F-recursiveness* when we take $m = 0$.

Theorem 2.13. The following statements are equivalent:
(a) P is F-recursive,
(b) P is F_0-recursive,
(c) P is recursively radically reckonable.

(b) → (a) follows immediately from the definitions. (a) → (c) is also easy. We use the fact that we only make use of finitely many formulae as premises for a logical inference, and so the logical inferences are r.e.

(c) → (b) makes use of the strong semantical form of the Kleene-Craig-Vaught theorem [3, 1]: For each recursively axiomatizable theory T there exists a finitely axiomatizable extension T' of T with additional predicates such that every infinite model of T can be enriched, without enlarging the domain of the model, to form a model of T'. Note: the Φ's used in (a) and (b) will not necessarily be the same. It may be that the Φ in (a) cannot be realized for any possible enrichment of E alone.

Corollary 2.14. The relativized notions are also equivalent. (In F_0 and F-recursiveness the constants $a_1, ..., a_n$ are added to $R_1, ..., R_m$ as monadic singletons $\{a_1\}, ..., \{a_n\}$ or as an n-adic singleton $\{(a_1, ..., a_n)\}$).

Examples. (1) Take $E = N$ with the one-place relation $\{O\}$ and the successor relation $\{(x, y) \mid y = x + 1\}$. Then to be recursively radically reckonable in this system is just to be recursive – but, for example, + is not first-order explicitly definable on $\langle N, O, \text{successor}\rangle$. On the other hand, to be recursively reckonable from $(+, \times)$ on N is also the same as being recursive, but we can get all arithmetic predicates from explicit definition over $\langle N, +, \times\rangle$. As for implicit definability, we know for instance that some hyper-arithmetic non-arithmetic predicates are implicitly first-order definable (with no extra predicates) over $\langle N, O, \text{successor}\rangle$. This illustrates the difference between F-recursiveness and other notions of definability. Actually, F-recursiveness is an instance of Kreisel's general notion of invariant definability [4].

(2) Take the system $\langle\text{reals}, +, \times\rangle$. Then in the particular case in which $P \subseteq R$ (i.e., P is a monadic predicate on the reals) we get that P is recursively radically reckonable in $\langle R, +, \times\rangle$ if P and its complement are each the union of a recursive sequence of algebraic (closed or open) intervals. The situation for non-monadic P is too complicated to state briefly.

(3) Let $E = N^N$ and let the set of operations on E be given by $S = \{\text{successor}$ (i.e., a monadic predicate on N^N), K, L (the inverses of the recursive pairing function), $\circ$ (where $(\xi \circ \eta)(x) = \eta(\xi(x))$)$\}$.

Then the recursively radically reckonable predicates for this system are the hyper-arithmetic predicates (on N^N). We obtain a similar result if we replace N^N by the set H of all hyper-arithmetic elements of N^N (we use Spector's result [7] on quantification over H): on H, the predicates recursively reckonable in the product $\circ$ are the predicates hyper-arithmetic (on H). But since H is denumerable the hyper-arithmetic predicates on H are also the predicates $\forall$-recursive in $\circ$.

3. Numberings $\alpha : N \xrightarrow{1-1} E$ have been considered so far. The objects we are really interested in are the equivalence classes of 'recursively indistinguishable' numberings.

Definition 3.1. A *numerotage* is an equivalence class of numberings under the equivalence relation: $\alpha \equiv \alpha'$ iff $\alpha^{-1}\alpha'$ is recursive.

We can extend our notions to the case where E is finite by considering any function defined on a finite set to be recursive. Thus, on a finite E there will be just one numerotage. More generally, we have:

Definition 3.2. Assume α, α' are enumerations $N \to E$ of a denumerable set E. We say:

(1) α is (recursively) *reducible to* α' iff there is a recursive function θ such that $\alpha = \alpha'\theta$,

(2) α, α' are *inter-reducible* if α is reducible to α' and α' is reducible to α.

An equivalence class determined by inter-reducibility of enumerations is an *enumerage*,

(3) α is *isogenic to* α' iff there is a recursive numbering (i.e. a permutation) of N, θ say, such that $\alpha = \alpha'\theta$.

There are cases in which notions (1) and (2) are equivalent, but this is not true in general.

Definition 3.3. Consider the following diagram:

$$\begin{array}{ccccc} N \times \cdots\cdots \times & N & \xrightarrow[\varphi]{} & N \\ \alpha_1 \downarrow \qquad & \alpha_p \downarrow & & \beta \downarrow \\ E_1 \times \cdots\cdots \times & E_p & \xrightarrow[\Phi]{} & F \end{array}$$

(1) We say φ *represents* Φ with respect to $\alpha_1, \ldots, \alpha_p, \beta$ iff

$$(\forall x_1, \ldots, x_p \in N)\,(\beta\varphi(x_1, \ldots, x_p) = \Phi(\alpha_1, x_1, \ldots, \alpha_p x_p))$$

(i.e., if the diagram commutes).

We say Φ is (recursively) *representable* with respect to $\alpha_1, \ldots, \alpha_p, \beta$. (The φ need not be unique since β need not be one-one.)

(2) φ is *intrinsic* iff

$$(\forall x_1, \ldots, x_p, x'_1, \ldots, x'_p \in N)\,(\alpha_1 x_1 = \alpha_1 x'_1 \wedge \ldots \wedge \alpha_p x_p =$$

$$\alpha_p x'_p \to \beta\varphi(x_1, \ldots, x_p) = \beta\varphi(x'_1, \ldots, x'_p))\,.$$

So φ is intrinsic iff φ represents some Φ.

We see immediately that the notion of recursively representable with respect to $\alpha_1, \ldots, \alpha_p, \beta$ is invariant under inter-reducibility of these enumerations – and so is a property of enumerages rather than enumerations.

When $E_1 = \cdots = E_p = F = E$ and $\alpha_1 = \cdots = \alpha_p = \beta = \alpha$, we can say that Φ is representable with respect to α (or with respect to the enumerage $\overline{\alpha}$).

We write $\mathrm{Rep}(\alpha)$ (or $\mathrm{Rep}(\overline{\alpha})$) for all such Φ representable with respect to α (or $\overline{\alpha}$).

Theorem 3.4. A set of functions $M = \text{Rep}(\overline{\alpha})$ for some $\overline{\alpha}$ iff M satisfies:
(1) M is countable,
(2) all constant functions and projection functions are in M, and
(3) M is closed under substitution.

It follows immediately from $M = \text{Rep}(\overline{\alpha})$ that M satisfies (1)–(3). And given an M satisfying (1)–(3), it is easy to show that $M \subseteq \text{Rep}(\overline{\alpha})$. The last part is more difficult. Given the denumerable M we build up $\overline{\alpha}$ (or rather the enumeration α) in such a way that all members of M are representable with respect to α but only members of M are representable with respect to α. The construction is Baire-like. At the nth step we work with the nth recursive function (of any number of places). It will follows from the way in which α is constructed that the nth recursive function either is not intrinsic or recursively represents something obtained by superposition from M. (Contrast the 1–1 case where *not* all structures on E were $\exists$-recursive.)

This theorem suggests that we might refine the problem by imposing new conditions on $\overline{\alpha}$.

We will say that a relation $R \subseteq E^p$ is recursive with respect to α iff $\alpha^{-1}(R) = \{(x_1, ..., x_p) \mid R(\alpha x_1, ..., \alpha x_p)\}$ is recursive. Similarly R is r.e. iff $\alpha^{-1}(R)$ is r.e.

Even the simplest relations may not be recursive with respect to a given α. In particular the equality relation is not in general recursive with respect to α.

Definition 3.5. (1) $\overline{\alpha}$ is *pure* iff = is recursive with respect to α.
(2) $\overline{\alpha}$ is *semi-pure* iff = is r.e. with respect to α.

The definitions make sense since both notions are invariant under inter-reducibility. If E is infinite, there is a correspondence between pure enumerages on E and numerotages on E. This is because $\overline{\alpha}$ is pure iff $\overline{\alpha} = \overline{\alpha}'$ for some numbering α'. If E is finite then there is just one semi-pure enumerage on E and this is also pure.

All the general definitions have meaning and interest for finite E. For instance if $\overline{\overline{E}} = 2$ there is still a richness of enumerages. An enumeration of E becones a 1-place predicate on N and our reducibility is the classical many-one reducibility.

Given E (where $2 \leqslant \overline{\overline{E}} \leqslant \aleph_0$) write $\text{Enum}(E)$ = the set of enumerages on E. So $\overline{\overline{\text{Enum}(E)}} = 2^{\aleph_0}$. The quotient relation of reducibility by inter-reducibility given an ordering on $\text{Enum}(E)$. This ordering is analogous to that on Turing degrees. For example we can prove that it is an upper semi-lattice which is not a lattice, and that (following Spector [8]) an increasing denumerable sequence of elements has no least upper bound. But if E is infinite Enum(E) has many minimal elements – unlike the Turing degrees for which **0** is the unique minimum. For example any semi-pure enumerage is minimal, since if α is semi-pure and α' is reducible to α, then α is reducible to α'. The semi-pure enumerages do not exhaust the minimal elements since there are at most countably many of these, and using Baire-like arguments we can obtain $2^{\aleph_0}$ minimal elements.

There are other interesting kinds of elements of Enum(E). For instance:

Definition 3.6. $\overline{\alpha}$ is *homogeneous* iff $\alpha_1, \alpha_2 \in \overline{\alpha}$ implies that α_1 is isogenic to α_2.

Examples. (a) H. Rogers Jr. [6] showed that on the set of partial recursive functions the enumerage consisting of all standard 'numberings' is homogeneous.

(b) It is easy to show that for a group G to possess a semi-pure enumerage compatible with the group operation is equivalent to it being 'recursively presentable' – that is, is the quotient of a free group by the normal subgroup generated by an r.e. subset. If G has a finite set of generators then there exists at most one semi-pure $\overline{\alpha}$ compatible with the group operation. If G is recursively presentable but not $\exists$-recursive, then the semi-pure $\overline{\alpha}$ compatible with the group operation is homogeneous. We can form a category in which the objects are the enumerages on all finite or denumerable sets and the morphisms $(E, \overline{\alpha}) \to (E', \overline{\alpha}')$ are those mappings $\theta : E \to E'$ which are representable with respect to $\overline{\alpha}, \overline{\alpha}'$. We use N with the structure given by the recursive functions as a fixed auxiliary set, in the same way, for instance, as we use the field K in the category of vector spaces over K.

There are several types of problem arising from the category of enumerages.

There are general category problems. For example, it is easy to define the following operations: direct product of finitely many enumerages, homomorphic image of an enumerage under a mapping, quotient of an enumerage by an equivalence relation, and induced enumerage – that is, the restriction of an enumerage $\overline{\alpha}$ on E to some subset E' of E. This last notion is not always defined. One can show that a necessary (but not sufficient) condition for it to be defined is that E' be r.e. with respect to $\overline{\alpha}$, in the weak sense that $E' = \varphi(N)$ for some $\varphi : N \to E$ recursive with respect to $\overline{\alpha}$. And a sufficient (but not necessary) condition is that E' be r.e. with respect to $\overline{\alpha}$, in the strong sense that $\alpha^{-1}(E')$ is r.e.

Another approach is to look at interesting sub-categories. We say that an enumerage $\bar{\alpha}$ on E is *autogenic* if the set of all $\bar{\alpha}$-recursive mappings of N into E can be enumerated by an $\bar{\alpha}$-recursive mapping of N^2 into E (where by $\bar{\alpha}$-recursive we mean recursively-representable with respect to $(\bar{1}, \bar{\alpha})$). Then, for instance, most classical results on standard numberings of partial recursive functions can be extended to the class of autogenic enumerages.

References

[1] W. Craig and R.L. Vaught, Finite axiomatizability using additional predicates, J. Symbolic Logic 23 (1958) 289–308.

[2] R. Fraissé, Une notion de récursivité relative, in: Infinitistic methods, Proceedings of the symposium on foundations of mathematics, Warsaw, 1959 (Pergamon Press, 1961) pp. 323–328.

[3] S.C. Kleene, Finite axiomatizability of theories in the predicate calculus using additional predicate symbols. Two papers on the predicate calculus, Mem. Am. Math. Soc. 10 (1952) 27–68.

[4] G. Kreisel, Model theoretic invariants: applications to recursive and hyper-arithmetic operations, in: The theory of models, Proceedings of the 1963 international symposium at Berkeley (North-Holland Publ. Co., 1965) pp. 190–205.

[5] H. Rogers Jr., Gödel numberings of partial recursive functions, J. Symbolic Logic 23 (1958) 331–341.

[6] C. Spector, Recursive well-orderings, J. Symbolic Logic 20 (1955) 151–163.

[7] C. Spector, On degrees of recursive unsolvability, Ann. Math. 64 (1956) 581–592.

MEASURABLE CARDINALS

J.R. SHOENFIELD
Duke University, Durham, N.C., USA

These lectures are an introduction to the role of large cardinals in the solution of problems in set theory not solvable by conventional methods. The reader is assumed familiar with the development of ZFC (Zermelo-Fraenkel set theory with the Axiom of Choice) through elementary cardinal arithmetic and with the elements of model theory. Some notation: σ, ν and τ are ordinals; κ, λ, and ρ are infinite cardinals; $\mathrm{PS}(x)$ is the power set of x; $|x|$ is the cardinal of x; y^x is the set of mappings from x to y.

Why can large cardinals solve problems which cannot be solved in ZFC? We recall that every set belongs to some $R(\sigma)$, where $R(\sigma)$ is defined (using transfinite induction) by

$$R(0) = 0\,,$$

$$R(\sigma+1) = \mathrm{PS}(R(\sigma))\,,$$

$$R(\sigma) = \bigcup_{\tau<\sigma} R(\tau) \qquad \text{for } \sigma \text{ a limit number.}$$

If we add an axiom guaranteeing the existence of a new ordinal σ, we also guarantee the 'presence' of the sets in $R(\sigma)$. These new sets may serve to settle old problems.

The sets which usually concern mathematicians are in $R(\sigma)$ for small σ; e.g., most of them are in $R(\omega+\omega)$. Can introducing large σ settle problems about these sets? It can, as has been known for a long time. The introduction of an inaccessible cardinal enables us to prove the consistency of ZFC, which cannot be proved in ZFC itself. This consistency is a statement about a set of natural numbers (viz., the set of Gödel numbers of theorems of ZFC). This is not a very exciting result, however; we already believed that ZFC was consistent. We would like to be able to settle more interesting problems, like the continuum hypothesis. Here inaccessibles do not help.

Despite this unfortunate situation, Gödel [6] suggested in 1946 that it might be possible to settle all problems of set theory (formalizable in the language of ZFC) by means of large cardinal axioms. Today we at least know that some interesting problems can be settled by large cardinal axioms. However, we still have not discovered a large cardinal axiom which seems likely to settle the continuum hypothesis.

To avoid getting lost in a jungle of definitions, we shall confine ourselves almost entirely to one type of large cardinal, the measurable cardinal. This will suffice to show what type of results have been obtained.

Let κ be an uncountable cardinal. A *measure* on κ is a mapping μ of $PS(\kappa)$ into $\{0, 1\}$ such that:

(a) If $[x_i \mid i \in I]$ is a pairwise disjoint family of subsets of κ, and $|I| < \kappa$, then

$$\mu(\bigcup_{i \in I} x_i) = \sum_{i \in I} \mu(x_i) ;$$

(b) $\mu(\kappa) = 1$; $\mu(\{\sigma\}) = 0$ for $\sigma < \kappa$.

We say κ is *measurable* if there is a measure on κ.

From the point of view of measure theory, our measures are quite special. Every subset of κ is measurable and has measure 0 or 1. Singletons have measure 0. Finally, the requirement (a) is stronger than countable additivity*. On the other hand, (a) and (b) suffice to prove:

(1) $a \subset b \subset \kappa \rightarrow \mu(a) \leqslant \mu(b)$;

(2) $\mu(a) = 0 \leftrightarrow \mu(\kappa - a) = 1$;

(3) $|a| < \kappa \rightarrow \mu(a) = 0$.

It is not immediately apparent that a measurable cardinal is large. However, this is easily settled.

Theorem 1 (Banach-Ulam [31]). If κ is measurable, then κ is inaccessible.

Proof. Let μ be a measure on κ. By definition, $\kappa > \omega$. If κ is singular, κ is the

* This is not too vital. If (a′) is (a) with "$|I| < \kappa$" replaced by "I is countable", then the existence of a κ and a μ satisfying (a′) and (b) implies the existence of a measurable cardinal [31].

union of $<\kappa$ sets having cardinal $<\kappa$. By (3) and (a), $\mu(\kappa) = 0$, a contradiction. Thus κ is regular.

It remains to derive a contradiction from the assumption $\lambda < \kappa \leqslant 2^\lambda$. Identify κ with a subset of 2^λ. For $\sigma < \lambda$, let

$$A_{\sigma,i} = [f \mid f \in \kappa \;\&\; f(\sigma) = i] \;.$$

Since κ is the disjoint union of $A_{\sigma,0}$ and $A_{\sigma,1}$, we can choose i_σ so that $\mu(A_{\sigma,i_\sigma}) = 1$. Setting

$$A = \bigcap_{\sigma<\lambda} A_{\sigma,i_\sigma} \;,$$

$\mu(A) = 1$ by (a). This is impossible, since the only f which can belong to A is defined by $f(\sigma) = i_\sigma$ for all σ.

The Banach-Ulam theorem suggests the problem: is the first inaccessible cardinal measurable? The solution of this problem by Hanf in 1960 (using unpublished results of Tarski) was the beginning of modern large cardinal theory. We shall present his solution in a version which is suitable for proving other facts about measurable cardinals.

We will need some facts about models. For the moment, we are only concerned with models for the language of set theory. Hence a model will be of the form $\mathfrak{A} = \langle A, \in_A \rangle$, where A is the universe of the model and $\in_A$ is the $\in$-relation of the model.

It is tacitly understood that all our set-theoretic results are to be proved in ZFC. Since model-theoretic results are used to prove set-theoretic results, they must also be proved in ZFC. There is no problem in doing this when the universe of the model is a set; but this is not always the case. For example, the standard model $\langle V, \in \rangle$ (where V is the class of all sets and $\in$ is the usual $\in$-relation) fails to have this property. We can still deal with $\langle A, \in_A \rangle$ in ZFC if both $x \in A$ and $x \in_A y$ are definable in ZFC. This is the case with the standard model $\langle V, \in \rangle$; $x \in V$ can be defined to mean $x{=}x$, and $x \in y$ is already a formula of ZFC. The details of how to treat such models in ZFC are tedious, but straightforward; we generally omit them.

If $\mathfrak{A} = \langle A, \in_A \rangle$ is a model, we use a superscript $\mathfrak{A}$ to mean "in the model $\mathfrak{A}$". Thus $\omega_1^{\mathfrak{A}}$ is the element of A which is designated by the defined symbol ω_1.

A defined predicate P is *absolute* for $\mathfrak{A}$ if

$$P^{\mathfrak{A}}(a_1,\dots, a_n) \leftrightarrow P(a_1,\dots, a_n)$$

for $a_1, \ldots, a_n \in A$. A defined operation F is *absolute* for $\mathfrak{A}$ if

$$F^{\mathfrak{A}}(a_1, \ldots, a_n) = F(a_1, \ldots, a_n)$$

for $a_1, \ldots, a_n \in A$.

A model $\mathfrak{A} = \langle A, \in_A \rangle$ is an *$\in$-model* if $\in_A$ is the usual $\in$-relation restricted to A. In this case, we generally identify $\mathfrak{A}$ and A. A *transitive* model is an $\in$-model A such that $(\forall x \in A)(x \subseteq A)$.

We use *absolute* to means absolute for transitive models of ZFC. Most of the predicates and operations defined in ZFC through the development of ordinals are absolute. Exceptions are PS and R; for these we have (for transitive A)

$$PS^A(x) = PS(x) \cap A ,$$

$$R^A(\sigma) = R(\sigma) \cap A .$$

Predicates and operations concerning cardinals are generally not absolute.

A binary relation E is *well-founded* if: (a) there is no sequence $\{a_n\}_{n \in \omega}$ such that $a_{n+1} E a_n$ for all n; (b) for each y, $[x \mid xEy]$ is a set. The important fact about a well-founded relation E is that we may do proofs and definitions by transfinite induction over E much as we do over the relation $<$ among ordinals.

A model $\langle A, \in_A \rangle$ is *well-founded* if $\in_A$ is well-founded. Obviously every $\in$-model is well-founded. A model $\mathfrak{A}$ is *extensional* if the axiom of extensionality holds in $\mathfrak{A}$. It is easy to see that transitive models are extensional.

Collapsing theorem (Mostowski [17]). If $\mathfrak{A}$ is well-founded and extensional, then $\mathfrak{A}$ is isomorphic to a transitive model B; both B and the isomorphism are uniquely determined.

The idea of the proof is to observe that the isomorphism l must satisfy

$$l(x) = [l(y) \mid y \in_A x] ,$$

and then to use this to define l by transfinite induction over $\in_A$. If $\mathfrak{A}$ is an $\in$-model, this equation becomes

$$l(x) = [l(y) \mid y \in x \cap A]$$

This leads to the following result.

Corollary. If A is an extensional $\in$-model and x is a transitive subset of A, then the collapsing isomorphism of A maps each element of x into itself.

Now suppose that μ is a measure on κ. Our object is to construct the ultrapower V^κ/μ and then collapse it. We sketch the construction of V^κ/μ, emphasizing the difficulties caused by the fact that V is not a set. We shall use *almost* to mean *except on a set of μ-measure* 0 and abbreviate *almost everywhere* to A.E.

Let V^κ be the class of mappings from κ to V. For $f, g \in V^\kappa$, define

$$f \sim g \quad \text{iff} \quad f(\sigma) = g(\sigma)\,\text{A.E.}$$

This is clearly an equivalence relation. Since the equivalence classes are not sets, we must define a new kind of equivalence class. Hence we set

$$\tilde{f} = [g \mid g \sim f \;\&\; g \in R(\sigma)]$$

where σ is chosen as small as possible so as to make $\tilde{f}$ non-empty. The basic property

$$\tilde{f} = \tilde{g} \quad \text{iff} \quad f \sim g$$

of equivalence classes is easily proved.

We let V^κ/μ be the class of all $\tilde{f}$. We define

$$\tilde{f} E \tilde{g} \leftrightarrow f(\sigma) \in g(\sigma)\,\text{A.E.}$$

It is easily checked that this depends only on $\tilde{f}$ and $\tilde{g}$, not on f and g. The model with universe V^κ/μ and $\in$-relation E is the *ultrapower* of V over μ; we also designate it by V^κ/μ.

Fundamental theorem (Łoś). If P is a defined predicate, then

$$P^{V^\kappa/\mu}(\tilde{f}_1, \ldots, \tilde{f}_n) \leftrightarrow \mu([\sigma \mid P(f_1(\sigma), \ldots, f_n(\sigma))]) = 1\,.$$

For each x, let c_x be the constant mapping from κ to $\{x\}$, and let $c(x) = \tilde{c}_x$. Putting c_{x_i} for f_i in the fundamental theorem, we get

$$P^{V^\kappa/\mu}(c(x_1),\ldots, c(x_n)) \leftrightarrow P(x_1,\ldots, x_n)\,.$$

This may be stated as: c is an elementary embedding of V into V^κ/μ. In particular, V^κ/μ is elementary equivalent to V and hence is a model of ZFC. It is therefore extensional.

Now we show that V^κ/μ is well-founded. First suppose that $\widetilde{f}_{n+1} E \widetilde{f}_n$ for all n. Setting

$$A_n = [\sigma \mid f_{n+1}(\sigma) \in f_n(\sigma)]\ ,$$

we have $\mu(A_n) = 1$. Hence $\mu(\cap A_n) = 1$ by countable additivity. Taking $\sigma \in \cap A_n$, $f_{n+1}(\sigma) \in f_n(\sigma)$ for all n, contradicting the well-foundedness of $\in$.

Now consider $[\widetilde{f} \mid \widetilde{f} E \widetilde{g}]$. If $\widetilde{f} E \widetilde{g}$, then $f(\sigma) \in g(\sigma)$ A.E. By changing f on a set of measure 0 (so that $\widetilde{f}$ is unchanged), we may suppose $f(\sigma) \in g(\sigma) \cup \{0\}$ for all σ. Hence

$$[\widetilde{f} \mid \widetilde{f} E \widetilde{g}] \subset [\widetilde{f} \mid f \in (\mathrm{Range}(g) \cup \{0\})^\kappa]\ ;$$

so $[\widetilde{f} \mid \widetilde{f} E \widetilde{g}]$ is a set.

We may therefore apply the collapsing theorem to get an isomorphism l of V^κ/μ and a transitive model M. Setting $x^* = l(c(x))$, we have the diagram:

$$V \xrightarrow{c} V^\kappa/\mu \xrightarrow{l} M \qquad (V \xrightarrow{*} M)$$

Remark. So far we have used the countable additivity of μ, but not the full κ-additivity.

We now study the elementary embedding $*$ of V into M. We have

$$\sigma \text{ an ordinal} \to \sigma^* \text{ an ordinal}^M \to \sigma^* \text{ an ordinal}$$

because *ordinal* is absolute. Similarly,

$$\sigma < \tau \to \sigma^* <^M \tau^* \to \sigma^* < \tau^*\ ,$$

Thus $*$ maps ordinals into ordinals in a strictly increasing way; so $\sigma \leqslant \sigma^*$.

Lemma 1. Every ordinal is in M.

Proof. We have $\sigma^* \in M$ and $\sigma = \sigma^*$ or $\sigma \in \sigma^*$. Since M is transitive, $\sigma \in M$.

Lemma 2. If $\sigma < \kappa$, then $\sigma^* = \sigma$.

Proof. We use transfinite induction on σ. Suppose that $\sigma < \sigma^*$. By Lemma 1, $\sigma = l(\tilde{f})$ for some f. Then $l(\tilde{f}) \in l(c(\sigma))$; so $\tilde{f} \tilde{E} c(\sigma)$; so $f(\tau) \in \sigma$ for almost all τ. Hence κ is the union of a set of measure 0 and the $< \kappa$ sets $A_\nu = [\tau \mid f(\tau) = \nu]$ for $\nu < \sigma$. By κ-additivity, some A_ν has measure 1. Then $f(\tau) = \nu$ for almost all τ; so $\tilde{f} = c(\nu)$; so $\sigma = l(\tilde{f}) = \nu^* = \nu$ by induction hypothesis. This contradicts $\nu < \sigma$.

Next we note that

$$\kappa^* = [l(\tilde{f}) \mid f \in \kappa^\kappa] \ .$$

For any element of κ^* is $l(\tilde{f})$ for some $f \in V^\kappa$. From $l(\tilde{f}) \in \kappa^*$ we get $\tilde{f} \tilde{E} c(\kappa)$ and hence $f(\sigma) \in \kappa$ A.E. Changing f on a set of measure 0, we may suppose $f \in \kappa^\kappa$. Conversely, $f \in \kappa^\kappa$ implies $\tilde{f} \tilde{E} c(\kappa)$ and hence $l(\tilde{f}) \in \kappa^*$.

Now for $f \in \kappa^\kappa$ and $\sigma < \kappa$, $l(\tilde{f}) = \sigma$ iff $l(\tilde{f}) = \sigma^*$ by Lemma 2; hence iff $\tilde{f} = c(\sigma)$; hence iff $f(\tau) = \sigma$ A.E. Thus $l(\tilde{f}) < \kappa$ iff f is almost constant. This implies that

$$(1) \qquad \kappa^* - \kappa = [l(\tilde{f}) \mid f \in \kappa^\kappa \ \& \ f \text{ is not almost constant}] \ .$$

Let i be the identity mapping of κ into κ. Since singletons have measure 0, i is not almost constant. Thus $\kappa^* - \kappa \neq 0$. From this and $\kappa \leqslant \kappa^*$:

Lemma 3. $\kappa < \kappa^*$.

From Lemma 3, $\kappa \in \kappa^* - \kappa$; and κ is clearly the smallest ordinal in $\kappa^* - \kappa$. Combining this with (1):

Lemma 4. If $f \in \kappa^\kappa$, then $l(\tilde{f}) = \kappa$ iff: (a) f is not almost constant; (b) for $g \in \kappa^\kappa$, $g(\sigma) < f(\sigma)$ A.E. implies that g is almost constant.

We say that μ is *normal* if $l(\tilde{i}) = \kappa$. By Lemma 4, this is equivalent to

$$(\forall g \in \kappa^\kappa)\,(g(\sigma) < \sigma \text{ A.E.} \rightarrow g \text{ is almost constant}) \ .$$

Theorem 2 (Scott [10]). If κ is measurable, then there is a normal measure on κ.

Outline of proof. Let μ be any measure on κ, and choose f so that $l(\widetilde{f}) = \kappa$. Set

$$\mu'(x) = \mu(f^{-1}(x))$$

for $x \subset \kappa$. Using Lemma 4, μ' is a normal measure on κ.

From now on, we assume that μ is normal. Using the fundamental theorem:

$$\begin{aligned} P^M(\kappa) &\leftrightarrow P^M(l(\widetilde{i})) \\ &\leftrightarrow P^{V^\kappa/\mu}(\widetilde{i}) \\ &\leftrightarrow \mu([\sigma \mid P(i(\sigma))]) = 1. \end{aligned}$$

Thus:

$$(2) \qquad P^M(\kappa) \leftrightarrow \mu([\sigma \mid \sigma < \kappa \;\&\; P(\sigma)]) = 1\,.$$

Lemma 5. If λ is inaccessible and A is a transitive model of ZFC containing λ, then λ is inaccessible^A.

Proof. Left to the reader.

Now we can prove that measurable cardinals are very large.

Theorem 3 (Hanf-Tarski [30]). If κ is measurable, then there are κ inaccessibles less than κ.

Proof. Let Inac(λ) mean that λ is inaccessible. By the Banach-Ulam theorem, Inac(κ). By Lemmas 1 and 5, Inac$^M(\kappa)$. By (2), $[\lambda \mid \lambda < \kappa \;\&\; \mathrm{Inac}(\lambda)]$ has measure 1 and therefore cardinality κ.

There are numerous extensions of Theorem 3, of which we give one. An inaccessible λ is *hyperinaccessible* if

$$|\,[\rho \mid \rho < \lambda \;\&\; \mathrm{Inac}(\rho)]\,| = \lambda\,;$$

equivalently, since λ is regular, if $[\rho \mid \rho < \lambda \;\&\; \mathrm{Inac}(\rho)]$ is cofinal in λ. From the latter definition, we see that Lemma 5 extends to hyperinaccessibles. Since Theorem 3 implies that κ is hyperinaccessible, we can repeat the proof of Theorem 3 to obtain:

Theorem 3a. If κ is measurable, then there are κ hyperinaccessibles less than κ.

We will now apply measurable cardinals to constructible sets. The basic properties of the class L of constructible sets which we need are:

(L1) $L = [C(\sigma) \mid \sigma \text{ an ordinal}]$, where C is a certain absolute operation (which is the F of [4]);

(L2) L is a transitive model of ZFC containing all the ordinals.

We define

$$C^*(\sigma) = [C(\tau) \mid \tau < \sigma] .$$

Now let A be a transitive model of ZFC; we shall investigate L^A. By the absoluteness of C and *ordinal*,

$$L^A = [C^A(\sigma) \mid \sigma \text{ an ordinal}^A]$$

$$= [C(\sigma) \mid \sigma \in A] .$$

If A contains all the ordinals, it follows that $L^A = L$. Otherwise the first ordinal not in A will be called the *index* of A, and will be designated by $\mathrm{Ix}(A)$. Using the transitivity of A, we get

$$\sigma \in A \leftrightarrow \sigma < \mathrm{Ix}(A) .$$

It follows that

$$L^A = [C(\sigma) \mid \sigma < \mathrm{Ix}(A)] = C^*(\mathrm{Ix}(A)) .$$

The sentence $V{=}L$ is called the *axiom* of *constructibility*. By the above, $L^L = L = V^L$; so $V{=}L$ is true in L.

Lemma 6. If A is a transitive model of ZFC containing all the ordinals, then $L \subset A$.

Proof. $L = L^A \subset A$.

This lemma shows that L is the smallest model satisfying (L2).

Theorem 4 (Scott [22]). If there is a measurable cardinal, then $V \neq L$.

Proof. Let κ be the first measurable cardinal, and let the notation be as above. From Lemmas 1 and 6, $L \subset M$; so it will suffice to prove $V \neq M$. Assume $V = M$. Then κ is measurable^M. By (2), $[\sigma \mid \sigma < \kappa \;\&\; \sigma \text{ measurable}]$ has measure 1, which is absurd.

We can extend the method to prove that if μ is a measure on κ, then $\mu \notin L$. Thus we know that $R(\kappa+2)$ contains a non-constructible set. We propose now to show that $R(\sigma)$ for quite small σ contains a non-constructible set. For this we introduce some new methods.

Let P be an n-ary relation on κ. A subset H of κ is *homogeneous* for P (or a *set of indiscernibles* for P) if the statements $P(\sigma_1,\ldots,\sigma_n)$, where $\sigma_1,\ldots,\sigma_n \in H$ and $\sigma_1 < \ldots < \sigma_n$, are all true or all false.

Lemma 7. Let μ be a normal measure on κ, and let $\mu(x_\sigma) = 1$ for all $\sigma < \kappa$. Then $[\sigma \mid (\forall \tau < \sigma)(\sigma \in x_\tau)]$ has measure 1.

Proof. Left as an exercise.

Theorem 5 (Rowbottom [21]). Let μ be a normal measure on κ, P an n-ary relation on κ. Then there is a subset H of κ such that $\mu(H) = 1$ and H is homogeneous for P.

Proof. We use induction on n. If $n=0$, take $H=\kappa$. Now assume the theorem for n, and let P be $(n+1)$-ary. Set

$$P_\tau(\sigma_1,\ldots,\sigma_n) \leftrightarrow P(\tau,\sigma_1,\ldots,\sigma_n).$$

For each $\tau < \kappa$, choose H_τ of measure 1 to be homogeneous for P_τ. Let $H = [\sigma \mid (\forall \tau < \sigma)(\sigma \in H_\tau)]$. By Lemma 7, $\mu(H) = 1$. Let i_τ be the constant value of the characteristic function of P_τ on increasing sequences in H_τ. Choose i and D so that $\mu(D) = 1$ and $i_\tau = i$ for $\tau \in D$. Then $\mu(H \cap D) = 1$. To show $H \cap D$ is homogeneous for P, let $\tau, \sigma_1,\ldots,\sigma_n \in H$ with $\tau < \sigma_1 < \ldots < \sigma_n$. Then $\sigma_i \in H_\tau$ and $\tau \in D$; so

$$P(\tau, \sigma_1, \ldots, \sigma_n) \leftrightarrow P_\tau(\sigma_1, \ldots, \sigma_n)$$
$$\leftrightarrow i_\tau = 1$$
$$\leftrightarrow i = 1 \,.$$

Since the right side is independent of $\tau, \sigma_1, \ldots, \sigma_n$, we have the desired result.

In the following, German letters represent models for arbitrary first order languages whose universes are set. The universe of $\mathfrak{A}$ will be designated by A, and similarly for other letters.

Recall that a submodel $\mathfrak{A}$ of $\mathfrak{B}$ is an *elementary* submodel (written $\mathfrak{A} \prec \mathfrak{B}$) if

$$P^{\mathfrak{B}}(b_1, \ldots, b_n) \leftrightarrow P^{\mathfrak{A}}(b_1, \ldots, b_n)$$

for all defined predicates P and all $b_1, \ldots, b_n \in B$.

The reflection principle tells us that, given $P_1, \ldots, P_n$ defined in set theory and a set a, we can find a set A such that $a \subset A$ and

$$P_i^A(x_1, \ldots, x_n) \leftrightarrow P_i(x_1, \ldots, x_n)$$

whenever $x_1, \ldots, x_n \in A$. It follows that it will do no harm if we assume that $A \prec V$, i.e., that

$$(3) \qquad P^A(x_1, \ldots, x_n) \leftrightarrow P(x_1, \ldots, x_n)$$

for *all* P and all $x_1, \ldots, x_n \in A$. For in any proof, we can use (3) only for a finite number of P; and for these P's there is an A making (3) true.

We let $l(\mathfrak{A})$ be the number of symbols in the language of $\mathfrak{A}$. An *expansion* of $\mathfrak{A}$ is a model $\mathfrak{A}'$ obtained from $\mathfrak{A}$ by assigning meanings to new non-logical symbols without changing the universe of the model.

Lemma 8. For each model $\mathfrak{A}$ there is an expansion $\mathfrak{A}'$ of $\mathfrak{A}$ such that $l(\mathfrak{A}) = l(\mathfrak{A}')$ and every submodel of $\mathfrak{A}'$ is elementary.

Idea of proof. Add symbols for a set of Skolem functions for $\mathfrak{A}$, and repeat this process ω times.

If $P(x)$ is a unary predicate defined in the language of $\mathfrak{A}$, we write $P^{\mathfrak{A}}$ for $[a \mid a \in A \;\&\; P^{\mathfrak{A}}(a)]$.

Theorem 6 (Rowbottom [21]). Let μ be a normal measure on κ. Let $\mathfrak{A}$ be a model such that $l(\mathfrak{A}) < \kappa$ and $\kappa \subset A$. Then there is a model $\mathfrak{B} \prec \mathfrak{A}$ such that $\mu(\kappa \cap B) = 1$ and either $|P^{\mathfrak{B}}| \leqslant l(\mathfrak{A})$ or $|P^{\mathfrak{B}}| \geqslant \kappa$ for each P defined in the language of $\mathfrak{A}$.

Proof. By Lemma 8, we may suppose that every submodel of $\mathfrak{A}$ is elementary. Let $\lambda = l(\mathfrak{A})$. Then there are λ P's. Since $\lambda < \kappa$, Theorem 5 and the κ-additivity of μ show that there is an H of measure 1 which is homogeneous for all $P^{\mathfrak{A}}$. Let $\mathfrak{B}$ be the smallest submodel of $\mathfrak{A}$ such that $H \subset B$. We need only prove that $|P^{\mathfrak{B}}| \leqslant \lambda$ or $|P^{\mathfrak{B}}| \geqslant \kappa$.

For each term $t(x_1,..., x_n)$ of the language of $\mathfrak{A}$, let

$$B_t = [t^{\mathfrak{A}}(\sigma_1,..., \sigma_n) | \sigma_1 < ... < \sigma_n \ \& \ \sigma_1,..., \sigma_n \in H] \ .$$

Then $B = \bigcup_t B_t$, and there are at most λ such B_t's. Hence our result will follow if we prove: (a) $P^{\mathfrak{B}}$ is a union of B_t's; (b) $|B_t| = 1$ or $|B_t| \geqslant \kappa$.

To prove (a), note that $P^{\mathfrak{B}}(b) \leftrightarrow P^{\mathfrak{A}}(b)$ for $b \in B$ (since $\mathfrak{B} \prec \mathfrak{A}$). Hence for $\sigma_1,..., \sigma_n$ as above,

$$\begin{aligned} P^{\mathfrak{B}}(t^{\mathfrak{A}}(\sigma_1,..., \sigma_n)) &\leftrightarrow P^{\mathfrak{A}}(t^{\mathfrak{A}}(\sigma_1,..., \sigma_n)) \\ &\leftrightarrow Q^{\mathfrak{A}}(\sigma_1,..., \sigma_n) \end{aligned}$$

where Q is independent of the σ_i. By choice of H, the right side is independent of the σ_i. This proves (a).

For (b), we define

$$Q(x_1,..., x_n, y_1,..., y_n) \leftrightarrow t(x_1,..., x_n) = t(y_1,..., y_n)$$

and consider two cases.

Case 1. $Q^{\mathfrak{A}}$ is false for increasing $2n$-tuples in H. Since $\mu(H) = 1$, we have $|H| = \kappa$; so H may be split into κ increasing n-tuples. Applying t to these gives κ distinct members of B_t.

Case 2. $Q^{\mathfrak{A}}$ is true for increasing $2n$-tuples in H. Suppose $\sigma_1,..., \sigma_n$ and $\sigma'_1,..., \sigma'_n$ are increasing n-tuples in H. Using $|H| = \kappa$ again, we can choose an increasing n-tuple $\tau_1,..., \tau_n$ in H with $\max(\sigma_n, \sigma'_n) < \tau_1$. Then

$$\begin{aligned} t^{\mathfrak{A}}(\sigma_1,..., \sigma_n) &= t^{\mathfrak{A}}(\tau_1,..., \tau_n) \\ &= t^{\mathfrak{A}}(\sigma'_1,..., \sigma'_n) \ . \end{aligned}$$

Hence $|B_t| = 1$.

To apply this result to constructible sets, we need the following result of Gödel [4]:

(4) $\mathrm{PS}(C^*(\lambda)) \cap L \subset C^*(\lambda^+)$.

Lemma 9. If κ is inaccessible, then $R(\sigma) \cap L \subset C^*(\kappa)$ for all $\sigma < \kappa$.

Outline of proof. Using (4), prove by induction on σ that

$$\sigma < \kappa \to (\exists \lambda < \kappa)\,(R(\sigma) \cap L \subset C^*(\lambda)) .$$

Theorem 7 (Rowbottom [21]). If κ is a measurable cardinal and $\omega \leqslant \sigma < \kappa$, then $|R(\sigma) \cap L| \leqslant |\sigma|$.

Proof. Choose A so that $\kappa \subset A$ and $A \prec V$. For each ordinal $\tau \leqslant \sigma$, introduce a symbol $\hat{\tau}$; and expand A to a model of this language by setting $\hat{\tau}^A = \tau$. Then $l(A) = |\sigma|$. Hence by Theorem 6, there is a $B \prec A$ such that $\mu(B \cap \kappa) = 1$ and $|P_B| \leqslant |\sigma|$ or $|P_B| \geqslant \kappa$ for all P.

Since $B \prec V$, B is a model of ZFC; so B is extensional. Let l be the collapsing isomorphism of B with a transitive model D. By the corollary to the collapsing theorem, $l(\sigma) = \sigma$. Hence $\hat{\sigma}^D = l(\hat{\sigma}^B) = l(\hat{\sigma}^A) = l(\sigma) = \sigma$.

Set $P(x) \leftrightarrow x \in R(\hat{\sigma})$. Then for $x \in D$, $P^D(x) \leftrightarrow x \in R^D(\sigma) \leftrightarrow x \in R(\sigma) \cap D$; so $P^D = R(\sigma) \cap D$. Since D is isomorphic to B, $|R(\sigma) \cap D| \leqslant |\sigma|$ or $|R(\sigma) \cap D| \geqslant \kappa$. But since $\sigma < \kappa$ and κ is inaccessible, we have $|R(\sigma)| < \kappa$. Hence $|R(\sigma) \cap D| \leqslant |\sigma|$.

It will now suffice to prove that $R(\sigma) \cap L \subset D$. By Lemma 9, it suffices to prove $C^*(\kappa) \subset D$. But $L^D = C^*(\mathrm{Ix}(D))$; so it suffices to prove $\kappa \leqslant \mathrm{Ix}(D)$.

Being an ordinal is absolute for B (since $B \prec V$) and D (since D is transitive). Since $\mu(B \cap \kappa) = 1$, there are κ ordinals in B and hence in D. This implies that $\kappa \leqslant \mathrm{Ix}(D)$, as required.

As an example, it follows from the theorem that $R(\omega+\omega) \cap L$ is countable. But $R(\omega+\omega)$ contains all sets of natural numbers, all sets of sets of natural numbers, etc.

Further results on the relation of measurable cardinals and constructible sets have been obtained by Gaifman [3] and Silver [24]. A reasonable summary of all these results is: if there is a measurable cardinal, then $V=L$ is as false as it possibly can be.

Having setting $V=L$, we look at the continuum hypothesis (CH) and the generalized continuum hypothesis (GCH). Here we shall see that measurable cardinals give no information. More precisely, let ZFM be ZFC + $(\exists\kappa)$ (κ is measurable). We show that if ZFM is consistent, then so are ZFM + GCH and ZFM + $\neg$CH.

Gödel proved that ZFC + GCH was consistent by using the model L. This won't do here, since L is not a model of ZFM. (This follows from Theorem 4, since L is a model of $V{=}L$.) We must therefore expand L somewhat.

For each set a, there is a class L_a consisting of the sets constructible from a. It has the following basic properties:

(L1′) $L_a = [C_a(\sigma) \mid \sigma$ an ordinal$]$, where C is a certain absolute operation with two arguments;

(L2′) L_a is a transitive model of ZFC containing all the ordinals;

(L3′) $a \cap L_a \in L_a$;

(L4′) $(\forall\tau < \sigma)(C_{a\cap M}(\tau) \subset M) \to C_a(\sigma) = C_{a\cap M}(\sigma)$.

We give a brief sketch of the proof for these familiar with [4]. To Gödel's operations $\mathcal{G}_1 - \mathcal{G}_8$, we add another defined by $\mathcal{G}(x, y) = a \cap x$. We then define C_a and L_a like C and L.

The proof of (L1′) and (L2′) is like that of (L1) and (L2). As in [4], we prove that every subset of L_a is included in a member of L_a. In particular, $a \cap L_a \subset b \in L_a$. By transistivity of L_a, $b \subset L_a$; so

$$a \cap L_a = a \cap b = \mathcal{G}(b, b) \in L_a .$$

Suppose (L4′) is false, and choose σ minimal so that $C_a(\sigma) \neq C_{a\cap M}(\sigma)$. Since $\mathcal{G}$ is the only operation involving a, we must have

$$C_a(\sigma) = C_a(\tau) \cap a ,$$

$$C_{a\cap M}(\sigma) = C_{a\cap M}(\tau) \cap a \cap M$$

for some $\tau < \sigma$. Now $C_{a\cap M}(\tau) \subset M$; so

$$C_{a\cap M}(\sigma) = C_{a\cap M}(\tau) \cap a = C_a(\tau) \cap a$$

by choice of σ. This contradicts $C_a(\sigma) \neq C_{a\cap M}(\sigma)$.

Let M be a transitive model of ZFC. Now L_a^M makes sense only if $a \in M$. We shall assume that $a \cap M \in M$ and evaluate $L_{a\cap M}^M$.

First we claim that

$$\sigma \in M \to C_a(\sigma) = C_{a \cap M}(\sigma) .$$

By (L4′), we need only prove that if $\tau < \sigma$, then $C_{a \cap M}(\tau) \subset M$. But $\tau \in M$, so

$$(C_{a \cap M}(\tau))^M = C_{a \cap M}(\tau)$$

and hence $C_{a \cap M}(\tau) \subset M$ by the transitivity of M.

It follows that

$$\begin{aligned} L^M_{a \cap M} &= [C_{a \cap M}(\sigma)^M \mid \sigma \text{ an ordinal}^M] \\ &= [C_{a \cap M}(\sigma) \mid \sigma \in M] \\ &= [C_a(\sigma) \mid \sigma \in M] . \end{aligned}$$

This gives the following result.

Lemma 10. Let M be a transitive model of ZFC, and let $a \cap M \in M$. If M contains all the ordinals, then $L^M_{a \cap M} = L_a$. Otherwise, $L^M_{a \cap M} = C^*_a(\mathrm{Ix}(M))$.

Taking $M = L_a$ and setting $a' = a \cap L_a$, $(L_{a'})^{L_a} = L_a$. This means that $V{=}L_{a'}$ is true in L_a.

Let μ be a normal measure on κ, and let $\mu' = \mu \cap L_\mu$. By the above, L_μ is a model of ZFC and $V{=}L_{\mu'}$. It is easy to check that "μ' is a normal measure on κ" holds in L_μ.

It follows that if ZFM is consistent, then so is the theory $\mathrm{ZFL}_{\kappa,\mu}$ obtained from ZFC as follows: we add two constants μ and κ and the axioms "μ is a normal measure on κ" and $V{=}L_\mu$. Thus our consistency result for ZFM + GCH is a consequence of the following.

Theorem 8 (Silver [25]). GCH is a theorem of $\mathrm{ZFL}_{\kappa,\mu}$.

We need some preliminary results. Throughout, $\mathrm{ZFL}_{\kappa,\mu}$ is our basic theory.

Lemma 11. Let B be a set such that $B \prec V$ and $\mu(B \cap \kappa) = 1$, and let l be the collapsing isomorphism of B onto D. Then $\kappa^D = \kappa$ and $\mu^D = \mu \cap D$.

Proof. First note that $B \prec V$ implies that B is extensional and well-founded; so l exists. Now 'ordinal' is absolute for B and D (since $B \prec V$ and D is transitive). Hence l maps the ordinals in B onto the ordinals in D. A similar argument using the absoluteness of $<$ shows that l is strictly increasing. Thus l^{-1} maps $\mathrm{Ix}(D)$ into the ordinals in a strictly increasing manner. Hence for $\sigma \in B, l^{-1}(\sigma) \geqslant \sigma$ and so $\sigma \geqslant l(\sigma)$. In particular, $l(\sigma) \leqslant \sigma$ for almost all $\sigma < \kappa$. We claim that $l(\sigma) = \sigma$ for almost all σ. For if not, $l(\sigma) < \sigma$ A.E.; so by normanilty, l is almost constant. This is absurd, since l is one-one.

Since $B \prec V, \kappa^B = \kappa$. Then $l(\kappa) \leqslant \kappa$. For almost all $\sigma < \kappa, \sigma \in B$ and $l(\sigma) = \sigma$ and $l(\sigma) < l(\kappa)$. Thus there are $\geqslant \kappa$ ordinals in D which are $< l(\kappa)$; so $l(\kappa) \geqslant \kappa$ and hence $l(\kappa) = \kappa$. Thus $\kappa^D = l(\kappa^B) = l(\kappa) = \kappa$.

To prove $\mu^D = \mu \cap D$, it suffices to show that if $x \subset \kappa$ and $x \in D$, then $\mu^D(x) = \mu(x)$. Now $\mu^D(x) = l(\mu^B(l^{-1}(x))) = l(\mu(l^{-1}(x)))$; so it suffices to show $\mu(x) = \mu(l^{-1}(x))$. Setting $y = l^{-1}(x)$, this becomes $\mu(l(y)) = \mu(y)$. But this follows from the fact that $l(y) = [l(\sigma) \mid (\sigma \in y \cap B)]$ and that for almost all $\sigma, \sigma \in B$ and $l(\sigma) = \sigma$.

Now we turns to Theorem 8, and prove $2^\lambda = \lambda^+$. First we suppose $\lambda \geqslant \kappa$ (this case is due to Solovay). It will suffice to prove $\mathrm{PS}(\lambda) \subset C^*_\mu(\lambda^+)$, since this gives

$$2^\lambda = |\mathrm{PS}(\lambda)| \leqslant |C^*_\mu(\lambda^+)| = \lambda^+ .$$

Let $x \subset \lambda$. Choose A so that $\lambda \cup \{x\} \subset A$ and $A \prec V$. By the Löwenheim-Skolem theorem, choose B so that $\lambda \cup \{x\} \subset B \prec A \prec V$ and $|B| \leqslant \lambda$. Let l be the collapsing isomorphism of B onto D. Since $\lambda \cup \{x\}$ is transitive, $l(x) = x$; so $x \in D$.

Now $\kappa \subset \lambda \subset B$; so by Lemma 11, $\kappa^D = \kappa$ and $\mu^D = \mu \cap D$. Hence by Lemma 10, $(L_\mu)^D = L^D_{\mu \cap D} = C^*_\mu(\mathrm{Ix}(D))$. But $V = L_\mu$ holds in $B \prec V$ and hence in D; so $D = C^*_\mu(\mathrm{Ix}(D))$. Since $|D| = |B| \leqslant \lambda$ and $\mathrm{Ix}(D) \subset D$, $|\mathrm{Ix}(D)| \leqslant \lambda$. Thus $D \subset C^*_\mu(\lambda^+)$. Since $x \in D$, we have $x \in C^*_\mu(\lambda^+)$.

Now we turn to the case $\lambda < \kappa$. We let $\mathrm{Od}_\mu(x)$ be the first σ such that $x = C_\mu(\sigma)$. We show that

$$x \subset \lambda \;\&\; \sigma = \mathrm{Od}_\mu(x) \rightarrow |\mathrm{PS}(\lambda) \cap C^*_\mu(\sigma)| \leqslant \lambda .$$

This will imply that $|\mathrm{PS}(\lambda)| \leqslant \lambda^+$. For otherwise, we would get a contradiction by taking σ to be the λ^+th ordinal which is $\mathrm{Od}_\mu(x)$ for some $x \subset \lambda$.

Choose A so that $\kappa \cup \{x\} \subset A \prec V$. For $y \in \lambda \cup \{\lambda, x\}$, introduce a symbol $\hat{y}$; and expand A by setting $\hat{y}^A = y$. Then $l(A) = \lambda$. Choose B by Theorem 6 so that $B \prec A \prec V, \mu(B \cap \kappa) = 1$, and $|P_B| \leqslant \lambda$ or $|P_B| \geqslant \kappa$ for all P. Let l be the collapsing isomorphism of B onto D.

Since $\lambda \cup \{\lambda, x\}$ is transitive, $l(y) = y$ for $y \in \lambda \cup \{\lambda, x\}$. But then

$$\hat{y}^D = l(\hat{y}^B) = l(\hat{y}^A) = l(y) = y\ .$$

Define P by $P(z) \leftrightarrow z \subset \hat{\lambda}$. Then $P^D(z) \leftrightarrow z \subset \lambda$ for $z \in D$; so $P^D = \mathrm{PS}(\lambda) \cap D$. Hence $|\mathrm{PS}(\lambda) \cap D| \leqslant \lambda$ or $|\mathrm{PS}(\lambda) \cap D| \geqslant \kappa$. But $\lambda < \kappa$ and κ is inaccessible; so $|\mathrm{PS}(\lambda)| = 2^\lambda < \kappa$. Thus $|\mathrm{PS}(\lambda) \cap D| \leqslant \lambda$.

It will therefore suffice to prove $C_\mu^*(\sigma) \subset D$. Using Lemmas 9 and 10 as above, $D = C_\mu^*(\mathrm{Ix}(D))$; so we need only prove $\sigma \leqslant \mathrm{Ix}(D)$. If not, then $\sigma = \mathrm{Od}_\mu(x)$ shows that $x \notin C_\mu^*(\mathrm{Ix}(D)) = D$. But $\hat{x}^D = x$; so $x \in D$.

There is another proof of Theorem 8 by Kunen using ultra products, and another proof of the consistency of ZFM + GCH by Jensen [9] using forcing. A great deal of further information about L_μ is contained in Kunen [11] and Paris [19].

Now we turn to the consistency of ZFM + $\neg$CH. The idea is to extend Cohen's proof [1] of the consistency of ZFC + $\neg$CH. We must therefore outline the Cohen method.

Let M be a countable transitive model of ZFC. (We obtain such a model by choosing a set $A \prec V$; using the Löwenheim-Skolem theorem to get a countable $B \prec A$; and then collapsing B.) The Cohen method enables us to extend M to a model having certain properties. We consider only a special but important case.

Let A and B be fixed sets in M. Let C be the set of all mappings from a finite subset of A to B. Note that $C \in M$. We call C the set of *conditions*. We use a, b, c for elements of M and p, q, r for conditions.

A subset D of C is *dense* if $\forall p \exists q\ (p \subset q\ \&\ q \in D)$. A mapping f in B^A is *generic* if for every dense set D such that $D \in M$, there is a $p \in D$ such that $p \subset f$ (as a set of ordered pairs).

Exercise. If $|A| \geqslant \omega$ and $|B| \geqslant 2$, then no generic mapping is in M. (Hint: For every f, $[p\,|\,p \not\subset f]$ is dense.)

Existence theorem. For every p, there is a generic f such that $p \subset f$.

Outline of proof. Let $D_0, D_1, \ldots$ be the dense sets in M, and extend p step by step to include a member of each D_i.

Now fix a generic f. Define a relation $\in_f$ on M by

$$a \in_f b \leftrightarrow \exists p (p \subset f\ \&\ \langle a, p \rangle \in b)\ .$$

Then $\in_f$ is well-founded; so we may define a mapping K_f on M by

$$K_f(b) = [K_f(a) \mid a \in_f b] .$$

We then set

$$M[f] = [K_f(b) \mid b \in M] .$$

It is easy to show that for each a there is an $\hat{a}$ such that $K_f(\hat{a}) = a$ for all f. Moreover, the operation $\hat{}$ is definable in M. It follows that $M \subset M[f]$. It is also easy to define an $F \in M$ such that $K_f(F) = f$ for all f; so $f \in M[f]$.

Fundamental theorem. If f is generic, then $M[f]$ is a countable transitive model of ZFC including M and containing f.

We shall not give the proof of the fundamental theorem; but we will sketch some notions used in the proof which are essential for further developments.

We add to the language of ZFC a constant a for each $a \in M$. The resulting language is called the *forcing language*. It is used to describe $M[f]$, with the variables ranging through $M[f]$ and the constant a designating $K_f(a)$. Thus the language is independent of f; but the meaning of the language depends upon f.

Let Φ be a sentence in the forcing language. We write $\vdash_{M[f]} \Phi$ if Φ is true in $M[f]$. We say that p *forces* Φ, and write $p \Vdash \Phi$, if $\vdash_{M[f]} \Phi$ for every generic f such that $p \subset f$.

There are three basic lemmas on forcing.

Extension lemma. If $p \Vdash \Phi$ and $p \subset q$, then $q \Vdash \Phi$.

Truth lemma. If f is generic, then $\vdash_{M[f]} \Phi$ iff $\exists p(p \subset f$ and $p \Vdash \Phi)$.

Definability lemma. Let $\Phi(x_1,..., x_n)$ be a formula of ZFC. Then there is a formula $\Psi(y, z, x_1,..., x_n)$ of ZFC such that

$$p \Vdash \Phi(a_1,..., a_n) \quad \text{iff} \quad \psi^M(C, p, a_1,..., a_n) .$$

The definability lemma shows that in defining sets in M we may make use of forcing. We often use this fact tacitly.

The fundamental theorem is useful only in conjunction with other results connecting M and $M[f]$. To illustrate, we sketch the construction of a model in which $2^{\aleph_0} \geqslant \aleph_2$. Take $A = \omega \times \aleph_2^M$, $B = \{0, 1\}$. Choose a generic f. For $\sigma < \aleph_2^M$,

$$x_\sigma = [i \mid i \in \omega \;\&\; f(i, \sigma) = 0]$$

is a subset of ω in $M[f]$. It is easy to show (using the genericity of f) that these sets are distinct. Thus in $M[f]$ there are $\geqslant \aleph_2^M$ subsets of ω. To complete the proof we need to know that $\aleph_2^M = \aleph_2^{M[f]}$. This is true, but the proof is non-trivial.

Now suppose our basic set theory is ZFM. Then M will be a model of ZFM. We need to know that $M[f]$ is also a model of ZFM. This will not always be true; but it will be if A and B are small.

Theorem 9 (Lévy-Solovay [14]). If κ is measurable^M and $|A|^M < \kappa$ and $|B|^M < \kappa$, then κ is measurable$^{M[f]}$.

Proof. A simple calculation (in M) shows that $|C|^M < \kappa$.

Suppose that in M, μ is a measure on κ. For $x \subset \kappa$ and $x \in M[f]$, we define

$$\mu'(x) = 0 \quad \text{if } \exists a(x \subset a \text{ and } \mu(a) = 0)\,,$$

$$\mu'(x) = 1 \quad \text{if } \exists a(a \subset x \text{ and } \mu(a) = 1)\,.$$

It is clear that $\mu' \in M[f]$. We claim that in $M[f]$, μ' is a measure on κ. We shall show that: (a) if $x \subset \kappa$ and $x \in M[f]$, then $\mu'(x)$ is defined; (b) in $M[f]$, the union of $< \kappa$ sets of μ'-measure 0 has μ'-measure 0. The desired conclusion then follows easily.

For (a), let a designate x (i.e., let $K_f(a) = x$). For $p \in C$ define

$$d_p = [\sigma \mid \sigma < \kappa \;\&\; p \Vdash \hat{\sigma} \in a]\,.$$

Then $d_p \in M$, and the function mapping p into d_p is in M, so that $\bigcup_{p \subset q} d_q$ is in M. We have

$$(5) \qquad p \subset f \rightarrow d_p \subset x \subset \bigcup_{p \subset q} d_q\,.$$

For if $\sigma \in d_p$, then $p \Vdash \hat{\sigma} \in a$ and hence $\vdash_{M[f]} \hat{\sigma} \in a$ by the truth lemma. But this means that $\sigma \in x$. If $\sigma \in x$, then $\vdash_{M[f]} \hat{\sigma} \in a$; so by the truth lemma, there is a $q \subset f$ which forces $\hat{\sigma} \in a$. We may suppose $p \subset q$; otherwise we replace q by $p \cup q$ and use the extension lemma. Hence $\sigma \in d_q \subset \bigcup_{p \subset q} d_q$.

In view of (5), it will suffice to prove $\exists p(p \subset f \text{ and } p \in D)$ where

$$D = [p \mid \mu(d_p) = 1 \vee \mu(\bigcup_{p \subset q} d_q) = 0] .$$

Since f is generic, it suffices to prove that D is dense and in M. Clearly $D \in M$. Let $p \in C$; we must find $q \in D$ with $p \subset q$. If $\mu(\bigcup_{p \subset q} d_q) = 0$,

$p \in D$; so we may take $q = p$. Now let $\mu(\bigcup_{p \subset q} d_q) = 1$. Since $|C| < \kappa$ and μ is

κ-additive in M, $\mu(d_q) = 1$ for some q with $p \subset q$. Then $q \in D$.

For (b), let g be, in $M[f]$, a mapping from some $\sigma < \kappa$ into PS(κ). We must show that either $\mu'(g(\tau)) = 1$ for some τ or $\mu'(\bigcup_{\tau < \sigma} g(\tau)) = 0$. Let $d_{p,\tau}$ be

like d_p above, but with x replaced by $g(\tau)$. Corresponding to (5), we have

$$p \subset f \rightarrow d_{p,\tau} \subset g(\tau) \subset \bigcup_{p \subset q} d_{p,\tau} ;$$

so it will suffice to prove $\exists p(p \subset f \,\&\, p \in D)$, where

$$D = [p \mid (\exists \tau < \sigma)\, (\mu(d_{p,\tau}) = 1) \vee \mu(\bigcup_{\tau < \sigma} \bigcup_{p \subset q} d_{q,\tau}) = 0] .$$

Again this reduces to showing that D is dense. Let $p \in C$. If

$$\mu(\bigcup_{\tau < \sigma} \bigcup_{p \subset q} d_{q,\tau}) = 0 ,$$

then $p \in D$. Otherwise (noting $\sigma < \kappa$ and $|C| < \kappa$ in M) there is a $\tau < \sigma$ and a q such that $p \subset q$ and $\mu(d_{q,\tau}) = 1$. Then $q \in D$.

Theorem 9 applies to the case $A = \omega \times \aleph_2^M$, $B = \{0, 1\}$; for $|A| = \aleph_2$ and $|B| = 2$ in M, while κ is inaccessible in M. Thus we obtain the consistency of ZFM + $\neg$CH.

Theorem 9 generalizes to more general sets of conditions (not discussed here), provided always that $|C|^M < \kappa$. This enables us to show, e.g., that Souslin's hypothesis cannot be settled in ZFM. It does not enable us to show that ZFM is consistent with $2^\kappa > \kappa^+$ (where κ is measurable); for to obtain this in $M[f]$ we must take C so that $|C|^M > \kappa$.

Theorem 9 (in the generalized form) is sometimes stated: measurability is preserved by mild Cohen extensions. It has been shown that almost all the large cardinal properties which have been considered are preserved by mild Cohen extensions; so these large cardinals do not suffice to settle CH.

It has been known for a long time that the existence of a measurable cardinal settles many mathematical problems; see [10] for examples. In all of these cases, the existence of a measurable cardinal makes the situation more complicated. Mathematicians would be more inclined to accept measurable cardinals if these cardinals had 'nice' mathematical consequences. We shall now examine some such consequences.

Since ω^ω is the product of ω copies of ω, we can give it a topology as follows: give ω the discrete topology and then give ω^ω the product topology. (With this topology, ω^ω is homeomorphic to the irrationals.) We use α, β, γ for elements of ω^ω.

We identify $(\omega^\omega)^2$ with ω^ω by identifying (α, β) with the γ defined by $\gamma(2n) = \alpha(n), \gamma(2n+1) = \beta(n)$. For $X \subset \omega^\omega$, $\mathrm{Proj}(X)$ is the projection of X, considered as a subset of $(\omega^\omega)^2$, on its first coordinate.

We say a subset X of ω^ω is:

(a) Σ^1_0 if it is open;

(b) Σ^1_{n+1} if it is $\mathrm{Proj}(Y)$ for some Π^1_n set Y;

(c) Π^1_n if $\omega^\omega - X$ is Σ^1_n;

(d) Δ^1_n if it is both Σ^1_n and Π^1_n.

Note that this is a proper definition by induction on n. A set which is Σ^1_n or Π^1_n for some n is called *projective*. The classification of projective sets given by (1) – (4) is called the *projective hierarchy*. For orientation, we remark that, by a theorem of Souslin, Δ^1_1 sets are identical with Borel sets.

It is known that sets low in the projective hierarchy have certain nice properties. Solovay has shown that in ZFM we can prove sets a little higher in the hierarchy have these nice properties. The nice properties are: (1) being measurable for all reasonable measures; (2) being countable or including a perfect subset; (3) having the property of Baire. We shall consider only (3).

A set X in ω^ω has the *property of Baire* if there is an open set Y such that $X \triangle Y$ (which is by definition $(X-Y) \cup (Y-X)$) is of the first category. This property is similar in many ways to being measurable. (See [13] for this and much further information about projective sets.)

Lusin proved that every Σ^1_1 or Π^1_1 set has the property of Baire (see [13]).

This is the best possible result in ZFC; for results of Gödel show that if $V=L$, then there is a Δ_2^1 set which does not have the property of Baire. We prove in ZFM that every Σ_2^1 or Π_2^1 set has the property of Baire. This is again best possible, since results of Silver show that if $V=L_\mu$, then there is a Δ_3^1 set which does not have the property of Baire.

We need the following result, whose proof is like that of Theorem 7.

Theorem 7a (Rowbottom [21]). If κ is a measurable cardinal, $\omega \leqslant \sigma < \kappa$, and $\alpha \in \omega^\omega$, then $|R(\sigma) \cap L_\alpha| \leqslant |\sigma|$.

Corollary. If κ is a measurable cardinal and $\alpha \in \omega^\omega$, then $\mathrm{PS}(\omega) \cap L_\alpha$ is countable.

Since ω^ω has a countable base, it has only $2^{\aleph_0} = |\omega^\omega|$ open sets; so there is a mapping σ_0 of ω^ω onto the set of Σ_1^0 sets. We can actually define such a σ_0 in set theory in a natural way. We then define

$$\pi_0(\alpha) = \omega^\omega - \sigma_0(\alpha)\ ,$$

$$\sigma_1(\alpha) = \mathrm{Proj}(\pi_0(\alpha))\ ,$$

$$\pi_1(\alpha) = \omega^\omega - \sigma_1(\alpha)\ ,$$

and so on. Then $\sigma_n(\pi_n)$ maps ω^ω onto the set of $\Sigma_n^1(\Pi_n^1)$ subsets of ω^ω.

Absoluteness theorem. Let M be a transitive model of ZFC containing all the ordinals. Then for $\alpha \in M$, $\pi_2(\alpha)^M = \pi_2(\alpha) \cap M$ and $\sigma_2(\alpha)^M = \sigma_2(\alpha) \cap M$.

Proof. See [23].

Theorem 10 (Solovay). If there is a measurable cardinal, then every Π_2^1 or Σ_2^1 set has the property of Baire.

Proof. We consider a Π_2^1 set X; the proof for Σ_2^1 sets is the same. Let $X = \pi_2(\alpha)$. We want to do forcing with $M=L_\alpha$, $A=\omega$, $B=\omega$. The first difficulty is that M is not countable (and is not even a set). However, the countability of M was used only in the proof of the existence theorem; and there we only needed to know that there were only countably many dense sets in M. Hence it will suffice to show

(6) $\mathrm{PS}(C) \cap M$ is countable.

Now it is clear that C is countable. In fact, we can define in set theory an absolute operation H mapping ω one-one onto C. Since $H^M = H$ for arguments in M, H maps sets in M onto sets in M. It follows that $|\mathrm{PS}(\omega) \cap M| = |\mathrm{PS}(C) \cap M|$; so (6) follows from the corollary to Theorem 7a.

Let G be the set of generic mappings. Then $G \subset \omega^\omega$. We shall prove that: (a) $\omega^\omega - G$ is of the first category; (b) $X \cap G$ is open in G. The theorem will follow. For by (b), $X \cap G = Y \cap G$ for some open Y. Then $X \triangle Y \subset \omega^\omega - G$ is of the first category; so X has the property of Baire.

To prove (a), note that

$$\omega^\omega - G = \bigcup_D Z_D$$

where D runs through the dense sets in M and

$$Z_D = [\beta \mid (\forall p \in D) \neg (p \subset \beta)] \ .$$

By (6) there are only countably many D's. Hence it suffices to show that Z_D is closed and nowhere dense. This is left to the reader.

To prove (b), let $\beta \in X \cap G$; we must obtain a neighborhood N of β in ω^ω so that $N \cap G \subset X$. Every member of α is in $L \subset L_\alpha$; so $\alpha = \alpha \cap L_\alpha \in L_\alpha$. Since $\alpha \in M[\beta]$, the absoluteness theorem gives $\pi_2(\alpha)^{M[\beta]} = X \cap M[\beta]$. Thus $\beta \in \pi_2(\alpha)$ is true in $M[\beta]$; so $\vdash_{M[\beta]} F \in \pi_2(\hat{\alpha})$. (Recall $F \in M$ is chosen so that $K_\beta(F) = \beta$.) By the truth lemma, there is a $p \subset \beta$ such that $p \Vdash F \in \pi_2(\hat{\alpha})$.

Let $N = [\gamma \mid p \subset \gamma]$. This is a neighborhood of β; so we must show $\gamma \in N \cap G \rightarrow \gamma \in X$. Since $p \subset \gamma$ and $p \Vdash F \in \pi_2(\hat{\alpha})$, we have $\vdash_{M[\gamma]} F \in \pi_2(\hat{\alpha})$; so $\gamma \in \pi_2(\alpha)^{M[\gamma]}$. As above, $\pi_2(\alpha)^{M[\gamma]} = \pi_2(\alpha) \cap M[\gamma]$; so $\gamma \in X$.

We now turn to another "nice" property of subsets of ω^ω which has been of much interest in set theory recently. For each $X \subset \omega^\omega$, we define a game $G(X)$. It is played by two players, I and II. First I picks $\alpha(0)$; then II picks $\alpha(1)$; then I picks $\alpha(2)$; then II picks $\alpha(3)$; etc. The game ends after ω steps. If $\alpha \in X$, I wins; otherwise, II wins.

A *strategy* (for I or II) is a rule which gives a play at each step depending on the results of the previous steps. It is a *winning* strategy if a player who plays by it can never lose. We say X is *determinate* if either I or II has a winning strategy for $G(X)$.

It can be proved in ZFC that there is an indeterminate set [18]. On the other hand, simple sets can be shown to be determinate. For example, suppose that X is closed. Suppose that II does not have a winning strategy in $G(X)$; we show that I does. If I plays badly, II may have a winning strategy after n moves; i.e., there may be a set of rules by which he can play from that point on which will guarantee that he wins. The strategy of I is to play so that II never has a winning strategy. It is easy to show by induction on n that he can play this way at his nth move. Now suppose I loses while playing this strategy. Then $\alpha \notin X$. Since X is closed, there is an n such that any function beginning with $\alpha(0),\ldots, \alpha(2n)$ is not in X. But this means that II has a winning strategy after the move $\alpha(2n)$.

It is not known whether one can prove in ZFC that every Δ^1_1 set is determinate. (See [2] for partial results.) However, nothing better can be proved; for Mycielski [18] has shown that if $V=L$, then there is a Π^1_1 set which is indeterminate.

Theorem 11 (Martin [15]). If there is a measurable cardinal, then every Π^1_1 set is determinate.

Proof. Let μ be a normal measure on κ. Let X be Π^1_1. By [13], there is a mapping F from ω^ω to $\mathrm{PS}(Q)$ (where Q is the set of rationals) such that:

(a) $\alpha \in X \leftrightarrow F(\alpha)$ is well-ordered;

(b) $[\alpha \mid r \in F(\alpha)]$ is open for all $r \in Q$.

Let $Q = \{r_0, r_1, r_2, \ldots\}$ and set

$$F(x_0, x_1,\ldots, x_n) = [r \mid r \in \{r_0, r_1,\ldots, r_n\}$$
$$\& \forall \alpha(\alpha(0) = x_0 \;\&\; \alpha(1) = x_1 \;\&\; \ldots \;\&\; \alpha(n) = x_n \rightarrow r \in F(\alpha))] \, .$$

Then the finite sets $F(\alpha(0), \alpha(1),\ldots, \alpha(n))$ expand with increasing n and, by (b), have $F(\alpha)$ as their union.

We describe a game $G'(X)$. Again I picks $\alpha(0)$, II picks $\alpha(1)$, I picks $\alpha(2)$, etc. But now when I picks $\alpha(2n)$, he must also pick a one-one mapping H_n of $F(\alpha(0),\ldots, \alpha(2n))$ into κ which (if $n \geqslant 1$) extends H_{n-1}. Then $H = \bigcup_n H_n$ is a one-one mapping of $F(\alpha)$ into κ. If $\alpha \in X$ and H is order preserving, then I wins; otherwise II wins.

If H is order preserving, then $\alpha \in X$ by (a), so I wins. Hence if I loses, he is lost irrevocably after some finite number of plays. The proof above that closed sets are determinate shows that either I or II has a winning strategy for $G'(X)$. If I has a winning strategy for $G'(X)$, he certainly has one for $G(X)$. We complete the proof by showing that if II has a winning strategy for $G'(X)$, he has one for $G(X)$.

The winning strategy for II in $G'(X)$ gives $\alpha(2n+1)$ as a function of $\alpha(0),\ldots,\alpha(2n)$ and H_n. We may assume that it is actually a function of $\alpha(0),\ldots,\alpha(2n)$ and Range(H_n). For if H_n is not order preserving, II will certainly win; so he may assume that H_n is the unique one-one order preserving mapping of $F(\alpha(0),\ldots,\alpha(2n))$ onto Range(H_n).

Next we claim that there is a $P \subset \kappa$ such that $\mu(P) = 1$ and such that $\alpha(2n+1)$ is also independent of Range(H_n) *provided* that Range$(H_n) \subset P$. This follows from Theorem 5; we leave the details to the reader.

With the assumption that Range$(H_n) \subset P$, $\alpha(2n+1)$ depends only on $\alpha(0),\ldots,\alpha(2n)$. Hence we have a strategy for II in $G(X)$. We show it is a winning strategy. Suppose not. Then some game in which II plays this strategy gives an $\alpha \in X$. Then $F(\alpha)$ is well-ordered. Since $|F(\alpha)| \leqslant \omega$ and $|P| = \kappa$, there is a one-one order preserving mapping H of $F(\alpha)$ into P.

Now we show how I can win in $G'(\alpha)$ against II's winning strategy. He picks $\alpha(2n)$ and picks H_n to be the restriction of H to $F(\alpha(0),\ldots,\alpha(2n))$. Since Range$(H_n) \subset P$, II will pick $\alpha(2n+1)$ each time. Clearly I will then win.

Again this result is best possible; for if $V{=}L_\mu$, then there is a Δ^1_2 set which is indeterminate.

Summing up, we can settle in ZFM some, but by no means all, of the problems left open in ZFC. This suggests that we try to find axioms giving even larger cardinals and use them to solve further problems. Plenty of axioms are known; but no one has succeeded yet in solving more problems with them. One difficulty is finding an extension of the absoluteness theorem. If we change π_2 to π_3, it becomes false; but further hypotheses on the model M might make it true again.

In view of this, we shall not discuss these very large cardinals. Instead we consider a large cardinal which has not yet been shown to be larger than a measurable cardinal.

Let $\kappa \leqslant \lambda$. A *κ-ring* on λ is a set $X \subset \mathrm{PS}(\lambda)$ such that $\lambda \in X$ and X is closed under complements (in λ) and unions of $< \kappa$ sets. A *κ-additive measure* on λ is a mapping μ of X into $\{0, 1\}$, where X is a κ-ring on λ, such that:

(a) if $[x_i \mid i \in I]$ is a set of pairwise disjoint sets in X, and $|I| < \kappa$, then $\mu(\bigcup_{i\in I} x_i) = \sum_{i\in I} \mu(x_i)$;

(b) $\mu(\lambda) = 1$.

If also $X = \mathrm{PS}(\lambda)$, then μ is total. Thus a measure on κ is a total κ-additive measure on κ which is 0 on singletons.

A cardinal κ is (*strongly*) *compact* if $\kappa > \omega$, κ is regular, and for every $\lambda \geqslant \kappa$, every κ-additive measure on λ can be extended to a total κ-additive measure. Note that this definition (unlike the definition of measurable) refers to arbitrarily large cardinals.

Lemma 11. Every compact cardinal κ is measurable.

Proof. For $x \subset \kappa$, let $\mu(x) = 0$ if $|x| < \kappa$, $\mu(x) = 1$ if $|\kappa - x| < \kappa$. Since κ is regular, this is a κ-additive measure. If we extend it to a total κ-additive measure, we obtain a measure on κ.

In trying to settle the converse, it is reasonable to assume the existence of compact cardinals. Thus we have the problem: is "every measurable cardinal is compact" provable, refutable, or independent in ZFC + $\exists\kappa$ (κ compact)? This problem is still unsettled. We have only a partial result when the theory is weakened to ZFM.

Theorem 12 (Vopěnka-Hrbáček [32]). If there is a compact cardinal, then $V \neq L_a$ for all *a*.

Corollary. If ZFM is consistent, then $\exists\kappa$ (κ compact) is not provable in ZFM. Hence $\forall\kappa$ (κ measurable $\rightarrow \kappa$ compact) is not provable.

Proof. If not, we could prove $\exists\kappa$ (κ compact) in $\mathrm{ZFL}_{\kappa,\mu}$; so $\mathrm{ZFL}_{\kappa,\mu}$ would be inconsistent.

Note that Theorem 12 results from Theorem 4 by strengthening the hypothesis and the conclusion. Its proof is similar to that of Theorem 4. This suggests we could strengthen it much as we did Theorem 4. This is done in [32].

Now we prove Theorem 12. We suppose that κ is compact and $V=L_a$, and derive a contradiction. Let $\rho = \max(\kappa, |\mathrm{TC}(a)|$, where $\mathrm{TC}(a)$ is the smallest transitive set including a; and let $\lambda = \rho^+$. For $x \subset \lambda$, set $\mu(x) = 0$ if $|x| < \lambda$ and $\mu(x) = 1$ if $|\lambda - x| < \lambda$. This is a κ-additive measure on λ. Extend it to a total κ-additive measure on λ, which we still designate by μ.

Define V^λ/μ, c, M, and l as in the proof of Theorem 4. (Of course μ is not λ-additive; but only the countable additivity is neede here.) We write p in place of $*$, so $p(x) = l(c(x))$.

Let

$$(V^\lambda/\mu)_1 = [\widetilde{f} \mid f \in V^\lambda \ \& \ |\mathrm{Range}(f)| < \lambda] \ .$$

This is a submodel of V^λ/μ, and includes the image of V under c. We claim that the fundamental theorem holds for $(V^\lambda/\mu)_1$; i.e.,

$$P^{(V^\lambda/\mu)_1}(\widetilde{f}_1,\ldots,\widetilde{f}_n) \leftrightarrow \mu([\sigma \mid P(f_1(\sigma),\ldots,f_n(\sigma))]) = 1 \ .$$

(provided $\widetilde{f}_1,\ldots,\widetilde{f}_n \in (V^\lambda/\mu)_1$). The proof is like that of the fundamental theorem for V^λ/μ, with a slight modification in one case: the proof of $\leftarrow$ when P is $\exists y Q$. From the right side, we see that for almost all σ, there is an $f(\sigma)$ such that $Q(f(\sigma), f_1(\sigma),\ldots, f_n(\sigma))$. Now we may assume that $|\mathrm{Range}(f_i)| < \lambda$ for all i. Hence there are $< \lambda$ n-tuples $f_1(\sigma),\ldots,f_n(\sigma)$. It follows easily that f may be chosen so that $|\mathrm{Range}(f)| < \lambda$. The proof then continues in the usual way.

One consequence is (as before) that c is an elementary embedding of V into $(V^\lambda/\mu)_1$; so the latter is a model of ZFC. Another is that $(V^\lambda/\mu)_1$ is an elementary submodel of (V^λ/μ). For if $\widetilde{f}_1,\ldots,\widetilde{f}_n \in (V^\lambda/\mu)_1$, then the two fundamental theorems give the same conditions for

$$P^{V^\lambda/\mu}(\widetilde{f}_1,\ldots,\widetilde{f}_n)$$

and

$$P^{(V^\lambda/\mu)_1}(\widetilde{f}_1,\ldots,\widetilde{f}_n) \ .$$

Hence the identity mapping I from $(V^\lambda/\mu)_1$ to V^λ/μ is an elementary injection.

Now $(V^\lambda/\mu)_1$, as a submodel of V^λ/μ, is well-founded; so we have a collapsing isomorphism l_1 of $(V^\lambda/\mu)_1$ onto a transitive M_1. Set $p_1(x) = l_1(c(x))$. We then have the diagram

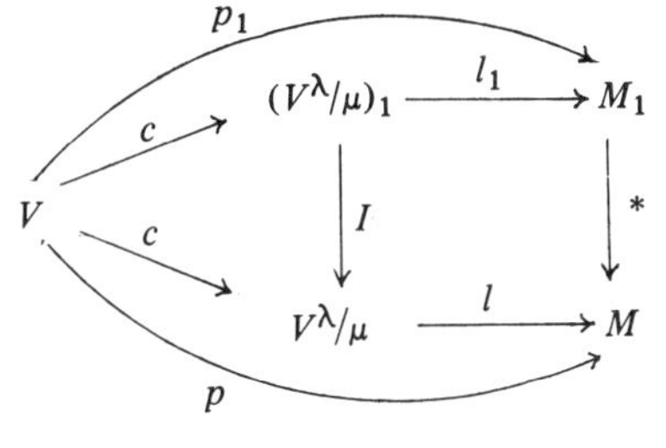

where $*$ is chosen to make the diagram commute.

We prove three facts:

(a) $p(a) = p_1(a)$;

(b) $p(\rho) = p_1(\rho)$;

(c) $p(\lambda) \neq p_1(\lambda)$.

First we show that these give the desired contradiction. Since $V=L_a$ holds in V, $V=L_{p(a)}$ holds in M. Then, noting M contains all the ordinals, $M=L_{p(a)}$. Similarly, $M_1 = L_{p_1(a)}$. Hence by (a), $M=M_1$.

Next, $\lambda = \rho^+$ in V, so

$$p(\lambda) = p(\rho)^+ \text{ in } M,$$

$$p_1(\lambda) = p_1(\rho)^+ \text{ in } M_1.$$

Using $M=M_1$ and (b), these give $p(\lambda) = p_1(\lambda)$, contradicting (c).

Lemma 12. If $f \in V^\lambda$ and $|\mathrm{TC}(\mathrm{Range}(f))| < \lambda$, then $l(\tilde{f}) = l_1(\tilde{f})$.

Proof. We use E-induction on $\tilde{f}$. If $\tilde{g} E \tilde{f}$, then $g(\sigma) \in f(\sigma)$ A.E. Hence $g(\sigma) \in \mathrm{TC}(\mathrm{Range}(f))$ A.E. By changing g without changing $\tilde{g}$, we may suppose that $\mathrm{Range}(g) \subset \mathrm{TC}(\mathrm{Range}(f))$; so $\mathrm{TC}(\mathrm{Range}(g)) \subset \mathrm{TC}(\mathrm{Range}(f))$; so $|\mathrm{TC}(\mathrm{Range}(g))| < \lambda$. Thus $\tilde{g} \in (V^\lambda/\mu)_1$, and, by the induction hypothesis, $l(\tilde{g}) = l_1(\tilde{g})$.

Now

$$l(\tilde{f}) = [l(\tilde{g}) \mid \tilde{g} E \tilde{f}] \ ,$$

$$l_1(\tilde{f}) = [l_1(\tilde{g}) \mid \tilde{g} E \tilde{f} \ \& \ \tilde{g} \in (V^\lambda/\mu)_1] \ .$$

Hence by the above, $l(\tilde{f}) = l_1(\tilde{f})$.

Corollary. If $|\mathrm{TC}(x)| < \lambda$, then $p(x) = p_1(x)$.

From the corollary, we obtain (a) and (b).

Now we examine $p_1(\lambda)$. Each element of $p_1(\lambda)$ is in M_1, so is $l_1(\tilde{f})$ for some $f \in V^\lambda$ with $|\mathrm{Range}(f)| < \lambda$. Since $l_1(\tilde{f}) \in p_1(\lambda) = l_1(c(\lambda))$,

$\tilde{f} E c(\lambda)$; so $f(\sigma) \in \lambda$ A.E. Changing f on a set of measure 0, we have Range$(f) \subset \lambda$. Since $|\text{Range}(f)| < \lambda$ and λ is regular, there is a $\tau < \lambda$ such that Range$(f) \subset \tau$. Then TC(Range$(f)) \subset \tau$; so $|\text{TC}(\text{Range}(f))| < \lambda$, and hence $l_1(\tilde{f}) = l(\tilde{f})$. By Lemma 12. Also $|\text{Range}(f)| < \lambda$ implies $\mu(\text{Range}(f)) = 0$. Thus every element of $p_1(\lambda)$ is $l(\tilde{f})$ for an $f \in \lambda^\lambda$ such that $\mu(\text{Range}(f)) = 0$.

Let i be the identity mapping of λ into λ. Then $i(\sigma) \in c_{\lambda}(\sigma)$ for all σ; so $\tilde{i} E c(\lambda)$; so $l(\tilde{i}) \in p(\lambda)$. If we assume (c) is false, we get $l(\tilde{i}) \in p_1(\lambda)$; so $l(\tilde{i}) = l(\tilde{f})$ with f as above. This gives $\tilde{i} = \tilde{f}$ and hence $\sigma = f(\sigma)$ A.E. But this clearly contradicts $\mu(\text{Range}(f)) = 0$.

Theorem 12 suggests that compact cardinals are not always measurable. We quote without proof another result confirming this impression.

Theorem 13 (Kunen [11]). If there is a compact cardinal, then for every λ there is a transitive model M of ZFC such that $\lambda \in M$ and

$$\vdash_M \text{There are } \lambda \text{ measurable cardinals.}$$

In view of these results, we conjecture that "every measurable cardinal is compact" is not provable in ZFC + $\exists\kappa$ (κ compact). We would hesitate to guess whether it is refutable or independent.

We return to measurable cardinals to consider briefly the problem: do measurable cardinals exist? To some, this question is meaningless; but we believe that Gödel [5] has given convincing reasons for thinking that it is both meaningful and important.

We cannot hope to prove the existence of measurable cardinals in ZFC. What reason could we have for assuming their existence as an axiom? The best reason would be that the axiom was what Gödel terms "intrinsically necessary"; that is, a consideration of the notion of set and of measurable cardinal would convince us that it must be true. Now what is intrinsically necessary is to some extent a matter of judgment. It is very doubtful if anyone would claim that, on the basis of what we know today, the existence of measurable cardinals is intrinsically necessary. This may change as we come to understand more about reflection principles.

However, there are other grounds for the acceptance of axioms. To quote Gödel: "There might exist axioms so abundant in their verifiable consequences, shedding so much light upon a whole field, and yielding such powerful methods for solving problems ... that, no matter whether or not they are intrinsically necessary, they would have to be accepted at least in the same sense as any well-established physical theory." At the time he wrote this, Gödel did not feel that this applied to the existence of measurable cardinals.

The results obtained since then, some of which are described above, have changed the situation somewhat; and new results may change it further. We venture the rash prediction that within a few years, the existence of measurable cardinals will come to be accepted as an axiom of set theory.

We would like to thank Paul Bacsich and Gordon Monro, whose notes on these lectures served as a basis for this article.

References

We include only papers referred to in the article and a few especially important papers on the subject.

[1] P.J. Cohen, Set theory and the continuum hypothesis (W.A. Benjamin, 1966).
[2] Morton Davis, Infinite games of perfect information, in: Advances in game theory, ed. M. Dresher (Princeton, 1964).
[3] H. Gaifman, Measurable cardinals and constructible sets, Notices Am. Math. Soc. 11 (1964) 771.
[4] K. Gödel, The consistency of the axiom of choice and of the generalized continuum hypothesis with the axioms of set theory, Ann. Math. Studies No. 3 (Princeton University Press, Princeton, N.J., 1940).
[5] K. Gödel, What is Cantor's continuum problem? in: Philosophy of mathematics, ed. P. Benacerraf and H. Putnam (1964) pp. 258–213.
[6] K. Gödel, Remarks before the Princeton bicentennial conference on problems in mathematics, in: The undecidable, ed. M. Davis (1965) pp. 84–88.
[7] W. Hanf, Incompactness in languages with infinitely long expressions, Fund. Math. 53 (1964) 309–324.
[8] W. Hanf and D. Scott, Classifying inaccessible cardinals, Notices Am. Math. Soc. 8 (1961) 445.
[9] R. Jensen, Measurable cardinals and the GCH, in: Notes from the symposium on set theory at Los Angeles (1967).
[10] H. Keisler and A. Tarski, From accessible to inaccessible cardinals, Fund. Math. 53 (1964) 225–308.
[11] K. Kunen, Inaccessibility properties of cardinals, Thesis, Stanford University (1968).
[12] K. Kunen, Some results on measurable cardinals, Notices Am. Math. Soc. 16 (1969) 322.
[13] C. Kuratowski, Topologie (1948).
[14] A. Lévy and R. Solovay, Measurable cardinals and the continuum hypothesis, Israel J. Math. 5 (1967) 234–238.
[15] D. Martin, Measurable cardinals and analytic games, to appear.
[16] D. Martin and R. Solovay, A basis theorem for Σ_3^1 sets of reals, Ann. Math. 89 (1969) 138–159.
[17] A. Mostowski, An undecidable arithmetical statement, Fund. Math. 36 (1949) 143–164.

[18] J. Mycielski, On the axiom of determinateness, Fund. Math. 53 (1964) 205–224.
[19] J. Paris, Boolean extensions and large cardinals, Thesis, University of Manchester (1969).
[20] K. Prikry, Changing measurable into accessible cardinals, Thesis, University of California (1968).
[21] F. Rowbottom, Large cardinals and small constructible sets, Thesis, University of Wisconsin (1964).
[22] D. Scott, Measurable cardinals and constructible sets, Acad. Polon. Sci., Ser. Sci. Math. Astron. et Phys. 7 (1961) 145–149.
[23] J. Shoenfield, The problem of predicativity, in: Essays in the foundations of mathematics, Jerusalem (1961) pp. 226–233.
[24] J. Silver, Some applications of model theory in set theory, Thesis, University of California (1965).
[25] J. Silver, The consistency of the GCH with the existence of a measurable cardinal, in: Notes from the symposium on set theory at Los Angeles (1967).
[26] R. Solovay, A model of set-theory in which every set of reals is Lebesgue measurable, to appear.
[27] R. Solovay, On the cardinality of Σ_2^1 sets of reals, in: Foundations of mathematics, Symposium papers commemorating the sixtieth birthday of Kurt Gödel (Springer-Verlag, 1969) pp. 58–73.
[28] R. Solovay, A non-constructible Δ_3^1 sets of integers, Trans. Am. Math. Soc. 127 (1967) 50–75.
[29] R. Solovay, Real-valued measurable cardinals, to appear.
[30] A. Tarski, Some problems and results relevant to the foundations of set theory, in: Logic, methodology and philosophy of science, Proceedings of the 1960 International Congress, Stanford (1962).
[31] S. Ulam, Zur Masstheorie in der allgemeinen Mengenlehre, Fund. Math. 16 (1930) 140–150.
[32] P. Vopěnka and K. Hrbáček, On strongly measurable cardinals, Bull. Acad. Polon. Sci., Ser. Math., Astron. et Phys. 14 (1966) 587–591.

PART II

INVITED PAPERS ON SET THEORY

INFINITARY METHODS IN THE MODEL THEORY OF SET THEORY*

Jon BARWISE
Yale University, New Haven, Conn., USA

1. Introduction

This paper is a contribution to the study of end extensions of models of ZF, Zermelo-Fraenkel set theory. This study was begun by Keisler and Morley in [5] and continued in Keisler-Silver [6]. One of the most striking results to date is the following:

Theorem (Keisler-Morley). *Every countable model of* ZF *has a proper elementary end extension.*

The main result of this paper is the following:

Theorem. Every countable model of ZF *has a proper end extension which is a model of* ZF + V=L.

On the other hand, we show that if ZF is consistent then there are uncountable models of ZF with no end extensions satisfying ZF + V=L.

The results just mentioned are proved in Section 3. The necessary preliminaries are given in Section 2. Related topics are discussed in the remaining sections. The title of the paper refers to the fact that all the results of this paper are proved using methods and results from infinitary logic. Some of the result of Section 5 on collapsing cardinals were announed in [2].

2. Preliminaries

We assume that the reader is familiar with the hierarchy put on the for-

* This research was partially supported by N.S.F. Grant GP-8625.

mulas of the language of set theory by Lévy in [7]. Given structures $\mathcal{M} = \langle M, E\rangle$ and $\mathcal{N} = \langle N, F\rangle$ (where E and F are binary relations) with $\mathcal{M}$ a substructure of $\mathcal{N}$, and given a subset X of M, we write

$$\mathcal{M} \prec_p \mathcal{N} \qquad [wrt\ X]$$

if for any Σ_p formula $\varphi(v_1 \dots v_n)$ and any $a_1 \dots a_n \in X$,

$$\mathcal{M} \models \varphi[a_1 \dots a_n] \quad \text{iff } \mathcal{N} \models \varphi[a_1 \dots a_n] .$$

We write $\mathcal{M} \prec_p \mathcal{N}$ if $\mathcal{M} \prec_p \mathcal{N}$ $[wrt\ M]$ and $\mathcal{M} \equiv_p \mathcal{N}$ if $\mathcal{M} \prec_p \mathcal{N}$ $[wrt\ \emptyset]$. We write, as usual, $\mathcal{M} \prec \mathcal{N}$ if $\mathcal{M} \prec_p \mathcal{N}$ for all $p < \omega$ and $\mathcal{M} \equiv \mathcal{N}$ if $\mathcal{M} \equiv_p \mathcal{N}$ for all $p < \omega$.

We say that the extension $\mathcal{N}$ of $\mathcal{M}$ is an *end extension* of $\mathcal{M}$, and write $\mathcal{M} \subseteq_e \mathcal{N}$, if aFb and $b \in M$ implies $a \in M$ for all $a, b \in N$. We write $\mathcal{M} \subset_e \mathcal{N}$ if $\mathcal{N}$ is a proper end extension of $\mathcal{M}$.

Given a model $\mathcal{M} = \langle M, E\rangle$ of ZE we use a superscript $\mathcal{M}$ to denote the value of a term or formula in the model $\mathcal{M}$. For example, $\omega_1^{\mathcal{M}}$ is that element $a \in M$ such that

$$\mathcal{M} \models \text{“}a \text{ is the first uncountable ordinal”}.$$

Given an $a \in M$ we write $a_E = \{b \in M \mid bEa\}$ so that $a_E = a$ if M is a transitive set and $E = \in_M$, i.e., $E = \in \cap M^2$. We say that an element $a \in M$ is definable if there is a (finite) formula $\varphi(x)$ such that

$$a = \text{the unique } b \text{ such that } \mathcal{M} \models \varphi[b] .$$

$D(\mathcal{M})$ is the substructure $\langle M_1, E \restriction M_1\rangle$ of $\mathcal{M}$ whose domain M_1 is the set of definable elements of $\mathcal{M}$. We say that $\mathcal{M}$ is *pointwise definable* if $D(\mathcal{M}) = \mathcal{M}$, that is, if every element of M is definable in $\mathcal{M}$.

By ZF we mean the axioms of Zermelo-Fraenkel set theory. AC is the axiom of choice, V = L the axiom of constructibility, and V = OD is the axiom “all sets are ordinal definable” (cf. Myhill and Scott [8]). Throughout this paper we use Q for a set of finite sentences in the language of set theory with only $\in$ as nonlogical symbol.

A model $\mathcal{M} = \langle M, E\rangle$ of ZF is *standard* if M is transitive and $E = \in_M$. $\mathcal{M}$ is an ω-model if $\langle \omega^{\mathcal{M}}, E \restriction \omega^{\mathcal{M}}\rangle$ is isomorphic to $\langle \omega, \in_\omega\rangle$.

One of the main results used below is the following:

Shoenfield-Levy Lemma. If θ is a Σ_1^{ZF} formula with no free variables except x and y then the following is provable in ZF:

$$y < \omega_1^{\mathrm{L}} \wedge \exists x \theta(x, y) \rightarrow \exists x [x \in L \wedge Od(x) < \omega_1^{\mathrm{L}} \wedge \theta(x, y)]$$

where ω_1^{L} is the least ordinal not constructibly denumerable.

The only essential strengthening of this result over Theorem 43 in Lévy [7] is the elimination of the axiom DC. That this is possible was pointed out to us by K. Kunen, using a simple forcing argument. To be a little more specific, one shows that Lemma 40 of [7] is true in every countable model $\mathcal{M}$ of ZF and hence is provable in ZF. To do this one goes to a Cohen extension $\mathcal{M}'$ of $\mathcal{M}$ where the set s of Lemma 40 is countable, but where there are no new constructible sets, and then applies Lemma 39 of [7].

We also assume in the following that the reader has some familiarity with the infinitary language $\mathcal{L}_A$ where A is an admissible set. The results we need are contained in pp. 228–239 of [1]. Readers familiar with other treatments of this topic should still be able to understand our proofs.

Some final notation and terminology. Cardinals are initial ordinals. $P(x)$ is the power set of x, $|x|$ is the cardinality of x and $TC(x)$ is the transitive closure of x. The sets R_α are defined by

$$R_\alpha = \bigcup_{\beta<\alpha} P(R_\beta)$$

using transitive induction. The set-theoretic rank of x, $rk(x)$, is the least α such that $x \in R_{\alpha+1}$.

3. End extensions satisfying V=L.

The Theorem stated in the Introduction follows from Theorem 3.3. Theorem 3.1. is a special case of 3.3. We present it first for expository reasons.

Let Q be a theory in the language of set theory. We say that Q is *absolute for L* if for every axiom φ of Q, the relativization $\varphi^{(L)}$ of φ to the constructible sets is a consequence of Q. Thus, for example, ZF and ZF + "there exists an inaccessible cardinal" are absolute for L whereas ZF + "there exists a measurable cardinal" is not.

Theorem 3.1. Let Q *be a r.e. theory which contains* ZF *and which is absolute* *for L. Let $\mathcal{M}$ be a countable standard model of* Q. *There is an end extension $\mathcal{N}$ of $\mathcal{M}$ which is a model of* Q + V=L. *Moreover, $\mathcal{N}$ can be chosen so that*:

(1) *$\mathcal{N}$ is pointwise definable*

(2) *For every $a \in M$,*

$$\mathcal{N} \models a \text{ is countable}.$$

Proof. Since $\mathcal{M} = \langle M, E\rangle$ is standard we can assume that M is a transitive set and $E = \in_M$. Let α be the least ordinal not in M. We define T to be the set of the following sentences of the infinitary language $\mathcal{L}_M$ (where $\bar{a}$ is the constant symbol $c_{(0,a)}$ of [1] and c is c_0):

(i)	φ	for each $\varphi \in$ Q + V=L
(ii)	c is a countable ordinal	
(iii)	$c > \bar{\beta}$	for each $\beta < \alpha$
(iv)	$\forall v_0\, [v_0 \in \bar{a} \leftrightarrow \bigvee_{b\in a} v_0 = \bar{b}]$	for each $a \in M$
(v)	$\forall v_1 \bigvee_{\varphi(v_0)} \forall v_0\, [\varphi(v_0) \leftrightarrow v_0 = v_1]$.	

The disjunction in (v) is taken over all finite formulas $\varphi(v_0)$ with just one free variable v_0. The theorem will follow if we can show that T has a model. Let $\theta_0(x)$ be a Σ_1 formula such that

$$T = \{\varphi \in M : \mathcal{M} \models \theta_0(\varphi)\},$$

and let Λ be the logically false sequent $\vdash (c \neq c)$. If T has no model then there is a derivation $\mathcal{D}$ of Λ from T and by the absoluteness lemma 2.10 of [1], $\mathcal{M}$ is a model of the Σ_1^{ZF} formula $\exists x \theta_1(x)$ where $\theta_1(x)$ is:

> x is a derivation of Λ all of whose nonlogical axioms φ satisfy $\theta_0(\varphi)$.

If $L^{\mathcal{M}}$ is the submodel of $\mathcal{M}$ of constructible sets, then $L^{\mathcal{M}} \models \exists x \theta_1(x)$ by the Shoenfield-Lévy Lemma. Also note that $L^{\mathcal{M}} \models$ Q + V=L since Q is absolute for L. Let β be the unique ordinal such that $\langle L_\beta, \in\rangle$ is isomorphic to $D(L^{\mathcal{M}})$. Then L_β is a pointwise definable model of Q + V=L + $\exists x \theta_1(x)$ since $D(L^{\mathcal{M}}) \equiv L^{\mathcal{M}}$. Now since $L_\beta \models \exists x \theta_1(x)$, $L_\beta \models \exists x\, [|\mathrm{TC}(x)| < \omega_1 \wedge \theta_1(x)]$ by

Theorem 36 of Levy (or use the Shoenfield-Levy lemma again). Let $\mathcal{D}$ be a derivation in L_β such that

$$L_\beta \models [|TC(\mathcal{D})| < \omega_1 \wedge \theta_1(\mathcal{D})] .$$

Let γ be the least ordinal not in $TC(\mathcal{D})$ so that $L_\beta \models \gamma < \omega_1$. Now on the one hand, $T_0 = \{\varphi \in TC(\mathcal{D}) : L_\beta \models \theta_0(\varphi)\}$ has no model since $\mathcal{D}$ is a proof that it is contradictory. On the other hand, T_0 only asserts (iii) for $\beta < \gamma$ and (iv) for $a \in TC(\mathcal{D})$. Thus, L_β is a model of T_0 if we assign a to $\bar{a}$ for all $\bar{a}$ mentioned in T_0 and assign γ to c. This contradiction establishes the theorem.

We now turn to the generalization of 3.1 to non-standard models $\mathcal{M}$. Let $\mathcal{M} = \langle M, E\rangle$ be a model of the axiom of extensionality. Let B be the set of those $a \in M$ such that there is no infinite sequence $a_0, \ldots, a_n, \ldots$ of elements of M with $a_0 = a$ and $a_{n+1} E a_n$ for each $n < \omega$. That is, B is the set of those $a \in A$ for which the following inductive definition is defined:

$$i(a) = \{i(b) : b \in a_E\} .$$

The function i is an isomorphism between $\langle B, E \cap B^2\rangle$ and $\langle A, \in_A\rangle$ for a transitive set A. As in [1], we call this A the standard part of $\mathcal{M}$, $A = sp(\mathcal{M})$, and we let $osp(\mathcal{M})$ be the least ordinal not in A. We can identify B with A without confusion.

Assume now that $\mathcal{M}$ is a model of ZF. In $\mathcal{M}$ there are objects x which satisfy:

$$\mathcal{M} \models x \text{ is a formula} ,$$

where "x is a formula" is the Σ_1 definition of Lemma 2.4 in [1]. We call such an x an $\mathcal{M}$-formula. The most obvious difficulty in extending Theorem 3.1 is that some of these $\mathcal{M}$-formulas are not really formulas at all. If x is an $\mathcal{M}$-formula and $x \in sp(\mathcal{M})$ then x really is a formula, of course. But consider the $\mathcal{M}$-sentence denoted by:

$$\forall x\, [x \in \bar{a} \leftrightarrow \bigvee \{x = \bar{b} : bEa\}] .$$

If $a \in M - sp(\mathcal{M})$ then this sentence is not in $sp(\mathcal{M})$ even though it is a perfectly good sentence. A moment's reflection shows that any $\mathcal{M}$-formula with $r^{\mathcal{M}}(\varphi) < osp(\mathcal{M})$ can be considered a real formula of $L_{\infty\omega}$. Here $r(\varphi)$ is defined as usual by: $r(\varphi) = 0$ if φ is atomic; $r(\varphi) = r(\psi) + 1$ if φ is $\neg\psi$, $\forall v\psi$ or $\exists v\psi$; and $r(\varphi) = sup\ \{r(\psi) + 1 : \psi \in \Phi\}$ if φ is $\wedge\Phi$ or $\vee\Phi$.

We need similar remarks for derivations. Let us define the length of a derivation $\mathcal{D}$, $l(\mathcal{D})$, in the obvious way:

$$l(\mathcal{D}) = sup\{l(\mathcal{D}')+1 : \mathcal{D}' \text{ an subderivation of } \mathcal{D}\},$$

and let the *complexity* of $\mathcal{D}$, $c(\mathcal{D})$ be the maximum of $l(\mathcal{D})$ and $sup\{r(\varphi) : \varphi \text{ occurs in } \mathcal{D}\}$.

If T is a definable class (over $\mathcal{M}$) of $\mathcal{M}$-formulas and $\mathcal{D}$ is an $\mathcal{M}$-derivation from T, then we can consider $\mathcal{D}$ to be a real derivation if $c^{\mathcal{M}}(\mathcal{D}) < osp(\mathcal{M})$; a real derivation in the sense that the sequent S proved by $\mathcal{D}$ really is a logical consequence of the axioms used in $\mathcal{D}$. This is seen by induction on $l(\mathcal{D})$.

With these few remarks the proof of 3.1 can be given, with only minor modifications, assuming only that $(\omega_1^L)^{\mathcal{M}}$ is standard instead of assuming that all of $\mathcal{M}$ is standard.

To estbalish the general result, however, we need some more remarks. Let us again assume that $\mathcal{M} \models$ ZF as above. Let T be a set of $\mathcal{M}$-formulas definable over $\mathcal{M}$ by a formula $\theta_0(x)$. Corresponding to Lemma 2.10 of [1], we have the following:

Lemma 3.2. If $\mathcal{D}$ is a derivation from a set T_0 of sentences where $T_0 \subseteq T$ of an $\mathcal{M}$-sequent S then there is an $\mathcal{M}$-derivation $\mathcal{D}'$ from T of *S with $c^{\mathcal{M}}(\mathcal{D}') < osp(\mathcal{M})$.*

Proof. The only case which differs from the proof of 2.10 in [1] is the case where $\mathcal{D} = (\vdash\bigwedge, I, f, \Gamma \vdash \Delta \cup \{\bigwedge\Delta_0\})$, where $\Gamma \vdash \Delta \cup \{\bigwedge\Delta_0\}$ is an $\mathcal{M}$-sequent as well as a sequent and hence has $r(\bigwedge(\Gamma \cup \Delta \cup \{\bigwedge\Delta_0\})) < osp(\mathcal{M})$. By induction on $l(\mathcal{D})$, for every $\varphi \in \Delta_0$ there is an $\mathcal{M}$-derivation $\mathcal{D}_\varphi$ of $\Gamma \vdash \Delta \cup \{\varphi\}$ such that $c(\mathcal{D}_\varphi) < osp(\mathcal{M})$. Let β be the least ordinal of $\mathcal{M}$ such that

$$\mathcal{M} \models \forall\varphi \in \Delta_0 \, \exists \mathcal{D} [\mathcal{D} \text{ is a derivation from } T \text{ of}$$

$$\Gamma \vdash \Delta \cup \{\varphi\} \wedge c(\mathcal{D}) < \beta] .$$

By the above, $\beta < osp(\mathcal{M})$. Now we may choose a function $g \in \mathcal{M}$ such that

$$\mathcal{M} \models \forall\varphi \in \Delta_0 \, [g(\varphi) \text{ is a non-empty set of derivations of}$$

$$\Gamma \vdash \Delta \cup \{\varphi\} \text{ from } T \text{ and all } \mathcal{D} \in g(\varphi) \text{ have}$$

$$c(\mathcal{D}) < \beta] .$$

(*Note*: we are using the fact that $\mathcal{M} \models$ ZF, not just $\mathcal{M} \models$ KP.) Then $\mathcal{D}' = (\vdash\wedge, I, g, \Gamma \vdash \Delta \cup \{\wedge \Delta_0\})$ is an $\mathcal{M}$-derivation of the required sequent with $c^{\mathcal{M}}(\mathcal{D}') < osp(\mathcal{M})$.

Theorem 3.3. Let Q *be an r.e. theory which contains* ZF *and which is absolute for* L. *Let* $\mathcal{M}$ *be a countable model of* Q. *There is an end extension* $\mathcal{N}$ *of* $\mathcal{M}$ *which is a model of* Q + V=L *and such that for every* $a \in M$,

$$\mathcal{N} \models a \text{ is countable}.$$

If $\mathcal{M}$ *is an* ω*-model then* $\mathcal{N}$ *can be chosen to be pointwise definable.*

Proof. Although it is not really recessary, it simplifies matters a little to assume that every ordinal $\alpha < osp(\mathcal{M})$ is constructibly countable in $\mathcal{M}$. This suffices by remarks made above. We won't worry at present about making $\mathcal{N}$ pointwise definable. Since the theory Q + V=L is r.e. it can be defined by

$$\varphi \in \mathrm{Q} + \mathrm{V{=}L} \quad \text{iff } \langle L_\omega, \in\rangle \models \exists n \theta_0(\varphi, n)$$

for some Δ_0 formula θ_0. Let T be the following set of $\mathcal{M}$-sentences:

(i) $\varphi \wedge (\bar{n} = \bar{n})$ if $\langle L_\omega, \in\rangle \models \theta_0(\varphi, n)$

(ii)–(iv) as in 3.1.

Let $\theta_1(x)$ be the Σ_1^{ZF} formula

$$\text{“}x \text{ is an } \mathcal{M}\text{-sentence”} \wedge (\theta^1 \vee \ldots \vee \theta^4)$$

where $\theta^k(x)$ is defined by (k) below, $k = 1, \ldots, 4$.

(1) x is of the form $\varphi \wedge (\bar{n} = \bar{n})$ for some natural number n and $\theta_0(\varphi, n)$
(2) x is "c is a countable ordinal"
(3) x is of the form $(c > \bar{\beta})$ for some ordinal β
(4) x is of the form $\forall v[v \in \bar{a} \leftrightarrow \vee \{v = b : bEa\}]$ for some a.

Then $\theta_1(x)$ satisfies:

$$T = \{x \in M : \mathcal{M} \models \theta_1(x) \quad \text{and} \quad r^{\mathcal{M}}(x) \text{ is standard}\}.$$

If T has no model then there is a derivation $\mathcal{D}$ from T of Λ. Applying 3.2, there is a standard ordinal α of $\mathcal{M}$ such that $\mathcal{M} \models \exists x \theta_2(x, \alpha)$ where $\theta_2(x, y)$ is:

> "x is *a* derivation of Λ all of whose nonlogical axioms φ satisfying $\theta_1(\varphi)$ and $c(x) = y$".

Recalling that α is constructibly countable in $\mathcal{M}$ we apply the Shoenfield-Levy lemma inside $\mathcal{M}$ and conclude that $L^{\mathcal{M}} \models \exists x \theta_2(x, \alpha)$ and hence, as before, $L^{\mathcal{M}} \models \exists x[|TC(x)| < \omega_1 \wedge \theta_1(x, \alpha)]$. Choose a $\mathcal{D} \in L^{\mathcal{M}}$ such that $L^{\mathcal{M}} \models [|TC(\mathcal{D})| < \omega_1 \wedge \theta_1(\mathcal{D}, \alpha)]$ and let

$$T_0 = \{x \in L^{\mathcal{M}} : L^{\mathcal{M}} \models \theta_1(x) \wedge x \in TC(\mathcal{D}) \wedge r(x) \leqslant \alpha\}.$$

We see that T_0 is an inconsistent set of sentences since $c(\mathcal{D}) \leqslant \alpha$ and $\mathcal{D}$ proves that T_0 is inconsistent. But, as before, $L^{\mathcal{M}}$ is a model of T_0 if we assign a to $\bar{a}$ for all $\bar{a}$ mentioned in T_0 and assign the least ordinal of $\mathcal{M}$ greater than those mentioned in T_0 to c. If $\mathcal{M}$ is an ω-model, then one can modify the proof by allowing axiom (v) of 3.1 to occur in T. (We need $\mathcal{M}$ to be an ω-model since otherwise (v) will not be an $\mathcal{M}$-sentence, its rank being $\omega+1$.)

At the time of our Manchester talk, we knew 3.2 only for ω-models. The proof is a little simpler in this case since the function $c(\mathcal{D})$ can be replaced by $l(\mathcal{D})$. A few days after proving the full 3.2 we learned by letter that K. Kunen had also seen how to extend our earlier result to get the full result.

After our talk in Manchester, several people called our attention to Gaifman's abstract [3] where the following is announced. Every countable model of ZF + "there exists arbitrarily large inaccessibles" has an end extension satisfying ZF + V=L. Using 3.2 we can strengthen this by getting an end extension satisfying ZF + V=L + "there exist arbitrarily large inaccessibles". Of course we can also apply 3.2 to theories like ZF, ZF + GCH, ZF + ¬Consis(ZF), ZF + "there is no standard model of ZF" and the like. We also note that the last mentioned theory shows we cannot strengthen 3.1 by requiring the model $\mathcal{N}$ to be wll founded, even though $\mathcal{M}$ is. On the other hand, we could strengthen 3.1 (though not 3.2) by weakening the requirement that Q be r.e. For example, if Q is Π^1_1, Π^1_1 in 0, etc., the theorem still holds. What is needed is that Q be definable by a Σ_1 formula without parameters over $\mathcal{M}$. Finally, we show that the assumption that $\mathcal{M}$ is countable is essential for 3.1 and 3.2. If $\mathcal{M}$ were a standard model with uncountably many ordinals such that $\mathcal{M} \models$ ZF + "there is a non-constructible subset of ω" then of course no end extension of $\mathcal{M}$ could satisfy ZF + V=L, since the ordinal constructing any subset of ω would already lie in $\mathcal{M}$. We can strengthen this argument to prove the following:

Theorem 3.4. Let Q *be any theory containing* ZF *such that* Q + V≠L *is consistent. There is an uncountable model* $\mathcal{M}$ *of* Q *such that no end extension of* $\mathcal{M}$ *is a model of* ZF + V=L.

Proof. Suppose, on the contrary, that every uncountable model of Q has an end extension satisfying ZF + V=L. We will show that Q $\vdash$ V=L by showing that V = L is true in every countable model $\mathcal{M} = \langle M, E\rangle$ of Q. It suffices to show that every $x \in M$ which is transitive and infinite in $\mathcal{M}$ is constructible in $\mathcal{M}$. Let a be a cardinal (i.e. initial ordinal) of $\mathcal{M}$ such that

$$\mathcal{M} \models rk(x) < a \, .$$

Let $\mathcal{M}' = \langle M', E'\rangle$ be an elementary extension of $\mathcal{M}$ such that

$$(1) \qquad \langle a_E, E \upharpoonright a_E\rangle \subseteq_e \mathcal{M}' \, ,$$

$$(2) \qquad |a_{E'}| = \omega_1 \, .$$

Such an $\mathcal{M}'$ exists by Theorem 2.2 of Keisler-Morley [5]. Note that as a consequence of $\mathcal{M} \prec \mathcal{M}'$ and (1) we have $x_E = x_{E'}$. Now let $\mathcal{M}' \subseteq_e \mathcal{N}$ where $\mathcal{N} \models$ ZF + V=L. There is an ordinal b (in the sense of $\mathcal{N}$) such that:

$$\mathcal{N} \models \text{"}x \text{ is the } b\text{th constructible set and } |b| = |x|\text{"} \, .$$

But then b_F is countable since there is a function mapping b_F one-one onto $x_F = x_{E'} = x_E$ and x_E is countable. Since $|a_F| = \omega_1$ we have

$$\mathcal{N} \models b < a \, ,$$

so $b \in M'$, and hence

$$\mathcal{M}' \models \text{"}x \text{ is the } b\text{th constructible set"} \, ,$$

since $\mathcal{M}'$ and $\mathcal{N}$ are models of ZF and $\mathcal{M}' \subseteq_e \mathcal{N}$, and the predicate "$x$ is the yth constructible set" is Δ_1 relative to ZF (or even KP). But then $\mathfrak{M}' \models$ "x is constructible" so $\mathfrak{M} \models$ "x is constructible".

4. Pointwise definable end extensions

Let $\mathfrak{M}$ be a countable ω-model for ZF. We know from Section 3 that there is a pointwise model $\mathfrak{N}$ of ZF such that $\mathfrak{M} \subset_e \mathfrak{N}$. The question which we discuss here is: How similar can we make $Th(\mathfrak{M})$ and $Th(\mathfrak{N})$? For example, can we find such an $\mathfrak{N}$ such that $\mathfrak{M} \equiv \mathfrak{N}$? The following lemma shows various ways in which this is too much to expect.

Lemma 4.1. Let $\mathfrak{M} = \langle M, E\rangle$ *and* $\mathfrak{N} = \langle N, F\rangle$ *be structures.*

(a) *If* $\mathfrak{M} \models$ ZF *and* $\mathfrak{M}$ *is pointwise definable then* $\mathfrak{M} \models$ V=OD.

(b) *If* $\mathfrak{M}$ *and* $\mathfrak{N}$ *are pointwise definable and* $\mathfrak{M} \equiv \mathfrak{N}$ *then there is a unique isomorphism of* $\mathfrak{M}$ *onto* $\mathfrak{N}$.

(c) *If* $\mathfrak{M}$ *and* $\mathfrak{N}$ *are pointwise definable models of* ZF *and* $\mathfrak{M} \equiv \mathfrak{N}$ *then* $sp(\mathfrak{M}) = sp(\mathfrak{N})$.

(d) *If* $\mathfrak{M}$ *is a standard model of* ZF *and* $\mathfrak{M} \subseteq_e \mathfrak{N}$ *where* $\mathfrak{N}$ *is pointwise definable then* $\mathfrak{M} \equiv \mathfrak{N}$ *implies* $\mathfrak{M} = \mathfrak{N}$.

Proof. (a) follows from the reflection principle and the definition of being ordinal definable. To probe (b), assume that $\mathfrak{M}$ and $\mathfrak{N}$ are pointwise definable and $\mathfrak{M} \equiv \mathfrak{N}$. Any isomorphism f of $\mathfrak{M}$ onto $\mathfrak{N}$ must satisfy:

$$\mathfrak{M} \models \varphi[a] \qquad \text{iff } \mathfrak{N} \models \varphi[f(a)]$$

for all formulas $\varphi(x)$ such that $\mathfrak{M} \models \exists! x\varphi(x)$. But since $\mathfrak{M}$ and $\mathfrak{N}$ are pointwise definable this condition uniquely defines a function from M onto N which is clearly an isomorphism. This isomorphism takes the well founded part of $\mathfrak{M}$ onto the well founded part of $\mathfrak{N}$ and hence $sp(\mathfrak{M})$ is isomorphic to $sp(\mathfrak{N})$. But if x and y are transitive sets and $\langle x, \in_x\rangle$ is isomorphic to $\langle y, \in_y\rangle$ then $x = y$. Hence (c) is true. To prove (d), let $\mathfrak{M}_0 = sp(D(\mathfrak{M}))$. Since $\mathfrak{N} \models$ V = OD and $\mathfrak{M} \equiv \mathfrak{N}$, $\mathfrak{M} \models$ V = OD so there is a definable well ordering $\mathfrak{M}$ according to Myhill-Scott [8]. Hence $D(\mathfrak{M}) \prec \mathfrak{M}$ so $\mathfrak{M}_0 \equiv \mathfrak{M}$ and $\mathfrak{M}_0$ is pointwise definable. Now since $\mathfrak{M}_0 \equiv \mathfrak{N}$ we have

$$\mathfrak{M}_0 = sp(\mathfrak{M}_0) = sp(\mathfrak{N})$$

by (c), and we also know that $\mathfrak{N}$ is well founded by (*b*). Since $\mathfrak{M} \subseteq_e \mathfrak{N}$ and $\mathfrak{M}$ is standard, it suffices to show that $\mathfrak{M} = sp(\mathfrak{N})$, since nothing outside $\mathfrak{M}$ can be collapsed to something inside $\mathfrak{M}$. But $\mathfrak{M}_0 \subseteq \mathfrak{M} \subseteq sp(\mathfrak{N})$ so the conclusion follows from the equality displayed above.

The following shows that even though we cannot in general require $Th(\mathfrak{M}) = Th(\mathfrak{N})$, we can push the point of divergence for the two theories out as far as we want.

Theorem 4.2. Let Q *be any r.e. (or even* Π^1_1*) theory containing* ZF + V=OD *and let* $\mathfrak{M}$ *be a countable* ω*-model of* Q. *For any* $p < \omega$ *there is a pointwise definable model* $\mathfrak{N}$ *of* Q *such that*

(1) $\mathfrak{M} \subset_e \mathfrak{N}$

(2) $\mathfrak{M} \equiv_p \mathfrak{N}$.

Proof. Let $\mathfrak{M} = \langle M, E\rangle$ be a countable ω-model of the r.e. theory Q. We identify the well founded part of $\mathfrak{M}$ with $sp(\mathfrak{M})$ be usual. Let Φ be the set of finite sentences $\varphi \in \Sigma_p \cup \Pi_p$ (considered as $\mathfrak{M}$-sentences) without parameters which are true in $\mathfrak{M}$. We need to see that the following set T of $\mathfrak{M}$-sentences has a model:

(i) $\bigwedge Q$

(ii) $\forall v_0 [v_0 \in \bar{a} \leftrightarrow \bigvee \{v_0 = \bar{b} : bEa\}$ all $a \in M$

(iii) $c \neq \bar{a}$ all $a \in M$

(iv) $\forall v_1 \bigvee_{\varphi(v_0)} \forall v_0 [\varphi(v_0) \leftrightarrow v_0 = v_1]$

(v) $\bigwedge \Phi$.

Let $\mathfrak{M}_0$ be the result of collapsing the well founded part of $D(\mathfrak{M})$ to $sp(D(\mathfrak{M}))$ so that:

(a) $\mathfrak{M}_0$ is isomorphic to $D(\mathfrak{M})$ and hence $\mathfrak{M}_0$ is pointwise definable

(b) $\mathfrak{M}_0 \equiv \mathfrak{M}$.

We see that (b) holds since $\mathfrak{M} \models$ ZF + V=OD implies $D(\mathfrak{M}) \prec \mathfrak{M}$. The set T is definable over $\mathfrak{M}$ by a Σ_1 formula with parameter Φ. The main point in the proof is that the element Φ is definable in $\mathfrak{M}$ and by a formula which also defines Φ in $\mathfrak{M}_0$, as follows from Theorem 17 of Lévy [7]. The remainder of the proof is very similar to the proof of 3.3, and so is omitted. The proof in the case where Q is π_1^1 is similar, but uses Lemma 3.3 of [1].

We do not know whether 4.2 holds for non-ω-models.

5. Collapsing cardinals

In 3.3 we collapsed all cardinals of $\mathcal{M}$ to ω in $\mathcal{N}$. This is a special case of the following:

Theorem 5.1. Let Q *be an r.e. theory containing* ZF + AC, *let* $\mathcal{M} = \langle M, E\rangle$ *be a countable model for* Q *and let a be an infinite cardinal of* $\mathcal{M}$. *There is a model* $\mathcal{N}$ *of* Q *such that*:

(1) $\mathcal{M} \subset_e \mathcal{N}$

(2) *For every* $x \in M$ *and every* $b \in a_E \cup \{a\}$

(a) $\mathcal{N} \models |x| \leqslant a$

(b) *if* $\mathcal{M} \models |x| = b$ *then* $\mathcal{N} \models |x| = b$.

Proof. Let us prove the special case where $\mathcal{M}$ is standard, leaving it to the reader to make the few necessary changes required for the full result. The set T of sentences of $\mathcal{L}_M$ is the following, where $d = \{b \in M : \mathcal{M} \models b$ is a cardinal $\leqslant a\}$:

(i) φ for all $\varphi \in$ Q

(ii) as in 3.1(iii)

(iii) as in 3.1(iv)

(iv) $|c| = \bar{a}$ and c is an ordinal

(v) $\bigwedge_{b \in d} [\bar{b}$ is a cardinal$]$.

This set T is definable by a Σ_1 formula with parameter d. If there is a derivation $\mathcal{D}$ of Λ from T then using 2.10 of [1] and Theorem 36 of Lévy [7] there is a derivation $\mathcal{D}' \in M$ of Λ from T such that $\mathcal{M} \models |TC(\mathcal{D}')| \leqslant a$. But $\mathcal{D}'$ can use axiom (ii) only for $\beta < \gamma$ where $|\gamma| = a$ in $\mathcal{M}$. This is a contradiction.

Corollary 5.2. Let $\mathcal{M} = \langle M, E\rangle$ *be a countable model of* Q, *where* Q *is* r.e.

and contains ZF + AC, *and let a be an infinite cardinal of* $\mathfrak{M}$. *There is a model* $\mathfrak{N} = \langle N, F \rangle$ *of* Q *such that*:

(1) $\mathfrak{M} \subset_e \mathfrak{N}$

(2) *If* $\mathfrak{M} \models$ (*b is a cardinal* $\leqslant a$) *then* $\mathfrak{N} \models$ (*b is a cardinal*).

(3) *For* $b \in N$, *if* $\mathfrak{N} \models$ (*b is a cardinal* $> a$) *then* $|b_F| = \omega_1$.

Proof. This follows from 5.1 and Theorem 2.2 of Keisler and Morley [5].

6. A theorem of Friedman

In this section we use the compactness theorem of [1] to prove the following theorem of Friedman [3]. For the axioms of KP see p. 232 of [1].

Theorem. Let Q *be a theory containing* KP *and suppose that* Q *is* Π^1_1 *in* X *for some* $X \subseteq \omega$. *Let* α *be the least ordinal not recursive in* X. *Then one of the following hold*:

(1) For some Δ^1_1 in X subset Q_0 of Q there is a $\beta < \alpha$ such that $osp(\mathfrak{M}) < \beta$ for all models $\mathfrak{M}$ of Q_0,

(2) *There is a model* $\mathfrak{M}$ *of* Q *such that* $osp(\mathfrak{M}) = \alpha$.

Proof. Let $A = L_\alpha(X)$ and suppose that (1) fails. Let T be the following set of sentences of $\mathcal{L}_A$, where **B** is a unary relation symbol:

(1) φ for all $\varphi \in Q$

(2) $\forall x [x < \bar{\beta} \leftrightarrow \bigvee_{\gamma < \beta} x = \bar{\gamma}]$ for all $\beta < \alpha$

(3) $Ord(c) \wedge c > \bar{\beta}$ for all $\beta < \alpha$

(4) $\forall x [\mathbf{B}(x) \leftrightarrow \bigvee_{n \in X} x = \bar{n}]$

(5) $\forall y$ [$y \leqslant c \rightarrow y$ is not admissible with respect to the predicate **B**].

Now if (1) fails then T has a model by the compactness theorem of [1]. So let $\langle M, E, B \rangle$ be a model of T. If the well founded part of $\mathfrak{M} = \langle M, E \rangle$ is identified with $sp(\mathfrak{M})$ then $B = X$. $\mathfrak{M} \models$ Q by (1). By (2), $osp(\mathfrak{M}) \geqslant \alpha$ and by (3)–(5), $\alpha \geqslant osp(\mathfrak{M})$ so $osp(\mathfrak{M}) = \alpha$.

Thus, for example, if ZF has an uncountable standard model than every countable admissible ordinal α is $osp(\mathcal{M})$ for some model $\mathcal{M}$ of ZF.

References

[1] J. Barwise, Infinitary logic and admissible sets, J. Symbolic Logic 34 (1969) 226–252.
[2] J. Barwise, Collapsing cardinals in models of set theory (abstract), J. Symbolic Logic, to appear.
[3] H. Friedman, On the ordinals in models of set theories, to appear.
[4] H. Gaifman, Two results concerning extensions of models of set theory, Notices Am. Math. Soc. (1968) 947.
[5] H.J. Keisler and M. Morley, Elementary extensions of models of set theory, Israel J. Math. (1968) 49–65.
[6] H.J. Keisler and J. Silver, End extensions of models of set theory, to appear.
[7] A. Levy, A hierarchy of formulas in the model theory of set theory, Mem. Am. Math. Soc., No. 57.
[8] J. Myhill and D. Scott, Ordinal definability, to appear.

ON SEMISETS

Petr HÁJEK
Czechoslovak Academy of Sciences, Prague, Czechoslovakia

This is the third paper concerning the theory of semisets and based on the book [3], which has still not appeared (but will very probably appear soon). The first one [1] was written by the author of the present paper and reflected an early version of [3]. The second one [2] was written by B. Balcar and A. Sochor and summarizes some important results of [3]. The present paper is devoted (1) to the intuitive derivation of the axioms of the theory of semisets and (2) to the presentation of several metamathematical results on the theory of semisets and connections between the theory of semisets and set theory in terms of equiprovability and conservative extensions.

1. Axioms on semisets

The set-theorist is interested in sets but often speaks about classes. He thinks of classes as collections of sets. Sets are also collections of sets but there are some collections of sets that are not sets, e.g. the class of all sets. One restricts himself usually to classes that are (say) comprehensive, i.e. intersect every set in a set: $(\forall a)(\exists b)(\forall u)(u \in b \equiv . u \in a \;\&\; u \in X)$. But let us ask if non-comprehensive classes can be imagined. (And if they are worthy of study.) First, observe that non-comprehensive classes exist if and only if there are subclasses of sets that are not sets. (Subclass of a set which is not a set is a non-comprehensive class; conversely, if X is a non-comprehensive class then there is a set a such that $a \cap X$ is not a set and $a \cap X$ is a subclass of a.) Hence we define a semiset to be a subclass of a set. All sets are semisets, semisets that are not sets are exactly all non-comprehensive classes that are included in sets. We shall now consider two examples showing us how semisets (that are not sets) can be imagined. Using these examples we shall formulate some reasonable axioms on semisets.

(a) *Transitive models*

Let M be a transitive model of ZF containing all the ordinals (think of the class L of all constructible sets). Call elements of M "sets", subclasses of M "classes", in particular, subsets of elements of M become "semisets". In general, not every subset of an element of M is an element of M, hence we have semisets that are not sets ("proper semisets"). Now we ask what statements about sets, semisets and classes are true in this interpretation. To be concrete, choose the following language: Capital letters X, Y, ... are variables for classes, $\in$ denotes the membership relation. Sets can be defined as classes that are elements of other classes; x, y, ... are variables for sets. Then semisets can be defined as subclasses of sets and we shall use variables σ, ρ, ... for semisets. First*, we see that the axiom of extensionality holds in our interpretation:

$$\text{(I)} \qquad X = Y \equiv (\forall u)\,(u \in X \equiv u \in Y)\,.$$

Secondly, the pairing axiom for sets holds:

$$\text{(II)} \qquad (\forall x, y)\,(\exists z)\,(\forall u)\,(u \in z \equiv .\; u = x \vee u = y)\,.$$

For every normal formula φ (i.e. formula in that only set variables are quantified) we can verify the axiom

$$\text{(III)} \qquad (\exists X)\,(\forall x_1 \ldots x_n)\,(\langle x_1 \ldots x_n \rangle \in X \equiv \varphi)\,.$$

(It is well-known that it is sufficient to assume only seven particular cases of this schema and all the other cases are provable from them.) One can call the theory with axioms (I) – (III) the theory of classes.

The axiom if infinity

(IV) there is an infinite set

(expressed in any usual form) is true in the interpretation. For every restricted formula φ (i.e. formula in that all quantifiers are of the form $(\exists x \in y)$, $(\forall x \in y)$) the comprehension axiom holds:

$$\text{(V)} \qquad (\forall a)\,(\exists b)\,(\forall x)\,(x \in b \equiv .\; x \in a \;\&\; \varphi) \quad (b \text{ not in } \varphi)$$

* To be precise, the introduction of new variables for sets requires that there are some sets, i.e. the axiom $(\exists X, Y)\,(X \in Y)$, but this is surely true in our interpretation.

It can be shown that it is sufficient to assume only six particular cases of this schema and the other cases follow.

Finally, we need some axiom on semisets. We do not have the axiom of replacement as a single statement on classes. E.g. imagine a 1–1 mapping of the set ω of all natural numbers onto a non-constructible subset of ω. This is a class in our interpretation, moreover a 1–1 mapping, the domain is a set but the range is a semiset that is no set. Surely, if a class is a 1–1 mapping and the domain is a semiset then the range is a semiset. But we can verify a little more. Imagine a 1–1 mapping F from L into the power-class of L. This is not a class in our interpretation but it can be "coded" as a class. Namely, define a relation R by $\langle y, x\rangle \in R \equiv y \in F(x)$. Then R is a class of the interpretation, for every $x \in D(F)$ we have $F(x) = R''\{x\}$ and R has the following properties from the point of view of the interpretation:

(regularity) for every x, $R''\{x\}$ is a semiset,

(Nowhere-constantness) $x, y \in D(R)$ and $x \neq y$ implies $R''\{x\} \neq R''\{y\}$.

Call a relation which is regular and nowhere constant an *exact functor*. Then we have:

(VI) If X is an exact functor then the domain of X is a semiset iff the range of X is a semiset.

The axioms (I) – (VI) are axioms of the theory of semisets (denotation: TSS). In fact, (TSS) is finitely axiomatized by 18 axioms. In Axiom (VI), the implication "semiset $(D(X)) \rightarrow$ semiset $(W(X))$" can be read: "the union of "semiset many" semisets is a semiset"; the converse implication can be used: "there are not "proper class many" subsemisets of any semiset". It can be easily shown that if we add the axiom saying that every semiset is a set to (TSS) then (1) the schema (V) becomes provable and (2) we obtain a theory equivalent to the Gödel-Bernays set theory with the axioms of groups A, B, C.

In (TSS) we can develop a reasonable theory of ordinals and also prove that the power set and the union of a set are sets. In particular, we can define for every ordinal α the set of all sets of rank α (denote it by r_α) and formulate the axiom of regularity

(D) $$V = \bigcup_{\alpha \in On} r_\alpha$$

(or, equivalently in TSS, $(\forall x \neq 0)\,(\exists y \in x)\,(x \cap y = 0)$). Define in TSS real classes as classes that intersect every set in a set (we called such classes comprehensive on the beginning). In (TSS, D), (or in TSS with a weaker axiom of regularity, see [2]), we can prove all axioms of the Gödel-Bernays set theory (with axiom groups A, B, C) with quantifiers restricted to real classes. This justifies the name "real" – real classes form a model of set theory, the other classes being "imaginary". As a consequence we obtain that in (TSS, D) (or as above) the replacement schema for all formulas with set variables only is provable and, consequently, all theorems of the Zermelo-Fraenkel set theory ZF are provable in (TSS, D). (Recall that all theorems of TSS are theorems of the Gödel-Bernays set theory). One very interesting open problem is whether the replacement schema for all formulas containing set variables only is provable in (TSS), i.e. if TSS → (ZF without regularity).

(b) *The ultraproduct*

We restrict ourselves to a particular example. Let z be a non-principal ultrafilter on ω. As usual, we define for every function $f : \omega \to \omega$ the set $\widetilde{f}$ of all functions g of the smallest possible rank such that the set $\{x \in \omega; f(x) = g(x)\}$ belongs to z. V^z is the class of all $\widetilde{f}$'s. Furthermore, we define $\widetilde{f}E\widetilde{g}$ to be equivalent to the condition $\{x \in \omega; f(x) \in g(x)\} \in z$. It is well-known that the relation E forms a model of ZF (with choice). Note that it is the same whether we deal with the elements of V^z or with their E-extensions. In other words, if we define $\widetilde{\widetilde{f}} = \{\widetilde{g}; \widetilde{g}E\widetilde{f}\}$ and $\widetilde{\widetilde{f}} \in^* \widetilde{\widetilde{g}} \equiv \widetilde{f}E\widetilde{g}$ then the class of all $\widetilde{\widetilde{f}}$'s with the relation $\in^*$ forms a model isomorphic to the former ome. Every $\widetilde{\widetilde{f}}$ is a particular subset of V^z. Now take all subsets of V^z for semisets and all subclasses of V^z for classes in the sense of a new interpretation. For $X, Y \subseteq V^z$ define $X \in^* Y \equiv (\exists f)\,(X = \widetilde{\widetilde{f}}\ \&\ \widetilde{f} \in Y)$. This definition coincides with the former one for subclasses of V^z of the form $\widetilde{f}, \widetilde{g}$ and all axioms of TSS hold in this interpretation. We obtain semisets that are not sets, e.g. the family of all $c(n)$ $(n \in \omega)$, $c(n)$ being $\widetilde{f}$ for the function f such that $f(x) = n$ for every $x \in \omega$. Denote the last semiset by σ and, if ω^* denotes the set of all natural numbers in the sense of the interpretation, put $\rho = \omega^* - \rho$. It can be proved in TSS and therefore holds in the interpretation that every non-empty set of ordinal numbers has a first element, but ρ is a non-empty semiset of ordinal numbers without any first element. Hence it cannot be proved in TSS that every non-empty semiset of ordinals has a first element. Denote the axiom "every non-empty semiset of ordinals has a first element" by (St) (the axiom of stndardness). Evidently, (St) is true in the interpretation described in (a). (St) is a very important additional axiom of TSS, both consistent with and independent from TSS(+D) and both (TSS, St) and (TSS, ⌝St) seem to be interesting theories. (Vopěnka and one of his students have studied a certain theory of semisets stronger than (TSS, ¬St) where an interesting "non-standard analysis" can be developed.)

(c) *Transitive models again*

Recall the interpretation of the notions of (TSS) described in (a). Assuming the axiom of choice in the form saying that there is a 1–1 mapping of the universe onto On (the class of all ordinal numbers) we conclude that there is a 1–1 mapping of the class M onto its power classe. Consequently, the following is true in the interpretation: There is an exact functor R such that, for every semiset σ, there is an x such that $\sigma = R''\{x\}$. (Say, there is a 1–1 coding of all semisets by (some) sets.) It follows that every semiset is "many-one-reducible" to R : if $\sigma \subseteq a$ where a is a set, take $f(y) = \langle y, x_0 \rangle$ for all $y \in a$ and a fixed x_0 with $\sigma = R''\{x_0\}$. It leads us to the following definition in TSS:

A semiset σ is said to be dependent on a class Z (denotation: $\mathrm{Dep}(\sigma, Z)$) iff there is a function f (which is a set) such that $\sigma = (f^{-1})''Z$. Furthermore, a class Z is said to be a total support iff all semisets are dependent on it.

The first axiom of support (S1) reads that there is a total support. We have just seen that (S1) is consistent with TSS(+D). The following is an important consequence of (S1): Call a formula seminormal if only set variables and semiset variables are quantified in it. Every seminormal formula is normal in TSS, (S1). Consequently, the schema (III) holds for every seminormal formula in TSS, (S1). (E.g., we can prove that there is the class of all absolute cardinals, i.e. of all ordinals α such that there is no 1–1 mapping (which is a semiset) of α onto a smaller ordinal.)

Particular total supports of two kinds are important: first, small total supports, i.e. total supports that are semisets, and, secondly, supports bearing a certain structure. Denote the axiom saying that there is a total support which is a semiset by (S3) (in accordance with [3]). To formulate an axiom of the second kind define a complete Boolean algebra to be a real class B with two functions $C : B \to B$ and $F : P(B) \to B$ such that C, F are real classes and the axioms for a complete Boolean algebra are satisfied (C being the complement function and F the meet function; every subset of B has to have its meet). Call a class $Z \subseteq B$ (Z not necessarily real) a complete ultrafilter on B if Z is an ultrafilter and, for every subset $a \subseteq Z$, the meet of a is in Z. The axiom (S4) of a Boolean total support reads that there is a total support which is a complete ultrafilter on a complete Boolean algebra. The axiom (S6) of a total Boolean semiset support reads that there is a total support which is a complete ultrafilter on a complete Boolean algebra which is a set.

Assume $V = L$ in set theory and let b be a complete Boolean algebra (a set).

Consider the Boolean valued model $V^{(b)}$ (see [4] or [5]) of the set theory. Inside of this model the following holds: define semisets as subsets of L and all the other notions of TSS as in (a). Then (S6) holds. (The diagonal function d with $d(u) = \check{u}$ for $u \in b$ is the desired support.) Hence the reader can "imagine" Boolean supports.

The following is provable in (TSS, D):

(i) $\quad$ S4 $\equiv$. S1 & St ,

(ii) $\quad$ S6 $\equiv$. S3 & S4 .

We can strengthen (S6) to an axiom saying that there is a total support which is a complete ultrafilter on a so and so defined Boolean algebra. (E.g. on the algebra of all regular open sets of the Cantor discontinuum.) Every particular case of such axiom is consistent with (TSS, D) (For an exact formulation see Section 2).

2. Equiprovability results; relations between the theory of semisets and the theory; some consistency results

Let TS denote the Gödel-Bernays set theory with axiom groups A to C. ZF means the Zermelo-Fraenkel set theory with regularity but without choice.

(1) In TSS, let r_α denote the set of all sets of rank α and put $Ker = \bigcup_{\alpha \in On} r_\alpha$. For every formula φ of TSS, let φ^{Ker} be the formula resulting from φ by replacing all quantifiers of the form $(\exists X)$, $(\exists \sigma)$ by $(\exists X \subseteq Ker)$, $(\exists \sigma \subseteq Ker)$ respectively and all quantifiers of the form $(\exists x)$ by $(\exists x \in Ker)$ (the same for $\forall$). Then the following holds for any closed φ:

$$\text{TSS} \vdash \varphi^{Ker} \quad \text{iff TSS, D} \vdash \varphi .$$

(2) For any closed formula φ containing set variables only, the following are equivalent:

(i) $\quad$ TSS, D $\vdash \varphi$,

(ii) $\quad$ TS, D $\vdash \varphi$,

(iii) $\quad$ ZF $\vdash \varphi$.

The proof of (i) ≡ (ii) uses the interpretation of TS, D in TSS, D described at the end of Section 1a; (ii) ≡ (iii) is well-known. Hence TS, D is a conservative extension of TSS, D w.r.t. formulas with set variables only. Consequently, the theories TS, TSS, ZF are equiconsistent (because TSS and TSS, D are by (1) and also TS, (TS, D) are).

(3) Existence of complete ultrafilters. The Gödel's form of the axiom of choice E reads in TSS: there is a function F which is a real class and such that $(\forall x \neq 0)\,(F(x) \in x)$. Let $\beta(b)$ be a formula with set variables only and with exactly one free variable b such that TSS, E $\vdash (\exists b)\beta(b)\ \&\ (\forall b)\,(\beta(b) \rightarrow b$ is a complete Boolean algebra). We denote by (S_β) the axiom "there is a complete Boolean algebra b such that $\beta(b)$ and such that there is a complete ultrafilter on b which is a total support". (Evidently, TSS $\vdash (S_\beta) \rightarrow (S6)$.) For any closed formula φ with set variables only,

$$\text{TSS, E} \vdash \varphi \quad \text{iff TSS, E}, (S_\beta) \vdash \varphi\,.$$

Hence (S_β) extends TSS, E conservatively w.r.t. formulas with set variables only. In particular, for every formula β as above, the axiom (S_β) is consistent with TSS, E.

(4) Here we describe (except others and not in the most general form) the relation between TSS and the set theory. Let φ be a closed formula with set and semiset variables only. φ^L denotes the formula resulting from φ by restricting all quantifiers of the form $(\exists\sigma)$ to subsets of L and quantifiers of the form $(\exists x)$ to elements of L. Let $\beta(b)$ be a formula as above. The following are equivalent:

(i) TSS, $V = L, (S_\beta) \vdash \varphi$

(ii) TS, D, $(S_\beta)^L \vdash \varphi^L$ *

(ii) TS, $V = L \vdash [\![\varphi^L]\!]_\beta = 1$

(iv) ZF, $(S_\beta)^L \vdash \varphi^L$

(v) ZF, $V = L \vdash [\![\varphi^L]\!]_\beta = 1$

* Note that TS, D $\vdash$ [If (σ is a total support)^L then $V = L_\sigma$] and hence TS, D, $(S_\beta) \vdash$ E.

This means that TSS presents an axiomatization of the practice of Boolean valued models. For statements φ on constructible sets and their subsets (thought as statements on sets and semisets) it is the same to prove $[\![\varphi^L]\!]_\beta = 1$ in the set theory or to prove φ in TSS, $V = L$, (S_β). Having proved either the former formula in the former theory or the latter one in the latter one one has demonstrated the consistency of φ with the set theory. (E.g. suppose that $\beta(b)$ says that b is an atomless complete Boolean algebra. To prove $[\![V \neq L]\!]_\beta = 1$ in ZF, $V = L$ it suffices to prove "there is a semiset that is not a set" in TSS, $V = L$, (S_β). But the last statement is evident in the last theory – the complete ultrafilter σ on an atomless Boolean algebra cannot be a set.)

(5) A hierarchy for statements on semisets. Call a normal formula φ with set and semiset variables only a Δ_0-formula, if φ is a Δ_0-formula call any formula of the form $(\exists\sigma)\varphi$ a Σ_1-formula, further call $(\forall\sigma)(\exists\rho)\varphi$ a Π_2-formula etc. The following holds true: Every closed formula with set and semisets variables only is equivalent to a Δ_2-formula in TSS, D, (S6). Moreover, for every closed formula ψ with set variables only, there is a closed Δ_2-formula φ such that

$$\mathrm{TS, D, (S6)}^L \vdash \psi \equiv \varphi^L \ .$$

In words, every statement on sets is equivalent in TS, D, $(S6)^L$ (or in ZF, ~~(S6)~~ $(S6)^L$) to a statement on constructible sets and their subsets in that only two variables for subsets of L are quantified. The last statement can be chosen both in $\forall\exists$-form and in $\exists\forall$-form.

Using (4) we obtain the following consequence for ZF ($\beta(b)$ being as above): For any closed formula ψ,

$$\mathrm{ZF}, V = L \vdash [\![\psi]\!]_\beta = 1 \quad \text{if } \mathrm{ZF}, (S_\beta)^L \vdash \psi \qquad \text{(the same for TS)} \ .$$

(6) Complete ultrafilters again. Let now $\beta(b)$ be a formula with set variables only and with exactly one free variable b such that TSS, E $\vdash \beta(b) \rightarrow b$ is a complete Boolean algebra. Let CU_β be the statement $(\exists b)\beta(b) \rightarrow$ $(\exists b, \sigma)$ ($\beta(b)$ & σ is a complete ultrafilter on b). By (3) (more precisely, by a proof of (3)), (CU_β) is consistent with TSS, E for every single β. But, moreover, the whole schema of axioms (CU_β) (for all β's with the assumed properties) is consistent with TSS, E and the schema $(\mathrm{CU}_\beta)^L$ is consistent with TS, E. (To show that every finite number of axioms (CU_{β_1}), ..., (CU_{β_n}) is consistent with TSS, E, one has only to find a formula $\beta(b)$ such that it is provable in TSS, E that (CU_β) implies all the (CU_{β_i})'s.)

As a consequence we obtain the following result concerning set theory:

if ψ is a closed formula with set variables only, then the consistency of ψ with TS, D, E (or with ZF with choice) can be shown using Boolean valued models if its consistency with (TS, D, E, schema$(CU_\beta)^L$) (or with ZF, choice, schema$(CU_\beta)^L$) can be shown by defining a transitive model containing all the ordinals. (Concerning the implication $\rightarrow$, let $\beta(b)$ be a formula describing a complete Boolean algebra used in the construction of a Boolean valued model for ψ. In (TS, D, E, schema(CU_β)) take a $b \in L$ such that $\beta^L(b)$ and a set $\sigma \subseteq L$ such that $(\sigma$ is a complete ultrafilter on $b)^L$. Then consider L_σ – it is a transitive model for ψ.)

(7) There is a single statement of TSS stronger then the whole schema (CU_β), namely the axiom (AS) saying that every complete Boolean algebra (which is a set) bears a complete ultrafilter. B. Balcar and the author of the present paper independently showed that (AS) is consistent with TSS, D, E. (The proof of this fact is not contained in [3] and will appear elsewhere.) In contradistinction to (6), $(AS)^L$ is inconsistent with TS (all cardinals would collaps). TSS, D, E, (AS) seems to be an interesting consistent theory where the continuum is a "proper class" in the sense that, for any α, we have at least $\aleph_\alpha$ many subsemisets of the set of all natural numbers.

Remark. A method using the theory of semisets for proving the consistency of statements contradicting the axiom of choice can be found in [3] but it is too long to describe here.

I hope that the reader found in Section 1 an answer to the questions concerning how semisets can be imagined and what axioms can be postulated for them. Furthermore I hope that he could see some relations between the theory of semisets and set theory (set theories) in Section 2. But the question still remains: what is the theory of semisets good for? Historically, the theory of semisets was thought to be an intermediate auxiliary theory good for the axiomatization of the techniques of Boolean valued models (see above). But it seems to both authors of the theory of semisets that this theory is an independent reasonable axiomatic theory. On the one hand (point (2) above) TSS, D extends conservatively ZF w.r.t. the formulas with set variables only. (Moreover, having proved some metatheorems on TSS, the proofs of statements on sets in TSS, D are no longer than in ZF.) It follows that it is equivalent to prove the consistency of a statement on sets with set theory or with the theory of semisets. On the other hand, two kinds of additional axioms consistent with TSS can be added to it: (a) axioms corresponding to the situation in Boolean valued models (support axioms) – then we obtain an axiomatization of this situation (cf. (4), (5) above) – and (b) axioms contradicting this situation, e.g. the axiom of non-standardness $\neg$(St) or the axiom (AS) on supports on all algebras, and then one has other applications. (E.g. the possibility of having non-standard analysis in terms of semisets; also some properties of the hierarchy described in (5) above in an extension of TSS where Boolean supports are not assumed could be interesting.)

To close this paper I should like to thank Professor G. Kreisel for his interest in the theory of semisets and for several very interesting discussions (in 1969 at Baden-Baden) about the notion of semisets.

References

[1] P. Hájek, The theory of semisets, in: Proc. UCLA Summer Institute on set theory, to appear.
[2] B. Balcar and A. Sochor, Syntactic models of the set theory. The general theory of semisets (Centro Internazionale Matematico Estivo, Varenna, 1968).
[3] P. Vopěnka and P. Hájek, Theory of semisets (book, to appear).
[4] P. Vopěnka, General theory of ∇-models, Comm. Math. Unic. Carol. 8 (1967).
[5] D. Scott and R.M. Solovay, Boolean valued models for set theory, in: Proc. UCLA Summer Institute on set theory, to appear.

COMPUTABILITY OVER THE CONTINUUM

Peter G. HINMAN
University of Michigan, Mich., USA

and

Yiannis N. MOSCHOVAKIS [1)]
University of California, Los Angeles, Calif., USA

Classical descriptive set theory was concerned with classifying sets of real numbers according to the complexity of their definitions from a point of view which considered individual reals as "given" and thus not subject to analysis. Modern hierarchy theory, on the other hand, as an outgrowth of recursion theory, takes as given only the (potentially) infinite sequence of natural numbers and analyses from this point of view the complexity of functions and relations of both natural and real numbers. The notions of computability which we study here are in the style of the latter theory – they are all extensions of ordinary recursion theory – but, following the example of the classical theory, each has as its main additional feature that the continuum is taken as "given" in one of several ways. The notions we discuss by no means exhaust the list of possible ones, but they seem to us the most interesting [2)].

We denote the set of natural numbers by ω and the set of functions from ω into ω by ${}^{\omega}\omega$. For $k, l \in \omega$, let $\omega^{k,l} = \omega^k \times ({}^{\omega}\omega)^l$. We shall study functions from subsets of $\omega^{k,l}$ into either ω or $\omega \cup {}^{\omega}\omega$ (partial functions) and subsets of $\omega^{k,l}$ (relations). The notion of such functions being recursive is well known; we shall adopt essentially the formulation of [Kl 1] in terms of schemata for an inductive definition of the set of computations. All of our

1) The preparation of this paper was supported in part by NSF Grants #GP-11542 and GP-22937.

2) This paper includes the results scheduled to appear in a paper entitled Post's theorem in the analytic hierarchy, announced by the second author in [Mo 2], and that paper will not appear. Some of the results of the first author were announced in [Hi].

other notions results from the adjunction of further schemata derived from various intuitive ways in which the continuum may be "given". The first of these is a schema which allows a search through ${}^{\omega}\omega$ to find those α (if any) which have a certain semi-computable property. This is analogous to the least-number operator of ordinary recursion theory and was motivated by a desire to have an analogue of Post's theorem for the analytic hierarchy (Theorems 1.12 and 1.26). A second group of schemata corresponding more closely to the classical point of view introduces each α in ${}^{\omega}\omega$ as a given function and allows computable functions to assume values in ${}^{\omega}\omega$.

Our main results concern two analogues of hyperarithmetic computability over ω obtained by treating quantification over the continuum as computable. We show in Section 2 that these two notions lead to the same integer-valued functions which, in accord with the results of the Appendix, we call *hyper-projective*.

Section 3 contains several further characterizations of the class of hyper-projective functions. In the Appendix we compare our notions of computability with the abstract theory of [Mo 2].

We shall in general give only brief sketches of proofs of our results. In most cases these outlines could be easily (but laboriously) filled in by anyone familiar with the techniques of [Kl 1] or [Mo 2]. We have attempted, however, to make the sense of our results as independent as possible of these details and intend that the paper be accessible also to readers lacking these prerequisites.

1. First-order computability

We use letters $a, b, ..., t$ as variables over ω; $\alpha, \beta, ..., \epsilon$ as variables over ${}^{\omega}\omega$; and $x, y, z, ...$ as variables whose domain may be either ω or ${}^{\omega}\omega$. The list $x_1, x_2, ..., x_k$ is abbreviated by x and $x_1, x_2, ..., x_k, y$ by (x, y). Most of our notions of computability will be relative to a fixed list $\Phi = (\Phi^1, \Phi^2) = \Phi^1_1, ..., \Phi^1_{r_1}, \Phi^2_1, ..., \Phi^2_{r_2}$ of single-valued functions, where Φ^1_i maps ω into ω and Φ^2_j maps ${}^{\omega}\omega$ into ω. We list here for reference all of the schemata we shall use. They will be explained as they are used.

(1.01i) $\varphi(\mathrm{x}) \simeq \Phi^1_i(\psi(\mathrm{x})).$ $(1 \leqslant i \leqslant r_1)$

(1.02j) $\varphi(\mathrm{x}) \simeq \Phi^2_j(\lambda m \cdot \psi(m, \mathrm{x})).$ $(1 \leqslant j \leqslant r_2)$

(1.1) $\varphi(m, \mathrm{x}) \simeq m + 1.$

(1.2) $\varphi(\mathrm{x}) \simeq m$.

(1.3) $\varphi(m, \mathrm{x}) \simeq m$.

(1.4) $\varphi(\mathrm{x}) \simeq \chi(\psi(\mathrm{x}), \mathrm{x})$.

(1.5) $\varphi(0, \mathrm{x}) \simeq \psi(\mathrm{x})$,

$\varphi(m+1, \mathrm{x}) \simeq \chi(m, \varphi(m, \mathrm{x}), \mathrm{x})$.

(1.6) $\varphi(\mathrm{x}) \simeq \psi(\mathrm{x}^i)$, where $(x_1, \ldots, x_k)^i = (x_i, x_1, \ldots, x_i, \ldots, x_k)$.

(1.7) $\varphi(m, \alpha, \mathrm{x}) \simeq \alpha(m)$.

(1.8) $\varphi(z, \mathrm{x}, \mathrm{y}) \simeq \{z\}(\mathrm{x})$.

(2) $\varphi(\mathrm{x}) \simeq \chi(\nu^1\alpha[\psi(\alpha, \mathrm{x}) \to 0])$.

(3) $\varphi(\mathrm{x}) \simeq 0 \iff \exists\alpha[\psi(\alpha, \mathrm{x}) \simeq 0]$.

(4) $\varphi(\mathrm{x}) \simeq \nu^0 m[\psi(m, \mathrm{x}) \to 0]$.

(5.1) $\varphi(\mathrm{x}) \simeq \alpha$.

(5.2) $\varphi(\alpha, \mathrm{x}) \simeq \alpha$.

(5.3) $\varphi(\mathrm{x}) \simeq \lambda m \cdot \psi(m, \mathrm{x})$.

(6) $\varphi(\mathrm{x}) \simeq \nu^1\alpha[\psi(\alpha, \mathrm{x}) \to 0]$.

(7) $\varphi(\mathrm{x}) \simeq 0$, if $\forall\alpha\exists m[\psi(\alpha, \mathrm{x}) \simeq m]$ and $\exists\alpha[\psi(\alpha, \mathrm{x}) \simeq 0]$;

$\simeq 1$, if $\forall\alpha[\psi(\alpha, \mathrm{x}) \simeq 1]$.

(8) $\varphi(\mathrm{x}) \simeq 0$, if $\exists\alpha[\psi(\alpha, \mathrm{x}) \simeq 0]$;

$\simeq 1$, if $\forall\alpha[\psi(\alpha, \mathrm{x}) \simeq 1]$.

These schemata are to be interpreted in the manner of [Kl 1, I §1] or [Mo 2, §2]. That is, to each schema is correlated an index z and a clause for an inductive definition of a relation $\{z\}(\mathrm{x}) \simeq y$. For example, to schema (1.4) we might assign the index $z = \langle 4, k, l, z_0, z_1 \rangle$ and the clause

$$\exists n[\{z_0\}(\mathrm{x}) \simeq n \quad \text{and} \quad \{z_1\}(n, \mathrm{x}) \simeq m] \Rightarrow \{z\}(\mathrm{x}) \simeq m .$$

Similarly schema (1.02j) could be assigned the index $z = \langle 0, 2, j, k, l, z_0 \rangle$ and the clause

$$\exists \alpha \forall m[\{z_0\}(m, \mathrm{x}) \simeq \alpha(m) \ \textit{and} \ \ \Phi_j^2(\alpha) = n] \Rightarrow \{z\}(\mathrm{x}) \simeq n .$$

As our proofs will not be detailed, we shall not need to make such an assignment of indices explicit, and hence we merely assume in the following that some fixed assignment has been made. Indices will be in ω or ${}^\omega\omega$ depending on which schemata are involved.

We denote by $\{z\}^{\Phi}_{(p_0,\dots,p_n)}(\mathrm{x}) \simeq y$ the relation obtained by use of schemata (p_0), ..., (p_n), where (1) means (1.01) – (1.8), etc. The superscript Φ will be omitted whenever it is clear from this context. A function φ defined on a subset of $\omega^{k,l}$ is called *partial* $(p_0, \dots, p_n)$*-computable* in Φ iff for some z, $\varphi(\mathrm{x}) \simeq \{z\}^{\Phi}_{(p_0,\dots,p_n)}(\mathrm{x})$. A relation R is $(p_0, \dots, p_n)$*-computable* in Φ iff its characteristic function is, and *semi-*$(p_0, \dots, p_n)$*-computable* iff it is the domain of a function partial $(p_0, \dots, p_n)$-computable in Φ.

The schemata (1) are just those of [Kl 1] for arguments of types 0 and 1. Hence the functions partial (1)-computable in Φ are just the functions partial recursive in Φ. Functions obtained by use of (1.01) – (1.7) are called *primitive computable* in Φ and in indexed form are denoted by $\{e\}^{\Phi}_{\mathrm{pr}}$.

In order to discuss schema (2), we need to introduce the notion of a multiple-valued function φ from $\omega^{k,l}$ into ω. Formally we take φ to be an ordinary function from $\omega^{k,l}$ into the power set of ω and write "$\varphi(\mathrm{x}) \to m$" for "$m \in \varphi(\mathrm{x})$" and "$\varphi(x)\downarrow$" for "$\varphi(x) \neq \emptyset$". Composition of two such functions is defined in the natural way:

$$\chi(\psi(\mathrm{x})) \to m \Longleftrightarrow \exists n[\psi(\mathrm{x}) \to n \quad \textit{and} \quad \chi(n, \mathrm{x}) \to m] .$$

Then all of the schemata (1) have natural interpretations as applied to multiple-valued functions. We refer the reader to [Mo 2, I §1] for a more complete discussion.

To explain the motivation for schema (2) we recall that in ordinary recursion theory, if $\psi(m, \mathbf{x})$ is partial recursive, there exists a partial recursive function ν, a *selection function*, such that

$$\exists m[\psi(m, \mathbf{x}) \simeq 0] \Rightarrow \nu(\mathbf{x}) \downarrow \quad \textit{and} \quad \psi(\nu(\mathbf{x}), \mathbf{x}) \simeq 0 \ .$$

Thus ν selects an element from any non-empty recursively enumerable subset of ω. The definition of ν makes use of the natural well-ordering of ω and the least-number operator, and its existence is due to the fact that ordinary recursion theory takes the natural number sequence as "given". Unless we wish to accept some well-ordering of ${}^{\omega}\omega$ as "given" (which we do not), it is not in general possible to define a schema which selects a single element from any non-empty semi-computable subset of ${}^{\omega}\omega$. The best we can do is to introduce a multiple-valued function ν^1 which takes on *all* the elements of the set as values. Thus the interpretation of schema (2) is

$$\exists \alpha[\psi(\alpha, \mathbf{x}) \to 0 \ \textit{and} \ \ \chi(\alpha) \to m] \Rightarrow \varphi(\mathbf{x}) \to m \ .$$

Note that even if ψ and χ were single-valued, φ might be multiple-valued.

Our first result is that if $\Phi^2 = \emptyset$ – that is, there are no type-2 functions of relativization – then schema (2) adds very little to (1). Towards this we first find a normal form for partial (1,2)-computable functions. Indices for such functions may be taken in ω.

Theorem 1.1. If $\Phi^2 = \emptyset$, then there exists a relation T primitive computable in Φ such that

$$\{e\}^{\Phi}_{(1,2)}(\mathbf{x}) \to m \iff \exists n T(e, \mathbf{x}, m, n) \ .$$

Proof. We define $T(e, \mathbf{x}, m, n)$ to mean "n is a code for a computation which establishes $\{e\}_{(1,2)}(\mathbf{x}) \to m$". For example, if e is an index for an application of schema (1.4) with e_0 an index for ψ and e_1 an index for χ, let

$$T(e, \mathbf{x}, m, n) \iff T(e_0, \mathbf{x}, (n)_0, (n)_1) \ \text{ and } \ T(e_1, ((n)_0, \mathbf{x}), m, (n)_2).$$

The crucial point which permits computations to be finite is that since $\Phi = \emptyset$, in applications of schema (2) ψ and χ depend only on a finite part of their arguments. Hence, if e is an index for schema (2) with e_0 and e_1 as above we can set

$$T(e, \mathbf{x}, m, n) \Longleftrightarrow T(e_0, ((n)_0 * 0^\omega, \mathbf{x}), 0, (n)_1)$$

and $$T(e_1, (n)_0 * 0^\omega, m, (n)_2)\ ,$$

where $(n)_0 * 0^\omega$ is the element of ${}^\omega\omega$ where first values are the components of the sequence coded by $(n)_0$ and whose remaining values are 0.

The other cases are similar and give a definition of T by recursion on n. It is then straightforward to prove by induction on n that

$$T(e, \mathbf{x}, m, n) \Rightarrow \{e\}_{(1,2)}(\mathbf{x}) \to m$$

and by an induction corresponding to the inductive definition of (1,2)-computations that

$$\{e\}_{(1,2)}(\mathbf{x}) \to m \Rightarrow \exists n T(e, \mathbf{x}, m, n)\ .$$

Corollary 1.2. If $\Phi^2 = \emptyset$, *then there exists a function* φ *partial* (1)*-computable in* Φ *such that for each e,* $\lambda\mathbf{x} \cdot \varphi(e, \mathbf{x})$ *is a single-valued branch of* $\{e\}_{(1,2)}$. *That is*:

(a) $\{e\}_{(1,2)}(\mathbf{x}) \downarrow \Longleftrightarrow \varphi(e, \mathbf{x}) \downarrow$;

(b) $\varphi(e, \mathbf{x}) \simeq m \Rightarrow \{e\}_{(1,2)}(\mathbf{x}) \to m$.

In particular, every single-valued function partial (1, 2)*-computable in* Φ *is partial* (1)*-computable in* Φ.

Proof. Take $\varphi(e, \mathbf{x}) \simeq (\text{least } n \cdot T(e, \mathbf{x}, (n)_0, (n)_1))_0$. This is partial (1)-computable, since by [Kl 1, XVII and XXXI], partial (1)-computability coincides with ordinary partial recursiveness when $\Phi^2 = \emptyset$.

Corollary 1.3. If $\Phi^2 = \emptyset$, *then for any relation R, R is* (*semi*)-(1, 2)*-computable in* Φ *iff R is* (*semi*)-(1)*-computable in* Φ.

Thus, when $\Phi^2 = \emptyset$, a relation is semi-(1, 2)-computable in Φ iff it is Σ^0_1 in Φ^1 (definable in existential form with matrix primitive computable in Φ^1) and (1, 2)-computable in Φ iff it is Δ^0_1 in Φ^1. This situation changes drastically with the introduction of type-2 functions or relativization. Use of schema (1.02) opens the possibility that the inductive definition of the set of computations be transfinite or, from another point of view, that an individual computation tree be infinite. This is reflected in the use of functions, rather than numbers as in 1.1, to code computations in 1.8 below. We first review some facts about (1)-computability. Our use here of the first recursion theorem may be justified by appeal either to [Kl 1, II, LXIV] or [Mo 3, §5].

Lemma 1.4. If Φ^2 *consists entirely of functions with graphs* $\Delta_1^1(\Delta_{n+2}^1)$ *in* Φ^1, *then every relation semi-*(1)*-computable in* Φ *is* $\Pi_1^1(\Delta_{n+2}^1)$ *in* Φ^1.

Proof. Immediate from [Kl 1, I, XXVI and XXVII].

Lemma 1.5. If Φ^2 *contains at least one total function, then every relation* Π_1^1 *in* Φ^1 *is semi-*(1)*-computable in* Φ.

Proof. If $R(\mathrm{x}) \Longleftrightarrow \forall\alpha\exists m S(x, \alpha(m))$ with S primitive computable in Φ^1, let Φ be a total function in Φ^2 and

$$F(\zeta, \mathrm{x}, s) \simeq 0, \quad \text{if } S(\mathrm{x}, s);$$

$$\simeq 0 \cdot \Phi(\lambda n \cdot \zeta(\mathrm{x}, s * \langle n \rangle)), \quad \text{otherwise.}$$

By the first recursion theorem, the smallest partial function ψ such that $\psi(\mathrm{x}, s) \simeq F(\psi, \mathrm{x}, s)$ is partial (1)-computable in Φ. It is routine to check that $R(\mathrm{x}) \Longleftrightarrow \psi(\mathrm{x}, \emptyset) \simeq 0$.

Lemma 1.6. Every function partial (1)*-computable in* Φ *is also partial* (1, 2)*-computable in* Φ.

Proof. Although this result seems intuitively obvious, a complete proof is surprisingly complex. We must construct a primitive computable function π such that

$$\{e\}_{(1)}(\mathrm{x}) \simeq m \Longleftrightarrow \{\pi(e)\}_{(1,2)}(\mathrm{x}) \simeq m\ .$$

We define a function π_p in terms of an index p and then apply the second recursion theorem (valid for (1, 2)-computability by the usual proof) to choose an index q such that $\pi_q = \{q\}_{\mathrm{pr}}$. Examples of this method may be found in [Kl 1, II, LXII] and [Mo 2, I, Lemma 31] and we shall not give any further details. In future proofs we shall refer to this method as *index transfer*.

Lemma 1.7. (a) *If* Φ^2 *contains at least one total function, then every relation* Σ^1_2 *in* Φ^1 *is semi-*(1, 2)*-computable in* Φ.

(b) *Every relation* Σ^1_{n+2} *in* Φ^1 *is semi-*(1, 2)*-computable in* Φ *and the characteristic function of some relation* Π^1_{n+1} *in* Φ^1.

Proof. (a) Suppose $R(\mathrm{x}) \Leftrightarrow \exists\alpha S(\mathrm{x}, \alpha)$ with S Π^1_1 in Φ^1. By 1.5 there exists a ψ partial (1)-computable in Φ such that $S(\mathrm{x}, \alpha) \Leftrightarrow \psi(\alpha, \mathrm{x}) \to 0$. By 1.6, ψ is also partial (1, 2)-computable in Φ. If $\chi = \lambda\alpha \cdot 0$,

$$R(\mathrm{x}) \Leftrightarrow \chi(\nu^1\alpha[\psi(\alpha, \mathrm{x}) \to 0]) \to 0 \,.$$

For (b) the proof is similar with ψ now the characteristic function of S.

Towards establishing a converse to this result, we first find a normal form for (1, 2)-computability in this case. Let WO be the set of γ in ${}^\omega\omega$ such that $\{\langle m, n\rangle \mid \gamma(\langle m, n\rangle) = 0\}$ is a well-order, and let $\gamma' < \gamma$ mean that the order coded by γ' is an initial segment of that coded by γ. It is well known that WO is Π^1_1 and the relation '$<$' is arithmetic.

Theorem 1.8. If Φ^2 *contains at least one total function, then there exists a relation* T *semi-*(1)*-computable in* Φ *such that*

$$\{e\}^{\Phi}_{(1,2)}(\mathrm{x}) \to m \Leftrightarrow \exists\delta T(e, \mathrm{x}, m, \delta) \,.$$

Proof. We define $T(e, \mathrm{x}, m, \delta)$ to mean "δ is a code for a computation which establishes $\{e\}_{(1,2)}(\mathrm{x}) \to m$". Except for the two cases given below, the clauses of the definition are very similar to those used in the proof of 1.1. However, here they are interpreted as clauses in the definition of a function Θ_t partial (1)-computable in Φ in terms of an index t. An application of the second recursion theorem yields a t such that $\Theta_t = \{t\}_{(1)}$ is the characteristic function of the T we want.

If e is an index for schema (1.02), with e_0 an index for ψ, we want

$$T(e, \mathrm{x}, m, \delta) \Leftrightarrow \delta_0 \in \mathrm{WO}$$

and $\quad \forall n[T(e_0, (n, \mathrm{x}), \delta_1(n), \delta_{n+2}) \quad \text{and} \quad \delta_{n+2,0} < \delta_0]$

and $\quad m = \Phi(\delta_1)\,,$

where $\delta_n = \lambda p \cdot \delta(\langle n, p\rangle)$. Using Lemma 1.5 it is not hard to see that the expression on the right is semi-(1)-computable as a relation in e, $\mathbf{x}$, m, δ, and t. If e is an index for schema (2), with e_0, e_1 indices for ψ, χ, set

$$T(e, x, m, \delta) \Longleftrightarrow \delta_0 \in \mathrm{WO}$$

and $$T(e_0, (\delta_1, \mathbf{x}), 0, \delta_2) \text{ and } T(e_1, \delta_1, m, \delta_3)$$

and $$\delta_{2,0} < \delta_0 \text{ and } \delta_{3,0} < \delta_0 \,.$$

Finally, it is straightforward to prove by induction on the ordinal of δ_0 that

$$T(e, \mathbf{x}, m, \delta) \Rightarrow \{e\}_{(1,2)}(\mathbf{x}) \to m \,,$$

and by an induction corresponding to the inductive definition of (1, 2)-computations that

$$\{e\}_{(1,2)}(\mathbf{x}) \to m \Rightarrow \exists \delta T(e, \mathbf{x}, m, \delta) \,.$$

Corollary 1.9. If Φ^2 contains at least one total function and consists entirely of functions with graphs Δ^1_{n+2} in Φ^1, then every relation semi-(1, 2)-computable in Φ is Σ^1_{n+2} in Φ^1.

Proof. Immediate from 1.4 and 1.8.

Corollary 1.10. For any γ and any relation R:
(a) *If Φ^2 is any total type-2 function with graph Δ^1_2 in γ, then R is Σ^1_2 in γ iff R is semi-(1, 2)-computable in γ and Φ^2,*
(b) *R is Σ^1_{n+2} in γ iff R is semi-(1, 2)-computable in γ and the characteristic function of some relation Π^1_{n+1} in γ.*

Proof. Immediate from 1.7 and 1.9.

Lemma 1.11. A relation R is (1, 2)-computable in Φ iff both R and $\sim R$ are semi-(1, 2)-computable in Φ.

Proof. If $R(\mathbf{x}) \Longleftrightarrow \varphi_0(\mathbf{x}) \to 0$ and $\sim R(\mathbf{x}) \Longleftrightarrow \varphi_2(\mathbf{x}) \to 0$, let $\psi(\alpha, \mathbf{x}) = \varphi_{\alpha(0)}(\mathbf{x})$ and $\chi(\alpha) = \alpha(0)$. Then

$$\varphi(\mathbf{x}) = \chi(\nu^1 \alpha[\psi(\alpha, \mathbf{x}) \to 0])$$

Theorem 1.12 (Post's Theorem). *For any γ and any relation R:*

(a) *If Φ^2 is any total type-2 function with graph Δ^1_2 in γ, then R is Δ^1_2 in γ iff R is* (1, 2)*-computable in γ and Φ^2.*

(b) *R is Δ^1_{n+2} in γ iff R is* (1, 2)*-computable in γ and the characteristic function of some relation Π^1_{n+1} in γ.*

Proof. Immediate from 1.10 and 1.11.

A natural assumption for a computation theory over the continuum would be that the relation of equality on ${}^{\omega}\omega$ is computable. If Φ consists only of the characteristic function of this relation, then the relations (semi)-(1, 2)-computable in Φ are exactly the $(\Sigma^1_2)\Delta^1_2$ relations.

We consider next the effect of replacing schema (2) by the apparently weaker schema (3). We shall see that for results concerning semi-computability we lose nothing by this replacement, but Post's Theorem no longer holds. (1, 3)-computability has, of course, the advantage of dealing only with single-valued functions.

Theorem 1.13. (a) *Every function partial* (1, 3)*-computable in Φ is also partial* (1, 2)*-computable in Φ.*

(b) *If Φ^2 is empty or contains at least one total function, then the relation $\{e\}_{(1,2)}(\mathrm{x}) \to m$ is semi-*(1, 3)*-computable.*

Proof. (a) is proved by the method of index transfer using the technique of the proof of 1.7 to imitate schema (3) by use of schema (2). (b) is immediate from the normal form Theorems 1.1 and 1.8.

Corollary 1.14. If Φ^2 is empty or contains at least one total function, then for any relation R, R is semi-(1, 2)*-computable in Φ iff R is semi-*(1, 3)*-computable in Φ.*

It follows that 1.10 holds also for (1, 3)-computability. On the other hand, 1.11 and 1.12 fail badly here:

Lemma 1.15. For any function φ partial (1, 3)*-computable in Φ, there exists a function ψ partial* (1)*-computable in Φ such that*

$$\varphi(\mathrm{x}) \simeq m \Rightarrow \psi(\mathrm{x}) \simeq m\ .$$

In particular, the (*total*) *functions and relations* (1, 3)*-computable in Φ*

coincide with the (total) functions and relations (1)*-computable in* Φ.

Proof. The first part is proved by an index transfer identical to that for 1.13(a) except that schema (3) is interpreted here by the function $\varphi(\mathrm{x}) \simeq 0$. The second part follows immediately.

The full strength of schema (2) is restored by the addition of schema (4), since this is all that is really needed for the proof of 1.11.

Theorem 1.16. If Φ^2 *is empty or contains at least one total function, then the functions partial* (1, 3, 4)*-computable in* Φ *coincide with those partial* (1, 2)*-computable in* Φ.

We do not know whether this result, or others with the same hypothesis, are true without the hypothesis.

With the introduction of the schemata (5), we consider two further ways in which the continuum may be "given". First, we now allow computable functions to take values in ${}^{\omega}\omega$ and hence need (5.2) to supplement (1.3). Schema (5.1) (in conjunction with (1.4) and (1.7)) introduces each member of ${}^{\omega}\omega$ as a computable function. Since an index for schema (5.1) must code the value, indices for computabilities using (5) will be functions rather than numbers. Schema (5.3) may be thought of as a kind of comprehension operator with the interpretation

$$\forall m[\psi(m, \mathrm{x}) \simeq \alpha(m)] \Rightarrow \varphi(\mathrm{x}) \simeq \alpha .$$

In the presence of (5.3), schema (1.4) takes on a new importance, since $\chi(\lambda m \cdot \psi(m, \mathrm{x}), \mathrm{x})$ is defined at most when $\lambda m \cdot \psi(m, \mathrm{x})$ is a total function. In particular, we have for (1, 5) computability full substitution for type-1 arguments, which fails for (1)-computability [Kl 1, II, LVI], unless Φ^2 contains at least one total function. However in this case:

Theorem 1.17. (a) *Every function partial* (1)*-computable in* Φ *is partial* (1, 5)*-computable in* Φ.

(b) *If* Φ^2 *contains at least one total function, then there exists a function* π *primitive computable in* Φ *such that*

$$\{\epsilon\}_{(1,5)}(\mathrm{x}) \simeq y \Leftrightarrow \forall m[\{\pi(\epsilon)\}_{(1)}(\epsilon, \mathrm{x}, m) \simeq y(m)] \ ,$$

where in case $y \in \omega, y(m) = y$.

(c) *If* Φ^2 *contains at least one total function, then every number-valued function* φ *partial* (1, 5)*-computable in* Φ *is partial* (1)*-computable in* Φ *and* ϵ, *the index of* φ.

Proof. The proof of (a) is very similar to that if 1.6 and (c) follows immediately from (b). The crucial case in the definition of π is when ϵ is an index for schema (1.4), say $\epsilon = \langle 4, k, l, \epsilon_0, \epsilon_1\rangle$. The desired relation is

$$\{\pi(\epsilon)\}_{(1)}(\epsilon, \mathrm{x}, m) \simeq \{\epsilon_1\}_{(1,5)}(\{\epsilon_0\}_{(1,5)}(\mathrm{x}), \mathrm{x})(m)$$

$$\simeq \{\pi(\epsilon_1)\}_{(1)}(\epsilon_1, \lambda m\, \{\pi(\epsilon_0)\}_{(1)}(\epsilon_0, \mathrm{x}, m), \mathrm{x}, m)\,.$$

When the first argument of the right side is total, the index $\pi(\epsilon)$ can be computed as in [Kl 1, II, LX]. Otherwise, the right side is undefined and we can ensure that the left side is also by choosing π such that

$$\{\pi(\epsilon)\}_{(1)}(\epsilon, \mathrm{x}, m)\downarrow \Longleftrightarrow \Phi^2(\lambda m\{\pi(\epsilon_0)\}_{(1)}(\epsilon_0, \mathrm{x}, m))\downarrow\,,$$

where Φ^2 is any total number of Φ.

Corollary 1.18. If Φ^2 *contains at least one total function, then a relation R is* (*semi*)-(1, 5)*-computable in* Φ *iff R is* (*semi*)-(1)*-computable in* Φ *and some* ϵ.

Proof. Immediate from 1.17(a) and (c).

Lemma 1.19 (Transitivity). *If* φ *is partial* (1, 5)*-computable in* Φ, Ψ *and each member of* Ψ *is partial* (1, 5)*-computable in* Φ, *then* φ *is partial* (1, 5)*-computable in* Φ.

Proof. By the method of index transfer using the fact that partial (1, 5)-computability is preserved under substitution for arguments of type-1.

Theorem 1.20. The relations semi-(1, 5)*-computable in* $\emptyset$ *are exactly the* Π^1_1 (= Π^1_1 *in some* ϵ) *relations.*

Proof. If R is Π^1_1 in ϵ, then by 1.5 R is semi-(1)-computable in $\lambda\alpha \cdot 0$ and ϵ, hence by 1.18 semi-(1, 5)-computable in $\lambda\alpha \cdot 0$ and by 1.19 semi-(1, 5)-computable in $\emptyset$. The argument is reversible by use of 1.4 at the last step.

One advantage in using functions with values in ${}^{\omega}\omega$ is that the search operator can now be introduced by the simpler schema (6). Computabilities involving this schema will, of course, again lead to multiple-valued functions.

Lemma 1.21. (a) *Every Σ^1_2 relation is semi-(1, 5, 6)-computable in $\emptyset$.*
(b) *Every Σ^1_{n+2} relation is semi-(1, 5, 6)-computable in the characteristic function of some Π^1_{n+1} relation.*

Proof. Similar to that of 1.7 using 1.20.

Theorem 1.22. There exists a relation T semi-(1, 5)-computable in Φ such that

$$\{\epsilon\}_{(1,5,6)}(\mathrm{x}) \to y \Longleftrightarrow \exists\delta T(\epsilon, \mathrm{x}, y, \delta)\,.$$

Proof. The proof is quite similar to that of Theorem 1.8 and we mention only two of the cases that are new here. WO is semi-(1, 5)-computable in Φ by 1.20. If ϵ is an index for schema (1.4), ϵ_0 an index for ψ with value in ${}^{\omega}\omega$, and ϵ_1 an index for χ, set

$$T(\epsilon, \mathrm{x}, y, \delta) \Longleftrightarrow \delta_0 \in \mathrm{WO} \quad \textit{and} \quad T(\epsilon_0, \mathrm{x}, \delta_1, \delta_2)$$

and $$T(\epsilon_1, (\delta_1, \mathrm{x}), y, \delta_3) \quad \textit{and} \quad \delta_{2,0} < \delta_0 \quad \textit{and} \quad \delta_{3,0} < \delta_0\,.$$

If ϵ is an index for schema 5.3,

$$T(\epsilon, \mathrm{x}, \gamma, \delta) \Longleftrightarrow \delta_0 \in \mathrm{WO}$$

and $$\forall m[T(\epsilon_0, (m, \mathrm{x}), \gamma(m), \delta_{m+1}) \quad \textit{and} \quad \delta_{m+1} < \delta_0]\,.$$

Corollary 1.23. If Φ^2 consists entirely of functions with graphs in Δ^1_{n+2} then every relation semi-(1, 5, 6)-computable in Φ is in Σ^1_{n+2}.

Proof. Immediate from 1.4, 1.18 and 1.22.

Corollary 1.24. For any relation R:
(a) *R is Σ^1_2 iff R is semi-(1, 5, 6)-computable in $\emptyset$.*
(b) *R is Σ^1_{n+2} iff R is semi-(1, 5, 6)-computable in the characteristic function of some relation in Π^1_{n+1}.*

Proof. Immediate from 1.21 and 1.23.

Lemma 1.25. A relation R is (1, 5, 6)*-computable in* Φ *iff both R and* $\sim R$ *are semi-*(1, 5, 6)*-computable in* Φ.

Proof. Identical to that of 1.11.

Theorem 1.26 (Post's Theorem). *For any relation R*:

(a) *R is* Δ_2^1 *iff R is* (1, 5, 6)*-computable in* $\emptyset$.

(b) *R is* Δ_{n+2}^1 *iff R is* (1, 5, 6)*-computable in the characteristic function of some relation in* Π_{n+1}^1.

This result together with Theorem 1.12 and the remark following it provide further evidence that Δ_2^1 rather than Δ_1^1 is the proper analog of Δ_1^0 when ${}^{\omega}\omega$ is the underlying space.

2. Quantification over the continuum

In [Kl 1, II, §11.26] Kleene proposed the hyperanalytic sets as a natural extension of the analytic hierarchy into the transfinite. Since these are the sets recursive in the type-3 functional ${}^3\mathbf{E}$ (in our notation, (1, 7)-computable in $\emptyset$), which embodies quantification over ${}^{\omega}\omega$, there is an obvious analogy with the hyperarithmetic sets, which by [Kl 1, I, XLVIII] are just those recursive in ${}^2\mathbf{E}$, the functional which embodies quantification over ω. It is now known that there are two major flaws in this analogy. First, the hyperarithmetic sets are just those in Δ_1^1, whereas the hyperanalytic sets are a proper subclass of Δ_1^2. Second, the class of relations semi-recursive in ${}^2\mathbf{E}$ (= Π_1^1) is closed under existential quantification over ω, whereas the class of relations semi-recursive in ${}^3\mathbf{E}$ (semi-(1, 7)-computable in $\emptyset$) is not closed under existential quantification over ${}^{\omega}\omega$ [Mo 1, Corollary 10.2]. We consider in this section two revisions of Kleene's analogy which eliminate the second flaw. No way is yet known to eliminate the first.

The first revision arises by adjoining schemas (5) and (6) to (1) and (7). It is easy to see by the method of proof of 1.7 that if $R(\alpha, \mathrm{x})$ is semi-(1, 5, 6, 7)-computable, so is $S(\mathrm{x}) \Longleftrightarrow \exists\alpha R(\alpha, \mathrm{x})$. Furthermore, it follows from the results of [Mo 2, II] together with those of the Appendix below that this class has most of the other nice properties of the class of hyperarithmetic relations. The only disadvantage of this method of repairing the analogy is that it requires the use of multiple- and function-valued functions. Our second revision, which replaces schema (7) by schema (8), avoids this unpleasant feature without sacrificing any structure. It is immediate from the form of schema (8) that the semi-(1, 8)-computable relations are closed under existential quantification over ${}^{\omega}\omega$ and that this class is a proper extension of the class of hyperanalytic relations. In fact, (1, 8)-computation may be regarded as Kleene recursion in the (extended) type-3 functional ${}^{3}\mathrm{E}^{\#}$ whose value on a *partial* function Θ from ${}^{\omega}\omega$ into ω is given by:

$$
\begin{aligned}
{}^{3}\mathrm{E}^{\#}(\Theta) &\simeq 0, \quad \text{if} \quad \exists\alpha[\Theta(\alpha)\simeq 0] \;; \\
&\simeq 1, \quad \text{if} \quad \forall\alpha[\Theta(\alpha)\simeq 1] \;; \\
&\text{undefined, otherwise.}
\end{aligned}
$$

It is not hard to see that a similar extension of ${}^{2}\mathrm{E}$ to ${}^{2}\mathrm{E}^{\#}$ does not affect the class of relations recursive or semi-recursive. Hence the class of relations recursive in ${}^{3}\mathrm{E}^{\#}$ is a very natural analogue of the hyperarithmetic relations.

The goal of the rest of this section is to show that except for the differences already mentioned, these two approaches are the same (Corollaries 2.5, 2.7, 2.10 and 2.11). For convenience we shall assume that Φ consists entirely of total, single-valued functions. For each k, l, let $\sigma^{k,l}$ mapping $\omega^{k,l}$ into ${}^{\omega}\omega$ be some standard coding function. We write $\bar{\mathbf{x}}$ for $\sigma(\mathbf{x})$ and $\langle x_1, ..., x_n\rangle$ for $\sigma(x_1, ..., x_m)$. For any Φ, let $\Phi^{=}$ denote Φ together with the characteristic function of $\{\langle\alpha,\beta\rangle \mid \alpha=\beta\}$. When Φ is denoted by a complicated expression we write $\{e\}(\Phi, \mathbf{x})$ instead of $\{e\}^{\Phi}(\mathbf{x})$.

Definition 2.1. H^{Φ} is the smallest subset H of ${}^{\omega}\omega$ such that for all $\mathbf{x}$, α, and e:

(a) $\langle\alpha, 0\rangle \in H$;

(b) $\forall\alpha[\langle\langle\gamma,\alpha\rangle, \{e\}_{\mathrm{pr}}(\Phi^{=},\gamma,\alpha)\rangle \in H] \Rightarrow \langle\gamma, \langle 0, e\rangle\rangle \in H$;

(c) $\exists\alpha[\langle\langle\gamma,\alpha\rangle, \{e\}_{\mathrm{pr}}(\Phi^{=},\gamma,\alpha)\rangle \in H] \Rightarrow \langle\gamma, \langle 1, e\rangle\rangle \in H$.

We write $m \in H^{\Phi}(\mathbf{x})$ for $\langle\bar{\mathbf{x}}, m\rangle \in H^{\Phi}$ and usually omit Φ.

Lemma 2.2. There exist functions Θ_1 and Θ_2 primitive computable in Φ such that

(a) $\{\epsilon\}_{(1,5,6,7)}(\mathrm{x}) \to y \Longleftrightarrow \Theta_1(\epsilon, \mathrm{x}, y) \in H(\epsilon, \mathrm{x}, y)$,

(b) $\{e\}_{(1,8)}(\mathrm{x}) \simeq m \Longleftrightarrow \Theta_2(e, \mathrm{x}, m) \in H(\mathrm{x})$.

Proof. The method is very similar to that used for Theorem 6 of [Mo 2, II] and we shall not give the details. The point is that all clauses of the inductive definitions of the left-hand relations except those corresponding to schemata (7) and (8) are primitive recursive in Φ. Hence the resulting relation can be reduced to H, which is complete for sets defined by such clauses.

Next, we assign ordinals to the elements of H in the natural way:

$$|\langle\gamma, 0\rangle| = 0 ;$$

$$|\langle\gamma, \langle 0, e\rangle\rangle| = \sup\{|\langle\langle\gamma, \alpha\rangle, \{e\}_{\mathrm{pr}}(\Phi^{=}, \gamma, \alpha)\rangle| + 1 : \alpha \in {}^{\omega}\omega\} ,$$

$$|\langle\gamma, \langle 1, e\rangle\rangle| = \inf\{|\langle\langle\gamma, \alpha\rangle, \{e\}_{\mathrm{pr}}(\Phi^{=}, \gamma, \alpha)\rangle| + 1 : \alpha \in {}^{\omega}\omega\} .$$

Let

$$\kappa = \sup\{|\langle\gamma, m\rangle| : \langle\gamma, m\rangle \in H\}$$

and

$$|\langle\gamma, m\rangle| = \kappa \quad \text{for} \quad \langle\gamma, m\rangle \notin H .$$

Theorem 2.3. There exist functions η_1 partial (1, 5, 6, 7)*-computable in* Φ *and η_2 partial* (1, 8)*-computable in Φ such that for i* = 1, 2:

(a) $|\langle\gamma, m\rangle| < \kappa$ *and* $|\langle\gamma, m\rangle| \leqslant |\langle\delta, n\rangle| \Rightarrow \eta_i(\gamma, \delta, m, n) \simeq 0;$

(b) $|\langle\delta, n\rangle| < |\langle\gamma, m\rangle| \Rightarrow \eta_i(\gamma, \delta, m, n) \simeq 1$.

Proof. The construction of η_1 follows that of [Mo 2, II, Theorem 7]. For η_2 we treat the trivial cases in the same way. Let $\mathbf{E}_0^{\#}$ be ${}^3\mathbf{E}^{\#}$ and $\mathbf{E}_1^{\#}(\Theta) \simeq 1 \dot{-} \mathbf{E}_0^{\#}(\lambda\alpha[1 \dot{-} \Theta(\alpha)])$. Then the four substantive clauses in the recursive definition of η_2 are given by

$$\eta_2(\gamma, \delta, \langle i, d\rangle, \langle j, e\rangle) \simeq \mathbf{E}^{\#}_{1 \dot{-} i}(\lambda\alpha \cdot \mathbf{E}_j^{\#}(\lambda\beta \cdot \eta_2(\langle\gamma, \alpha\rangle, \langle\delta, \beta\rangle,$$

$$\{d\}_{\mathrm{pr}}(\Phi^{=}, \gamma, \alpha), \{e\}_{\mathrm{pr}}(\Phi^{=}, \delta, \beta)))) .$$

The proof is completed by an application of the second recursion theorem and proof of (a) and (b) by induction on $\min\{|\langle\gamma, m\rangle|, |\langle\delta, n\rangle|\}$.

Theorem 2.4. H is both semi-$(1, 5, 6, 7)$*-computable in* Φ *and semi-*$(1, 8)$*-computable in* Φ.

Proof. Following the method of [Mo 2, II, §13], let

$$R_1(\epsilon) \Longleftrightarrow \{\epsilon\}^{\Phi}_{(1,5,6,7)}(\epsilon) \to 0\,, \qquad \text{and}$$

$$R_2(e, \alpha) \Longleftrightarrow \{e\}^{\Phi}_{(1,8)}(e, \alpha) \simeq 0\,.$$

By Lemma 2.2, there exist functions χ_1 and χ_2 primitive computable in Φ such that

$$R_1(\epsilon) \Longleftrightarrow \chi_1(\epsilon) \in H(\epsilon)\,, \qquad \text{and}$$

$$R_2(e, \alpha) \Longleftrightarrow \chi_2(e, \alpha) \in H(\alpha)\,.$$

Let

$$\nu_1 = \sup\{|\langle\epsilon, \chi_1(\epsilon)\rangle| : R_1(\epsilon)\}\,, \qquad \text{and}$$

$$\nu_2 = \sup\{|\langle\alpha, \chi_2(e, \alpha)\rangle| : R_2(e, \alpha)\}\,.$$

Suppose $\nu_1 < \kappa$. Then for some $\langle\gamma, m\rangle \in H$, $|\langle\gamma, m\rangle| = \nu_1$ and we have by 2.3

$$R_1(\epsilon) \Longleftrightarrow \eta_1(\epsilon, \gamma, \chi_1(\epsilon), m) \simeq 0\,, \qquad \text{and}$$

$$\sim R_1(\epsilon) \Longleftrightarrow \eta_1(\epsilon, \gamma, \chi_1(\epsilon), m) \simeq 1\,.$$

Hence R_1 would be $(1, 5, 6, 7)$-computable in Φ, which leads to a contradiction by a standard diagonal argument. Hence $\nu_1 = \kappa$. A similar argument shows $\nu_2 = \kappa$. Thus

$$\begin{aligned}\langle\gamma, m\rangle \in H &\Longleftrightarrow \exists\epsilon[R_1(\epsilon) \;\; \textit{and} \;\; \eta_1(\gamma, \epsilon, m, \chi_1(\epsilon)) \to 0]\\ &\Longleftrightarrow \exists e \exists\alpha[R_2(e, \alpha) \;\; \textit{and} \;\; \eta_2(\gamma, \alpha, m, \chi_2(e, \alpha)) \simeq 0]\end{aligned}$$

Since R_1 is semi-$(1, 5, 6, 7)$-computable in Φ and R_2 is semi-$(1, 8)$-computable in Φ, this establishes the desired conclusion.

Corollary 2.5. Every function partial (1, 8)*-computable in* Φ *is also partial* (1, 5, 6, 7)*-computable in* Φ.

Proof. Let χ be partial (1, 5, 6, 7)-computable in Φ such that $\langle\gamma, m\rangle \in H \Leftrightarrow \chi(\gamma, m) \to 0$, and $\psi(\alpha) = \alpha(0)$. Then

$$\{e\}^{\Phi}_{(1,8)}(\mathrm{x}) \simeq \psi(\nu^1\alpha[\chi(\bar{\mathrm{x}}, \Theta_2(e, \mathrm{x}, \alpha(0))) \to 0]) .$$

Corollary 2.6. The relation $\{\epsilon\}^{\Phi}_{(1,5,6,7)}(\mathrm{x}) \to y$ *is semi-*(1, 8)*-computable in* Φ.

Proof. Immediate from 2.2 and 2.4.

Corollary 2.7. A relation is semi-(1, 5, 6, 7)*-computable in* Φ *iff it is semi-*(1, 8)*-computable in* Φ *and some* ϵ.

Since (1, 8)-computability has no search schema, it is not immediate that the same relations are computable in the two senses. This follows, however, from the fact that (1, 8)-computability, like ordinary recursion theory, has a selection function.

Theorem 2.8. There exists a function ν *partial* (1, 8)*-computable in* Φ *such that for any e, m, and* x,

$$\exists m[\{e\}_{(1,8)}(m, \mathrm{x}) \simeq 0] \Rightarrow \nu(e, \mathrm{x})\downarrow \qquad \textit{and}$$

$$\{e\}_{(1,8)}(\nu(e, \mathrm{x}), \mathrm{x}) \simeq 0 .$$

Proof. Essentially the same as that of [Mo 1, Theorem 7].

Corollary 2.9. A relation R is (1, 8)*-computable in* Φ *iff both R and* $\sim R$ *are semi-*(1, 8)*-computable in* Φ.

Corollary 2.10. A relation is (1, 5, 6, 7)*-computable in* Φ *iff it is* (1, 8)*-computable in* Φ *and some* ϵ.

Corollary 2.11. There exists a function φ *partial* (1, 8)*-computable in* Φ *such that for each* ϵ, $\lambda\mathrm{x} \cdot \varphi(\epsilon, \mathrm{x})$ *is a single-valued branch of the number-valued part of* $\{\epsilon\}_{(1,5,6,7)}$. *That is*

(a) $\exists m[\{\epsilon\}_{(1,5,6,7)}(\mathrm{x}) \to m] \Leftrightarrow \varphi(\epsilon, \mathrm{x})\downarrow$,

(b) $\varphi(\epsilon, \mathrm{x}) \simeq m \Rightarrow \{\epsilon\}_{(1,5,6,7)}(\mathrm{x}) \to m$.

In particular, every single-number-valued function partial (1, 5, 6, 7)*-computable in* Φ *is partial* (1, 8)*-computable in* Φ *and some* ϵ.

3. The class of hyperprojective functionals

The reader familiar with [Mo 2] will recognize that the notions of computability discussed above are closely related to the notions of prime, search and hyperprojective computability over an abstract domain B. The precise relationship is quite complicated because of the extension from B to B^* in the development of [Mo 2]. (In the specific case $B = {}^{\omega}\omega$, this means that the theories of [Mo 2] are defined on some proper extension of ${}^{\omega}\omega$.) We state two theorems in the Appendix which make this relationship precise. Here we summarize some characterizations of the important class of *hyperprojective* functions on the continuum which follows from the results of [Mo 2] – [Mo 6], [B-G-M], once the class is identified. We shall omit proofs, since they will be obvious to these familiar with the abstract theory and incomprehensible to others.

Let $\Phi = \Phi^1, \Phi^2$ consists entirely of total, single-valued functions, let $\varphi(\mathrm{x})$ be partial, single-valued with range in ω. It follows from Corollaries 2.5, 2.11 and Theorems A and B of the Appendix that

φ *is* (1, 5, 6, 7)*-computable* (*in* Φ)

iff φ *is* (1, 8)*-computable in some* ϵ (*and* Φ)

iff φ *is hyperprojective in* $(\Phi), \Psi_1, \Psi_2$,

where Ψ_1, Ψ_2 are the operations of function application and "comprehension" defined in the Appendix. Since it is natural to assume Ψ_1, Ψ_2 as given for the continuum, we call these partial functions simply *hyperprojective* (*in* Φ) (*over the continuum*). Relations of the form

$$R(\mathrm{x}) \Longleftrightarrow \varphi(x) \quad \textit{is defined}$$

with a *hyperprojective* (*in* Φ) φ are called *semihyperprojective* (*in* Φ) and sets or relations whose characteristic functions are hyperprojective (in Φ) are called hyperprojective (in Φ).

We shall concentrate on the hyperprojective functions, relations and sets, taking $\Phi = \emptyset$ and leaving the relativization to the reader. It will also be convenient to have a name for the class of hyperprojective total unary functions on ${}^{\omega}\omega$ to ω,

$$\mathbf{HP} = \{F : F \text{ maps } {}^{\omega}\omega \text{ into } \omega, F \text{ hyperprojective}\} .$$

We call functions on ${}^{\omega}\omega$ to ω *functionals* to distinguish them from *functions* on ω to ω. Most of our characterizations aim to extend to **HP** known properties of the class of hyperarithmetic functions,

$$\mathrm{HA} = \{\alpha : \alpha \text{ maps } \omega \text{ into } \omega, \alpha \text{ hyperarithmetic}\} .$$

The sets of (pure) *types* over ω are defined inductively,

$$T_0 = \omega ,$$

$$T_{n+1} = {}^{(T_n)}\omega = \text{all unary functions on } T_n \text{ to } \omega .$$

For each $n \geqslant 0$, let $\mathcal{L}_{n+1}$ be *the language of* $n + 1$*-order arithmetic* with variables $\alpha_0^i, \alpha_1^i, \ldots$ ranging over T_i for each $i \leqslant n$, symbols $0, ', +, \cdot, =$ and the obvious formation rules and interpretation. $\mathcal{L}_1$ is the ordinary language of arithmetic and $\mathcal{L}_2$ is often called *the language of analysis*. It is convenient to choose unsuperscripted variables for the first three types,

$$n, k, m, n_1, \ldots \quad \text{etc. for} \quad \alpha_0^0, \alpha_1^0, \alpha_2^0, \ldots \quad \text{(number variables)}$$

$$\alpha, \beta, \gamma, \alpha_1, \ldots \quad \text{etc. for} \quad \alpha_0^1, \alpha_1^1, \alpha_2^1, \ldots \quad \text{(function variables)}$$

$$F, G, H, F_1, \ldots \quad \text{etc. for} \quad \alpha_0^2, \alpha_1^2, \alpha_2^2, \ldots \quad \text{(functional variables)} .$$

We often allow constants denoting members of some T_i in these languages, in which case we talk of *formulas with parameters* (of the appropriate types).

A. *Inductive definability*

Let $\mathcal{L}_2(S)$ be the extension of the language of analysis by the symbol "$\in$" and the variable "S" over subsets of ${}^{\omega}\omega = T_1$. If $\varphi(\alpha, S)$ is any formula with parameters from ${}^{\omega}\omega$ and only α, S free, it defines an operator

$$\Gamma_{\varphi}(S) = \{\alpha : \varphi(\alpha, S)\}$$

on subsets of ${}^{\omega}\omega$. It is well known that if all occurrences of subformulas of $\varphi(\alpha, S)$ of the form "$\beta \in S$" are *positive*, then Γ_φ is *monotone* and has a least fixed point φ^∞,

$$\varphi^\infty = \bigcap\{S : \Gamma_\varphi(S) = S\} .$$

A relation $R(\mathbf{x})$ of number and functions variable is *inductively definable over* $\mathcal{L}_2$ if there is a positive $\varphi(\alpha, S)$ and an $\mathcal{L}_2$-definable function $f(\mathbf{x})$ such that

$$R(\mathbf{x}) \Longleftrightarrow f(\mathbf{x}) \in \varphi^\infty .$$

Theorem 3.1. A relation R is semihyperprojective iff R is inductively definable over $\mathcal{L}_2$; hence R is hyperprojective iff both R and $\sim R$ are inductively definable over $\mathcal{L}_2$.

For each ordinal η, put

$$\varphi^\eta = \Gamma_\varphi\Big(\bigcup_{\xi<\eta} \varphi^\xi\Big) ;$$

it is then easy to verify that

$$\varphi^\infty = \bigcup_\eta \varphi^\eta .$$

From cardinality considerations, it is obvious that for some λ,

$$\varphi^\infty = \varphi^\lambda ;$$

the least such λ is called *the order* of Γ_φ (or φ) and denoted by $o(\Gamma_\varphi)$. Put

$$\kappa = \text{supremum}\,\{o(\Gamma_\varphi) : \varphi(\alpha, S) \text{ is positive in } \mathcal{L}_2(S),$$
$$\text{with parameters from } {}^{\omega}\omega\} .$$

(The choice of name is not accidental, as it can be shown that this is precisely the κ defined after Lemma 2.2.) We can further refine Theorem 3.1 to

Theorem 3.2. Suppose R is semihyperprojective, so that for some positive

$\varphi(\alpha, S)$ in $\mathcal{L}_2(S)$ and some $\mathcal{L}_2$-definable f,

$$R(\mathrm{x}) \Longleftrightarrow f(\mathrm{x}) \in \varphi^\infty ;$$

R is hyperprojective iff there is some $\lambda < \kappa$ so that

$$R(\mathrm{x}) \Longleftrightarrow f(\mathrm{x}) \in \varphi^\lambda .$$

(In particular this holds if $\mathrm{o}(\Gamma_\varphi) < \kappa$.)

These are direct generalizations of classical results relating semihyperarithmetic ($= \Pi^1_1$) and inductively definable relations on ω. In that case the common closure ordinal of all positive inductive definitions over $\mathcal{L}_1$ can be identified as the Church-Kleene ω_1, so that κ is the second-order analog of ω_1. However, it is by no means the case that κ is the supremum of all "recursive" (or even $\mathcal{L}_2$-definable) well-orderings (or even partial well-orderings) of ${}^\omega\omega$; κ is much larger than that.

B. *Model-theoretic characterizations*

The basis fact about hyperarithmetic relations is the theorem of Kleene which identifies them with the Δ^1_1 relations and it is natural to seek its analogs of the second type.

Recall that a relation $R(\mathrm{x})$ is Σ^2_1 if it is definable by a formula $(\exists F)\varphi(F, \mathrm{x})$ of $\mathcal{L}_3$, where $\varphi(F, \mathrm{x})$ may have parameters from ${}^\omega\omega$ but has no functional quantifiers; $R(\mathrm{x})$ is Π^2_1 if $\sim R(\mathrm{x})$ is Σ^2_1 and Δ^2_1 if both $R(\mathrm{x})$ and $\sim R(\mathrm{x})$ are Σ^2_1.

Each hyperprojective relation is Δ^2_1, but the converse fails by a long shot; in fact each semihyperprojective relation is Δ^2_1 and these exhaust only a very small part of Δ^2_1. We must be a bit more subtle in formulating analogs of the Kleene theorem.

Suppose $\mathcal{N}$ is a collection of functionals (i.e. $\mathcal{N} \subseteq T_2$) and F a given functional. We say that F is *Δ^2_1-definable with basis $\mathcal{N}$ and parameters from $\mathcal{N}$* if there is a formula $\varphi(G, \alpha, n)$ of $\mathcal{L}_3$ with parameters from ${}^\omega\omega$ and $\mathcal{N}$ and no functional quantifiers such that for all α, n

$$F(\alpha) = n \Longleftrightarrow (\exists G)\varphi(G, \alpha, n) \Longleftrightarrow (\exists G \in \mathcal{N})\varphi(G, \alpha, n) .$$

(We say Δ^2_1 rather than Σ^2_1 since this equivalence implies

$$F(\alpha) = n \iff (\forall G)(\forall m)[\varphi(G, \alpha, m) \Rightarrow m = n]$$

$$\iff (\forall G \in \mathcal{N})(\forall m)[\varphi(G, \alpha, m) \Rightarrow m = n].)$$

For each $\mathcal{N}$ put

$$\mathcal{N}^+ = \{F : F \textit{ is } \Delta^2_1\textit{-definable with basis } \mathcal{N} \textit{ and parameters from } \mathcal{N}\}.$$

Clearly $\mathcal{N} \subseteq \mathcal{N}^+$ and in general $\mathcal{N} \subsetneqq \mathcal{N}^+$.

Put

$$\mathcal{P}_0 = \{F : F \textit{ is } \mathcal{L}_2\textit{-definable with parameters from } {}^\omega\omega\},$$

$$\mathcal{P}_{\eta+1} = \mathcal{P}^+_\eta ;$$

$$\mathcal{P}_\eta = \bigcup_{\xi<\eta} \mathcal{P}_\xi \textit{ if } \eta \textit{ is a limit ordinal}.$$

Theorem 3.3. (i) $\bigcup_\eta \mathcal{P}_\eta = \mathcal{P}_\kappa = \mathbf{HP}$.

(ii) *if* $\eta < \xi < \kappa$, *then* $\mathcal{P}_\eta \subsetneqq \mathcal{P}_\xi$.

This is the application to the case of the continuum of the chief result of [Mo 4] and generalizes to the second type the main result of [Kl 2]. It characterizes the class of hyperprojective functionals as *the smallest class of functionals fixed by the operation* $\mathcal{N} \to \mathcal{N}^+$. It also leads directly to Theorem 3.4 below which is (perhaps) the closest we can come in relating **HP** with "Δ^2_1".

A class $\mathcal{N}$ of functionals is a model of *the* Δ^2_1*-comprehension schema* if for each formula $\varphi(G, \alpha, n)$ of $\mathcal{L}_3$ with parameters from ${}^\omega\omega$ and $\mathcal{N}$ and no functional quantifiers,

$$(\forall\alpha)(\exists!\, n)(\exists G \in \mathcal{N})\varphi(G, \alpha, n) \Rightarrow (\exists F \in \mathcal{N})(\forall\alpha)(\exists G \in \mathcal{N})\varphi(G, \alpha, F(\alpha));$$

$\mathcal{N}$ is a model of the Σ^2_1*-functional comprehension schema* if for each such $\varphi(G, \alpha, n)$,

$$(\forall\alpha)(\exists n)(\exists G \in \mathcal{N})\varphi(G, \alpha, n) \Rightarrow (\exists F \in \mathcal{N})(\forall\alpha)(\exists G \in \mathcal{N})\varphi(G, \alpha, F(\alpha)).$$

Theorem 3.4. **HP** *is the smallest model of the* Δ^2_1*-comprehension schema; it is also the smallest model of the* Σ^2_1*-functional comprehension schema.*

The result says that although **HP** does not exhaust the Δ^2_1 functionals, it contains all those which can be proved to exist using true assertions about ${}^{\omega}\omega$ expressible by functional-quantifier-free formulas of $\mathcal{L}_3$ as well as the schemata of Δ^2_1-comprehension and Σ^2_1-*functional comprehension*. (Incidentally, it is an open and apparently deep problem whether the class of Δ^2_1 functionals satisfies either of these schemata.)

Theorem 3.3 characterizes all semihyperprojective relations because of the following lifting to type-2 of the Spector-Gandy theorem [Sp].

Theorem 3.5. A relation $R(\mathbf{x})$ is semihyperprojective iff there is a formula $\varphi(F, \mathbf{x})$ of $\mathcal{L}_3$ with parameters from ${}^{\omega}\omega$ and no functional quantifiers such that

$$R(\mathbf{x}) \Longleftrightarrow (\exists F \in \mathbf{HP})\varphi(F, \mathbf{x}) .$$

Another consequence of Theorem 3.3 is that it gives a very natural *hierarchy* of length κ on **HP**, similar to the classical hierarchy on HA.

The sequence $\{\mathcal{P}_\eta\}_{\eta<\kappa}$ is also reminiscent of the classical hierarchy of Borel functionals. In looking for analogies of this type we should keep in mind that **HP** is the second order analog of HA and not of the Borel functionals; the correct analog of the latter consists of *all objects in T_3 that are hyperprojective in some $F \in T_2$* which we do not discuss here. The properties that **HP** shares with the Borel functionals are the simple structure properties which the Borel functionals share with HA.

It is interesting, or at least amusing to examine some consequences of Theorem 3.3 in the context of the *axiom of determinacy*, which is inconsistent with the axiom of choice. The next theorem follows easily from the results of [Mo 5].

Theorem 3.6. Assume the axioms of Zermelo-Fraenkel set theory (without choice) and the axiom of determinacy. Then the class of hyperprojective subsets of ${}^{\omega}\omega$ is the smallest class of sets containing all open sets and closed under complementation, the taking of continuous images and unions of length less than κ. Moreover κ is a weakly inaccessible cardinal, in fact the κ'th *weakly inaccessible cardinal.*

This type of result is not entirely frivolous, since there is some hope that we may someday show (from reasonable axioms) the axiom of determinacy to hold in

$$L[{}^{\omega}\omega] = \text{smallest model of Zermelo-Fraenkel theory}$$

$$\text{containing } {}^{\omega}\omega \text{ and all ordinals} .$$

This has been conjectured by Solovay. Granting Solovay's conjecture, Theorem 3.6 gives an elegant characterization of **HP** within $L[{}^{\omega}\omega]$. (One should observe that $\mathbf{HP} \subseteq L[{}^{\omega}\omega]$.)

C. *Admissible sets*

The results of [B-G-M] give an entirely different characterization of **HP**.

Theorem 3.7. Let $A = ({}^{\omega}\omega)^{+}$ *be the smallest transitive set which is admissible (in the sense of Kripke-Platek) and such that* ${}^{\omega}\omega \in A$. *Then*

$$A = L_{\kappa}({}^{\omega}\omega) ,$$

$$\kappa = \text{least ordinal not in } A ,$$

and

$$\mathbf{HP} = \{F : F \in A\} .$$

This lifts to the second type the known characterization of HA as the set of functions in L_{ω_1}.

D. *The game quantifier*

Perhaps the most attractive kind of characterization of a class of relations $\mathcal{R}$ is one like this: $\mathcal{R}$ consists of all relations definable in a language $\mathcal{L}$ by formulas of such and such form. This is particularly attractive if $\mathcal{L}$ and the formulas in question are simple and natural. Such a characterization for the semihyperprojective relations is given in [Mo 6].

Let $\langle n_0, n_1, ..., n_k\rangle, \langle \alpha_0, ..., \alpha_k\rangle$ be fixed recursive codings of tuples in ω and ${}^{\omega}\omega$ respectively. The *Suslin quantifier* $\mathcal{S}$ is defined by

$$(\mathcal{S} n)\varphi(n, \mathrm{x}) \Longleftrightarrow (\forall n_0 \forall n_1 \forall n_2 ...)(\exists k)\varphi(\langle n_0, ..., n_k\rangle, \mathrm{x}) .$$

Kleene's normal form theorem characterizes the Π^1_1 relations as precisely those of the form $(\mathcal{S} n)\varphi(n, \mathrm{x})$, with $\varphi(n, \mathrm{x})$ $\mathcal{L}_1$-definable.

The Suslin quantifier may be applied to a function variable,

$$(\mathcal{S}\alpha)\varphi(\alpha, \mathrm{x}) \Longleftrightarrow (\forall\alpha_0 \forall\alpha_1 \forall\alpha_2 \ldots)(\exists k)\varphi(\langle\alpha_0, \ldots, \alpha_k\rangle, \mathrm{x}) ;$$

however it is trivial to verify that if $\varphi(\alpha, \mathrm{x})$ is $\mathcal{L}_2$-definable, then so is $(\mathcal{S}\alpha)\varphi(\alpha, \mathrm{x})$, so that $\mathcal{S}$ on ${}^\omega\omega$ has no more power than $\forall$.

We define the *game quantifier* by turning every other $\forall$ to $\exists$ in the infinite string defining $\mathcal{S}$, i.e.

$$(\mathcal{G} n)\varphi(n, \mathrm{x}) \Longleftrightarrow (\forall n_0 \exists n_1 \forall n_2 \ldots)(\exists k)\varphi(\langle n_0, \ldots, n_k\rangle, \mathrm{x}) ,$$

$$(\mathcal{G}\alpha)\varphi(\alpha, \mathrm{x}) \Longleftrightarrow (\forall\alpha_0 \exists\alpha_1 \forall\alpha_2 \ldots)(\exists k)\varphi(\langle\alpha_0, \ldots, \alpha_k\rangle, \mathrm{x}) .$$

(A precise definition of $\mathcal{G}$ is easily given in terms of an infinite sequence of Skolem functions or in terms of *closed games*, e.g. see [Mo 6].)

It is an easy exercise in the theory of Π^1_1 relations to show that the relations of the form $(\mathcal{G} n)\varphi(n, \mathrm{x})$ with $\varphi(n, \mathrm{x})$ $\mathcal{L}_1$-definable are precisely the Π^1_1 relations. The main result of [Mo 6] is that this lifts one type to a characterization of the semihyperprojective relations.

Theorem 3.8. A relation $R(\mathrm{x})$ is semihyperprojective iff there is a formula $\varphi(\alpha, \mathrm{x})$ of $\mathcal{L}_2$ with parameters from ${}^\omega\omega$ such that

$$R(\mathrm{x}) \Longleftrightarrow (\mathcal{G}\alpha)\varphi(\alpha, \mathrm{x}) .$$

This characterization of the semihyperprojective relations suggests a development of their theory that does not involve computability in objects of higher type, the way we often study "Π^1_1 relations" without bothering to identify them with "semihyperarithmetic", i.e. "semicomputable in ${}^2E$". There are direct and fairly simple proofs of the main properties of relations of the form $(\mathcal{G}\kappa)\varphi(\alpha, \mathrm{x})$, including the existence of a universal relation, the reduction property etc.

We hope that the characterizations of hyperprojective relations in terms of Kleene recursion in ${}^3\mathrm{E}^\#$ in Section 2 and the structure theorems of this section have convinced the reader of two things: this class is the correct analog in type-2 of the hyperarithmetic sets and it is an interesting class deserving the attention paid to it. It should be pointed out that our emphasis on type-2 is only for simplicity of notation, to avoid messy-looking superscripts. The abstract theory of hyperprojective relations on arbitrary structures applies directly to any type over ω and leads to the *hyper-order n-projective* relations having much the same structure properties as those we have outlined here.

Appendix. Abstract computability over the continuum

The reader familiar with [Mo 2] will recognize that the notions of computability discussed above are closely related to the notions of prime, search, and hyperprojective computability over an abstract domain B. In this section we shall specify precisely this relationship.

We consider the special case of abstract computability over $B = {}^{\omega}\omega$ (as promised in [Mo 2, I, Example 2]). Letters $u, v, w, \ldots$ will be used as variables over $({}^{\omega}\omega)^*$, the closure of ${}^{\omega}\omega \cup \{0\}$ under the formation of ordered pairs. Recall that ω is canonically embedded in $({}^{\omega}\omega)^*$ by $n + 1 \to \langle n, 0\rangle$.

Since in general there is no relationship assumed between B and ω, the schemata of [Mo 2] contain nothing corresponding to our schemata (1.7) and (5.3). Hence in considerations of abstract computability over ${}^{\omega}\omega$ we shall often want to relativize to the function

$$\begin{aligned}\Psi_1(u, v) &= u(v), \quad \text{if} \quad u \in {}^{\omega}\omega \quad \text{and} \quad v \in \omega\,;\\ &= 0, \quad \text{otherwise,}\end{aligned}$$

and the functional

$$\Psi_2(\Theta) = \lambda m \Theta(m)\,.$$

The theory in [Mo 2] was not relativized to an arbitrary functional but only to the specific functional $\mathbf{E}$; in order to relativize to Ψ_2 we add to the definition of $\{u\}_i(\mathrm{v}) \to w$ with $i = p, v, h$ the clause

C11′ If $\forall m \{u\}(m, \mathrm{v}) \to \alpha(m)$, then $\{\langle 11, n, u\rangle\}(\mathrm{v}) \to \alpha$.

(See Def. 9 and clause XV following it in §3 of [Mo 3] for a discussion of this method of introducing *uniformly computable* functionals.) Since partial functions from $\omega^{k,l}$ into ω are also partial functions from $(({}^{\omega}\omega)^*)^{k+l}$ into $({}^{\omega}\omega)^*$, all of the functions we have been studying are also functions appropriate to the present abstract context.

Theorem A. For any function φ:

(a_1) *φ partial* (1)*-computable in* $\Phi \Rightarrow \varphi$ *absolutely prime computable in* Φ, Ψ_1, *and* Ψ_2.

(b_1) *φ partial* (1, 2)*-computable in* $\Phi \Rightarrow \varphi$ *absolutely search computable in* Φ, Ψ_1, *and* Ψ_2.

(c_1) *φ partial $(1,8)$-computable in $\Phi \Rightarrow \varphi$ absolutely hyperprojective in Φ, Ψ_1, and Ψ_2.*

(a_2) *φ partial $(1,5)$-computable in $\Phi \Rightarrow \varphi$ prime computable in Φ, Ψ_1, and Ψ_2.*

(b_2) *φ partial $(1,5,6)$-computable in $\Phi \Rightarrow \varphi$ search computable in Φ, Ψ_1, and Ψ_2.*

(c_2) *φ partial $(1,5,6,7)$-computable in $\Phi \Rightarrow \varphi$ hyperprojective in Φ, Ψ_1, and Ψ_2.*

Furthermore, in (a_1) – (c_1), *if $\Phi^2 = \emptyset$ then Ψ_2 may be omitted on the right side.*

Proof. In each case the proof is by a laborious but straight-forward construction of an index transfer function π such that

$$\{z\}_{(1,\ldots)}(\mathbf{x}) \to y \quad \text{iff} \quad \{\pi(z)\}(\mathbf{x}) \to y$$

and $\{\pi(z)\}(\mathbf{u})$ is undefined for u not in $\omega^{k,l}$.

Let τ be an absolutely primitive computable coding map from $({}^{\omega}\omega)^*$ into ${}^{\omega}\omega$. We write $\bar{u}$ for $\tau(u)$ and assume that for u in $\{\emptyset\}^*$, $\bar{u}$ is primitive computable.

Lemma. There exist primitive computable functions $\pi_p, \pi_\nu, \pi_h, \rho_p, \rho_\nu$ and ρ_h such that:

(a_1) $\{u\}_p(\Phi, \Psi_1, \mathbf{v}) \simeq w \Longleftrightarrow \forall m[\{\pi_p(\bar{u})\}_{(1)}(\Phi, \bar{u}, \bar{\mathbf{v}}, m) \simeq \bar{w}(m)]$.

(b_1) $\{u\}_\nu(\Phi, \Psi_1, \mathbf{v}) \simeq w \Longleftrightarrow \forall m[\{\pi_\nu(\bar{u})\}_{(1,2)}(\Phi, \bar{u}, \bar{\mathbf{v}}, m) \simeq \bar{w}(m)]$.

(c_1) $\{u\}_h(\Phi, \Psi_1, \mathbf{v}) \simeq w \Longleftrightarrow \forall m[\{\pi_h(\bar{u})\}_{(1,8)}(\Phi, \bar{u}, \bar{\mathbf{v}}, m) \simeq \bar{w}(m)]$.

(a_2) $\{u\}_p(\Phi, \Psi_1, \Psi_2, \mathbf{v}) \simeq w \Longleftrightarrow \{\rho_p(\bar{u})\}_{(1,5)}(\Phi, \bar{u}, \bar{\mathbf{v}}) \simeq \bar{w}$.

(b_2) $\{u\}_\nu(\Phi, \Psi_1, \Psi_2, \mathbf{v}) \simeq w \Longleftrightarrow \{\rho_\nu(\bar{u})\}_{(1,5,6)}(\Phi, \bar{u}, \bar{\mathbf{v}}) \simeq \bar{w}$.

(c_2) $\{u\}_h(\Phi, \Psi_1, \Psi_2, \mathbf{v}) \simeq w \Longleftrightarrow \{\rho_h(\bar{u})\}_{(1,5,6,7)}(\Phi, \bar{u}, \bar{\mathbf{v}}) \simeq \bar{w}$.

Furthermore, in (a_1) – (c_1), *if Φ^2 contains at least one total function, then Ψ_2 may be added to the list of relativization functions on the left side.*

For φ a $k + l$-place function on $({}^{\omega}\omega)^*$, let $\varphi^{k,l}$ be the restriction of φ to $\omega^{k,l}$.

Theorem B. For any number-valued function φ *and any* ψ:

(a$_1$) φ *absolutely prime computable in* Φ *and* $\Psi_1 \Rightarrow \varphi^{k,l}$ *partial* (1)-*computable in* Φ.
(b$_1$) φ *absolutely search computable in* Φ *and* $\Psi_1 \Rightarrow \varphi^{k,l}$ *partial* (1, 2)-*computable in* Φ.
(c$_1$) φ *absolutely hyperprojective in* Φ *and* $\Psi_1 \Rightarrow \varphi^{k,l}$ *partial* (1, 8)-*computable in* Φ.

(a$_2$) φ *prime computable in* Φ, Ψ_1, *and* $\Psi_2 \Rightarrow \varphi^{k,l}$ *partial* (1, 5)-*computable in* Φ.
(b$_2$) φ *search computable in* Φ, Ψ_1, *and* $\Psi_2 \Rightarrow \varphi^{k,l}$ *partial* (1, 5, 6)-*computable in* Φ.
(c$_2$) φ *hyperprojective in* Φ, Ψ_1, *and* $\Psi_2 \Rightarrow \varphi^{k,l}$ *partial* (1, 5, 6, 7)-*computable in* Φ.

Furthermore, in (a$_1$) – (c$_1$), *if* Φ^2 *contains at least one total function, then* Ψ_2 *may be added to the list of relativization functions on the left side.*

References

[Hi] P.G. Hinman, Extensions of the hyperanalytic hierarchy, Abstract 663–721, Notices Am. Math. Soc., January, 1969.

[B-G-M] K.J. Barwise, R.O. Gandy and Y.N. Moschovakis, The next admissible set, to appear in J. Symbolic Logic.

[Kl 1] S.C. Kleene, Recursive functionals and quantifiers of finite types, I and II, Trans. Am. Math. Soc. 91 (1959) 1–52; 108 (1963) 106–142.

[Kl 2] S.C. Kleene, Quantification of number-theoretic functions, Comp. Math. 14 (1959) 23–40.

[Mo 1] Y.N. Moschovakis, Hyperanalytic predicates, Trans. Am. Math. Soc. 129 (1967) 249–252.

[Mo 2] Y.N. Moschovakis, Abstract first order computability, I and II, Trans. Am. Math. Soc. 138 (1969) 427–464, 465–504.

[Mo 3] Y.N. Moschovakis, Axioms for computation theories – first draft. This volume, p.

[Mo 4] Y.N. Moschovakis, Abstract computability and invariant definability, J. Symbolic Logic 34 (1969) 605–633.

[Mo 5] Y.N. Moschovakis, Determinacy and prewellorderings of the continuum, Mathematical logic and foundations of set theory, Amsterdam, 1970, pp. 24–62.

[Mo 6] Y.N. Moschovakis, The game quantifier, to appear.

[Sp] C. Spector, Inductively defined sets of natural numbers, in: Infinitistic Methods, Warsaw 1961, pp. 97–102.

ON THE GCH AT MEASURABLE CARDINALS

Kenneth KUNEN*
University of Wisconsin, Madison, Wisconsin, USA

Theorem. Suppose that there is a measurable cardinal, κ, such that at least one of the following holds:

(i) $2^\kappa > \kappa^+$.

(ii) *Every κ-complete filter over κ can be extended to a κ-complete ultrafilter.*

(iii) *There is a uniform κ-complete ultrafilter over κ^+.*

Then for all ordinals θ, there is a transitive proper class, N, such that

$$N \models [\text{ZFC} + \textit{there are } \theta \textit{ measurable cardinals}] \ .$$

We consider the proof from (i) to be the main result of this paper. That the conclusion follows from (ii) and (iii) will be proved by the same method and with little additional effort.

This theorem is an improvement over some results in §§9–10 of [2], which showed that the existence of Solovay's $0^\dagger$ followed from (i) and from (ii), and that the full conclusion of the theorem followed from the assumption that κ is strongly compact. Familiarity with [2] is assumed here throughout.

Both the statement of the theorem and the proof are understood to be formalized within Morse-Kelley set theory with the axiom of choice (see the appendix to Kelley [1]). We leave to the reader the simple modifications necessary to carry out everything within ZFC.

Note that we may forget completely about assumption (iii), since it implies that either (i) or (ii) hold. Thus, from now on we shall assume that κ is a measurable cardinal satisfying (i) or (ii). We shall, in fact, derive the conclusion of the theorem directly from Lemma 4. The proof of Lemma 4 from (i) uses Lemma 1, and from (ii) uses Lemma 3.

Lemma 1 combines some modifications, due to C.C. Chang and to

* The author is a fellow of the Alfred P. Sloan Foundation.

A. Hajnal, of Gödel's proof of GCH in L.

Lemma 1. If $a \subseteq \kappa^+$, $\mathcal{S} \subseteq \mathcal{P}(\kappa)$, $b \subseteq \mathrm{ORD}$, *and* $\bar{\bar{b}} < \kappa$, *then* $\mathrm{card}(\mathcal{P}(\kappa) \cap L[a, b, \mathcal{S}]) \leqslant \kappa^+$.

Proof. Let $\lambda = \bar{\bar{b}}$. By a Löwenheim-Skolem and collapsing argument,

$$\mathcal{P}(\kappa) \cap L[a, b, \mathcal{S}] \subseteq \bigcup \{L_\mu[a \cap \nu, x, \mathcal{S}] : \mu < \kappa^+ \wedge \nu < \kappa^+ \wedge x \subset \kappa^+ \wedge \bar{\bar{x}} = \lambda\},$$

which has power κ^+.

To prove Lemma 3, we need an unpublished combinatorial result, due to J. Ketonen, which we include with his permission.

Lemma 2 (Ketonen). *There is a family* $\mathcal{G} \subset {}^\kappa\kappa$ *such that* $\bar{\bar{\mathcal{G}}} = 2^\kappa$, *and such that whenever* $\lambda < \kappa$, g_ξ $(\xi < \lambda)$ *are distinct members of* $\mathcal{G}$, *and* ρ_ξ $(\xi < \lambda)$ *are any ordinals* $< \kappa$, *then* $\exists \eta < \kappa \forall \xi < \lambda [g_\xi(\eta) = \rho_\xi]$.

Proof. Let A_α $(\alpha < 2^\kappa)$ be almost disjoint subsets of κ. Let $f_\alpha : A_\alpha \to \kappa$ be such that $\forall \rho < \kappa [\mathrm{card}(\{\zeta \in A_\alpha : f_\alpha(\zeta) = \rho\}) = \kappa]$. Let s_η $(\eta < \kappa)$ enumerate $\{s \subset \kappa : \bar{\bar{s}} < \kappa\}$. Let $\mathcal{G} = \{g_\alpha : \alpha < 2^\kappa\}$, where $g_\alpha(\xi) = f_\alpha(\zeta)$ when $s_\xi \cap A_\alpha = \{\zeta\}$, and $g_\alpha(\xi) = 0$ if $\mathrm{card}(s_\xi \cap A_\alpha) \neq 1$.

Lemma 3. If (ii) *of the theorem holds and* $\delta < (2^\kappa)^+$, *then there is a* κ*-complete ultrafilter,* $\mathcal{U}$, *over* κ *such that* $i_{01}^{\mathcal{U}}(\kappa) > \delta$ *(where* $i_{01}^{\mathcal{U}} : V \to \mathrm{Ult}_1(V, \mathcal{U})$*).*

Proof. We may assume $\delta \geqslant 2^\kappa$. Let $\mathcal{G}$ be as in Lemma 2; let $\mathcal{G} = \{g_\alpha : \alpha \leqslant \delta\}$, in 1–1 enumeration. Let $\mathcal{U}$ be such that whenever $\alpha < \beta \leqslant \delta$, $\{\xi : g_\alpha(\xi) < g_\beta(\xi)\} \in \mathcal{U}$.

Lemma 4. If $b \subset \mathrm{ORD}$, $\bar{\bar{b}} < \kappa$, *and* $\delta < \kappa^{++}$, *then there is a transitive model,* M, *for* ZFC, *and a* κ*-complete ultrafilter,* $\mathcal{U}$, *over* κ, *such that* $b \in M$, $\mathcal{U} \cap M \in M$, $\mathrm{ORD} \subset M$, *and* $\delta < i_{01}^{\mathcal{U} \cap M}(\kappa) < \kappa^{++}$ *(where* $i_{01}^{\mathcal{U} \cap M} : M \to \mathrm{Ult}_1(M, \mathcal{U} \cap M)$*).*

Proof. If $2^\kappa = \kappa^+$, then (ii) holds, so we may take $M = V$ and apply Lemma 3. If $2^\kappa > \kappa^+$, let $\mathcal{U}$ be any κ-complete non-principal ultrafilter over κ. $i_{01}^{\mathcal{U}}(\kappa) > 2^\kappa \geqslant \kappa^{++}$, so there are $g_\alpha \in {}^\kappa\kappa$ for $\alpha \leqslant \delta$ such that $\{\xi : g_\alpha(\xi) < g_\beta(\xi)\} \in \mathcal{U}$ whenever $\alpha < \beta \leqslant \delta$. Let $a \subset \kappa^+$ be such that each $g_\alpha \in L[a]$. Let $M = L[a, b, \mathcal{U}]$. and apply Lemma 1.

Definition 1. For any ordinal θ, a θ-set is a set $b \subset$ ORD such that b has order type $\omega\cdot\theta$, and such that for each $\zeta \in b$, $\zeta > \sup(b \cap \zeta)$.

Definition 2. If b is a θ-set, $\xi < \theta$, and $n < \omega$,

(a) $\gamma(b, \xi, n)$ *is the* $\omega\cdot\xi+n$th *ordinal in* b *and* $\lambda(b, \xi) = \sup(\{\gamma(b, \xi, m) : m < \omega\})$.

(b) $\mathcal{G}(b, \xi) = \{x \subseteq \lambda(b, \xi) : \exists m \forall k > m [\gamma(b, \xi, k) \in x]\}$; $\vec{\mathcal{G}}(b) = \langle \mathcal{G}(b, \eta) : \eta < \theta \rangle$.

(c) $\Phi(b, \xi)$ *is the statement that* $\mathcal{G}(b, \xi) \cap L[\vec{\mathcal{G}}(b)]$ *is, in* $L[\vec{\mathcal{G}}(b)]$, *a normal ultrafilter on* $\lambda(b, \xi)$.

Definition 3. (a) $K_0 = \{\sigma : \sigma$ *is a cardinal* $\wedge \operatorname{cf}(\sigma) > \kappa \wedge \forall \tau < \sigma [\tau^\kappa < \sigma]\}$.

(b) $K_{\alpha+1} = \{\sigma \in K_\alpha : \operatorname{card}(K_\alpha \cap \sigma) = \sigma\}$.

(c) $K_\beta = \bigcap_{\alpha<\beta} K_\alpha$ *for limit ordinals* β.

We shall eventually show that if $\theta < \kappa$, b is a θ-set, and $b \subset K_{\omega\cdot\theta+1}$, then $\forall \xi < \theta \Phi(b, \xi)$. This will prove the theorem for $\theta < \kappa$. The full theorem will then follow immediately, since for any θ one may carry out the entire proof within $\mathrm{Ult}_{\theta+1}(V, \mathcal{U})$, where $\mathcal{U}$ is any κ-complete non-principal ultrafilter over κ.

Lemma 5. If M is a transitive model for ZFC, *$\mathcal{U}$ a κ-complete ultrafilter over κ, $\mathcal{U} \cap M \in M$, $\sigma \in K_0$, and $\mu < \sigma$, then $i_{0\mu}^{\mathcal{U}\cap M}(\sigma) = \sigma$.*

Lemma 6. If b is a θ-set, $\theta < \kappa$, and $\xi < \theta$ is such that $\forall \eta < \xi [\lambda(b, \eta) < \kappa^{++}]$ and $\forall n \forall \eta [\xi \leqslant \eta < \theta \to \gamma(b, \eta, n) \in K_0]$, then $\Phi(b, \xi)$.

Proof. Let $\delta = \sup(\{\lambda(b, \eta) : \eta < \xi\})$. $\delta < \kappa^{++}$. In the special case that $\delta < \kappa$, the lemma is standard and just uses the fact that κ is measurable. In the general case, let M and $\mathcal{U}$ be as in Lemma 4, and apply the proof for $\delta < \kappa$ within $\mathrm{Ult}_1(M, \mathcal{U} \cap M)$.

Definition 4. If $a \subset$ ORD and $X \subseteq L[a]$, then $H(a, X) = \{y \in L[a] : y$ is definable in $L[a]$ from elements of $a \cup X \cup \{a\}\}$.

Lemma 7. Let $a \subset K_\alpha$ be of order type α, where $\alpha < \kappa$. Say $a = \{\sigma_\xi : \xi < \alpha\}$, in increasing enumeration. Then for each $\xi < \alpha$, $\operatorname{card}(\sigma_\xi \cap H(a, K_{\xi+1} \sim \sigma_\xi)) \leqslant \kappa^+$.

Proof. We use induction on $\xi < \alpha$. Assume the lemma holds for $\eta < \xi$. Let

$\rho = \sup\{\sigma_\eta : \eta < \xi\}$. If $\rho = \sigma_\xi$, the induction step is trivial, so assume $\rho < \sigma_\xi$. Let $A = H(a, K_\xi \sim \rho)$. $\mathrm{card}(\rho \cap A) \leqslant \kappa^+$. Let T be the transitive collapse of A, j the isomorphism from A onto T, $\delta = j(\rho)$, and $b = j(a)$. $\delta < \kappa^{++}$, and $j(\tau) = \tau$ for $\tau \in K_{\xi+1} \sim (\rho + 1)$. Let M and $\mathcal{U}$ be as in Lemma 4. Lemma 7 for ξ will follow if we can show that for all $\pi \in \sigma_\xi \cap H(a, K_{\xi+1} \sim \sigma_\xi)$, $j(\pi) < i_{01}^{\mathcal{U} \cap M}(\kappa)$. Say π is definable in $L[a]$ from elements of $s \cup t \cup \{a\}$, where s is a finite subset of $a \cap \sigma_\xi$ and t is a finite subset of $K_{\xi+1} \sim \sigma_\xi$. Then $j(\pi)$ is definable in $L[b]$ from elements of $j(s) \cup t \cup \{b\}$. Now if $\pi < \sigma_\xi$, then $j(\pi)$ is fixed by $i_{1\pi}^{\mathcal{U} \cap M}$, since all elements of $j(s) \cup t \cup \{b\}$ are. Thus, $j(\pi) < i_{01}^{\mathcal{U} \cap M}(\kappa)$.

We now prove the theorem, using the remark following Definition 3. Suppose $\theta < \kappa$, b is a θ-set, and $b \subset K_{\omega\cdot\theta+1}$. Let $\xi < \theta$. Let $A = H(b, K_{\omega\cdot\theta} \sim \sup(\{\lambda(b, \eta) : \eta < \xi\}))$, and j be the isomorphism from A onto the transitive collapse of A. For $\xi \leqslant \eta < \theta$ and $n < \omega$, $j(\gamma(b, \eta, n)) = \gamma(b, \eta, n) \in K_0$, and for $\eta < \theta$ and $n < \omega$, $j(b, \eta, n) < \kappa^{++}$ by Lemma 7. Hence, by Lemma 6, $\Phi(j(b), \xi)$. It follows that $\Phi(b, \xi)$, since $A \prec L[b]$ and $\Phi(b, \xi)$ is a property of b and ξ expressible within $L[b]$. Thus, $\forall \xi < \theta \Phi(b, \xi)$.

References

[1] J.L. Kelley, General topology (Van Nostrand, Princeton, N.J., 1955).

[2] K. Kunen, Some applications of iterated ultrapowers in set theory, Ann. Math. Logic 1 (1970) 179–227.

PART III

INVITED PAPERS ON GENERALISATIONS OF RECURSION THEORY

AXIOMATIC RECURSIVE FUNCTION THEORY*

Harvey M. FRIEDMAN
Department of Philosophy, Stanford University, California, USA

Introduction

When we examine ordinary recursive function theory, a certain construction is seen to play a fundamental role not only in proofs of theorems but in actual statements of theorems (e.g. the Kleene 2nd recursion theorem): namely, the enumeration functions, as in the Kleene enumeration theorem. Further, certain basic properties of the particular enumerations constructed are used over and over again. The properties referred to here are summed up by the existence of functions S_n^m satisfying the so-called *S-m-n* theorem.

These considerations immediately suggest an axiomatic theory. An appropriate axiomatic theory was set down by H. Strong who, on reformulating results of E. Wagner and himself, considered systems of the form $(D,F,\phi_1,\phi_2,...)$ satisfying the following first-order conditions: D has at least two elements; F is a class of partial functions on D of many arguments; each ϕ_i is an $i+1$-ary enumeration function and lies in F; functions S_n^m exist in F; and (D,F) satisfies certain basic closure conditions. Structures satisfying these conditions are called BRFT's (basic recursive function theories).

We refer the reader to Strong [9] for an extensive summary and bibliography of work of Wagner and Strong. Remarks on the connection between Strong's formulation and the earlier Wagner formulation is given in Section 4.

Wagner and Strong discovered that a considerable portion of the statements of the theorems in ordinary recursive function theory can be expressed elegantly in the language of BRFT (often without mention of the constants ϕ_i), and that many of these theorems are true in all BRFT, and so, consequently, can be derived from the BRFT axioms.

In contrast, this paper is concerned with a metamathematical investigation of the BRFT axioms. In particular the emphasis is on two traditional metamathematical issues: (relative) categoricity and minimality results. The rela-

* This research was partially supported by NSF GP 13355.

tive categoricity results are found in Theorem 2.5 and Corollaries 3.4.1 and 3.5.1. The minimality results are found in Wagner and Strong's Theorem 2.1 and our improvement, Theorem 2.3, and Theorem 3.1.

In Section 1 we mainly exposit those parts of the Wagner-Strong theory that will be needed for the later proofs. These results are general; they hold for all BRFT. One result of ours, Theorem 1.1, establishes a kind of mutual interpretability between the axioms of BRFT and an elegant set of axions on monadic and binary partial functions, the latter not containing any *S-m-n* condition. Our Theorem 1.5 gives added meaning to our relative categoricity results, which are stated in terms of piecewise isomorphism.

In Section 2 we concentrate on BRFT $(\omega, F, \phi_1, \phi_2, ...)$ such that $\lambda n(n+1) \in F$. The special case of Theorem 2.5 when F is the collection of all partial recursive functions on ω of several arguments is of special interest [1]. This, along with Theorem 1.5 provide a theoretical explanation of why it is that the only properties of enumeration functions ever used in ordinary recursive function theory is the existence of S_n^m functions. Theorem 2.3, very roughly speaking, shows that any collection of partial functions on ω satisfying the Kleene enumeration theorem must contain all partial recursive functions. In Theorem 2.6, a forcing argument is used to characterize those classes of total functions present in some ω-BRFT.

In Section 3, a minimality theorem is obtained for the hyperarithmetic functions in Theorem 3.1 together with Theorem 3.3. In Theorems 3.4 and 3.5 we present a generalization of the relative categoricity result of Section 2 to BRFT on ordinals, in connection with the ordinal recursive function theory of Kripke and Platek. Here, a consideration appears which does not appear in Section 2; namely, nonprojectibility. Theorem 3.6 establishes that relative categoricity is sensitive to that consideration.

In Section 4 we make additional remarks and discuss open questions.

Section 1 is intended to be self-contained. For reference, consult Strong [9].

Section 2 presupposes knowledge of ordinary recursive function theory and the theory of forcing with finite conditions. For reference, see Rogers [8] and Feferman [1].

Section 3 presupposes knowledge of the ordinal recursive function theory of Platek and Kripke. For reference, see Platek [6] and Kripke [5].

Section 1

Whenever E_1, E_2 are expressions, by $E_1 = E_2$ we mean that both E_1, E_2

[1] Added in proof: We recently learned that this important special case of our Theorem 2.5 was proved independently by M. Blum.

are defined and are equal. By $E_1 \simeq E_2$ we mean that either $E_1 = E_2$ or else both E_1 and E_2 are undefined. Thus for monadic partial functions f, g, we have $f = g \equiv (\forall x)\ (f(x) \simeq g(x))$. We have that f is total if and only if $(\forall x)\ (f(x) \equiv f(x))$. We have the important transitivity, $(E_1 \simeq E_2\ \&\ E_2 \simeq E_3) \rightarrow E_1 \simeq E_3$. If any part of an expression, at given arguments, is undefined, then at those arguments the whole expression is undefined. Thus, if f is monadic, g is monadic, and f is undefined at 0 but defined at 1, and g is total, then $f(g(x))$ is defined at 1 but undefined at 0.

Definition 1.1. An *adequate system* is a pair (D,F) such that

(1) D has at least two elements,

(2) F is a collection of partial function on D of several arguments,

(3) F is closed under generalized composition; i.e., whenever g is an n-ary element of $F, f_1,\ldots,f_n$ are m-ary elements of F, then $\lambda x_1..x_m(g(f_1(x_1,\ldots,x_m), \ldots,f_n(x_1,\ldots,x_m))) \in F_m$,

(4) each U_n^m, $1 \leqslant n \leqslant m$, is in F, where $U_n^m = \lambda x_1\ldots x_m(x_n)$,

(5) each C_x^m, $x \in D$, $1 \leqslant m$, is in F, where $C_x^m = \lambda x_1\ldots x_m(x)$,

(6) $\lambda abcx$ (b if $x = a$; c if $x \neq a$) $= \alpha \in F$. Whenever (D,F) satisfies (2) we let F_n, $1 \leqslant n$, be the set of all n-ary element of F.

Definition 1.2. A *basic recursive function theory* (BRFT) is a system $(D,F,\phi_1,\phi_2,\ldots)$, which we will always write (D,F,ϕ_n), such that

(1) (D,F) is an adequate system,

(2) each $\phi_n \in F_{n+1}$, and enumerates F_n; i.e.,

$$F_n = \{\lambda x_1\ldots x_n(\phi_n(a,x_1,\ldots,x_n)) : a \in D\},$$

(3) for each m , $n \geqslant 0$ there is a total $S_n^m \in F_{m+1}$ such that for all $x,x_1,\ldots,x_m$ we have

$$\lambda y_1..y_n(\phi_{m+n}(x,x_1,\ldots,x_m,y_1,\ldots,y_n)) = \lambda y_1..y_n(\phi_n(S_n^m(x,x_1,\ldots,x_m),y_1,\ldots,y_n)).$$

Definition 1.3. An *enumerative system* (ES) is a pair (D,F) such that

(1) D has at least two elements,

(2) F is a collection of $\leqslant$ 2-ary partial functions on D,

(3) F is closed under generalized composition,

(4) $\lambda x(x) \in F$, and there is a pairong mechanism (P,f_1,f_2); i.e., $P \in F_2$, P 1–1 and total, $f_1, f_2 \in F_1$, and for all $x,y \in D$, $f_1(P(x,y)) = x, f_2(P(x,y)) = y$,

(5) each C_x^1, $x \in D$, is in F,

(6) for each $z, w \in D$, $z \neq w$, $E_{z,w} \in F_2$, where $E_{z,w} = \lambda xy$ (z if $x = y$; w if $x \neq y$),

(7) there is a $\phi \in F_2$ such that $F_1 = \{\lambda x_1(\phi(a,x_1)) : a \in D\}$.

We now prove a kind of mutual interpretability between ES and BRFT, the first half of which is due to Strong and Wagner, the second half due to us. This theorem is of independent interest since ES is a very elegant functorial condition on monadic and binary partial functions.

Theorem 1.1. Let (D,F,ϕ_n) be a BRFT. Then $(D,F_1 \cup F_2)$ is an ES. Let (D,G) be an ES. Then there is a unique adequate (D,H) with $H_1 = G_1$; furthermore, for this unique (D,H) there exists a BRFT (D,H,ψ_n), and $H_2 = G_2$.

Proof. Let (D,F,ϕ_n) be a BRFT. Choose $e \in D$ such that for all b,c,a,x, $\phi_4(e,b,c,a,x) \simeq \alpha(a,b,c,x)$. Now fix $y \neq e$. Then $(\lambda bc(S_2^2(e,b,c))$, $\lambda z(\phi_2(z,e,e))$; $\lambda z(\phi_2(z,e,y)))$ is a pairing mechanism for (D,F). Clearly $E_{z,w} = \lambda xy(\alpha(x,z,w,y))$. So $(D,F_1 \cup F_2)$ is an ES.

Let (D,G) be an ES. Let (P,f_1,f_2) be a pairing mechanism for (D,G), ϕ as in (7). Then the following notation will be used throughout this paper: Define $P_1 = \lambda x(x)$, $f_1^1 = \lambda x(x)$, $P_{k+1} = \lambda x_1 \ldots x_{k+1}(P(x_1, P_k(x_2,\ldots,x_{k+1})))$, $f_1^{k+1} = f_1$, $f_{i+1}^{k+1} = \lambda x(f_i^k(f_2(x)))$, $1 \leqslant i \leqslant k$. Then each $f_i^k \in G_1$, and $f_i^k(P_k(x_1,\ldots,x_k)) = x_i$. Furthermore, each P_k is total and 1–1. Define $H_n = \{\lambda x_1 \ldots x_n(g(P_n(x_1,\ldots,x_n))): g \in G_1\}$. We now check the adequacy of (D, H).

To check closure under generalized composition, let $h \in H_n$, $\beta_1,\ldots,\beta_n \in H_m$. Set $h(x_1,\ldots,x_n) \simeq g(P_n(x_1,\ldots,x_n))$, $\beta_i(x_1,\ldots,x_m) \simeq g_i(P_m(x_1,\ldots,x_m))$. Then $h(\beta_1(x_1,\ldots,x_m),\ldots,\beta_m(x_1,\ldots,x_m)) \simeq g(P_n(g_1(P_m(x_1,\ldots,x_m)),\ldots,$ $g_n(P_m(x_1,\ldots,x_m))))$. Set $f = \lambda x(g(P_n(g_1(x),\ldots,g_n(x))))$. Clearly f is obtained by generalized composition, from $P, g_1,\ldots,g_n, g$, and so is in G_1. So $\lambda x_1,\ldots,x_m(h(\beta_1(x_1,\ldots,x_m),\ldots,\beta_n(x_1,\ldots,x_m))) \in H_m$.

To check that H contains the projection functions, note that $U_n^m(x_1,\ldots,x_m) = f_n^m(P_m(x_1,\ldots,x_m))$.

To check that H contains each C_x^m, note that G_1 contains each C_x^1.

To check that $\alpha \in H$, we need to use ϕ. Choose z, w such that $\lambda a(\phi(z, a)) = f_1$, $\lambda a(\phi_1(w, a)) = f_2$. Then $z \neq w$. So $E_{z,w} \in G_2$. We must produce $g \in G_1$ such that $\alpha = \lambda abcx(g(P_4(a, b, c, x)))$. Choose $g = \lambda y(\phi(E_{z,w}(f_1^4(y), f_4^4(y))$, $P(f_2^4(y), f_3^4(y)))))$. Clearly $g \in G_1$.

Obviously $H_1 = G_1$, $H_2 \subset G_2$.

To show $H_2 = G_2$, let $h \in G_2$. Then $h(x_1, x_2) \simeq g(P(x_1, x_2))$, where $g = \lambda x(h(f_1(x), f_2(x)))$. Clearly $g \in G_1$.

Uniqueness of (D, H): In fact, we show that if (D, F) and (D, G) satisfy (1), (2), (3), (4) of Definition 1.1, and have a pairing mechanism, then $F_1 = G_1 \rightarrow F = G$.

To see this assume $F_1 = G_1$ and let (P, f_1, f_2) be a pairing mechanism for (D, F). Let $g \in G_m$. Let $h = \lambda x(g(f_1^m(x),\ldots, f_m^m(x)))$. Then since each $f_i^m \in F$, we have that $h \in G$. Hence $h \in F$. Note that $\lambda x_1 \ldots x_m(h(P_m(x_1,\ldots, x_m))) = g$. However, it is obtained by generalized composition from h, P, and projection functions, and so lies in F. So $G \subset F$. By symmetry, $F = G$. This completes the proof of uniqueness.

For the existence of a BRFT, (D, H, ψ_n), define $\phi_n(x, x_1,\ldots, x_n) \simeq \phi(x, P_n(x_1,\ldots, x_n))$. To see that ϕ enumerates H_n, let $h \in H_n$. Let $g \in G_1$ have $h(x_1,\ldots, x_n) \simeq g(P_n(x_1,\ldots, x_n))$. Let a be such that $\lambda y(\phi(a, y)) = g$. Then clearly $\phi_n(a, x_1,\ldots, x_n) \simeq h(x_1,\ldots, x_n)$. However, we cannot conclude the existence of S_n^m functions for these enumerations. So we will define new enumerations ψ_n in terms of the old ones by $\psi_n(x, x_1,\ldots, x_n) \simeq \phi_{n+1}(f_1(x), f_2(x), x_1,\ldots, x_n)$. To see that the ψ_n are still enumeration functions for (D, H), let $h \in H_n$. Then choose b such that $\phi_{n+1}(b, x, x_1,\ldots, x_n) \simeq h(x_1,\ldots, x_n)$. Then clearly $\psi_n(P(b, b), x_1,\ldots, x_n) \simeq h(x_1,\ldots, x_n)$.

To complete the existence, we have to find S_n^m functions for the ψ_n. Let $x, x_1,\ldots, x_m$ be fixed elements of D. We wish to define $S_n^m(x, x_1,\ldots, x_m)$. That is, we wish to find a in terms of $x, x_1,\ldots, x_m$ such that $\psi_n(a, y_1,\ldots, y_n) \simeq \psi_{n+m}(x, x_1,\ldots, x_m, y_1,\ldots, y_n)$. In other words, we wish to find b, c such that $\phi_{n+1}(b, c, y_1,\ldots, y_n) \simeq \psi_{n+m}(x, x_1,\ldots, x_m, y_1,\ldots, y_n)$ and then set $a = P(b, c)$. The first step is to associate an element of G_{n+1} with ψ_{n+m} by considering $h = \lambda y y_1 \ldots y_n(\psi_{n+m}(f_1^{m+1}(y),\ldots, f_{m+1}^{m+1}(y), y_1,\ldots, y_n))$. Choose d such that $h = \lambda y(\phi_{n+1}(d, y, y_1,\ldots, y_n))$. Then clearly $\phi_{n+1}(d, P_{m+1}(x, x_1,\ldots, x_m), y_1,\ldots, y_n) \simeq \psi_{n+m}(x, x_1,\ldots, x_m, y_1,\ldots, y_n)$. So we set $a = P_2(d, P_{m+1}(x, x_1,\ldots, x_m)) = P_{m+2}(d, x, x_1,\ldots, x_m)$. Thus we define $S_n^m = \lambda x x_1 \ldots x_m(P_{m+2}(d, x, x_1,\ldots, x_m))$. The choice of d was independent of $x, x_1,\ldots, x_m$.

In connection with Theorem 1.1 and its proof, we remark that the *Smn* condition on BRFT cannot be derived from the rest of the conditions. For by the recursion theorem for BRFT (see Theorem 1.6) it is easily seen that for any BRFT (D, F, ϕ_n) there is an $a \neq b$ such that $\lambda x(\phi_1(a, x)) = \lambda x(\phi_1(b, x))$. So if $D = \omega$, F is the class of partial recursive functions of many arguments on ω, ϕ_n are enumeration functions for (ω, F), and ϕ_1 is an enumeration without repitition (see Friedberg [2]), then (D, F, ϕ_n) is not a BRFT.

Definition 1.4. Let (D, F, ϕ_n) be a BRFT. Then we set $[x]_k = \lambda x_1 \ldots x_k$

$(\phi_k(x, x_1,..., x_k))$ for $k > 0$; $[x]_0 = x$. Let $F_0 = D$. Let $T_0 = D$, $T_{k+1} = \{x\colon [x]_{k+1}$ is total$\}$.

Definition 1.5. Let (D, F, ϕ_n) be a BRFT, $0 \leqslant r$, $0 \leqslant m_i$, $0 < k$, G a partial function from $F_{m_1} x ... x F_{m_k}$ into F_r. Such G are called functionals. Then we write PRep $((D, F, \phi_n), e, g)$ if and only if for all $x_1,..., x_n$, if $G([x_1]_{m_1},...,[x_k]_{m_k})$ is defined, then its value must be $[\phi_k(e, x_1,..., x_k)]_r$. We write TRep $((D, F, \phi_n), e, g)$ if and only if PRep $((D, F, \phi_n), e, G)$ & $e \in T_k$. We say G is partial representable if and only if $(\exists e)$ (PRep $((A, F, \phi_n), e, G)$).

The idea behind the notation is that PRep mean "partial represents", TRep mean "total represents".

Theorem 1.2. Let (D, F, ϕ_n) be a BRFT, $G\colon F_{m_1} \times ... \times F_{m_k} \to F_r$, $1 \leqslant r$, PRep $((D, F, \phi_n), e, G)$. Then there is a y such that TRep $((D, F, \phi_n,) y, G)$.

Proof. Consider $\lambda x_1 ... x_k y_1 ... y_r([[e]_k(x_1,..., x_k)]_r(y_1,..., y_r))[x]_{k+r}$. Define $f(x_1,..., x_k) = S^k_r(x, x_1,..., x_k)$. Choose $[y]_k = f$. Then TRep $((D, F, \phi_n), y, G)$.

Definition 1.6. Let (D, F, ϕ_n) be a BRFT. Let $h \in F_n$, $g_1,..., g_n \in F_m$, $0 \leqslant n$, $0 \leqslant m$. Then by $h(g_1,..., g_n)$ we mean that element of F_m obtained by generalized composition, remembering the convention $F_0 = D$.

Theorem 1.3. Let (D, F, ϕ_n) be a BRFT. Then

(a) each functional $H\colon F_{m_1} \times ... \times F_{m_k} \to F_r$ given by $H(x_1,..., x_k) = x_i$ is partial representable,

(b) each functional $H\colon F_{m_1} \times ... \times F_{m_k} \to F_r$ given by $H(x_1,..., x_k) = y$ is partial representable,

(c) if $H, G_1,..., G_k$ are partial representable functionals, $H\colon F_{m_1} \times ... \times F_{m_k} \to F_r$, $G_i\colon F_{n_1} \times ... \times F_{n_p} \to F_{m_i}$, then $G\colon F_{n_1} \times ... \times F_{n_p} \to F_r$ is partial representable, where $G(x_1,..., x_p) = H(G_1(x_1,..., x_p),..., G_k(x_1,..., x_p))$,

(d) if $H, G_1,..., G_r$ are partial representable functionals, $H\colon F_{m_1} \times ... \times F_{m_k} \to F_r$, $G_i\colon F_1 \times ... \times F_{m_k} \to F_p$, then $G\colon F_{m_1} \times ... \times F_{m_k} \to F_p$ is partial representable, where $G(x_1,..., x_k) = H(x_1,..., x_k)\,(G_1(x_1,..., x_k),..., G_r(x_1,..., x_k))$ in the sense of Definition 1.5.

Proof: (a), (b) are obvious. For (c), let PRep $(D, F, \phi_n), e, H)$, PRep $((D, F, \phi_n), q_i, G_i)$. Then if $[a]_p = [e]_k([q_1]_p,...,[q_k]_p)$ in the sense of Definition 1.6., then PRep $((D, F, \phi_n), a, G)$. For (d) let PRep $((D, F, \phi_n), e, H)$, PRep $((D, F, \phi_n), q_i, G_i)$. It suffices to find an $f \in F_{r+1}$ such that f is total and $[f(x_1,..., x_r, x_{r+1})]_k = [x_{r+1}]_r([x_1]_k,..., [x_r]_k)$. Towards this aim, choose a such that $[a]_{k+r+1} = \lambda x_1 \ldots x_r x_{r+1} y_1 \ldots y_k([x_{r+1}]_r([x_1]_k(y_1,\ldots,y_k),\ldots,[x_r]_k(y_1,\ldots,y_k)))$. Then choose $f = \lambda x_1 \ldots x_r x_{r+1}(S_k^{r+1}(a, x_1,\ldots, x_r, x_{r+1}))$.

We will make use of the following result later.

Theorem 1.4. If (D, F, ϕ_n), (D, F, ψ_n) are BRFT, then for each $0 < m$ there is an $f \in F_1$ such that f is total and for all $x, x_1,\ldots, x_m \in D$ we have $\phi_m(x, x_1,\ldots, x_m) \simeq \psi_m(f(x), x_1,\ldots, x_m)$.

Proof: Choose $a \in D$ with $[a]_{m+1} = \phi_m$. Take $f = \lambda x(S_m^1(a, x))$ in (D, F, ψ_n).

Definition 1.7. Let D, E be sets, G a total 1–1 monadic function from D onto E. Then we take $T(G, f)$, for f a partial n-ary function on D, to be given by $G(f(x_1,\ldots, x_n)) \simeq T(G, f)(G(x_1),\ldots, G(x_n))$. $T(G, f)$ is an abbreviation for "the translate" of f by G. When G is onto and total, $f(T(G, f))$ is a 1–1 total function from $\bigcup_k D^{(D^k)}$ onto $\bigcup_k E^{(E^k)}$ preserving application.

Definition 1.8. Let (D, F, ϕ_n) and (E, G, ψ_n) be BRFT's. They are called *piecewise isomorphic* if and only if for each m there is a 1–1 onto total function $H: D \to E$ such that for all $x \in D$, $T(H,[x]_m) = [H(x)]_m$, where the second brackets refers to (E, G, ψ_n). If we wish to use the [notation for two BRFT's at once, we use [for the first and { instead of [for the second, from now on.

After Rogers [7] we take

Definition 1.9. Let (D, F, ψ_n), (D, F, ψ_n) be BRFT's. Then they are called *piecewise Rogers isomorphic* if and only if for each a there is a 1–1 onto function $G: D \to D$ such that for all $x \in D$, $[x]_n = \{G(x)\}_n$, and $G \in F_1$. (D, F, ϕ_n) is *piecewise Rogers imbeddable* in (D, G, ψ_n) if we require G to be a 1–1 function, but not necessarily onto.

Definition 1.10. We write 1-isomorphic, Rogers 1-isomorphic, Rogers 1-imbeddable, just in case the respective definitions above hold for $n = 1$.

The notion in Definition 1.7, in contrast to those in Definition 1.8, is connected with the usual notion of model-theoretic isomorphism. For this reason, if two BRFT are piecewise isomorphic then they are elementarily equivalent with respect to an appropriate rich language, described below.

We may think of a BRFT (D, F, ϕ_n) as a relational structure

$(D, F_1, F_2, ...; A_1, A_2, ...; \phi_1, \phi_2, ...)$, where there are infinitely many disjoint domains $D, F_1, F_2, ...$, and infinitely many i+2-ary relations A_i, and infinitely many constants $\phi_i \in F_{i+1}$, with the stipulation that $A_i(f, x_1, ..., x_i, y)$ is well-formed just in case $f \in F_i$, $x_1, ..., x_i$, $y \in D$. The intention is that $A_i(f, x_1, ..., x_i, y)$ if and only if $f(x_1, ..., x_i) = y$. For each n let μ_n be the similarity type of many-sorted structures $(D, F_1, F_2, ...; A_1, A_2, ...; \phi_n)$. Let $\mathcal{L}_n$ be the language $L_{\infty,\infty}$ based on μ_n only. Finally let $\mathcal{L}$ be the least language closed under arbitrary conjunction and disjunction, and negation, and containing each sentence in each $\mathcal{L}_n$.

Theorem 1.5. If (D, F, ϕ_n), (E, G, ψ_n) are piecewise isomorphic, then $(D, F_1, ...; A_1, ...; \phi_1, ...) \equiv_{\mathcal{L}} (F, G_1, ...; A_1, ...; \psi_1, ...)$.

Proof. One merely verifies that for each m, and total 1–1 onto $H: D \to E$ with for all $x \in D$, $T(H, [x]_m) = \{H(x)\}_m$, there is an extension of H to a model-theoretic isomorphism from $(D, F_1, ...; A_1, ...; \phi_m)$ onto $(E, G_1, ...; A_1, ...; \psi_m)$. To see this, take $H^*(f) = T(H, f)$ for $f \in F$. Clearly f is total and 1–1, $H^*(\phi_m) = \psi_m$, H^* extends H. It suffices to prove $\mathrm{Rng}\,(H^*) = G \cup E$. To see this, note that $(E, \mathrm{Rng}\,(H^* \restriction F), H^*(\phi_n))$ is a BRFT. By the part of Theorem 1.1 concerning uniqueness, it now suffices to prove that $\mathrm{Rng}\,(H^* \restriction F_1) = G_1$. Clearly if $m = 1$ we are done. If $1 < m$ then let $f \in F_1$ and set $g = \lambda x_1 ... x_m (f(x_1))$. Then $H^*(g) \in G_m$. But $H^*(f) = \lambda x_1 (H^*(g)(x_1, x_1, ..., x_1))$, and so $H^*(f) \in G$. If $f \in G_1$ then set $g = \lambda x_1 ... x_m (f(x_1))$. Then $(H^*)^{-1}(g) \in F_m$. But $(H^*)^{-1}(f) = \lambda x_1 ((H^*)^{-1}(g)(x_1, x_1, ..., x_1))$, and so $f \in \mathrm{Rng}\,(H^* \restriction F_1)$. We are done.

We do not know whether 1-isomorphic, Rogers 1-isomorphic, Rogers 1-imbeddable are equivalent to, respectively, piecewise isomorphic, piecewise Rogers isomorphic, Rogers imbeddable.

The recursion theorem, in the language of BRFT, says that for each $0 < k$ and each total $f \in F_1$ there is an $x \in D$ with $[x]_k = [f(x)]_k$. Not only is this true in all BRFT, but the stronger effective version is true.

Theorem 1.6. Let (A, F, ϕ_n) be a BRFT, $0 < k$. Then there is an $R_k \in F_1$ such that for all $x \in T_1$ we have $[R_k(x)]_k = [\phi_1(x, R_k(x))]_k$.

Proof. Fix $e \in D$ such that for all $x, y, x_1, ..., x_k \in D$ we have $\phi_{k+2}(e, y, x, x_1, ..., x_k) \simeq \phi_k(\phi_1(y, S_k^1(x, x)), x_1, ..., x_k)$. Hence for all $y, x, x_1, ..., x_k \in D$ we have $\phi_{k+1}(S_{k+1}^1(e, y), x, x_1, ..., x_k) \simeq \phi_k(\phi_1(y, S_k^1(x, x)), x_1, ..., x_k)$. Clearly for all $y, x_1, ..., x_k \in D$ we have

$\phi_{k+1}(S^1_{k+1}(e, y), S^1_{k+1}(e, y), x_1,\ldots, x_k) \simeq \phi_k(\phi_1(y, S^1_k(S^1_{k+1}(e, y), S^1_{k+1}(e, y))), x_1,\ldots, x_k) \simeq \phi_k(S^1_k(S^1_{k+1}(e, y), S^1_{k+1}(e, y)), x_1,\ldots, x_k)$. So set $R_k = \lambda y(S^1_k(S^1_{k+1}(e, y), S^1_{k+1}(e, y)))$.

Theorem 1.7. Let (D, F, ϕ_n) be a BRFT and let $0 < m$. Then there is a total $\mathrm{FP}_m \in F_m$ such that whenever $H\colon F_m \to F_m$ and PRep $((D, F, \phi_n), e, H)$ we have that $H([\mathrm{FP}_m(e)]_m) = [e]_m$ or $H([\mathrm{FP}_m(e)]_m)$ is undefined.

Proof. Upon examination of the proof of Theorem 1.2, we see that it can be made "effective". Thus, let $H_1 : F_1 \to F_{m+1}$ by $H_1(g) = \lambda x_1 y_1 \ldots y_m ([g(x_1)]_m(y_1,\ldots, y_m))$. Then by Theorem 1.3, H_1 is partial representable. Hence by Theorem 1.2, $(\exists q)$ (TRep $(D, F, \phi_n), q, H_1$). Fix this q. Fix a with $[a]_2 = \lambda xy(S^1_m(x, y))$. Then whenever PRep $((D, F, \phi_n), e, H)$, we have TRep $((D, F, \phi_n), S^1_1(a,[q]_1(e)), H)$. Hence whenever PRep $((D, F, \phi_n), e, H)$, we have that $H([R_m(S^1_1(a,[q]_1(e)))]_m)$ is either undefined or is $[R_m(S^1_1(a,[q]_1(e)))]_m$ by Theorem 1.6. Hence set $\mathrm{FP}_m = \lambda e(R_m(S^1_1(a, \phi_1(q, e))))$.

FP stands for "fixed point".

Theorem 1.8. Let (D, F, ϕ_n) be a BRFT, and let $0 < m, 0 \in D$. Then H: $F_m \times F_m \to F_m$ given by $H(f, g) = \lambda x_1 \ldots x_m (f(x_1,\ldots, x_m)$ if $x_m = 0$; $g(x_1,\ldots, x_m)$ if $x_m \neq 0)$ is everywhere defined and partial representable.

Proof. If $[a]_m = f$, $[b]_m = g$, then $H(f, g) = \lambda x_1 \ldots x_m (\phi_m(\alpha(0,a, b, x_m), x_1,\ldots, x_m)) \in F_m$. Let $h(x, y, x_1,\ldots, x_m) \simeq \phi_m(\alpha(0, x, y, x_m), x_1,\ldots, x_m)$, and let $h = [e]_{m+2}$. Then $H(f, g) = [S^2_m(e, a, b)]_m$.

A word of caution about H above is in order. Note that $H(f, g)$ is not, in general, $\alpha(C^m_0, f, g, U^m_m)$, since the latter will always be undefined at any arguments at which f is undefined.

Section 2

In this section we first present a new minimality theorem for the partial recursive functions which we derive immediately from Wagner and Strong's original minimality theorem via Theorem 1.1. We will exposit a proof of Wagner and Strong's result. In so doing we will establish key lemmas necessary for the second part of this section, which concerns relative categoricity properties of BRFT; roughly speaking, any two BRFT on ω which contain

$\lambda x(x+1)$ (ω-BRFT's) and which have the same partial functions are piecewise isomorphic. However, rather than proceeding directly in the second part with the relative categoricity result, we first will develop a general sufficient condition for piecewise isomorphism, and use this to derive the relative categoricity result. In the third part we classify those sets of total functions on ω which are the total functions present in some ω-BRFT.

Definition 2.1. An ω-BRFT (ω-ES) is a BRFT (ES), (ω, F, ϕ_n) $((\omega, F))$, such that $\lambda x(x+1) \in F_1$. We will use the convention $n - m = 0$ if $n \leqslant m$.

Theorem 2.1. (Wagner and Strong). If (ω, F, ϕ_n) is an ω-BRFT then $\mathrm{PR}(\omega) \subset F$, where $\mathrm{PR}(\omega)$ is the class of partial recursive functions of many arguments on ω.

Proof. Fix (ω, F, ϕ_n) as an ω-BRFT. It suffices to prove

(a) whenever $f \in F_{n+1}$, $0 < n$, f total, then $\mu f = \lambda x_1 \ldots x_n(\mu y(f(y, x_1, \ldots, x_n) = 0)$ if $(\exists y)(f(y, x_1, \ldots, x_n) = 0)$; undefined otherwise$) \in F_n$.

(b) whenever $g \in F_m$, $h \in F_{m+2}$, $0 \leqslant m$, h, g total, and $f = \lambda x_1 \ldots x_{m+1}(g(x_1, \ldots, x_n)$ if $x_{m+1} = 0$; $h(x_1, \ldots, x_m, x_{m+1}, f(x_1, \ldots, x_{m+1} - 1))$ if $x_{m+1} \neq 0)$, then $f = \mathrm{PRIM}(g, h) \in F_{m+1}$.

The partial functionals PRIM, μ are defined only on total arguments.

Towards (a), we need to consider H: $F_{n+1} \to F_{n+1}$ given by $H(g) = \lambda y x_1 \ldots x_n$ $(0$ if $f(y, x_1, \ldots, x_n) = 0$; $S(g(S(y), x_1, \ldots, x_n))$ if $f(y, x_1, \ldots, x_n) \neq 0))$. By Theorem 1.3 and Theorem 1.7 and Theorem 1.8 we have that H has a fixed point, $h \in F_{n+1}$. We wish to prove $\lambda x_1 \ldots x_n(h(0, x_1, \ldots, x_n)) = \mu f$. If $\mu f(x_1, \ldots, x_n) = z$ then clearly $g(z, x_1, \ldots, x_n) = 0$, and hence $g(0, x_1, \ldots, x_n) = z$, since f is total. If $\mu f(x_1, \ldots, x_n)$ is undefined then clearly g is nowhere defined.

Towards (b), we first note that $\lambda x(x - 1) \in F$ since $\lambda x(x - 1) = \mu f$, where $f(x, y) = 0$ if $x = 0, y = 0$; 0 if $S(y) = x$; 1 otherwise. Then define H: $F_{m+1} \to F_{m+1}$ by $H(\beta) = \lambda x_1 \ldots x_{m+1}(g(x_1, \ldots, x_m)$ if $x_{m+1} = 0$; $h(x_1, \ldots, x_m, x_{m+1}, \beta(x_1, \ldots, x_{m+1} - 1))$ if $x_{m+1} \neq 0)$. Then by Theorem 1.3 and Theorem 1.7 and Theorem 1.8, there is a fixed point for H. But any fixed point for H is $\mathrm{PRIM}(g, h)$ and lies in F.

Upon examination of this proof, we can obtain a stronger theorem of this kind which will be useful later.

Theorem 2.2. Let (ω, F, ϕ_n) be an ω-BRFT, and let $0 < n$, $0 \leqslant m$. Then the partial functionals PRIM: $F_m \times F_{m+2} \to F_{m+2} \to F_{m+1}$ and μ: $F_{n+1} \to F_n$ are defined on and only on all total arguments and are partial representable in (ω, F, ϕ_n).

Proof. Consider $H^*(f, g) = \lambda y x_1 \dots x_n$ (0 if $f(y, x_1, \dots, x_n) = 0$; $S(g(S(y), x_1, \dots, x_n))$ if $f(y, x_1, \dots, x_n) = q$ for some $q \neq 0$; undefined otherwise). Then let PRep $((\omega, F, \phi_n), e, H^*)$ by Theorem 1.3. If $a \in T_{n+1}$ then $\mu([a]_{n+1}) = [S_n^1(\mathrm{FP}(S_1^1(e, a)), 0)]_n$.

Consider $H^{**}(g, h, \beta) = \lambda x_1 \dots x_{m+1}(g(x_1, \dots, x_m)$ if $x_{m+1} = 0$; $h(x_1, \dots, x_m, x_{m+1}, \beta(x_1, \dots, x_{m+1} - 1))$ if $x_{m+1} \neq 0)$. Let PRep $((\omega, F, \phi_n), e, H^*)$ by Theorem 1.3. If $a \in T_m$, $b \in T_{m+2}$ then PRIM $([a]_m, [b]_{m+2}) = [\mathrm{FP}(S_1^2(e, a, b))]_{m+1}$.

Theorem 2.3. If (ω, F) is an ω-ES then $\mathrm{PR}_1(\omega) \cup \mathrm{PR}_2(\omega) \subset F$.

Proof. Let (ω, F) be an ω-ES. Then by Theorem 1.1 there is a BRFT (ω, G, ϕ_n) such that $F_1 = G_1$, $F_2 = G_2$.

It is well known that Theorem 2.1 and Theorem 2.3 are best possible in the sense that there is a BRFT $(\omega, \mathrm{PR}(\omega), \phi_n)$.

Theorem 2.4. Let (D, F, ϕ_n) and (D, F, ψ_n) be two BRFT satisfying

(a) the inverse of every permutation of D that lies in F_1 lies in F_1

(b) the partial inverse functional $I: F_1 \to F_1$ given by $I(g) = g^{-1}$ if g is a permutation of D; undefined otherwise, is total representable in (D, F, ϕ_n).

(c) for each $0 < m$ there is a total function $\beta \in F_2$ such that whenever $a, b \in T_1$ we have that $[\beta(a, b)]_1$ is a permutation of D, and for all x either $\{[\beta(a, b)]_1(x)\}_m = \{[a]_1(x)\}_m$ or $[x]_m = [[b]_1([\beta(a, b)]_1(x))]_m$.

Then (D, F, ϕ_n) and (D, F, ψ_n) are piecewise isomorphic.

Proof. Let (D, F, ϕ_n), (D, F, ψ_n) be as in the hypotheses, and fix $0 < m, \beta$. We first note the existence of a permutation $f \in F_1$ such that $[x]_m = \{f(x)\}_m$. This is clear from Theorem 1.4, and (c). We fix this f. Next, note that $T^*: F_1 \times F_m \to F_m$ by $T^*(G, g) = \lambda x_1 \dots x_m(G(g(I(G)(x_1), \dots, I(G)(x_m))))$ is partial representable by (b). Hence let TRep $((D, F, \phi_n), p, T^*)$. Similarly let $T^{**}(G, g) = T^*(I(G), g)$, and let TRep $((D, F, \phi_n), q, T^{**})$. Then let r be such that $[r]_2 = f([p]_2)$. Let s be such that $[s]_2 = \lambda xy([q]_2(x, f^{-1}(y)))$. Now let $b(x) = \beta(S_1^1(r, x), S_1^1(s, x))$. Then by Theorem 1.7, there is an e such that $[e]_1 = [h(e)]_1$. Then $[e]_1 = [\beta(S_1^1(r, e), S_1^1(s, e))]_1$. Since $S_1^1(r, e), S_1^1(s, e) \in T_1$, we have that $[e]_1$ is a permutation. Furthermore, either $\{[e]_1(x)\}_m = \{[S_1^1(r, e)]_1(x)\}_m = \{[r]_2(e, x)\}_m = \{f([p]_2(e, x))\}_m = [[p]_2(e, x)]_m = T([e]_1, [x]_m)$, or $[x]_m = [[S_1^1(s, e)]_1([e]_1(x))]_m = [[s]_2(e, [e]_1(x))]_m = [[q]_2(e, f^{-1}([e]_1(x)))]_m = T(I([e]_1), [f^{-1}([e]_1(x))]_m) = T(I([e]_1), \{[e]_1(x)\}_m)$. Hence, in either case, $\{[e]_1(x)\}_m = T([e]_1, [x])$.

We wish to use Theorem 2.4 to prove that any two ω-BRFT are piecewise isomorphic.

Lemma 2.5.1. Let (ω, F, ϕ_n) be an ω-BRFT. Then the inverse of every permutation of ω that lies in F_1 lies in F_1, and the partial inverse functional I: $F_1 \to F_1$ given by $I(g) = g^{-1}$ if g is a permutation of ω; undefined otherwise, is total representable.

Proof: If f is a permutation of ω then $f^{-1} = \mu(\lambda xy(\alpha(x, 0, 1, f(y))))$. So by Theorem 2.2., I is partial representable. Hence I is total representable.

Lemma 2.5.2. Let (ω, F, ϕ_n) be an ω-BRFT, $0 < m$. Then there is an $f \in F_2$, f total, such that $[f(x, y)]_m = [y]_m$, and for each y, Rng $(\lambda x(f(x, y)))$ is infinite.

Proof. First we show that there is a total $g \in F_2$ such that whenever $[x]_2$ is total, $y \in \omega$, $z \in \omega$, $(\forall a < y)\,([[x]_2(a, z)]_m = [z]_m)$, then $[g(x, y, z)]_m = [z]_m$ and either $(\forall a < y)\,([x]_2(a, z) \neq g(x, y, z))$ or $y = g(x, y, z)$. Firstly let $H_1(h, y, z) = \lambda r(y$ if $r \in$ Rng $(h \restriction y)$; z if $r \notin$ Rng $(h \restriction y))$ if h total; undefined otherwise. Then H_1 can be seen to be partial representable by considering $H_2(h) = \lambda yr(H_1(h, y, z)(r))$ and applying Theorem 2.2. Let TRep $((\omega, F, \phi_n), e, H_1)$. Then set $H(h, x, y, z) = \lambda a_1 \ldots a_m([y]_m(a_1, \ldots, a_m)$ if $(\exists a < y)\,(h(a, z) = R_m([e]_3(x, y, z)))$; $[R_m([e]_3(x, y, z))]_m(a_1, \ldots, a_m)$ otherwise), if h is total; undefined otherwise. Then by Theorems 1.2, 1.3, 1.8, and 2.2, let TRep $((\omega, F, \phi_n), q, H)$. Then finally set $g(x, y, z) = [q]_4(x, x, y, z)$.

Next let $g = [k]_3$, and let $h(x) = S^1_2(k, x)$. Then there is a p such that $[p]_2 = [h(p)]_2$. Clearly $[h(p)]_2$ is total since $k \in T_3$. Now $[h(p)]_2(y, z) = [S^1_2(k, p)]_2(y, z) = [k]_3(p, y, z) = g(p, y, z) = [p]_2(y, z)$. But since $p \in T_2$, we must have, for all y, z, if $(\forall a < y)\,([[p]_2(a, z)]_m = [z]_m)$ then $[g(p, y, z)]_m = [[p]_2(y, z)]_m = [z]_m$ and either $(\forall a < y)\,([p]_2(a, z) \neq g(p, y, z))$ or $y = g(p, y, z)$. Hence by induction, we have that for all y, z, $[[p]_2(y, z)]_m = [z]_m$ and either $(\forall a < y)\,([p]_2(a, z) \neq g(p, y, z))$ or $y = [p]_2(y, z)$. So clearly $[p]_2$ satisfies the conclusion of Lemma 2.5.2.

This argument is connected with the proof of the Roger's isomorphism theorem in [7].

Lemma 2.5.3. Let (ω, F, ϕ_n) be an ω-BRFT. Let $f, g \in F_2$, total, such that for each y, $\lambda x(f(x, y))$ and $\lambda x(g(x, y))$ have infinite range. Then there is a total $h \in F_2$ such that whenever $a, b \in T_1$, we have that $[h(a, b)]_1$ is a permutation of ω and for all z, either $[h(a, b)]_1(z) \in \text{Rng}(\lambda x(g(x, [a]_1(z))))$ or $z \in \text{Rng}(\lambda x(f(x, [b]_1([h(a, b)]_1(z)))))$.

Proof. By Theorem 2.2. it suffices to show that we can write a function β in terms of functions F, G, f, g and PRIM, μ, such that whenever F, G are total, f, g satisfy hypotheses, then β is a permutation of ω and for all z, $\beta(z) \in \text{Rng}(\lambda x(g(x, F(z))))$ or $z \in \text{Rng}(\lambda x(f(x, G(\beta(z)))))$. The expression is routinaly constructed, using the development of sequence numbers; one imitates the Cantor back and forth argument.

Theorem 2.5. Let (ω, F, ϕ_n), (ω, F, ψ_n) be ω-BRFT. Then (ω, F, ϕ_n) is piecewise isomorhpic to (ω, F, ψ_n).

Proof. In view of Lemma 2.5.1, it suffices to check clause (c) of the hypotheses of Theorem 2.4. Choose f satisfying Lemma 2.5.2 for (ω, F, ϕ_n), g satisfying Lemma 2.5.2 for (ω, F, ψ_n). Then use Lemma 2.5.3 for (ω, F, ϕ_n).

We now wish to characterize the sets of total functions of many arguments of ω-BRFT's. In doing so, we will abandon, for the moment, our general approach, and use freely notions, theorems, and methods from ordinary recursive function theory, and from the theory of generic sets. In particular, we will use the notation $f \leqslant_T g$ for total monadic $g: \omega \to \omega$, partial $f: \omega \to \omega$ of many arguments, read "f is partial recursive in g". We will also freely use a fixed standard BRFT $(\omega, \text{PR}(\omega), \phi_n)$, together with well-known theorems about such standard BRFT even though by our isomorphism Theorem 2.5 our proofs must be independent of the choice of $(\omega, \text{PR}(\omega), \phi_n)$.

We firstly wish to extend the notion $f \leqslant_T g$ to the case of arbitrary partial $g: \omega \to \omega$ of many arguments.

Definition 2.2. Usually one talks of an effective sequence number scheme for partial $f: \omega \to \omega$ whose domain is an initial segment of ω. We wish a scheme of numbers for all partial functions of many arguments whose domain is finite. Thus, we let N_k, D_k, B_k, $0 < k$, be a system of partial recursive functions such that

(a) each N_k, D_k is $k + 1$-ary, B_k is 1-ary and total

(b) $\{\lambda x(N_k(x, x_1,..., x_k)): x_1,..., x_k \in \omega\}$ = the set of all partial k-ary functions on ω with finite domain

(c) $N_k(x, x_1,..., x_k)$ is defined (undefined) if and only if $D_k(x, x_1,..., x_k) = 0$ (1)

(d) for each x, if $(\exists i)(1 \leqslant i \leqslant k \;\&\; B_k(x) \leqslant x_i)$ then $N_k(x, x_1,..., x_k)$ is undefined. We write $\text{Sub}_k(x, y)$ for $(\forall x_1,\ldots, x_k)(D_k(x, x_1,\ldots, x_k) = 0 \rightarrow N_k(x, x_1,\ldots, x_k) = N_k(y, x_1,\ldots,x_k))$. (Sub means "subset".) For f a partial k-ary function on ω, $0 < k$, we write $\text{Sub}_k(x, f)$ for $(\forall x_1,\ldots, x_k)(D_k(x, x_1,\ldots, x_k) = 0 \rightarrow f(x_1,\ldots, x_k) = N_k(x, x_1,\ldots, x_k)$.

Definition 2.3. A (k, l)-program, $0 < k$, $0 < l$, is a $k + 1$-ary partial recursive function G such that whenever $\text{Sub}_l(x, y)$ and $G(x, x_1,\ldots, x_k)$ is defined, we have $G(y, x_1,\ldots, x_k) = G(x, x_1,\ldots, x_k)$.

Definition 2.4. We write $\text{Comp}_{(k,l)}(f, g)$, for (k, l)-programs f, partial l-ary g, for that unique partial k-ary function h given by $h(x_1,\ldots, x_k) = y$ if and only if $(\exists a)(\text{Sub}_l(s, g)\ \&\ f(s, x_1,\ldots, x_k) = y)$.

Definition 2.5. We write $h \leqslant_{\text{T}} g$, for partial k-ary h, l-ary g, just in case $(\exists f)$ (f is a (k, l)-program & $\text{Comp}_{(k,l)}(f, g) = h$).

Lemma 2.6.1. Let g be l-ary, $0 < l$. Then there is a BRFT (ω, F, ψ_n) such that $F = \{f\colon f \leqslant_{\text{T}} g\}$.

Proof. In checking adequacy of (ω, F), the only clause presenting a problem is closure under generalized composition. Let $H \in F_n$, $h_1,\ldots, h_n \in F_m$, $0 < n$, $0 < m$. Then let $\text{Comp}_{(m,l)}(f_i, g) = h_i$, $\text{Comp}_{(n,l)}(f, g) = H$. Then $\text{Comp}_{(m,l)}(\lambda x x_1 \ldots x_m(f(x, f_1(x, x_1,\ldots, x_m),\ldots, f_n(x, x_1,\ldots, x_m))), g) = H(h_1,\ldots, h_n)$, and clearly $\lambda x x_1 \ldots x_m(f(x, f_1(x, x_1,\ldots, x_m),\ldots, f_n(x, x_1,\ldots, x_m)))$ is a (k, l)-program since f, f_i are.

From Theorem 1.1 we now only have to prove that $(\omega, F_1 \cup F_2)$ is an ES. Clearly there is a pairing system (P, f, g) for $(\omega, F_1 \cup F_2)$. We have only to produce a 2-ary enumeration function $\psi \in F_2$. Using the standard BRFT $(\omega, \text{FR}(\omega), \phi_n)$ chosen earlier, and the functions N_l, D_l, and B_l, we can find a total recursive monadic function G such that for all e, $[G(e)]_2$ is a $(1, l)$-program, and for all e, if $[e]_2$ is a $(1, l)$-program then $[G(e)]_2 = [e]_2$. Informally, $G(e)$ is the index of the $(1, l)$-program $f(x, x_1)$ computed in the following way: given x_1 compute $[e]_2(x, x_1)$ for various x in a criss-cross pattern, first computing a bit on $[e]_2(0, x_1)$, then a bit on $[e]_2(1, x_1)$, then a bit again on $[e]_2(0, x_1)$, etcetera. Of course, once $[e]_2(i, x_1)$ is found to converge we no longer spend any more time computing $[e]_2(i, x_1)$. Now we compute in the criss-cross pattern until an inconsistency arises (which we will define later). If no inconsistency ever arises, then set $f(x, x_1) = [e]_2(x, x_1)$. If an inconsistency arises at a certain stage s, then set $f(x, x_1) = [e]_2(x, x_1)$ if $[e]_2(y, x_1)$ has been found to converge before stage s, for some y with

$\mathrm{Sub}_l(y, x)$; undefined otherwise. An inconsistency is said to arise at stage s just in case at stage s, $[e]_2(i, x_1)$ is seen to converge to, say, j, and there is some y such that, prior to s, $[e]_2(y, x_1)$ is seen to converge to a value other than j, and $(\exists r)(\mathrm{Sub}_l(y, r) \,\&\, \mathrm{Sub}_l(i, r))$.

Now set $f(x, x_1, x_2) = [G(x_1)]_2(x, x_2)$. Then clearly f is a $(2, l)$-program. Then set $\psi = \mathrm{Comp}_{(2,l)}(f, g)$. Then $\psi(x_1, x_2) = (\mathrm{Comp}_{(1,l)}([G(x_1)]_2, g))(x_2)$, and we are done.

Theorem 2.6. Let Y be a collection of total functions on ω of several arguments. Then the following three are equivalent:

(a) $(g \in Y \,\&\, f \leqslant_T g \,\&\, f$ total$) \to f \in Y$, and Y is countably infinite

(b) there is an ω-BRFT (ω, F, ψ_n) with Y = set of total elements of F

(c) there is a partial g on ω such that $Y = \{\text{total } f\colon f \leqslant_T g\}$.

Proof. In view of Lemma 2.6.1, and Theorem 2.2, it suffices to prove that (a) $\to$ (c). This implication is of independent interest for ordinary recursive function theory.

Let I be any total 2-ary function such that $Y = \{f\colon f \text{ is total } \& \ (\exists n)(f \leqslant_T \lambda m(I(n, m)))\}$ and $n \leqslant r \to \lambda m(I(n, m)) \leqslant_T \lambda m(I(r, m))$. Consider the forcing language based on the terms arithmetical in I, and the generic constant $a \subset \omega$. Let $a \subset \omega$ be generic, using finite conditions, with respect to this forcing language. Finally let $g(n, m) = I(n, m)$ if $n \in a$; undefined otherwise. Firstly, it is well known that a is infinite. Hence for all $f \in Y$, there is an n such that $f \leqslant_T \lambda m(f(n, m))$. Hence, $f \in Y \to f \leqslant_T g$.

Note that g is given by a term in the forcing language. Now fix $f \leqslant_T g$, f k-ary, total. We wish to prove $f \in Y$.

Let h be such that $\mathrm{Comp}_{(k,2)}(h, g) = f$. Then h is partial recursive, and so is given by a term not mentioning a. Then some condition $p = \{x_1, \ldots, x_r \in a; y_1, \ldots, y_s \notin a\}$ forces "$\mathrm{Comp}_{(k,2)}(h, g)$ is total", where p is compatible with a; i.e., p is true of a. We claim that for any $b, c, x_1, \ldots, x_k$, if

(1) $(\forall i)(\forall m)((1 \leqslant i \leqslant r \,\&\, D_2(b, x_i, m) = 0) \to N_2(b, x_i, m) = I(x_i, m))$ $\&\ (\forall j)(\forall m)((1 \leqslant j \leqslant r \,\&\, D_2(c, x_i, m) = 0) \to N_2(c, x_i, m) = I(x_i, m))$

(2) $(\forall i)(\forall m)((1 \leqslant j \leqslant s \to (D_2(b, y_j, m) = 1 \,\&\, D_2(c, y_j, m) = 1)))$

(3) $h(b, x_1, \ldots, x_k)$ and $h(c, x_1, \ldots, x_k)$ are defined, we must have $h(b, x_1, \ldots, x_k) = h(c, x_1, \ldots, x_k)$. If this is so, then $f(x_1, \ldots, x_k) = y$ if and only if $(\exists b)(b$ satisfies (1)–(3) above $\&\ h(b, x_1, \ldots, x_k) = y)$; consequently $f \leqslant_T \lambda m(I(\max(x_1, \ldots, x_r, y_1, \ldots, y_s)(m)))$, and hence $f \in Y$.

To prove the claim, let b, c be as in the hypotheses. Choose $\alpha < \omega$ such that for all $\alpha \leqslant \beta < \omega$, all $m \in \omega$, we have that $N_2(b, \alpha, m)$ and $N_2(c, \alpha, m)$

are undefined. Then let a^* be any generic set compatible with p and such that $a^* \cap \beta = \{x_i\colon 1 \leqslant i \leqslant r\}$. Then we must have $\mathrm{Comp}_{(k,2)}(h, g^*)$ is total, where $g^*(n, m) = I(n, m)$ if $n \in a^*$; undefined otherwise. But then there is a d such that $h(d, x_1, \ldots, x_k)$ converges and $\mathrm{Sub}_2(d, g^*)$. So, by the choice of a^*, d is compatible with both b and c. Hence b, d have a common extension, c, d have a common extension. Hence $h(d, x_1, \ldots, x_k) = h(c, x_1, \ldots, x_k) = h(b, x_1, \ldots, x_k)$.

As is usual from proofs involving forcing, one can obtain cardinality information from the proof.

Corollary 2.6.1. If Y is a collection of total functions on ω of several arguments such that $(g \in Y \;\&\; f \leqslant_{\mathrm{T}} g \;\&\; f \text{ total}) \to f \in Y$, and Y is countably infinite, then there are continuumly many distinct F such that (ω, F, ψ_n) is a BRFT and Y = set of total elements of F.

Proof. This follows immediately from the proof on Theorem 2.6 using the following well-known fact from the theory of forcing: there are continuumly many distinct sets generic with respect to the forcing language based on the terms arithmetical in Y.

Section 3

In the first part of this section we will establish a minimality theorem for the hyperarithmetic functions. More precisely, we prove that every partial hyperarithmetic function is present in every ω-BRFT in which existential quantification is partial representable, and that this result is best possible. In the second part of this section we will estbalish relative categoricity results connected with the Kripke-Platek ordinal recursive function theory.

Definition 3.1. We let LO (f) mean

(1) f is a total binary function from ω into $\{0, 1, 2\}$

(2) the binary relation $f(n, m) = 0$ constitutes a linear ordering with field $\{i\colon f(i, i) = 1\}$. In case LO (f) we write f_n for $\lambda ab\,(f(a, b)$ if $f(a, n) = f(b, n) = 0$; 2 otherwise). Thus we have $(\forall f)\,(\forall n)\,(\mathrm{LO}(f) \to \mathrm{LO}(f_n))$.

LO means "linear ordering".

We will assume familiarity with the jump operator, which sends each total function on ω of several arguments into a monadic total function. We will write $J(f)$ for the jump of f, for total f. We will let $\langle a, b\rangle$ abbreviate $2^a 3^b$.

We now give the definition of relative hyperarithmeticity ($f \leqslant_{\mathrm{h}} g$), for partial f of many arguments, total monadic g.

Definition 3.2. We write HY (f, g, h) just in case

(a) LO (f) and g is total and monadic and h is total and binary

(b) for all n, m we have $h(n, m) = \langle g(m), J(\lambda xy(h(x, y)$ if $f(x, n) = 0$; 0 otherwise)) $(m)\rangle$.

HY (f, g, h) means "h is a hierarchy based on jump on the ordering f relative to g".

Definition 3.3. We write WO (f) for LO (f) & $\sim (\exists g)(\forall n)(f(g(n \dotplus 1), g(n)) = 0)$.

Definition 3.4. We let $\beta \leqslant_{\mathrm{h}} g$ be

(a) β is a partial function of many variables

(b) g is total and monadic

(c) $(\exists f)(\exists h)(f \leqslant_{\mathrm{T}} g$ & WO (f) & HY (f, g, h) & $\beta \leqslant_{\mathrm{T}} h)$.

Definition 3.5. Let (D, F, ϕ_n) be a BRFT with $0, 1 \in D$. Then we let E be the partial functional $E: F_1 \to D$ given by $E(f) = 0$ if f is total and $(\exists x)(f(x) = 0)$; 1 if f is total and $\sim (\exists x)(f(x) = 0)$; undefined if f is not total. We call (D, F, ϕ_n) an E-BRFT if and only if $0, 1 \in D$ and F is partial representable in (D, F, ϕ_n). We call (D, F, ϕ_n) an ω-E-BRFT if and only if (D, F, ϕ_n) is both an ω-BRFT and an E-BRFT.

Lemma 3.1.1. Let (ω, F, ϕ_n) be an ω-E-BRFT. Then $J^*: F_2 \to F_1$ given by $J^*(f) = J(f)$ if f is total; undefined otherwise, is partial representable in (ω, F, ϕ_n).

Proof. It is well known that the jump can be written in terms of projection functions, constant functions, α, PRIM, μ, and E. Then apply Theorem 2.2.

Lemma 3.1.2. Let (ω, F, ϕ_n) be an ω-E-BRFT, $f \in F_2$, WO (f), $g \in F_1$, g total. Then there is an $h \in F_2$ such that HY (f, g, h).

Proof. Let $H: F_2 \to F_2$ be given by $H(\beta) = \lambda nm(\langle g(m), J^*(\lambda xy(\beta(x, y)$ if $f(x, n) = 0$; 0 otherwise)) $(m)\rangle)$. Then it is easily seen using Lemma 3.1.1. and Theorem 1.8 that H is partial representable. Hence by Theorem 1.7 there is an $h \in F_2$ with $H(h) = h$. It suffices to prove that h is total. Note that the relation $f(n, m) = 0$ well-orders $\{i\colon f(i, i) = 1\}$. We will prove that for each n,

$\lambda m(h(n, m))$ is total. It is clear that if $n \notin \{i: f(i, i) = 1\}$ then $\lambda m\, (H(h)\,(n,m))$, and consequently $\lambda m\, (h(n, m))$ is total. So it now suffices to assume $f(n, n) = 1$ and $(\forall r)\,(f(r, n) = 0 \rightarrow \lambda m(h(r, m))$ is total), and conclude that $\lambda m(h(n, m))$ is total.

Under these assumptions, we see that $\lambda xy\,(h(x, y)$ if $f(x, n) = 0$; 0 otherwise) is total. Hence $\lambda m\,(H(h)\,(n, m))$ is total, and hence $\lambda m\,(h\,(n, m))$ is total.

Theorem 3.1. Let (ω, F, ϕ_n) be an ω-E-BRFT, $g \in F_1$, g total, $\beta \leqslant_{\mathrm{h}} g$. Then $\beta \in F$.

Proof. Immediate from Theorem 2.2, Lemma 3.1.1, and Lemma 3.1.2.

We now prove that Theorem 3.1 is best possible. As in Gandy et al. [3], we view an ω-model as a collection of subsets of ω.

Lemma 3.2.1. Let g be monadic, total, and let $\beta \not\leqslant_h g$, β partial with k arguments. Then there is an ω-model of 2nd-order arithmetic which contains $\{\langle n, m\rangle: g(n) = m\}$ and omits $\{\langle\langle n_1 ... \langle n_{k-1}, n_k\rangle ...\rangle, m\rangle: \beta(n_1,..., n_k) = m\}$.

Proof. This follows immediately from the straightforward relativized version of Theorem 1 of Gandy et al. [3].

Lemma 3.2.2. Let $Y \subset P(\omega)$ be an ω-model of 2nd-order arithmetic. Then for each total monadic g such that $\{\langle n, m\rangle: g(n) = m\} \in Y$, there is an ω-E-BRFT (ω, F, ϕ_n) such that $g \in F_1$ and whenever $\beta \in F_k$ we have $\{\langle\langle n_1,..., \langle n_{k-1}, n_k\rangle ...\rangle, m\rangle: \beta(n_1,..., n_k) = m\} \in Y$.

Proof. It is well known that, for any monadic total g, there is an ω-E-BRFT (ω, F, ϕ_n) such that $g \in F_1$ and F is the collection of all partial functions whose graph is Π^1_1 relative to g. Furthermore, this assertion, when suitable translated into the language of 2nd-order arithmetic, can easily be seen to be provable in 2nd-order arithmetic.

Theorem 3.2. Let g be monadic and total. Then a necessary and sufficient condition that a partial β of many arguments be present in every ω-E-BRFT in which g is present is that $\beta \leqslant_{\mathrm{h}} g$.

Proof. Immediate from Lemma 3.2.1, Lemma 3.2.2, and Theorem 3.1.

Of course, as remarked before, we have the following.

Theorem 3.3. For each monadic total g there is an ω-E-BRFT (ω, F, ϕ_n) such that the total elements of F are *exactly* the total $\beta \leqslant_{\mathrm{h}} g$, and the elements of F are exactly those partial functions whose graph is Π_1^1 relative to g.

We now wish to generalize the work of Section 2 on relative categoricity for ω-BRFT to relative categoricity for α-BRFT, for admissable ordinals α. Unfortunately we were not able to obtain the most satisfactory generalizations, and were forced to add conditions on our definition of α-BRFT.

Definition 3.6. Let g be a k-ary function from ordinals to ordinals. Define $L^g(0) = \emptyset$, $L^g(\alpha+1) = \{\text{graph}\,(g) \cap \alpha^{k+1}\} \cup \{x : x \subset L^g(\alpha)$ and x is first-order definable, with parameters, over $(L^g(\alpha), \epsilon)\}$, $L^g(\lambda) = \bigcup_{\alpha<\lambda} L^g(\alpha)$. If β is admissible and g is a total function on β and f is a partial function on β, then we write $f \leqslant_\beta g$ just in case the graph of f is Σ definable over $L^g(\beta)$ with the graph of g allowed as a parameter, as well as any finite list of elements of $L^g(\beta)$.

Definition 3.7. A β-BRFT is a BRFT (β, F, ϕ_n) such that

(a) whenever $g \in F$ is total, f partial, $f \leqslant_\alpha g$, we have $f \in F$

(b) whenever $f \in F_k$ there is a total $g \in F_{k+2}$ such that $f(x_1,\ldots, x_k) = y \equiv (\exists z \in \beta)\,(g(x_1,\ldots, x_k, y, z) = 0)$.

Definition 3.8. An admissible β-BRFT is a β-BRFT (β, F, ϕ_n) such that no 1–1 element of F with unbounded domain is onto an ordinal $<\beta$. A nonprojectible β-BRFT is a β-BRFT is a β-BRFT such that no 1–1 element of F with unbounded domain has bounded range.

The standard examples of admissible β-BRFT are found in the β-recursive functions on β. Other examples are based on the partial functions on β whose graph is Σ_m on $L(\beta)$, providing $L(\beta)$ satisfies Σ_m-Reflection.

We wish to use Theorem 2.4 to generalize Theorem 2.5 for nonprojectible β-BRFT.

Lemma 3.4.1. Let β be admissible. Then there is a pairing mechanism (P, f_1, f_2) such that f_1, f_2 are total monadic functions on β, P a total 1–1 binary function on β, and all three are β-recursive.

Proof. This is well known.

We now establish the analogue to Lemma 2.5.1.

Lemma 3.4.2. Let (β, F, ϕ_n) be a β-BRFT. Then the inverse of every permutation of β that lies in F_1 lies in F_1, and the partial inverse functional I: $F_1 \to F_1$ given by $I(g) = g^{-1}$ if g is a permutation of β; undefined otherwise, is total representable.

Proof. The first part is immediate from (a) of Definition 3.7. For the second part, find a 4-ary total g such that $\phi_1(x, y) = z$ if and only if $(\exists a)(g(x, y, z, a) = 0)$. Let (P, f_1, f_2) be as in Lemma 3.4.1. Then clearly $\lambda xz(\mu b(g(x, f_1(b), z, f_2(b)) = 0)) \leqslant_\beta g$, and hence lies in F_2. Let $h = f_1(\lambda xz(\mu b(g(x, f_1(b), z, f_2(b)) = 0)))$. Then $h \in F_2$, and whenever $[x]_1$ is a permutation of β then $\lambda z(h(x, z)) = ([x]_1)^{-1}$. Hence, letting $[a]_2 = h$, $[c]_1 = \lambda x(S_1^1(a, x))$, we have TRep $((\beta, F, \phi_n), c, I)$. We have actually proved more: $(\forall f \in F_1)(\exists g \in F_1)(f(x) \simeq f(g(f(x))) \;\&$ $\mathrm{Dom}(g) = \mathrm{Rng}(f))$.

We now establish the analogue to Lemma 2.5.2.

Lemma 3.4.3. Let (β, F, ϕ_n) be a nonprojectible β-BRFT, $0 < m$. Then there is an $f \in F_2$, f total, such that $[f(x, y)]_m = [y]_m$, and for each y, $\mathrm{Rng}(\lambda x(f(x, y)))$ is unbounded in β.

Proof. First we show that there is a total $g \in F_3$ such that whenever $[x]_2$ is total, $y \in \beta$, $z \in \beta$, $(\forall a < y)([[x]_2(a, z)]_m = [z]_m)$, then $[g(x, y, z)]_m = [z]_m$ and either $(\forall a < y)([x]_2(a, z) \neq g(x, y, z))$ or $y = g(x, y, z)$. Firstly let $\phi_1(x, y) = z \equiv (\exists a)(G(x, y, z, a) = 0)$, G total. Then $r \in \mathrm{Rng}([x]_1 \restriction y)$ $\equiv (\exists a)(\exists z)(\exists b < y)(G(x, b, z, a) = 0)$, and if $x \in T_1$, then $r \notin \mathrm{Rng}$ $([x]_1 \restriction y) \equiv (\exists a)(\exists z)(\exists b < y)(G(x, b, z, a) = p$ for some $p \neq 0)$. Hence H_1 is partial representable where $H_1(h, y, z) = \lambda r(y$ if $r \in \mathrm{Rng}(h \restriction y)$; z if $r \notin \mathrm{Rng}(h \restriction y))$ if h total; undefined otherwise, using (a) of Definition 3.7. Let TRep $((\beta, F, \phi), e, H)$. Let $\phi_2(x, y, z) = w \equiv (\exists b)(G^*(x, y, z, w, b) = 0)$, G^* total. Then set $g(x, y, z) = y$
if $(\exists a < y)(\exists b)(G^*(x, a, z, R_m([e]_3(x, y, z))) = 0)$; $R_m([e]_3(x, y, z))$
if $(\exists a < y)(\exists b)(\exists c)(G^*(x, a, z, R_m([e]_3(x, y, z))) = c \;\&\; c \neq 0)$; undefined otherwise.

Next let $g = [k]_3$, and let $h(x) = S_1^1(k, x)$. Then there is a p such that $[p]_2 = [h(p)]_2$. Clearly $[h(p)]_2$ is total since $k \in T_3$. Now $[h(p)]_2(y, z) = [S_2^1(k, p)]_2(y, z) = [k]_3(p, y, z) = g(p, y, z) = [p]_2(y, z)$. But since $p \in T_2$, we must have, for all y, z, if $(\forall a < y)([[p]_2(a, z)]_m = [z]_m)$ then

$[g(p, y, z)]_m = [[p]_2(y, z)]_m = [z]_m$ and either $(\forall a < y)([p]_2(a, z) \neq g(p, y, z))$ or $y = [p]_2(y, z)$. Hence by transfinite induction, we have that for all y, z, $[[p]_2(y, z)]_m = [z]_m$ and either $(\forall a < y)([p]_2(a, z) \neq g(p, y, z))$ or $y = [p]_2(y, z)$. So clearly $[p]_2$ satisfies the conclusion of Lemma 3.4.3 provided that for each z, Rng $(\lambda y([p]_2(y, z)))$ is unbounded. Fix z. Suppose $\lambda y([p]_2(y, z)]$ is bounded by $b < \beta$. Then $\lambda y([p]_2(y, z)) \restriction \{c: b < c\}$ is 1–1, but has bounded range, which contradicts the nonprojectibility.

We now establish the analogue to Lemma 2.5.3.

Lemma 3.4.4. Let (β, F, ϕ_n) be an admissible β = BRFT. Let $f, g \in F_2$, total, such that for each y, $\lambda x(f(x, y))$ and $\lambda x(g(x, y))$ have unbounded range. Then there is a total $h \in F_2$ such that whenever $a, b \in T_1$, we have that $[h(a, b)]_1$ is a permutation of α and for all z, either $[h(a, b)]_1(z) \in$ Rng $(\lambda x(g(x, [a]_1(z))))$ or $z \in$ Rng $(\lambda x(f(x, [b]_1([h(a, b)]_1(z)))))$.

Proof. Let $\phi_1(x, y) = z \equiv (\exists a)(G(x, y, z, a) = 0)$, $G \in F_4$, G total. Define 3-ary partial functions K, L by simultaneous recursion on β, by $K(x, y, z) = w$ if and only if $(\forall a < z)(\exists r)(\exists s)(K(x, y, z) = r \mathbin{\&} L(x, y, z) = s)$ and either $(w < z \mathbin{\&} L(x, y, w) = z)$ or $(\forall a < z)(\exists r)((L(x, y, a) = r \mathbin{\&} z \neq r) \mathbin{\&}$ $(\exists b)(\exists c)(w = g(b, c) \mathbin{\&} c = f_1(\mu a(G(x, z, f_1(a), f_2(a)) = 0) \mathbin{\&} z \leqslant w \mathbin{\&}$ $(\forall a < b)(g(b, c) < z))$; $L(x, y, z) = w$ if and only if $(\forall a < z)(\exists r)(\exists s)(K(x, y, a) = r \mathbin{\&} L(x, y, a) = s)$ and either $(w \leqslant z \mathbin{\&} K(x, y, w) = z)$ or $(\forall a \leqslant z)(\exists r)((K(x, y, a) = r \mathbin{\&} z \neq r) \mathbin{\&}$ $(\exists b)(\exists c)(w = f(b, c) \mathbin{\&} c = f_1(\mu a(G(x, z, f_1(a), f_2(a)) = 0) \mathbin{\&} z < w \mathbin{\&}$ $(\forall d < b)(f(b, c) \leqslant z))$. We now wish to prove that $K, L \in F$.

Let $A = \lambda x_1 x_2 x_3 x_4(P_3(G(x_1, x_2, x_3, x_4), f(x_1, x_2), g(x_1, x_2)))$.

Define the 4-ary relation $P(x, y, z, w)$ on $L^A(\beta)$ as follows: $P(x, y, z, w) \equiv (\exists b)(x < b \mathbin{\&} y < b \mathbin{\&} z < b \mathbin{\&} w < b \mathbin{\&} L^A(b + 2) \models (\exists K)(\exists L)(\text{Dom}(K) = \text{Dom}(L) = z + 1 \mathbin{\&} K, L$ satisfy the above conditions when G, f_1, f_2, f, g are interpreted as their intersection with $L^A(b + 2)$, and $K(z) = w))$. Now P is Σ with parameter A, over $L^A(\beta)$. We claim that for all $x, y \in T_1$, we have $(\forall z)(\exists ! w)(P(x, y, z, w))$. To see this, fix $x, y \in T_1$. Suppose true for all $d < e$. Then set $h(d) = \mu b$ (b holds in the definition of $P(x, y, d, w)$), for $d < e$. Then clearly $h \leqslant_\beta A$, and so $h \in F$. By the proof of Lemma 3.4.2 and admissibility, we have that h is bounded. Then use the unboundedness of $\lambda b(f(b, c))$, $\lambda b(g(b, c))$ for each $c < \beta$, and $x, y \in T_1$ to conclude the claim for e.

Now clearly for all x, y we have $K(x, y, z) = w \equiv P(x, y, z, w)$, $L(x, y, z) = w \equiv P(x, y, w, z)$, and if $x, y \in T_1$ then $\lambda z(K(x, y, z))$ is a permutation of β. Hence $K \in F$. Let $[e]_3 = K$. Set $h = \lambda xy(S_1^2(e, x, y))$.

Theorem 3.4. Let $(\beta, F, \phi_n), (\beta, F, \psi_n)$ be nonprojectible β-BRFT. Then (β, F, ϕ_n) is piecewise isomorphic to (β, F, ψ_n).

Proof. In view of Lemma 3.4.2, it suffices to check clause (c) of the hypotheses of Theorem 2.4. Choose f satisfying Lemma 3.4.3 for (β, F, ϕ_n), g satisfying Lemma 3.4.3 for (β, F, ψ_n). Then use Lemma 3.4.4 for (β, F, ϕ_n).

Corollary 3.4.1. Let $(\beta, \mathrm{PR}(\beta), \phi_n), (\beta, \mathrm{PR}(\beta), \psi_n)$ be BRFT, where β is a nonprojectible ordinal and $\mathrm{PR}(\beta)$ is the collection of all partial β-recursive functions of many arguments on β. Then $(\beta, \mathrm{PR}(\beta, \mathrm{PR}(\beta), \phi_n)$ is piecewise isomorphic to $(\beta, \mathrm{PR}(\beta), \psi_n)$.

Since Lemma 3.4.4 does not need nonprojectibility, we have the following.

Theorem 3.5. Let $(\beta, F, \phi_n), (\beta, F, \psi_n)$ be admissible β-BRFT. Furthermore assume (*) there are total $f, g \in F_2$ such that $[f(x, y)]_m = [y]_m$, $\{g(x, y)\}_m = \{y\}_m$, and for each y, $\mathrm{Rng}(\lambda x(f(x, y)))$, $\mathrm{Rng}(\lambda x(g(x, y)))$ are unbounded in β. Then (β, F, ϕ_n) is piecewise isomorphic to (β, F, ψ_n).

Corollary 3.5.1. Let $(\beta, \mathrm{PR}(\beta), \phi_n), (\beta, \mathrm{PR}(\beta), \psi_n)$ be BRFT such that β is admissible. Furthermore assume (*) there are total $f, g \in \mathrm{PR}(\beta)$ such that $[f(x, y)]_m = [y]_m$, $\{g(x, y)\}_m = \{y\}_m$, and for each y, $\mathrm{Rng}(\lambda x(f(x, y)))$, $\mathrm{Rng}(\lambda x(g(x, y)))$ are unbounded in β. Then $(\beta, \mathrm{PR}(\beta), \phi_n)$ is piecewise isomorphic to $(\beta, \mathrm{PR}(\beta), \psi_n)$.

We now show that Corollary 3.5.1 is best possible.

Theorem 3.6. Let β be a projectible admissible ordinal. Then there are BRFT $(\beta, \mathrm{PR}(\beta), \phi_n), (\beta, \mathrm{PR}(\beta), \psi_n)$ such that the first satisfies condition (*) and the second does not; furthermore, the first is not piecewise isomorphic to the second.

Proof. Take $(\beta, \mathrm{PR}(\beta), \phi_n)$ to be the usual BRFT obtained from Gödel numbering. To define $(\beta, \mathrm{PR}(\beta), \psi_n)$, let f have bounded domain and be onto β, $f \in \mathrm{PR}(\beta)$. Then set $\psi_n = \lambda x x_1 \ldots x_n (\phi_n(f(x), x_1, \ldots, x_n))$. To prove that $(\beta, \mathrm{PR}(\beta), \psi_n)$ is a BRFT, it suffices to verify the *S-m-n* condition. For this, it is sufficient to prove that for each $m, n > 0$ there is a total $g \in F_1$ such that for all $x \in \beta$, $\{x\}_{m+n} = [g(x)]_{m+n}$. To do this, let $[e]_{m+n+1} = \psi_{m+n}$. Set $g = \lambda x(S^1_{m+n}(e, x))$. The second part of the theorem follows from the first. It suffices to prove that $(\beta, \mathrm{PR}(\beta), \phi_1)$ and $(\beta, \mathrm{PR}(\beta), \psi)$ satisfy distinct first-order sentences. Consider "$(\exists g)$ (g is total & $(\forall x)(g(x) \neq 0 \rightarrow (\forall y)(\phi_1(x,y)$

is undefined)) & $(\forall h)$ (h 1–1 and total $\to$ Rng $(h) \neq \{z: g(z) = 0\}$))". Then this sentence holds of the first and fails of the second BRFT.

Section 4

Theorem 1.1 gives a transparent necessary and sufficient condition on monadic and binary partial functions in order for they to be conservatively extended to a BRFT. A related problem considered here is to give a transparent necessary and sufficient condition on monadic and binary partial functions together with a distinguished binary partial function in order for this structure to be conservatively extended to a BRFT.

Theorem 4.1. Let D be a set, F a collection of partial functions on D of many arguments, $\phi \in F$, ϕ binary. Then a necessary and sufficient condition for there to be a BRFT (D, G, ϕ_n) with $G_1 = F_1$, $G_2 = F_2$, $\phi_1 = \phi$ is that

(1) (D, F) is an ES and ϕ is an enumeration function

(2) there is a total $g \in F_2$ such that for all $x, y \in D$, $\lambda z(\phi(g(x, y), z)) = \lambda z(x, z)(\lambda z(y, z))$

(3) there is a pairing mechanism (P, f_1, f_2) and a total $h \in F_1$ such that for all $x \in D$, $\lambda y(\phi(h(x), y)) = \lambda y(P(x, y))$.

Proof. The necessity follows from Theorems 1.1 – 1.3. For the sufficiency, set (D, G) to be adequate with $G_1 = F_1$, $G_2 = F_2$, by Theorem 1.1, and define $\phi_n(x, x_1, \ldots, x_n) \simeq \phi(x, P_n(x_1, \ldots, x_n))$. Then $\phi_1 = \phi$. By the proof of Theorem 1.1, the ϕ_n are enumeration functions. Now let $m, n > 0$. We must find a total function $S_n^m \in F$ such that $\phi_n(S_n^m(x, x_1, \ldots, x_m), y_1, \ldots, y_n) \simeq \phi_{m+n}(x, x_1, \ldots, x_m, y_1, \ldots, y_n)$. Thus, we wish to find a total S_n^m with $\phi_1(S_n^m(x, x_1, \ldots, x_m), P_n(y_1, \ldots, y_n)) \simeq \phi_{m+n}(x, x_1, \ldots, x_m, y_1, \ldots, y_n)$. It will be sufficient to find a total S_n^m with $\phi_1(S_n^m(x, x_1, \ldots, x_m), z) \simeq \phi_{m+n}(x, x_1, \ldots, x_m, f_1^n(z), \ldots, f_n^n(z))$. Consider $K = \lambda yz(\phi_{m+n}(f_1^{m+1}(y), f_2^{m+1}(y), \ldots, f_{m+1}^{m+1}(y), f_1^n(z), \ldots, f_n^n(z)))$. Finally consider $L = \lambda w(K(f_1^2(w), f_2^2(w)))$. Then for each y, $\lambda z(K(y, z)) = L(\lambda z(P(y, z)))$. Fix e with $\lambda x(\phi(e, x)) = L$. Then for each y, $\lambda z(K(y, z)) = \lambda z(\phi(e, h(y)), z))$. Finally set $S_n^m = \lambda x x_1 \ldots x_m(g(e, h(P_{m+1}(x, x_1, \ldots, x_m))))$.

Wagner's conditions (URS) are of real algebraic character, but transparency suffers. Wagner essentially considers nonassociative algebras consisting of a domain $D \cup \{*\}$ with the distinguished "undefined element" $*$ and an everywhere defined binary operation, $\cdot$, on $D \cup \{*\}$, and writes finitely many axions of the form $\exists\forall$. The algebras $(D, \cdot)$ satisfying Wagner's conditions are called URS's. The connection between URS and BRFT, established by Strong, is the following: If (D, F, ϕ_n) is a BRFT then $(D \cup \{*\}, \cdot)$ is a URS, where $* \cdot * = x \cdot * = * \cdot x = *, x \cdot y = \phi_1(x, y)$ if $\phi_1(x, y)$ is defined; $*$ otherwise, for $x, y \in D$. If $(D \cup \{*\}, \cdot)$ is a URS then there is a BRFT (D, F, ϕ_n) such that $\phi_1(x, y)$ is undefined if and only if $x \cdot y = *$, and $\phi_1(x, y) = x \cdot y$ if $x \cdot y \neq *$. See Strong [9] for an exposition.

It would be preferable to have a more elegant general sufficient condition for piecewise isomorphism than Theorem 2.4.

In Theorem 2.5, note that the isomorphisms constructed are actually in F. This is a general phenomenon as seen by the following.

Theorem 4.2. Let $(\omega, F, \phi_n), (\omega, F, \psi_n)$ be ω-BRFT, $0 < m$, and let G be a permutation of ω such that for all $x \in \omega$, $T(G, [x]_m) = \{G(x)\}_m$. Then $G \in F$.

Proof. Clearly $T(G, \lambda x_1 \ldots x_m (S(x_1))) = h \in F$. But then $\lambda x_1 \ldots x_m (h^{(x_1)}(G(0))) = \lambda x_1 \ldots x_m (G(x_1)) \in F$ by Theorem 2.2. Hence $G \in F$.

We do not know if Theorem 4.2 holds for β-BRFT; we suspect so.

We do not know a characterization of those F such that there exists an ω-BRFT (ω, F, ϕ_n). In addition, we do not know an analogue of Theorem 2.6 for ω-E-BRFT.

We conjecture that there is a minimality theorem for the collection of total functions on ω whose graph lies in the minimum β-model for Σ^1_2-AC (i.e., $\omega_1^{E_1}$-finite). A natural choice would be to replace E in Theorem 3.1 by $H: F_2 \to \{0, 1\}$ given by $H(f) = 0$ if WO (f)); 1 if f is total and $\sim$ WO (f); undefined otherwise. We suspect that there is a connection between minimality theorems in ω-BRFT, and the theory of recursive functionals of finite type on ω. In particular, we suspect a connection between our Theorem 3.2 and the result of Kleene [4] characterizing the functions recursive in his functional E.

We do not know of a more satisfactory generalization of our theorems about ω-BRFT to BRFT on ordinals. The major difficulty is generalizing Theorem 2.2 (or even Theorem 2.1); there the true finiteness of every ordinal $< \omega$ plays a definite role in the proofs. In particular, we know of no minimality theorem in BRFT on ordinals.

For projectible admissible β, we have not obtained a classification, up to piecewise isomorphism, of those BRFT $(\beta, \text{PR}(\beta), \phi_n)$. It would seem likely that such a classification is possible in terms of the obvious analogue of the notion of projectum to BRFT.

Platek and Kripke have discussed the notion of partial A-recursive function on A for arbitrary admissible A. But at present recursive function theory on arbitrary admissible sets is undeveloped as compared to the theory of partial β-recursive functions for admissible β. A major obstacle is that it seems unlikely that for arbitrary admissible A, there is a 2-ary partial A-recursive enumeration of the monadic partial A-recursive functions, although this is still an open question. However, leaving recursive function theory aside, arbitrary admissible sets may be a reasonable setting for a general recursive *set* theory. (the theory of recursively enumerable sets), since for any admissible A there is an A-r.e. enumeration of the A-r.e. sets. Thus, generalized recursive set theory may be a different matter from generalized recursive function theory. It would be interesting to find an axiomatic recursive set theory which would reflect that difference, and then to reinvestigate the kind of questions dealt with here.

References

[1] S. Feferman, Some applications of the notions of forcing and generic sets, Fundamenta Mathematicae (1965) LVI.

[2] R. Friedberg, Enumeration without duplication, J. Symbolic Logic (1958) 23.

[3] R. Gandy, G. Kreisel and W. Tait, Set existence, Bull. Acad. Polon. Sci. Ser. Sci. Math. Astron. Phys. 8 (1960).

[4] S. Kleene, Recursive functionals and quantifiers of finite types, I, Trans. Am. Math. Soc. (1959) 91.

[5] S. Kripke, Transfinite recursion on admissible ordinals, I. II (abstracts), J. Symbolic Logic (1964) 29.

[6] R. Platel, Foundations of recursion theory, Dissertation, Stanford University (1966).

[7] H. Rogers, Gödel numerings of partial recursive functions, J. Symbolic Logic (1958) 23.

[8] H. Rogers, Theory of recursive functions and effective computability (McGraw-Hill, 1968).

[9] H. Strong, Algebraically generalized recursive function theory, IBM J. Research and Development, 12 (1968).

SOME REASONS FOR GENERALIZING RECURSION THEORY

G. KREISEL
Stanford University, Stanford, Calif.

Introduction

The purpose of this article is to correct two common and almost equally frustrating errors: the first is to suppose that there is just *one* use of g.r.t. (and of course none that is ever mentioned is found to be right); the other is to suppose that there are so many (depending on unspecified purposes) that one doesn't even mention any of them. The fact is that there are a few, and we shall look at some of them critically. To be realistic we want to know to what extent our aims are feasible and so we refer to existing results on various generalizations in Section 2, excluding metarecursion theory which is considered in Section 3. But our principal purpose is the formulation and analysis of the aims of g.r.t. themselves. This is done in a general way in Section 1 and reviewed in Section 4 in the light of the formal work reported in 2 and 3.

The remainder of this introduction is intended to prepare the reader for an 'unusual' and, perhaps, difficult aspect of the exposition below: it treats as objects of study, notions and distinctions which are indeed used by logicians and mathematicians, but only conversationally. Below I weigh my words also in informal discussions, giving them the kind of attention to detail usually reserved for formal work. The reader should guard against mistaking this 'informal rigour' for pedantry since I constantly rely on the results of these discussions and refer to the distinctions made there.

(a) A striking feature of current mathematical style is this: aims (or even questions) are often not explicitly stated and certainly not analyzed; but, when the answers are available, a suitable wording of theorems in a carefully chosen order with a very brief comment can convey most artistically these aims without stating them; properly chosen definitions replace the distinctions that are brought out by an explicit analysis. In short, one lets 'results speak for themselves'. In these circumstances anyone among us may be in doubt whether the kind of detailed analysis of aims attempted in this paper is *ever* worthwhile. In this connection g.r.t. provides a possibly instructive object lesson because the relevant analysis (for the aims 1a and 1b) was spelt out well before the detailed development and one can see the relation between the first analysis, corrections of oversights and final results. Evidently *if* one is interested in this relation at all, the kind of 'unusual' attention to informal detail mentioned earlier is essential.

(b) Following ordinary mathematical usage except for a proviso in (c) below, we shall distinguish between two kinds of purposes (of generalizations): for *applications* and for *conceptual analysis*. The former concern the solution of problems stated in the vocabulary of current mathematics without reference to the generalization. The latter use the generalized notion to *state an explicit and faithful analysis* of concepts which we understand already. To take a hackneyed example, the notions of group theory help one analyze concepts of *symmetry* and topological notions allow one to analyze *stability* in mechanics. In the present paper we shall touch on more logical analysis (of the logical concepts of inductive definition, and of computation).

(c) The explanations in (b) are *rough*. First of all, there isn't really such a thing as *the* vocabulary of current mathematics. It certainly isn't the language of set theory! To put it paradoxically: mathematicians constantly recognize that some set-theoretic definition (of a concept not 'yet' formulated in set theory) is correct; this very act presupposes their understanding the defined concept too. Even more to the point: mathematicians understand problems which leave open in what terms they are to be answered, and recognize their solutions when they see them! for example, Higman's solution, by use of classical recursion theory, of the question: Which finitely generated groups are subgroups of some finitely presented group? Nobody told him just how the word 'which' was to be narrowed down.

Another ambiguity affects the explanation of the distinction under (b) when we really do not understand an abstract idea very well although we may use it freely in ordinary life or, say, philosophical discussion. Two such cases will concern us in 1(d), and its correlate 4(d), namely *functions with self application* and *kinds of evidence*. (It is often stressed, I think correctly, that at best we have here a mixture of concepts and that it is a sheer act of faith to suppose that further reflection and few distinctions will lead to a definitive answer. What is often overlooked is how 'unreasonably' reliable these suppositions are.) When I want to emphasize the problematic aspect of an idea, I shall speak of *illustrations* for it, and not of its analysis. As it happens, it may well be that these illustrations together with the distinctions made in 4(c) constitute, in a reasonable sense, the most useful part of the present text.

(d) Finally two remarks concerning relations to the literature are in order. This note is by no means self-contained, but relies on lots of references for precise and detailed statements of existing results. My aim was to tell the reader enough about such work to let him decide if he wants to pursue it and to tell him places in the literature where he can do so. (It has to be admitted that more detailed statements, even with full definitions, would have been no longer and no less agreeable to read, though they would have been harder to write).

The references given do *not* constitute an attempt at a history of this subject. (For example such a history would contain a good deal on the work, by Mostowski and his students, reported in [39], which has been superseded by more recent work.) I believe, history of (parts of) logic is interesting, but *only* if done seriously; specifically by sifting such evidence as the autobiographical remarks that various authors have chosen to publish; for example, when or whether they first thought of an idea or first read somebody's paper. Such remarks, even if believed by their authors (and of interest to them and theirs), do not constitute history but may, occasionally, provide useful raw material for it. For an apparently different point of view, see Montague [36, pp. 63, 64].

1. Purposes of g.r.t.

We consider four distinct purposes under (a)–(d) which will be re-examined in Section 4(a)–(d) resp.

(a) *Advancing other parts of logic and mathematics* (including classical recursion theory itself)

This concerns *applications* in the sense explained in (b) and (c) of the Introduction. I shall not be concerned with the familiar and important aim of 'economic' generalizations which include several existing branches of mathematics, e.g. descriptive set theory [1], without necessarily making much difference to any one of them.

To me the single most striking element in applications of (classical) recursion theory is its intimate connection with *finiteness*, both in model theory and group theory as in [17] mentioned already. So in applications of g.r.t. one expects the generalization of the notion of *finite* to be central. Now finiteness is such a familiar and basic idea that one is hardly conscious of using it; consequently, any generalization (which, of course, must make its use explicit) is bound t, have sophisticated features. Let me just mention two of them.

(i) Finiteness in its literal sense has of course lots of consequences: the close relation between *cardinal* and *ordinal* properties is probably best known. In addition, we also have *definability* properties, every subset of a finite set being definable for almost any notion of definability, some subsets of an infinite set being indefinable, again for almost any notion of definability. (NB. I mean here definitions in their ordinary 'classical' sense and not some kind of non-set-theoretic foundation. It just happens to be the case that the natural logical laws for propositions about definable objects have a lot in common with, say, constructive logic.) In generalizations one has to analyze *which* of these consequences of finiteness are relevant.

(ii) The very statement of a generalization must be expected to have a novel character, if *finite* is replaced by a suitable but unfamiliar *definability property*. For example, at one time mathematicians underestimated the interest of the finiteness theorem in general model theory, but they certainly understood it. Its proper generalization to ω-models requires the notion of hyperarithmetic set, and the illiterate model theorist or mathematician doesn't even understand its statement!

As a corollary, if the generalization is relevant at all it should be very useful! Indeed, a quite superficial knowledge of the concepts themselves, without any proofs, may be sufficient. For example, given a mathematical problem we can often reduce it quite trivially to the existence of a suitable kind of model for a set Σ of axioms; once one knows the appropriate g.r.t., it may be quite obvious that each generalized finite subset of Σ has such a model (though not obvious of Σ itself). Well, it won't be if one doesn't know the words, and so those who have not learnt the concepts of g.r.t. are at a genuine disadvantage. Speaking of useful superficial knowledge, according to J.P. Serre, a nodding acquaintance with classical recursion theory was sufficient to make good use of Higman's theorem [17] ; people wanted to know if certain groups G were subgroups of some finitely presented group; it turned out it was easy to check that G satisfied Higman's conditions (actually the easy 'partial' criterion for *countably* generated G was most useful).

Ultimately one would expect g.r.t. to advance classical recursion theory, by drawing attention to distinctions which are (too?) subtle here, but are magnified in g.r.t.; distinctions which are needed to keep technical details in their place.

(b) *Understanding the mathematical character of ordinary recursion theory*

Though obviously related to the last paragraph of (a), the emphasis is different: not so much on finding new results as on understanding existing ones in the classical case. To avoid unrealistic hopes, one must here distinguish between the (mathematically) elementary and coherent part of recursion theory developed from the notion of mechanical computation and the more difficult theory mainly of r.e. sets and their degrees. The latter was not guided by simple and clear logical or practical considerations; it goes back to imaginative but loose analogies (of Post) that led to various striking notions and problems; and the later development resembles, in many ways, *number theory*. Thus we may indeed expect a tidy theory of some simple parts corresponding, perhaps, to abelian group properties of the integers; but, remembering the difficulties Bourbaki has with number theory, one must be prepared for real obstacles to an adequate systematic understanding. (The quite explicit aim of [26, p. 318] was to use g.r.t. to manufacture analogues to recursion theory by use of, apparently, far fetched model theoretic invariants; evidently, this was done because one of the difficulties with understanding number theory is that there are no *obvious* analogues.)

Speaking for myself I was absolutely fascinated by the question: How could one bring order into 'difficult' recursion theory? (if one wanted to; I have no interest in the subject itself). Oddly enough, one obstacle in work on this question is a kind of blind spot produced by overlooking two simple, but important points.

(i) Though full generality is certainly aesthetically more satisfactory, a specific generalization (such as metarecursion theory) may be useful, provided it 'magnifies' enough the particular distinctions arising in the theorems one wants to understand.

(ii) How do we judge better understanding if not by new results? (Timid authors often squeeze a dull new result out of a marvellous new *proof* just to have an 'objective' improvement.) Why not leave the judgement to the experts? However reasonable one's doubts may be about possible disagreement or even systematic error, as a matter of intellectual experience there is the *fact* of reliability of the experts; cf. the end of (c) in the introduction for another example of 'unreasonable' reliability. Of course, it would be too much to expect not only *objective* judgement, but also an *explicit* criterion! Don't we all, at least occasionally, know more than we can say? (Actually, all this is perfectly familiar from generalizations in traditional mathematics; the experts consider, and agree on, the *correct* setting of a particular result or on the correct topology to use, of course *before* they have made applications.) For a nice illustration of all this, recall that, for a long time, the experts on r.e. sets concentrated on just a few problems which seemed wholly isolated to an outsider like myself. But quite recently Lachlan [30] showed that the solution of these few problems was the key to a whole *class* of problems, a class as natural as one could wish.

When it comes to *advancing* knowledge (not its exposition), being too skeptical can be as damaging as being too gullible.

(c) *Analysis of a general concept of computation*

In view of (b) in the introduction, the issue here can be put quite simply:

Is there such a thing as a general concept of computation? or the related question:

Is there such a thing as an extension of Church's Thesis to general (abstract) structures?

The answers determine the whole *shape* of the development, the *kind* of theorems that one tries to prove.

(i) If the answer is positive, we have to do with the *analysis* of a concept, specifically a phenomenlogical analysis of the kind given by Turing for mechanical computations of number theoretic functions. The mathematics has here much the same role as in the natural sciences: to *state* rival hypotheses and to help one deduce from them a consequence, an *experimentum crucis* which distinguishes between them; one will try to avoid artefacts and systematic error. Equivalence results do not play a special role, simply because one good reason is better than 20 bad ones, which may be all equivalent because of systematic error [1)].

(ii) If the answer is negative, the criteria of 1a and 1b above apply; it may still be possible to use some idea of 'general computation' in a metaphorical way for *applications* (which do not in turn refer to the metaphor). What one has to guard against is to imitate mechanically the basic development of classi-

1) The familiar emphasis on *stability* or *equivalence* results is not rooted in some kind of 'common sense', but in a *positivistic* philosophy of research which rejects the objectivity of problem (i). An equivalence result allows one to act in accordance with this doctrine without formally adopting it: the result allows one to *evade* the issue (for the time being).

cal recursion theory. Specifically, at least many of the basic theorems were used in connection with Turing's (genuine) analysis, not for their pure mathematical interest. It would be quite unreasonable to suppose that generalizations of these same theorems will be useful for applications of g.r.t.

As already mentioned I shall touch also on another logical concept (besides computation), but only in passing: inductive definitions in 2b.

(d) *Other uses*

The other uses of g.r.t. here considered are *illustrations* of two rather difficult logical ideas of the kind explained in (d) of the introduction. These two abstract ideas turn up quite naturally in the context of recursion theory; namely

(i) a formal theory of *self application*,

(ii) an axiomatic analysis of *kinds of evidence*.

(i) occurs in classical recursion theory because in a perfectly natural and non-trivial way, on a syntactic level instructions (rules) apply to themselves. By use of g.r.t., specifically g.r.t. on *sets*, we then derive formal theories exhibiting the same features of self application but with much greater expressive power. This is *only* an illustration: but it may help us pin point the genuinely problematic aspects of the abstract notion, by showing what is not problematic here.

(ii) comes up here in connection with the assertion that all mechanically computable number theoretic functions are recursive. Leaving aside the truth of the assertion (about which there is little doubt, in the sense made quite explicit in (i) of 4c below) we consider *refinements* analogous to those obtained in the 19th century for geometry. Corresponding to the assertion above (Church's thesis), one accepted the Euclidean character of space, but *classified* geometric propositions axiomatically, that is by analyzing just which properties of, or what knowledge about, Euclidean space was needed to derive a given geometric proposition.

2. Mathematical generalizations

Each of the subsections (a)–(d) below (loosely related to the subsections of Section 1) treats a group of g.r.t.'s. By definition, all g.r.t. specialize to or include the case of classical recursion theory, on the natural numbers with zero and successor. The various groups differ in generality in as much as some of them do *not* generalize all the following basic notions of classical recursion theory: recursive; finite (cardinal), finite (ordinal), finite (explicitly definable); bounded (= included in a finite set); r.e. (range of a recursive function); s.r. (semi-recursive or: generated by a recursive process). The definition of 'bounded' is used to apply to structures which are not ordered.

Note that the recent literature is very clear on finiteness, already discussed in Section 1, but less so (in fact, less than the literature in the fifties) on the distinction between r.e. and s.r. Here it is to be remarked, and this will be taken up in (d) below, that the usual literature takes the notion of *partial recursive function* as central, which is not included in the list above. In particular, an s.r. set is defined as the *domain* of a partial (recursive) function, here thought of as being determined by the recursive process which generates its graph (the set of pairs consisting of its arguments and values). The roles of domain and range are symmetric because they are generated simultaneously (and a 'search' operator is available to recover domain and range from the graph). In contrast, when an r.e. set is generated the domain of the process, the universe, is given in advance. The use of partial recursive functions for defining s.r. is reasonable because the generation of their graphs is typical of arbitrary recursive processes. For reference below, note the amusing curiosity:

$$\text{r.e.} = \Sigma_1 \text{ over the universe}$$
$$\text{s.r.} = \Sigma_1 \text{ over associated ordinals,}$$

namely the ordinals involved in constructing the (least) class generated by the recursive process considered. To be quite precise: not only the *size* of ordinals but the *complexity* of well-orderings may have to be considered, cf. Section 3.

(a) *Model theoretic treatment*

Here we have the most extensively developed g.r.t.; perhaps not surprisingly because the class of recursive functions first appeared in the literature in connection with Gödel's *entscheidungsdefinite Eigenschaften*. All the basic notions of recursion theory are taken into account in these g.r.t. The principal *uses* of these 'model theoretic' g.r.t. are applications, specifically to model theory (e.g. [23], [2], [3]) and recursion theory [12], and, as far as I know, not logical analysis of concepts.

(i) For general surveys, to which I have to add only very little below, see [23] and [25]. The central idea is that of *invariant* and *semi-invariant* definition; *explicit* ones going back to Gödel's above, *implicit* ones to Fraissé;

both yield generalizations of recursiveness. For a generalization of finiteness (in terms of definability) *absolute* invariance is used [23].

Barwise [2], [3] gives very polished expositions of this theory applied to *sets* with *membership*. Invariance over suitable transitive extensions (of a given structure) is treated. An important correction in [3] of earlier work concerns the somewhat *ad hoc* notion of an *admissible set*. Barwise introduces the notion of a *strict* Π_1^1*-admissible set*, shows that all *countable* admissible sets 'happen' to be strict Π_1^1-admissible; but in the uncountable case the two notions diverge. Barwise [3] gives a g.r.t. for strict Π_1^1-admissible sets, which has satisfactory applications for *infinitary languages*, while mere admissibility is not sufficient. Specifically, Gregory [16] has a convincing 'negative' result for the socalled generalized finiteness property (GFP):

> There is a necessarily uncountable admissible A and an A-s.r. subset X of axioms in the A-language such that each A-finite subset of X has a model but X does not.

(The result is convincing because, at present, GFP plays a central role in the particular applications considered.) Gregory also shows that GFP cannot be established by the methods of ZFC for all A which satisfy an interesting additional condition due to Barwise (BC):

$$A \text{ is admissible and } \forall X\,(X \text{ is } A\text{-finite} \leftrightarrow X \in A).$$

As far as I know it is open whether GFP is implied by BC by use of a reasonable *axiom of infinity*. The question is natural because axioms of infinity have long been known to be relevant to infinitary logic, cf. [25, p. 146, 1. 11].

Concerning further work on the relation between GFP and BC, see Mrs. Karp's contribution [18]. She constructs an admissible set A, which satisfies GFP, but not BC, nor the 'finiteness property' corresponding to BC. Specifically there is an A-s.r. set X of formulae which has no model, but for every subset Y of X, Y has a model if $Y \in A$.

The principal gap in the existing model theoretic g.r.t. here described is its preoccupation with *sets* (that is sets built up from the empty set by some cumulative operation, as in ordinary set theory); not even sets of individuals are treated.

(ii) Uniform inductive definitions, published (in the sense of [36]) by Feferman [8], present a very natural and probably most useful *refinement* of (i). The general idea of *semi-invariant implicit definition* is retained, but two new conditions are imposed on the definitions (both of which are satisfied in the classical recursion theory on the natural numbers; in particular, in connection with Post's canonical systems which serve as a model for [8]). Specifically, in (i) the use of auxiliary relations is permitted which need not in turn be invariantly defined; roughly speaking, the definitions of P in (i) take the form $\exists P_1 \ldots \exists P_n A(P, P_1, \ldots, P_n)$, where A is of first order over the structures considered, but no requirement (of invariance or even uniqueness) is imposed on the P_i, $1 \leqslant i \leqslant n$. The corresponding definitions of [8] for P are such that $P, P_1, \ldots, P_n$ are *simultaneously* (semi-invariantly) defined.

The second condition involves *monotonicity* or *persistence*, familiar from inductive definitions. (For a consistently model theoretic or extensional treatment it is natural to *start* with a 'mathematical' property like monotonicity; but Feferman is able to show that often a monotone predicate of sets can be defined by a positive formula, which is a useful syntactic condition [6]; for this purpose we need monotonicity in a class of structures; hence he speaks of 'uniform' inductive definitions.)

For g.r.t., though not stressed in [8], the single most important feature of inductive definitions is probably this. With any such definition an *ordinal* is associated in the usual way, and hence the following generalization of *finiteness* (*both in the ordinal sense* and in terms of *definability*):

> a set X (contained in the universe of the g.r.t.) is *finite* iff it occurs in the generating process determined by some inductive definition, at a stage below the ordinal of the recursion theory.

Consequently we have a *partial ordering* between these 'finite' sets or, at least, between pairs (α, X) and (β, Y) where α and β are inductive definitions:

$(\alpha, X) < (\beta, Y)$ iff $\alpha \equiv \beta$, X and Y occur in the generating process defined by α, and X occurs before Y.

Such a partial ordering may be useful in view of Moschovakis' axiomatization [38].

The ideas of [8] are clear and attractive; but the detailed information on this g.r.t. is about structures where it coincides with the g.r.t. in (i) (when the classes of structures over which invariance is required are properly matched up). In other words, in these cases one can get rid of the unattractive auxiliary predicates allowed in (i). But (ii) may well be more satisfactory, for the projected uses, where (i) and (ii) differ.

(b) *Fixed point operations* (Platek [42])

This development is closely related to Kleene's reworking of classical recursion theory in [21] with the recursion theorem in a central role. An obvious difference is that [42] deals with *general* structures, a subtler difference is that the basic objects of [42] are *extensional*, numberings of definitions being used in the proofs, not in the basic definitions (in contrast to Kleene's S9). I shall return to this matter in (c) below.

As it stands, [42] generalizes all the principal notions of classical recursion theory except: finite. (Obviously, [23, p. 202], for certain structures there are adequate definitions of *finite* in terms of the other notions.)

The potential applications are not too different from those of the model theoretic g.r.t. in (a), provided a proper generalization of finiteness is added, e.g. as in (a)(ii). For *understanding* classical recursion theory including [21] (cf. Section 1b) the concept of hereditarily monotone functions of finite type in [42] seems essential; for instance, it is needed in (a)(ii) for the proper choice of the class of structures over which the definitions are to be invariant. Another potential use of [42] is to the *analysis* of *inductive definitions.*

A curious technical gap of [42] is the restriction to structures with a (generalized) recursive equality relation at the ground type.

Let me go a little more into the relations of the g.r.t. in [42] with the model theoretic g.r.t. My own main interest here is in the differences, that is where one g.r.t. may be useful and the other is not; the similarities help to pinpoint such differences.

(i) A striking, but *per se* minor, difference is the use of all finite types in [42], explained very clearly in the introduction to [42]. In a model theoretic treatment higher type operations are 'hidden' in the logical symbolism, as is well-known from *functional* interpretations. The elimination, from definitions, of types higher than the type of the object defined is one of the principal theorems of [42]. It should be compared with the reduction of uniform inductive definitions in (a)(ii) to a simple syntactic form ('hiding' lower types) and, above all, to Tait's generalization of *cut-elimination* [56], used to eliminate detours via higher types. Evidently, logical consequence in (a) (for the class of structures considered) corresponds to consequence in the λ-calculus of [42]; but this particular matter is more naturally considered in (c) below.

(ii) Platek, in the introduction to [42], faces squarely what is undoubtedly the most delicate problem in setting up a g.r.t. for some given *concrete* collection of objects, such as sets or ordinals or functions (and, I think, more explicitly than the earlier literature). *What principles should guide us in the choice of basic relations and operations to be 'put on' our collection*? A more 'formalistic' way is to ask: what language is to be used? The corresponding question arises for the model theoretic treatment in the *choice of extensions* (over a given structure) which enters into the invariance conditions; and again in 4(c), for a phenomenological analysis when we have to ask how the objects are *given* (or: presented).

Practically speaking, it may be easier to see the answer in a specific case (e.g. transitive extensions for sets, end extensions for orderings) than to say why; recall 1(b)(ii). (And as long as we consider applications there is no practical need.) Platek, in the introduction to [42], makes the interesting suggestion of using operations which 'reflect' the build up of the structures considered. However, it must be admitted that this aim is not fulfilled by his two principal applications, to admissible ordinals and admissible sets. Specifically, in the case of *uncountable* ordinals α, it is not at all plausible that his principal operation (supremum) 'reflects' the building up of the ordinals; one needs a separate operation to generate uncountable *sets* of ordinals. In the case of a recursion theory on sets, the general theory is not used at all; instead we have the *ad hoc* definition: r.e. = Σ_1 on A where A satisfies familiar closure conditions (Platek's *admissible* sets), which, by [3] and [16], are not satisfactory for *uncountable* A, as already mentioned in 2(a)(i) above. – NB. Though these delicate questions arise in all g.r.t., it should not be forgotten that, for some *given* structure, the answer may be easier to *find* for Platek's development than for, say, the model theoretic g.r.t., or *vice versa*.

(iii) As to a *logical analysis* of the concept of inductive definition, Feferman's [8] and Platek's [42] are equally defective, if one thinks of inductive definitions for a *foundation* independent of familiar set theory (and nobody ever denied this). Thus [8] uses freely the classical logical operations, including *negation* while the central feature of an inductive definition is that we know what goes *into* the set defined, but not its complement; [8] derives positivity conditions only by *use* of ordinary model theoretic notions. In [42] the basic definitions are given *in terms of the full finite type structure* built up on the ground type considered; even if, *post hoc*, one were to analyze the closure conditions on the higher type objects needed for the proofs of the principal results, this would not be an *independent* foundation! The very statement of a closure condition contains variables whose range is supposed to be given if the closure conditions are to be understood in the ordinary 'classical' way. To be independent of familiar set theoretic concepts, that is to use inductive definitions *foundationally*, one has to face the question: what kinds of objects make up this range?

But it seems to me that a quantifier-free treatment, like that of [42], may be more help in answering the question than the corresponding model theoretic treatment (which would need a reinterpretation of the logical operations à la Kripke).

(c) *Formal calculations*

The 'formal calculations' that arise with g.r.t. may be either literally formal, as in a formalized predicate or equation calculus, or infinitary, for instance with some kind of ω-rule, Kripke's calculus [28] or the λ-calculus in [42].

Which comes first: the calculations or the functions that result from the calculations?

The answer is quite clear, if we go back to the two principal sources of classical recursion theory (which happen to produce extensionally equal notions in arithmetic): *invariant definability* (or, in Platek [42] and Feferman [8]: inductive definability) and *computations*. In the former case the functions come first, they are picked out from a class of models in (a) or from a hierarchy of types in (b); the calculi are chosen by suitable *completeness* properties. In the latter, the processes or rules come first, ideally determined by a phenomenological analysis of possible instructions (understandable by the means considered); the functions are defined in terms of these processes, and one may look for other systems of rules which compute the functions of the same class. Evidently one must expect a possible conflict of interest here: for a complete *description* of possible processes we want lots of rules; for *deciding* some specific question about the computable functions we often want few rules (provided only they generate all these functions). Here it is to be remarked that Kleene's well-known equation calculus, e.g. in [20], is very convenient for the second kind of purpose because it permits an *easy* analysis of deductions, but not for the first, in fact, its derivability relation has no good meaning; see e.g. [27, para. 4, p. 32] where the appropriate completeness property and an adequate equation calculus are given. From an algebraic point of view, [27] is superseded by Julia Robinson's remarkable formulation in [44].

Looking back at (a) and (b) above, the role of a complete formal calculus is plain enough. In the model theoretic treatment a calculus adequate for *logical consequence* is involved; I made some use of the completeness of the ω-rule in [23], and Barwise used a complete Gentzen-type calculus in [2]. As already mentioned, Platek put his λ-calculus to the traditional Gentzen-type use of avoiding detours (in his case: via higher types).

In view of the doubts about a general concept of computation (and, *a fortiori*, about a convincing analysis of this hypothetical concept) expressed in 1(c) and developed in 4(c), I shall confine myself to metaphorical uses of this idea which have led to various g.r.t., namely *applications of the class of 'computable' functions*.

(i) Some of these classes turn up in g.r.t. based on *definability* considerations too, for example, as mentioned in (b), Kleene's [21] and Platek's [42]; indeed [21] *led* to the extensional reformulation in [42]. Kleene's schemata explicitly operate on indices (of rules or instructions): so, *as they stand*, they are about rules and processes. Similarly, Kripke's equation calculi for 'admissible' ordinals α [28] 'compute' essentially the Σ_1 and Δ_1 sets of ordinals over the segment L_α in Gödel's constructible (ramified) hierarchy where L_α satisfies familiar axioms; see [28] or [23, p. 194(b)]; the role of Σ_1 in connection with generalizing r.e. *and* s.r. was mentioned already at the beginning of this section [2)].

Evidently, because of these equivalent descriptions in familiar terms, the calculi are not essential in a legalistic sense to any applications of the *class* of functions which they compute. But it may be significant that in such delicate constructions as Sacks' for maximal simple sets on suitable α, a concept which can be very simply *defined* in the Σ_1 notation, the *proofs* seem to use Kripke's equation calculus in a very natural way; just as, occasionally, one uses a suitable calculus for delicate refinements in logic.

[2)] Some of the simpler results in recursion theory are obtained by mere quantifier manipulation, for example, Post's theorem that a set is s.r. in some s.r. set iff it is Δ_2, as stressed, e.g., by Montague [36, p. 71]. In [36] a g.r.t. is described on structures built up from a ground type by iterating the power set operation limited to subsets of *uniformly restricted cardinality*, a version of Zermelo's original hierarchy with individuals. (Thus the cardinality restriction corresponds to first order definability in Kripke's g.r.t. on segments of the constructible hierarchy.) Σ_1 over the whole structure is taken to generalize s.r., Δ_1: recursive, r.e. is (presumably) the range of a recursive function with the ground type as its domain. No generalization of *finite* in terms of definability is attempted; a natural candidate is: recursive *and* of restricted cardinality. – As it stands, the g.r.t. of [36] excludes precisely those recursion theories (other than classical recursion theory) which have been most useful up to the present such as Barwise' use of *countable* admissible sets; just as the languages $L_{k,\lambda}$ with *cardinals* k, λ ($> \aleph_0$) omitted some of the most useful infinitary languages. Evidently, for further progress one has to look for structures where Montague's g.r.t. *differs* from others (and to apply it). Specifically, it might be interesting to describe in familiar terms the class of Σ_1 subsets $\subset N^N$, where Σ_1 ranges over the type structure hereditarily of cardinal $< 2^{\aleph_0}$, and the basic 'operations' on N^N are $\lambda n\,(n+1)$, inverse pairing functions, composition, and =. (Montague always includes =.) Do not assume the continuum hypothesis.

(ii) In contrast to (i), Moschovakis [37] studies two notions of 'abstract computability' for which, as far as I know, there are in general no equally familiar equivalent descriptions. Curiously, though the word 'computability' occurs in the title the computability notions of [37], *prime* and *search computability*, are not studied in much detail in [37]; in fact, *there is no description of the class of prime computable relations in any familiar structure*! One would expect these classes to be really interesting for very simple algebraic structures and, at the other extreme, in structures without a (prime computable) well ordering. The difference between the two computabilities corresponds pretty well to that between deterministic rules (of the kind one has in computations) and non-deterministic 'rules' (as in the usual systems of inference with detachment) where nothing is said about the *order* in which the principles are to be applied.

More logical uses of these notions will be taken up in 4(c).

Here it is to be remarked that in Part II of his paper [37], Moschovakis develops his theory of *hyper projective* sets (not mentioned in the title of [37]) which gives a very satisfactory generalization of Kleene's theory of hyperarithmetic sets. Though [37] does not present the theory of hyperprojective sets as a g.r.t., this can be done straightforwardly by means of Feferman's [8] described in 2(a)(ii) above. In any case, Moschovakis, jointly with Barwise and Gandy [4], have already related hyperprojectivity to Barwise' g.r.t. using strict Π_1^1-sets in [3].

(d) *Enumerated structures and an axiomatic theory*

Trivially, whenever we have a general definition (here: of a recursion theory) on a *class* of structures as in (a) or (b), we have an induced axiomatic 'theory' (but, of course, not necessarily recursively axiomatized) modulo a given *language* corresponding to the basic definitions: the set of those statements in the language which are true in all the structures considered. (Obvious languages were considered in [23, pp. 202–204] or [26, pp. 319, 320].) It is sometimes said that *the* axiomatization problem is to generate the set of valid statements. But this is a logician's parody of the role in *mathematics* of genuine axiomatic theories! Of course the problem is attractive because it is clear cut. Now, first of all, quite often the set of valid statements is not r.e., when there is no question of a literal axiomatization; but even when the set is r.e., a complete axiomatization is not necessarily useful; for example, if *very* few axioms yield the *bulk* of interesting results while, perhaps demonstrably, a complete axiomatization is bound to be complicated. In such a case the law of diminishing returns applies; concerning objectivity in judging partial axiomatizations, see 1(b). Naturally, and this comes up in (ii) below, once such 'partial' axiomatic theories have been discovered they may present interesting material for metamathematical study.

The enumerated *structures* and *definitions* of essential concepts are given in Malcev [35] and Lacombe's thesis [31] [3]. The *axiomatic theories* were introduced by Wagner [57], developed and studied by Strong [55] and, more recently, Friedman [11]. (His work and my knowledge of Strong's go back to an exposition by Mr. Richard Thompson in my seminar at Stanford, in autumn 1968.)

The basic notions of recursion theory (listed at the beginning of Section 2) all appear explicitly in these axiomatic theories *except* for: finite. There is an important *additional* relation, between the universe and its s.r. subsets, familiar from the s.r. enumeration of s.r. sets.

(i) The works of Malcev and Lacombe constitute a g.r.t. on *countable* structures, defined in terms of one-one numberings (numérotage) or many-one enumerations (énumérage). Any enumeration together with given relations of the structure obviously *induces* number theoretic relations. The enumerated structure is then called *recursic* if the number theoretic relations are recursive. Evidently since only the (extensional) notions of ordinary recursion theory are used, the theory is free of details about Gödel numberings and the like.

Both Malcev and Lacombe emphasized a particular application, namely where the enumerated structure consists of the partial recursive functions themselves with composition and some other simple operations (but not: extensional equality). From this point of view, it was natural to analyze properties of the *enumeration function* which implied the principal theorems of what has been called: pre-Friedberg recursion theory. Though *names* were given to such properties (e.g. Lacombe's 'autogenic' enumerations, obviously connected with a partial recursive enumeration of partial recursive functions) no attempt seems to have been made to set out these properties in the form of an ordinary axiomatic theory.

It seems fair to say that, because of this, it was not realized just how abstract the analysis was! Specifically, the fact that one was thinking of an enumeration of the structure by means of natural numbers was not used in the theory of partial recursive functions at all. One could just as well think of

[3] I know Malcev's work only from reviews. I have the impression that Lacombe shows more *finesse* (but it often happens that reviewers do not appreciate and hence do not communicate the finer points of a theory). This compensates for the distressingly affected style of Lacombe's thesis.

an enumeration by 'themselves' provided such a function is identified, not with its graph, but a presentation.

(ii) Wagner [57] also analyses the usual operations on partial recursive functions, but in a proper axiomatic style, also emphasizing the 'enumeration' relation. The language is very simple, the axioms are equations. In his original version he introduced explicitly an undefined element; consequently, when specialized to the classical case, his basic relation = cannot be realized by an r.e. predicate. This would be interesting to modify this.

Wagner and Strong [55] developed what might be called Post-recursion theory from these axioms. In view of the work under (i) (or, indeed, from inspection of the older literature), it was not surprising that *something like this* could be done; but (at least to me) it is remarkable just how much these few axioms yield.

Recently, Friedman [11] has developed this theory some steps further, in (at least, to me) most satisfactory ways.

First of all, he establishes an *isomorphism* result which again is not unexpected from (i) but was not previously established: if we take the partial recursive functions and consider any two enumerations satisfying the Wagner-Strong conditions the resulting structures are isomorphic. He describes it as a *relative* categoricity result, relative to keeping our class of *objects* fixed, varying only the enumeration. Naturally this result does *not* explain why so much recursion theory can be developed by *logical deduction* from the few axioms. Indeed, logically quite unrelated axioms in the language of Wagner-Strong might also be relatively categorical! This raises the (probably delicate) question:

> In what way, if at all, are the Wagner-Strong axioms *distinguished among relatively categorical ones* in the language considered?

The other novel element which, of course, depends on the existence of g.r.t. described in (a) – (c) above, is that the Wagner-Strong axioms hold for a large class of them. And, I think equally important, Friedman [11] shows that his isomorphism result does *not* hold for all of them, e.g. not for all admissible L_α: satisfyingly enough it holds for the subclass of 'nonprojectible' L_α which have already played an important role in the subject. So things fit together quite well.

Possibly the Wagner-Strong axioms are *necessary* in the precise sense that they are satisfied by g.r.t. which have the kind of uses described in Section 1.

(But cf. Friedman's remark at the end of [11] concerning *countable* admissible A without a definable well ordering.)

But it cannot be expected that they are sufficient. It will be interesting to see whether a few *additional* axioms in the Wagner-Strong language are sufficient for the bulk of, say, post-Friedberg recursion theory.

3. Metarecursion theory (a g.r.t. on the recursive ordinals)

It is useful to begin with an up-to-date summary by reference to the list of basic notions at the beginning of Section 2.

The generalizations of these notions were *defined* in [26, p. 322] in terms of model theoretic invariance discussed in 2a, but the technical development used principally one of two *equivalent* descriptions, in terms of Kripke's equation calculus [28] (the standard equivalence *proof*, which belongs in para. 3 of [26, p. 324], somehow having been omitted); and, to a lesser extent, notations and concepts of Kleene's hierarchy (the logical status of the notations being made clear in [26, p. 319, l. 21–31], but not in [49]). The equivalences with familiar concepts of Kleene's hierarchy are of importance not only for the proofs, but also for results, because some of the more important applications of metarecursion theory concern precisely these concepts.

An important element of metarecursion theory is the notion of *metafinite*, the generalization of *finite* in the sense of definability. This notion is taken as primitive in [26, p. 322], but turns out to be extensionally equivalent to: metarecursive and included in a proper initial segment of the recursive ordinals. (Thus this use of 'metafinite' is not essential to stating *results*.) But thinking of this conjunction as generalizing 'finite' was essential in the *development* of the subject, to guide (1) conjectures, (2) proofs and (3) the choice of notions. The following examples should be kept in mind.

(1) For reasonable conjectures in applications of the theory to Π_1^1 sets, that is to meta r.e. sets of natural numbers, one must remember this (where metafiniteness is relevant): these sets are generalizations of *bounded* r.e. sets (like r.e. sets whose elements are 0 or 1), since they are included in ω, which is a metafinite subset of the recursive ordinals.

The distinction between finite sets, given by a standard presentation, and bounded r.e. sets is familiar not only from the intuitionistic literature (cf. [26, p. 323]) but from Post's notion of hyper – and hyper hyper simple sets.

But the distinction is not always developed in ordinary recursion theory when it could be introduced in a natural and perhaps fruitful way.

(2) The notion of metafinite is used in *proofs* of theorems which generalize classical results where 'finite' is not even mentioned such as enumeration of r.e. sets without repetition or the splitting of non-recursive r.e. sets into disjoint ones; or generalizations which retain the literal meaning of 'finite', as in (the strongest version, in (a) below of) the maximal simple set; cf. Friedberg [10].

(3) In a more subtle way, one of the *reducibility relations*, namely $<_M$ ('M' for metarecursiveness), is tied to our emphasis on metafiniteness.

To explain this let me go over the subject of metareducibility; its discussion in [26] and [49] was unnecessarily complicated because of our ignorance of simple mathematical facts below. Nowadays it can be put as follows.

The various notions which are equivalent to Turing reducibility in classical recursion theory fall naturally into two groups: those concerned with *computation* and those concerned with *definability*. Now *if* one things of computations as operations from (and to) finite objects, and if the notion of metafiniteness is to be used *then* $<_M$ is the proper generalization: it involves a *uniform* metarecursive process from and to metafinite neighbourhoods. *All* the definability notions that arise turn out to be equivalent. They are denoted by $<_c$ where 'c' stands for 'calculus', i.e. definable in Kripke's equation calculus. The reason for saying 'definable' and not 'computable' (which also begins with a *c*) in turn involves the notion of metafinite! in the sense that we obtain $<_c$ only if 'calculation' trees which are *not* metafinite are admitted; cf. [49]. (To be precise, $<_c$ would conflict even with a weaker requirement on meta-computations', restricting only the ordinal *length* of the calculation trees by ω_1 and not their *complexity*, that is without requiring them to be metafinite, and thus uniformly of length $< \omega_1$.) Of course since both metarecursiveness and metafiniteness are explained in terms of invariant definability, so can $<_M$. But there seems to be no *natural* invariant definability relation which is equivalent to $<_M$.

NB. Since, by 1(c), the idea of meta-'computation' is metaphorical, the argument above is purely heuristic. *If* the metaphor is to be used, *then* the relation $<_M$ will be studied. We would also look for *applications* in the sense of the introduction. But since a related problem arises also for classical degree theory I prefer to discuss it in a general way at the end of the section. Note, however, that, e.g. by (2), not all post-Friedberg recursion theory has to do with degrees!

Digression on needless complications in the literature. In [26, bottom of p. 328] and particularly [23, p. 197] also two notions of relative invariant definability were emphasized. But (and this was not known at the time), in contrast to $<_c$ and $<_M$, they do turn out to be equivalent for the proper infinitary language. This was cleared up in my abstract [24] with help from Lacombe's [32]. (The equivalence between $<_c$ and this definability relation *was* known; it is proved by the same argument as the basic equivalence proofs which, as mentioned already, were omitted from [26, para. 3].)

Another kind of needless complication is due not to neglect of formal relations, but to certain expository principles of Sacks (who, with his students, is of course the person who has done most for metarecursion theory). He treated, with equal emphasis, many different relations. My own view is different, and applies quite generally. There are only a *few* to emphasize. But, for any one of them such as $<_M$ and $<_c$, *variants* present themselves for consideration both before and after the discovery of the notions. Thus, in our case, we had two *asymmetric* examples: socalled weak metarecursiveness (which, in contrast to $<_M$, operates from metafinite neighbourhoods to single values; it was introduced through an oversight in the abstract preceding [26]) and ω_1-computability (which, in constrast to $<_c$, bounds the ordinal length of computations; but only of individual computations in the whole set that constitutes an ω_1-reduction *to* a given predicate). This asymmetry actually leads to *intransitivity* as shown by Driscoll [5] for the former and Sacks (unpublished) for the latter. This type of situation is familiar, e.g. from Schwartz' attempt to generalize a definition of *area of a plane figure* (corresponding here to: ordinary reducibility) to *area of a surface*. Some ingenuity was needed to show that his definition was untenable even for the unit cylinder of unit length: This caused a surprise. (Ignorance or thoughtlessness is a *guarantee* for surprises.) Texts on analysis, quite properly, separate such results from the main body of development. Current literature on metarecursion theory does not always put its results in their place. (End of digression.)

An *economical presentation* of metarecursion theory must employ two familiar principles which are constantly used when a notion *splits up* in a generalization. First of all, for a given proposition P there may be a *strongest* ramification, that is, P applied to one of these generalizations is easily seen to imply P applied to any of the others ('is easily seen', because if the strongest version of P is true so are the others, and hence, logically, they are all equivalent). This situation may arise both when we do not know whether the distinct notions are extensionally equivalent and when we do; cf. (a) and (b) below. Lastly when, for a given P, we do not know an (obviously) strongest version, yet P holds for all, an *axiomatic* treatment may be desirable, as in (c) below.

(a) *Maximal simple set*

This topic does not involve any reducibility relation at all. The generalization of finite, which enters into the notion of *simple* set is crucial. It is evident that we have a *strongest* notion when, in the metarecursive generalization, the literal meaning of finite is retained. By [26] there exists a meta r.e. set of recursive ordinals which is maximal in this strongest sense, and a refinement, also in [26, p. 332], of a maximal meta r.e. set $\subset \omega$, i.e. a maximal Π_1^1 set (among Π_1^1 sets).

To the best of my knowledge certain natural questions that arise from this work have been curiously neglected in favour of teratoid theorems. (A similar remark applies also to other results in metarecursion theory.)

First of all, what do we get in ordinary recursion theory if we *project back* the result above on *bounded* meta r.e. maximal sets (i.e. Π_1^1 sets)? Or, even more simply in connection with understanding ordinary recursion theory; cf. 1(b): How would one set out the proof in the ordinary case *after* having thought through the metarecursive analysis? Speaking for myself I found it easier to remember the generalized proof because here one is forced to *say* explicitly what properties the various auxiliary functions must have, while in ordinary recursion theory they are so obvious from equations that one does not stop to make them explicit.

More objectively, as far as I know there is no discussion in the literature *whether the different generalizations of the notion of maximal set are extensionally inequivalent*! Of course, this general question may possibly be solved by closer study of *specific* maximal set constructions already in the literature! namely by Friedberg [10], Sacks [46], Yates [59]. As pointed out in [26, p. 329], (the unambiguous generalizations to metarecursion theory of) their definitions yield meta r.e. sets for which the *obvious* proofs establish maximality in progressively stronger senses.

Wouldn't it be satisfying, in view of 1(b)(ii), if (even extensionally speaking) the sets themselves were distinguished by maximality in different senses, corresponding (?) to the experts' impression that [10], [46], [59] constituted progress.

For reference in Section 4, note that there is a quite different analysis of proofs in the literature by Lachlan [30] and Owings [41] who consider *two types of maximal* sets, but only one of them occurs in the ordinary case. Thus proofs in the ordinary case can be analyzed in terms of whether they are applicable to both types, or only to the case which occurs here. Evidently this distinction is more manageable than the distinction in terms of different generalizations of 'finite', since the two types are exclusive, while a given set is maximal for all generalizations if it is maximal in the strongest sense.

(b) *Post's problem and reducibility*

Here it is known that $<_c$ and $<_M$ are extensionally inequivalent, e.g. when applied to the Π_1^1 set of ordinal notations (cf. [23, p. 204]), and $<_c$ is the coarsest reducibility relation, actually among *all* relations that have been considered. So incomparability for $<_c$ is strongest. Here, and in contrast to (a), the cases of *unbounded* and *bounded* meta r.e. sets have to be treated differently.

(i) Sacks [48] solves Post's problem positively for $<_c$, for *unbounded* meta r.e. sets (of recursive ordinals). The incomparable sets X_1 and X_2 are *subgeneric* which makes Sacks' solution *uniform* in the reducibility relations in a very delicate sense! (not merely in the brutal sense that incomparability for $<_c$ obviously implies incomparability for the others.) *Every* $X <_c X_1$ or $<_c X_2$ is also $<_M X_1$, resp. $<_M X_2$!

(ii) For Π_1^1 (bounded meta r.e.) sets, the situation is different.

Since $<_c$, restricted to Π_1^1 sets, is equivalent to relative hyperarithmeticity, Spector's observation [54] applies and shows that there are *no* $<_c$-incomparable Π_1^1 sets.

For $<_M$, Sacks [49] produces incomparable Π_1^1 sets (and announces ω_1-incomparable sets for this coarser, but teratoid relation discussed in the Digression above).

For reference in connection with applications of $<_M$, and in particular with its usefulness compared with $<_c$, (ii) should be kept in mind! We certainly want to apply metarecursion theory to Π_1^1 sets (e.g. to the set of consequences of a formula in ω-models); $<_M$ is at least a *candidate* for a possible application where some kind of incomparability between Π_1^1 sets is required; see d(ii) below.

(c) *Density theorems* for meta r.e. sets

I mention the theorems (see, in particular, Driscoll [5]) not primarily for their intrinsic interest, but because they illustrate two quite distinct interesting points.

First, a splendid use of metarecursion theory was made in this connection! namely the discovery of an error in [47] when Driscoll tried to generalize [47]; cf. the review of [47] by Robinson in J. Symbolic Logic 34 (1969)

294–295. Certainly it is not the first use of a generalization for such mathematical hygiene: it is generally said that a computational error in Siegel's [51] was found by use of Weil's generalization [58]. (The relevant footnote on p. 86 of [58] is, however, non-committal on this matter.)

Secondly, in contrast to (a) and (b), there is apparently no 'strongest' generalization of the classical theorem involved [47]. Thus, if one is thinking of an axiomatic theory of reducibility the density theorem may provide a good *test*; for, in contrast to (b) (i), there is now no trivial 'uniform' generalization by simply proving the strongest instance, and thus no need for delicate additional uniformity requirements.

Another, more 'logical', way of putting the matter is that density theorems are *conditional*. We have binary and ternary relations R and S between (meta) r.e. sets, say Y and Z, resp. X, Y and Z, and we establish

$$R(Y, Z) \to \exists X S(X, Y, Z)\,,$$

while in (a) and (b) the logical form is simply

$$\exists X T(X)\,,$$

where T may contain parameters other than X, the 'relativized' version.

It may be useful to keep this obvious difference in mind when one looks for significantly different extensions of Friedberg's priority method; cf. (e) below.

(d) *Discussion of possible applications*

To avoid all misunderstanding: I am not here concerned with inane discussions on whether people should work in a subject without applications, or how long they should work in it before looking for applications. Apart from their more obvious defects these discussions seem to make the romantic assumption that people can do anything they want if they only try! The issue is quite simply this: given, say, post-Friedberg recursion theory particularly concerning r.e. sets, what *more* can we do with it than what we have? We must be quite prepared for a good deal of effort to answer this, and it *may* turn out to be negative; in which case effort was needed to *show* that it was not worthwhile! In any case there is no practical problem for an outsider (like myself) who is not particularly attracted by the details of the subject: he leaves them in the hands of the experts and can be sure they will keep him informed.

(i) The single most important point to remember is that the subject goes back to a philosophically very dubious consideration of Post: formal systems generate r.e. sets; formal systems are fundamental to our thinking; so the 'classification' of r.e. sets is fundamental. Then he thrashed about for principles of classification.

Myhill's classical result on the recursive isomorphism between the sets of theorems of all the more usual formal systems makes it evident that much more structure on these sets, e.g. preserving some logical operations (and thus referring to meaning), is needed; structure which does not enter into Post's classification.

Again, Gentzen's work makes clear, beyond a shadow of doubt, that *proof theory* (which *is* concerned with our thinking) *begins where recursion theory ends*. It is not the r.e. character of the set of theorems, but the choice of formal rules, the *particular* enumeration which is needed for this analysis. The fact that *some* very general theorems like the first incompleteness theorem are easily derived from purely recursion theoretic ones only means that *these* do not depend on more delicate structure.

But granted all this, a general even if false logical conception can lead to interesting mathematics, just as a general but false physical theory may have an interesting mathematical content. Only certain conjectures based on the conception will be wrong.

(ii) In contrast to model theory, where Malcev, A. Robinson, Tarski and others looked out for applications, both systematically and energetically, people familiar with post-Post recursion theory did not. It is not unreasonable to suppose that they partly followed an instinct; they just did not expect too much. But the result is that we nowhere find applications put together, and the *difference is exaggerated*. Here I have to rely on my very incomplete knowledge, but let me give a few examples, which, incidentally, have not yet been generalized by means of metarecursion theory (to ω-logic in the first two cases):

Mrs. Pour-El [43] established a close connection between hyper simple sets and independent axiomatization; the concept of independence certainly does not mention 'hypersimple'.

Shoenfield [50] used a priority argument to answer a 'pure' definability question. (In fact, Lemma 1 is a purely recursion theoretic result, about *degrees*, from which the answer to the question follows quite easily.)

Specker [52] used incomparables fairly directly to show that Ramsey's theorem does not relativize to recursive sets. (This mentions recursiveness but again not degrees.) This constitutes a genuine application of degree theory despite the fact that Jockusch [63] has (recently) obtained sharper results by different methods.

These are not systematic results. But were applications of model theory so very systematic before people tried to make them so?

(iii) Trivially, however we look at current degree theory, there is something *lopsided* about it. If we have degrees of arbitrary sets, we have lots of sets and only *recursive* reducibility. If we have degrees of r.e. sets, we have the basic *asymmetry* between membership and non-membership for r.e. sets, and the *symmetric* use of membership and non-membership in Turing-reducibility. (This is why one sometimes considers the *field* of sets generated by r.e. sets, or one-one reducibility or, in connection with r.e. sets, *positive* existential definability.) Incidentally, this perfectly well known lopsidedness becomes more striking in g.r.t.!

But so what? Certainly one does not expect instantaneous simple results which are eye openers, the kind of thing some of us like particularly about *logic*. But within limits it is not wholly unreasonable to look for what mathematicians call 'deep' connections, between notions which, at first sight, do *not* fit together. Certainly what has been called the Dutch Garden of complex function theory is full of *prima facie* odd connections between the growth of functions at infinity and the values omitted in an angle. Or, for that matter, aren't additive problems about prime numbers equally lopsided since primes are distinguished by multiplicative conditions?

It is probably fair to say that these 'curious' questions do not belong to the socalled mainstream of mathematics and, perhaps, rightly so. But, viewing the matter with total detachment, I think there is something lopsided about the fairly general complaint about a lack of applications of the current theory of degrees.

With this out of the way, we can return to the beginning of this subsection and ask: Can we do better?

By 1(b), let us trust the experts, who emphasize the *methods of proof* over the results. Again by 1(b) we trust their judgement, not their analysis, and more simply we interpret this emphasis to mean: they haven't formulated particularly interesting results. I can't do better but I do have some *considerations* which, it seems to me, have been neglected so far.

(e) *Priority methods and its results*

Since, after all, the experts have not succeeded in formulating the essence of the method to their satisfaction, let us look at the *pattern of results*, and their *logical* character. (Of course I do not dispute such bromides as: the results are very attractive, because once you understand a very few notions, the problems are very simple to state, and difficult to prove, like in number theory.) As already mentioned at the end of (c), we have, in the first place, existential theorems

$$\exists X C(X)$$

with X ranging over r.e. sets and C of quite low logical complexity when written in the sort of language considered in [26, p. 320].

Friedberg emphasized (without considering explicitly the language of $C(X)$) that Post's theorem has an essentially *non-recursive character*: though the X is produced explicitly, a certain $\forall\exists$ combination in C cannot be realized by a recursive function. (This too involves Myhill's classical result in (d) (i).) It is a striking feature in that non-recursiveness is very common in logic or in *general* mathematical theorems on convergence say, but relatively rare with special ones; even more, though we may *suspect* such non-recursiveness (e.g. in Roth's theorem on the approximation of algebraic numbers with uniform bound-2-ϵ), he it is demonstrable.

Neglected question. Is Friedberg's proof non-recursive enough?

Putting first things first: if we only want the existential theorem, isn't there a radically more non-constructive argument which is much simpler? Afterwards we can see what more information proofs by the priority method give, or, more precisely, the various extensions needed for relativized or conditional existential theorems, cf. (c).

Second order neglected question (which is neglected in the 'neglected' question). What relations C arise in *theorems* proved by means of the priority method?

Naturally this question is difficult, cf. (b) of the Introduction. In which *language* should we formulate C to state the facts usefully? To what extent is logical complexity of C (in a given language) critical? or: Is it best to look for socalled synthetic conditions, as in the case of basis theorems, namely the existence of an X in a suitably chosen class (containing also non r.e. sets), where X satisfies $C(X)$? It is understood that this weaker condition should be *easy* to verify in applications.

Speaking for myself I had hoped that the added complexity of g.r.t., in particular metarecursion theory, would force one to be very explicit about each step, so that finally a proper choice of language forces itself upon one too! In addition, the insistence on subgeneric sets in (b) forced one to think of *forcing* in this connection. (Of course all this self-discipline seemed reasonable only because the free-and-easy way of the experts had not been successful.)

Let me only add (if this is of interest) that the project of a less constructive proof is not meant as a porte-manteau, but connected with a specific parallel.

Autobiographical remark. During my collaboration with Spector some ten years ago, I often meant to ask him about a famous 'priority construction' due to Hilbert and developed by Ackermann: the socalled ϵ-substitution method as explained in Hilbert-Bernays, but I never did. My own interest was primarily in this method. But perhaps some recursion theorist would be interested in its relation to Friedberg's construction. Admittedly there is a difference in that only finitely many requirements have to be 'met' in the substitution method. But now look at the similarities:

The requirements are *purely combinatorial*, expressed in the language of function application. (They arise from the socalled critical ϵ-formulae in various formal systems.)

The basic procedure consists in defining recursivelty (a finite number of) *sequences of finite partial functions*. (Each sequence is associated to an ϵ-matrix occurring in the critical formulae considered.)

As long as the requirements are not met at stage n, the definition of the $(n + 1)$th partial function in each of the sequences is determined by an *order of priorities*. (The order is given by the logical complexity of the corresponding ϵ-matrix.)

In all cases where the method is shown to converge, the crucial combinatorial property is that, for each argument, the value of any of the functions *can change only finitely often.* (The cases in question are critical formulae arising from predicate logic, and certain extensions of arithmetic.)

Most telling of all: the *problem* of convergence of this purely combinatorial procedure is very simple to state, but, in one sense or another, the proofs are surprisingly sophisticated! Hilbert pointed out, quite correctly, that the convergence implies trivially the consistency of the corresponding formal system and added, somewhat incorrectly, that therefore *only* this purely combinatorial (arithmetical) convergence problem had to be solved. Since Gödel's incompleteness theorems we know that the convergence problems cannot be solved by methods formalized in the corresponding systems. If one tries to give *explicit* (quantitative) solutions with relatively simple bounds on the rate of convergence, methods in the style of Gentzen are needed.

Now, as I have often pointed out (though not in a recursion-theoretic context), in contrast to the consistency problem itself, the convergence of Hilbert *particular* procedure is not obvious, but non-constructive *continuity* considerations give a quite easy convergence proof for the method applied to the systems for which it is known to converge. But *there is a 45-year old open problem*!

Ackermann [62] *formulated* the ϵ-substitution method for (the critical formulae arising in) classical analysis, and *proposed* a proof of convergence. Von Neumann [66] discovered a gap in the *argument*. But we don't know whether or not the procedure in fact converges; cf. [64, p. 168, 3.351].

4. Uses of g.r.t. and some problems

As mentioned at the beginning of Section 1, the subsections (a)–(d) below concern the aims described in 1(a)–1(d) resp. Here it is to be emphasized that the first two aims, in 1a and 1b, differ from the other two, both theoretically and in degree of development; 'theoretically' in that the former concern applications, and the latter concern the analysis of familiar concepts. While 4a and 4b can rely on a substantial if somewhat chaotic body of results in the literature, the aims 1c and 1d are undeveloped; in fact, 4c and 4d are intended to contribute by making explicit some simple but quite basic distinctions (which, as quotations from the literature show, are badly neglected). Such an heterogeneous mixture of subjects is often quite unattractive. But, on reflection, I believe that the explicit separation of this mixture is at least as necessary at the present stage as anything else in this paper. Also the various topics are sufficiently well separated not to interfere with each other. Indeed, for those with a somewhat literary way of thinking, the broad canvas filled by g.r.t. may be quite pleasing. In other words, it's the thought of heterogeneity which is disturbing rather than the actual experience of it (like a *mésalliance* of May and December?)

(a) *Advancing other parts of logic and mathematics*

By far the most important element of g.r.t. here is not in connection with computation (or, more generally, with constructivity), but with *definability*. It arises as follows.

Starting with a, perhaps quite trivial, existence theorem, a first improvement is to get some *explicitly defined* realization, as in descriptive set theory. A next step is a more *quantitative refinement*, often by use of some hierarchy of definitions indexed by ordinals. Experience shows that these hierarchies, or at least the more useful ones, are best replaced by *infinitary languages*. Each infinite formula, being a well founded tree, has in any case an *ordinal* associated with it and thus a classification by ordinals is induced. But, better still, we have a natural refinement by the *complexity* of the tree, not only its (ordinal) size. It is precisely here, in limiting the complexity of the formulae, that g.r.t. have been of use in the *choice of infinitary languages.*

There is no shadow of doubt that some of these definability refinements, in fact ordinal bounds, answer *natural* questions. One of the first was this: given a closed set F of reals, how long is the sequence of its derived sets? Evidently, some definability 'measure' on F has to be used. Suppose the set of complementary intervals to F with rational end points is Π_1^1 in the subset a of ω. Then an optimal bound for the length of the sequence is ω_1^a. (Thus there is a connection with admissible ordinals too [12].) More recently, better uses have been made of g.r.t. in connection with Hanf numbers, see e.g. [2].

What is in doubt is: just how useful are these refinements to the outsider, e.g. the 'working' mathematician. The refinement mentioned is not; since Cantor-Bendixson type theorems are marginal to *current* interests there is simply no good reason to learn the concepts used for stating the bounds, here: ω_1^a; the refinements would have been more 'palatable' if they had involved familiar ordinals like ω^ω or ϵ_0. After all there is a lot of other mathematics to learn; there would be an incentive if some spectacular application were made like Ax-Kochen's work on Artin's conjecture. (To be realistic it must also be remembered that current mathematical training is highly stylized; like any such training it makes it hard to understand related ideas which do not fit exactly into its frame. For example, in connection with the first uses of model theory, the fact remains, behind their bluff about 'triviality' or 'imprecision', that mathematicians simply had difficulty in understanding the concept of *first order formula* which was all that was needed here.) Of course discovering an area where a technical notion 'belongs', here: an area where definability refinements lead to progress in the judgment of the experts (cf. 1b(ii) above), may well require more imagination than the refinement itself.

As to specific proposals, let me modify and elaborate a little the remarks in 2(a) concerning Barwise's languages L_A associated with admissible sets A in [2] and the improvements in [3].

(i) By 2a the *ad hoc* condition of admissibility is not sufficient for uncountable A. Though good positive results were obtained by Barwise for strict Π_1^1-admissible sets A [3], as far as I know, *no applications of this theory are known for uncountable A*. None can be realistically expected from ordinary mathematics: its notions are usually either not of first order at all or else definable in L_A where A is some 'small' L_α. However, perhaps set theoretical problems may lend themselves to applications.

As should be obvious (or see [25, p. 149]), g.r.t. has a *selective*, not a *creative* role in the choice of infinitary languages. For given logical operations it will limit the complexity of their use or iteration, and thus match up the syntax and semantics for a smooth theory. A g.r.t. may help us *locate* a gap, e.g. in [2, p. 249, 3.12], Barwise shows that for recursively inaccessible τ the validity predicate is τ-recursive: so we could *add* some new logical operations and still have a τ-r.e. validity predicate. But g.r.t. does not tell us *what* to add. So to speak new *logical* ideas may be needed here, not 'combinatorial' ideas of g.r.t. The type of proposal I made in [25, p. 149 (iv)] has not so far led to any progress.

(ii) At the opposite extreme: is admissibility necessary? Barwise has a certainly interesting 'necessity' result in 3.4 of [2, p. 245]. It depends on an apparently harmless assumption that the set A considered satisfies Δ_0-separation. If we leave 'high theory' and look at the facts around us, we find the *recursive* language, that is the fragment of $L_{\omega_1\omega}$ with recursive structure, considered by Lopez-Escobar [33], corresponding to the collection A of hereditarily recursive sets; it is *not* admissible since it does not satisfy Δ_0-separation. Yet Lopez-Escobar establishes good (proof theoretic) closure properties such as the completeness of the *recursive* cut-free calculus, and the interpolation theorem.

We do not know genuine uses of this recursive language. But two points are clear. (i) For *constructive* proof theory Lopez-Escobar's language is too wide; more precisely, it is *incomparable*. The best developed example is the Feferman-Schütte ramified hierarchy restricted to predicative ordinals, which Feferman [7] has formulated most transparently by use of infinitary formulae. Even though (certain) non-recursive formulae are used, *not* all predicates definable in Lopez-Escobar's language are defined. As is well known, in constructive proof theory we do find 'desirable' ordinals like ω^ω or ϵ_0 mentioned earlier on in this subsection. (ii) So to speak at the other extreme, the *validity predicate* for recursive formulae (with the usual Gödel numbering) is as 'complicated' as for hereditarily hyperarithmetic formulae of $L_{\omega_1\omega}$; which do constitute an admissible language. Each is a complete Π^1_1 predicate and indeed, since the formulae are well founded, the syntactic predicate of being a recursive formula of $L_{\omega_1\omega}$ is itself a complete Π^1_1 predicate. Also the validity of interpolation for the recursive language is less striking than appears at first sight since, by what has just been said, each hyperarithmetic formula defines the same class of models as some recursive formula.

(iii) Another variant of the languages considered by Barwise is suggested by the 'principal gap' of model theoretic g.r.t. mentioned at the end of 2a(i). The variant corresponds to (ordinary) predicate calculus with an arbitrary set of symbols, constantly used when the method of diagrams is applied. Its formulae are hereditarily finite subsets built up from a collection of individuals, namely the symbols in question. Similarly one would consider infinite formulae supplied by a g.r.t. on other collections of sets built up from individuals. From the point of view of g.r.t. the idea is attractive because *relative* recursiveness mentioned in Section 3 might find a use. But, from the point of view of applications, the idea is not very promising. On the one hand, the method of diagrams can often be replaced by using single formulae in a many sorted calculus, as has been shown recently mainly by Feferman [6]. On the other hand, even for axiomatic theories formulated in ordinary predicate calculus, more delicate results (such as those needing arithmetization) depend essentially on the collection of symbols being r.e. and indeed on the definability of each symbol. So if the variant is useful, imagination will be needed to find such a use.

Finally it is perhaps not superfluous, after all these open problems, to remind the reader that there are plenty of *applications* mentioned at the beginning of 2a. Also he should look back to 3d(ii) for some applications of 'difficult' recursion theory. The applications of ordinary recursion theory to ordinary logic should extend to ω-logic by use of metarecursion theory. (Incidentally, the first pretty application of metafiniteness, i.e. having hyperarithmetic correspond to finite instead of then current: recursive, came from the generalization of a very simple result in predicate logic! A meta r.e. set, i.e., Π_1^1 set which is not equivalent in ω-logic to an hyperarithmetic set; on the old correspondence this would have contradicted Craig's observation that an r.e. set is equivalent to a recursive set of formulae. Amusingly, the generalization of Craig's result to sets of *finite* formulae in ω-logic trivializes; the proper language to use consists of metafinite formulae.)

(b) *Understanding ordinary recursion theory*

In contrast to (a) above, recursion theory is here an object of study, not a tool. We do not reject problematic notions as in 3d(iii), on the basis of impressions (which would be reasonable if the notions were only tools), but *investigate* them. A perfectly good use of the two metareducibility relations $<_M$ and $<_c$ discussed in Section 3, is simply to bring some order into the subject of Turing reducibility, for example by separating out theorems or lemmata which generalize to both $<_M$ and $<_c$, or only to one (or to neither). This would be the analogue to *one* important role of topological spaces: to help us organize proofs in ordinary (concrete) analysis according to the kinds of spaces to which they generalize. (Old expositions such as Goursat's *Cours d'analyse* could not be so explicit but, it has to be admitted, arrived at a corresponding organization.) Evidently, for this purpose it is essential not to have *too* many distinctions in the g.r.t. which would only create a chaos of their own; the reduction to only two relations $<_M$ and $<_c$ in Section 3 is important in this connection.

The organization below follows the two main principles of the relevant Section 2 of [26]. By p. 318 and p. 320, elaborated at the beginning of 2d above, we do *not* insist on axiomatic theories which are complete for the class of structures considered, that is some g.r.t. of 2a or 2b. By p. 323, a specific g.r.t. such as metarecursion theory can serve as an *aid* towards an axiomatic theory, for instance by giving us more phenomena to think about.

In view of the obviously marked difference between pre-Post and post-Friedberg recursion theory, it is not to be excluded that *different* axiomatizations would be best for understanding either; just as one has algebraic and topological analyses in familiar mathematics. At any rate, we shall *separate* them.

(i) Let me say at once that the (incomplete) *axiomatic theory* of Wagner-Strong in 2d(ii) would certainly have been taken into account if it had existed at the time of the abstract preceding [26] (or if Wagner's work [57] had been available when the paper [26] was written).

It is a remarkably elegant axiomatization of pre-Post and of a good deal of Post's recursion theory. The bulk of the theorems here have *intrinsic* character in that they are invariant under renumbering of the defining equations in the sense of Rogers [45], and the axiomatic theory makes this apparent by not referring to algorithmic details at all. But it would be an error (cf. [45]) to insist on this intrinsic characterization as a point of principle. First of all, we can find exceptions in recursion theoretic practice if we only look for them! cf. end of 4d. More subtly, the intrinsic fragment may have a more elegant and more *intelligible axiomatization* in a richer non-extensional language reminiscent of the use of analytic notions in number theory (cf. 1b) where, it appears, arithmetic laws about prime numbers cannot be *proved intelligibly* without such notions. This important need for a richer language is obscured by the popular misconception that *strong logical* principles are needed for the proofs (whereas, in fact, the proofs are easily formalized in *conservative extensions* of ordinary arithmetic).

As mentioned in 2d(ii), Friedman [11] has used g.r.t. as a help in the *metamathematics* of this axiomatic theory by appeal to the distinctions between the projectible universe of metarecursion theory and the non-projectible universe of ordinary recursion theory.

(ii) As to an *axiomatic theory of post-Friedberg recursion* theory it is wide open *whether the language of the Wagner-Post theory is adequate*; cf. end of 2d. True, when applied to ordinary recursion theory, all the basic notions listed at the beginning of Section 2 can be defined in the language; indeed in many formally inequivalent ways! (cf. [23, p. 202–204] where such matters were first discussed). But this leaves open whether there are elegant axioms in the Wagner-Strong language which permit the development of post-Friedberg recursion theory for any of these definitions.

Here it is to be remarked that another axiomatic treatment of a part of post-Friedberg recursion theory is *implicit* in the literature, namely those properties of r.e. sets used in Lachlan's decision method [30] mentioned in 1b and 3a. But here intelligibility and not simply objective truth is a principal issue; so not mere existence (of such an axiomatic theory), but its exact shape must be before us; indeed, as Bishop Berkeley (over) stressed, even this is not enough: we must *look* at it to see it.

Digression: subclasses of the class of recursive functions. The Wagner-Strong theory, and many g.r.t. in Section 2, have ordinary recursion theory as some kind of *minimal* model. It is not denied that this makes for an *easy* axiomatic theory; the question is whether we can do better. It is easy (to handle) because the class of recursive functions has excellent closure conditions and so axioms which have this class as some kind of minimal or even unique model have a chance of implying (logically) lots of interesting properties. This is perfectly familiar from the use of 'unnecessarily' strong axioms in informal expositions of familiar mathematics, such as the use of the principle of the least upper bound. If we look at *such* an exposition we shall be convinced of the 'need' for the strong axioms because they come in right at the beginning! As to the elimination of such axioms a *distinction* is to be made. If we have *applications* in mind, the guiding principles are usually clear enough; for example, to come back to recursion theory, any computational analysis of algorithms has to take their complexity into account which inevitably takes us to subclasses; similarly any epistemological analysis of how algorithms are recognized to be well defined leads to the kind of subclasses familiar from proof theory; cf. also 4a(i). But the matter is much subtler from a purely *mathematical* point of view. The choice of subclasses or weaker closure conditions requires the same kind of mathematical imagination which allowed mathematicians to see the use of 'algebraization' in analysis (instead of the least upper bound principle) or the role of finite fields in arithmetic. Perhaps most of us had better not try. The totalitarians will reject subclasses, those of a more philosophical turn of mind recognize that the interest of the class of recursive functions does not depend on rejecting smaller or larger classes. (Both types will leave the subject alone; the only ones who really go wrong are those who work in it without realizing its subtlety.) To conclude this digression, let me mention a typical difference between recursion theory and effective schemata like primitive recursion. While in the latter case the natural extension to *higher types* usually introduces new functions of lowest type (as in [14]), in the former they do not (as in the main result of [42]).

(iii) It must be admitted that the use of *metarecursion* theory for understanding the classical case (cf. [23, p. 191] or (a) and (c) of Section 3 above) has been neglected. Not (I believe) because of objective obstacles but because of current preoccupation with new results (however minor) instead of better understanding (of central facts). However, there are exceptions.

MacIntyre, on pp. 59–60 of his thoughtful dissertation [34], analyzed in terms of metarecursion theory the difference between two *minimal degree constructions* by Spector [53] and Sacks [46]. He finds certain *lemmata* in the two constructions *which permit a clear separation* in that their *statements* generalize unambiguously to metarecursion theory, but *one is true, the other false*. To be quite precise, MacIntyre slurs over one ambiguity: which of the metareducibilities $<_M$ or $<_c$ is taken to generalize Turing reducibility; he takes $<_M$ (instead of looking for a criterion of choice, analogously to his work just mentioned!); of course, minimal degree results for $<_c$ and $<_M$ are incomparable. Actually, since in ordinary recursion theory minimal degrees are *not* r.e., one's naive impression is that the more brutal relation $<_c$ may be more to the point. Be that as it may, in terms of $<_c$, the minimal hyperdegree construction of Gandy and Sacks [13] raises the following

Problem: The minimal hyperdegree, say h, of [13] is minimal for $<_c$ among all *bounded* sets, specifically sets $\subset \omega$. But $<_c$ is defined for all sets $\subset \omega_1$. Is h also minimal for $<_c$ in the latter class of sets?

Evidently, this question is a refinement of the problem whether there is *any* subset of ω which is minimal for $<_c$ among all sets of recursive ordinals.

MacIntyre's analysis above concerns two specific *constructions*. As a paradigm for analyzing the nature of a *problem* we can go back to the questions on Post's problem which presented themselves at the beginning of the subject. (It is easy to lose sight of such simple questions.) In the abstract preceding [26], but not in the paper [26] itself, the results were set out so as to suggest the questions: Is it significant for the existence of incomparable sets that

the universe is infinite,
the universe is in recursive correspondence with the set of computations,
the universe is well orderd by a recursive relation?

Of course the universe ω of ordinary recursion theory has all these properties. (Probably the last question is a bit simple-minded, and something else should be well ordered in place of the universe.) The results of Sacks' [49], mentioned in 3b(ii), give an interesting partial solution to the first two questions; it is *partial* without further analysis of the roles of $<_M$ and $<_c$, because the answers are *different* for these two reducibility relations. It cannot be denied that, in a subject full of talk about the *order of priorities*, the third question is more interesting, but metarecursion theory is not suitable for it since everything in sight is well-ordered by some metarecursive relation.

(iv) *Limitations of metarecursion theory*. To analyze this last question, I should very much like to see a solution to the following

Problem. Is there a countable admissible set A *without* an A-recursive well-ordering for which Post's problem has a positive solution?

The reasons for this formulation are clear enough. First of all, by Barwise's work, such an A has an eminently manageable g.r.t. satisfying not only the requirements of [2] but of [3]. Also such an A may have a perfectly manageable well ordering even if it is not A-recursive. In so fas as forcing techniques are used in Sacks' solution of Post's problem by means of subgeneric sets [48], well-ordering of A would not seem to be essential since recent expositions of forcing no longer limit themselves to work on constructible sets.

Evidently, even when metarecursion theory gives a useful analysis, a *significantly* better one may be possible by considering a larger class of g.r.t. Thus after I presented [23] at Berkeley back in 1963, Kripke immediately pointed out that Myhill's isomorphism result on creative sets does not generalize to metarecursion theory: there are both bounded and unbounded creative meta r.e. sets and they, of course, are not metarecursively isomorphic. Further there are only these two kinds of meta r.e. creative sets up to metarecursive isomorphism. This certainly *suggests* that Myhill's isomorphism theorem for creative sets has to do with the (classical) property that bounded r.e. sets are finite. But g.r.t. on countable admissible segments α gives here not only more (trivially), but, I think, significantly more: the property mentioned is sufficient. Equivalently it is sufficient for the generalization if α is not projectible (which ensures that every bounded α-r.e. set is α-finite), a property already referred to in 2d(ii).

Correction of the literature, e.g. [48]. If [26] is taken literally, none of the alleged generalizations of metarecursion theory to *uncountable* admissible ordinals has been proved and some are false. The *definitions* of [26] (called 'fundamental' on p. 322) are uncompromisingly model theoretic as in 2a(i), but the 'proofs' are given for Kripke's g.r.t. on admissible segments L_α: in general, this object is (extensionally) different from the g.r.t. on L_α in 2a(i) when α is uncountable. This difference affects not only the reducibility notions, which, by Section 3, are perhaps problematic, but also, for example, the notion of a maximum simple set; cf. 3a.

Such errors are easily spotted and corrected. What is much more difficult to decide is *whether g.r.t. on uncountable structures are useful for our present purpose at all*, that is for understanding the mathematics of ordinary recursion theory. (Recall an analogous question in 4a(ii) above.) An example would be a theorem, in a sufficiently rich language, such that one proof generalizes to some uncountable cases too, the other does not, *and* that the experts find this distinction illuminating; cf. 1b(ii).

It is perhaps superfluous to mention that a g.r.t. even on L_α is useless for studying whether Post's problem depends on a well-ordering of the universe, as in the Problem above in (iv).

Returning now to the opening paragraph of the present subsection it has to be admitted that we have no really convincing proposal for removing the most striking defect in our understanding of classical recursion theory, namely its appalling lack of elegance; with a metaphorical vocabulary of coloured markers, a vocabulary which in no way adds to clarity, but merely distracts attention from the lack of clarity. I, for one, would not underestimate the difficulty of finding a *real* improvement, since perhaps genuinely new concepts are needed here. We need only think of the early work on exceptional sets in the theory of Fourier series, full of ad hoc constructions. Nothing short of Cantor's theory of sets of points, with the concepts of perfect nowhere dense sets and the like, would allow us to say explicitly what the aims of these constructions were. Nor should it be assumed, where elegance and intelligibility of exposition are concerned, that it does not matter whether we do recursion theory on the natural numbers (with zero and successor) or on the hereditarily finite sets (with ϕ, $\{\ \}$ and $\cup$); it is not enough that either structure is recursive in the other. Indeed an elegant axiomatization, as in (i) above, does not by itself ensure an elegant axiomatic development of the actual state of the subject.

(c) *Analysis of the general concept of computation*

Here we are principally concerned with the aim which is described in the literature by: extending Church's thesis to higher types [21] or to abstract structures [37] ; not in a metaphorical, but in a literal sense explained in 1c. Now the plain fact is that our ordinary meaning of 'computation' is: finite mechanical computation, and hence, inasmuch as a faithful literal analysis is involved, the word 'computation' is out of place. This slip does not, by itself, cast doubt on the objectivity of the concept to be analyzed since there are plenty of *specific examples* and assertions in the literature to work on, to help one infer which concept has been 'misnamed'. (The ancients described the stars as holes in the sky; this was a wrong analysis, but does not stop us from knowing what objects they meant.) Indeed, the literature is locally often extremely thorough and acute.

But there is a serious defect (reflected, I believe, in the slipshod terminology), namely a *failure to state the informal issues carefully and to analyze rigorously the relevance of formal results.* In terms of the comparison with the use of mathematics in theoretical science mentioned in 1c, the defect is parallel to the familiar failure of mathematicians to 'remember the physics behind the mathematics'; or, more explicitly, the failure to make sure that the problems are solved under realistic conditions. I shall go into the exact use of formal work here (though it must be admitted that I have made some of the points already, possibly *ad nauseam*, in scattered reviews, not to speak of conversations on our present topic.) As pointed out in (a) of the introduction, in a future definitive exposition few words will be used to convey these same points. But, in plain empirical terms, experience over the last 10 years with our particular topic has shown that it took less time to settle formal problems than to discover adequacy conditions or even to discover oversights in proposed conditions. (In favourable cases it was enough to hint at correct conditions to open an impasse of several months.) Naturally the situation will change in this respect as soon as the subject is a *little* more developed.

(i) We begin with some questions about ordinary computability. They will allow us to see the main problem more clearly. But they are also of independent interest because of some *useful, yet neglected distinctions.* Turing's work might be called an analysis of

mechanical numerical computation and related processes;

'related processes' because, though Turing gave of course many *examples* of mechanical operations on objects other than the natural numbers or, more precisely, their presentation by numerals [4]), he himself did not give an *explicit description of the kinds of objects that come under the heading of* 'related processes'.

The examples carry *conviction*, so much so that a significant increase in degree of conviction can hardly be expected. What, however, can be very much increased is the *degree of precision* of Turing's analysis, *an explicit description of what we are talking about*. There seems to be a vast literature, both in logic and in computer science, of more explicit expositions. I am not competent to review it. But to the best of my knowledge, there is no analysis of the kind of *questions for which this greater precision is needed*. Evidently two elements are involved in Turing's analysis,

as already mentioned: the objects on which we operate,

the instructions or rules of computation.

Concerning the *objects*, the first problem is to know which mathematical objects, in particular which 'abstract structures' are suitable for a computational treatment. The issue comes up very naturally in connection with numerical computation. It is a commonplace that mathematical recursion theory 'relativizes'; or, as one usually says, it applies to 'computation' on the structure ω with 0 and successor and an *arbitrary* function f on ω. On the other hand, and this is the essence of Turing's analysis, for non-recursive f such a structure is *computationally inadmissible*; realistically speaking, there is no presentation system of ω for which (the diagramme of) such a structure can be mechanically computed. To answer the question:

> To which situations can Turing's analysis (in its intended sense) be extended?

we can use the notion: $\exists$*-recursive* (which Lacombe [31] introduced for this

[4]) Computations operate of course on presentations, not on the abstract objects themselves; indeed the standard answer to the question 'What do we lose when we abstract from a given *concrete* structure and consider only its 'imomorphism type' '? refers to nothing else but *effectiveness*. We may be able to operate effectively on one isomorphic copy of the structure, but not on another; at least, in general; cf. Lacombe's notions of $\forall$ and $\exists$ recursiveness in [31] to which we shall return below.

purpose); there is *some* [5] enumeration of the domain of the structure for which the induced diagramme is recursive in the ordinary sense. An important formal problem here is to *describe classes of abstract structures which are* $\exists$*-recursive*, since structures which are not $\exists$-recursive are certainly not suitable for computations (in the ordinary sense); some examples of such classes are in [31]. I should not have belaboured this point, were it not for the fact that some of the literature on 'abstract computability' misinterprets a mathematical generalization of the body of theorems in classical recursion theory to other structures as constituting a 'computation' theory. The nature of this error could also be illustrated by comparison with other familiar conceptual analyses, for example, when we extend the notion of *area* to a wide class of (planar) sets by use of *Borel-content* or *Lebesgue-measure*. Both these measures are defined for sets of points which have no geometric significance, which do not represent *geometric figures* (e.g. in the sense analyzed precisely in topological terms). It is perfectly true that, *provided* the intended notion of area applies at all, it is given by B- or, equivalently, L-measure, since the defining conditions for these measures are satisfied by the notion of area. But a *separate* and, in the present case, more sophisticated investigation is needed to pick out sets of points which do have geometric significance. It should not be necessary to add that mere absence of 'paradoxical' results concerning these measures on a set S of points is no guarantee for the geometric significance of S (since, as every good liar knows, false assertions may be plausible, and certainly consistent).

Turning now to the second element in Turing's analysis, namely the instructions, the evidence for Church's thesis, which only refers to *results*, to the functions computed, actually establishes more, a kind of *superthesis*: to each mechanical rule or algorithm is assigned a more or less specific programme, modulo trivial conversions, which can be seen to define the same computation *process* as the rule (as in footnote 5, there is a distinction between the objective identity of processes and our knowledge). The reader may wish to compare the thesis to Frege's empirical analysis of logical validity in terms of his formal rules; the superthesis would then correspond to an assignment of specific deductions, again modulo trivial conversions, to intuitive logical proofs. Here again the result, that is the theorem proved, does not de-

[5] Throughout the subject of computation theory we have the distinction between *existence* (of an instruction or, here enumeration with required properties) and our *knowledge*; this is familiar in connection with *means of proving* the termination of a procedure; here the question arises also with the *means of defining* the enumeration, naturally depending on the presentation (or: on our knowledge) of the objects to be enumerated.

termine the process, that is the proof (*a fortiori*, not the formal description of the process); in fact, not even in propositional logic: thus we have at least two obviously different proofs of the theorem $(p \wedge \neg p) \to (p \to p)$, one using $p \to p$ and $q \to (r \to q)$ with $q = p \to p$ and $r = p \wedge \neg p$, the other using $(p \wedge \neg p) \to s$ with $s = p \to p$. (Curry's well known connection between derivations in the positive implicational calculus and terms in his theory of combinators, e.g. pp. 313–314 of [61] allows one to define a convertibility or identity relation between such formal derivations from the familiar convertibility relation between terms; in privately circulated notes W.A. Howard has recently extended Curry's ideas to the whole of Heyting's predicate logic and also developed variants of it.) But I know of no *systematic work* conciously aimed at making Turing's work *on* (what was called) *the superthesis* more precise (explicit) though, inevitably, there must be a good deal in the literature of computer science which does so, perhaps incidentally.

Turing analyses the idea of a machine programme in terms of his concept of (what we call: Turing) machine. But the least we must do to make the evidence for the superthesis precise is to give an *explicit description of a general class of* (mechanical) *rules*, for which the reduction to machine programmes is asserted. The second step is to state *conversion procedures* to say, recursion equations or more detailed 'programmes' such that not only the functions (graphs) computed, but the computation processes are preserved. It may not be easy to be precise about identity of processes (so as to have a *useful* theory about them), but, computationally speaking, it is more important than many precise results. There is no question in my mind that, treated realistically, this second issue is the *principal open problem for which a more precise analysis* (than Turing's) *is required*. In view of successful recent work on definitional equality this problem may indeed be ripe for study at the present time.

(ii) While Turing's analysis concerns the question

What are mechanical computations?

the literature on 'extending' Church's thesis (as I interpret the aim behind it), wants to know:

Why do we use mechanical computations?

More specifically, what is it, about our intellectual or technical equipment or perhaps about the objects studied, that leads us to *confine* ourselves to such operations? There are clearly *empirical* and *a priori* answers. The former are quite well-known. For computing in the engineering sense of the word, the crucial point is this: we have *analogue* computers, which are physical systems not given as Turing machines; what makes mechanical computations empirically so important is that, nevertheless, according to the laws of current physics, known analogue computers may be expected to have recursive behaviour. (Digital computers are simply *designed* to realize mechanical rules.) For computing in the epistemological sense of the word, or following out instructions *effectively*, where we are sure from our understanding of the words used that we shall know what to do, we have not only formal rules (which, by definition, *are* mechanical) but also rules appealing to meaning as in intuitionistic mathematics. But *so far* the latter have been shown, in every specific case, to be equivalent to recursive rules, by recursive realizability or by other means [6)].

The *a priori* answer (admittedly based on intensive specialized experience in proof theory which, as I understand this subject, deals with operations on our knowledge) takes the following general form:

> Mechanical operations are appropriate (fitting) if the objects on which we operate are given in a finitistic (spatio-temporal) way.

As already mentioned in footnote 5, there is also a quite separate question of choosing appropriate means of proof. *If* one takes the view that our knowledge has, necessarily, finitistic character then of course, literally speaking, only mechanical operations will be fitting. It is of course this view which tells us that a precise representation of *proofs* or other mental constructions *must* be finite or 'formal'; this is not the natural way we look at proofs, in the perfectly ordinary sense of 'natural' since the possibility of formalization came as a startling discovery. If now we do not adopt this view but (following either Brouwer or Zermelo according to taste!) regard, fully analyzed, mental constructions as *infinite objects* then the question arises *what infinitistic operations are fitting*. – I insist here on this realistic question because possible metaphorical uses of infinitistic 'computations' have already been considered anyway. – The question, loosely formulated in 1c, concerning the existence

6) It should not be necessary to emphasize that this equivalence is a discovery. (I have discussed this matter elsewhere, perhaps *ad nauseam*.) Here it is perhaps sufficient to note that these proofs of equivalence are *not* always routine in contrast to familiar *routine* appeals to Church's thesis (justified by Turing's analysis), which asserts *grosso modo*: rules that strike us as being mechanical (in the ordinary sense of the word) define recursive functions.

of such a general concept of computation becomes the question whether, on examination, it will be possible to isolate a general notion from the mixture which we associate with the idea of

> operations proper to (or: implicit in) the nature of the objects,

and whether this (hypothetical) notion admits of a useful theory. *This question is wide open.* I have no idea whether it would be useful to pursue this question. But it would be a simple error of judgement not to recognize the massive evidence in favour of the hypothetical concept.

Here it is to be remarked that, in connection with metarecursion theory, I intentionally excluded any discussion of the notion of 'metacomputation' for conceptual analysis: the applications of the notion were interesting enough to state formal problems. There was no need to bring in controversial matters. But it will not have escaped the reader's notice that some of the questions in, say, 4b(ii), have obvious *heuristic* relevance to our present topic; for example questions concerning a *well-ordering* of the computations to be considered, would be expected to connect with the question:

> is our hypothetical concept well determined only if the computations used can be built up in an inductive manner?

(The *domain of objects* need not be so built up; after all we have a good computation theory on the number-theoretic functions.)

Digression on the glamour of conceptual analysis. Turing's work stands out in recent times as a *faithful analysis of a basic logical concept,* which is both convincing and useful. As has often been observed, there was no more hope of such an analysis of the concept of *mechanical computation* than, for instance, of the concepts of *proof* or *definition* (which we also understand). The latter do not seem to permit an explicit analysis, at least not in terms of any current concepts. By the nature of the case, an explicit analysis of *kinds of definition* as supplied, e.g., by g.r.t. (or kinds of proof) cannot be equally satisfactory. But it must not be forgotten that it is not merely the *faithfulness* of the analysis, to our intended concept, which is at issue but its usefulness for understanding our experience. Few concepts that have been analyzed can be more useful than that of computation though I suppose the discovery of geometry, that is a correct analysis of how we *think* of space around us (not physical or visual space) remains outstanding here. Needless to say, it is not claimed that the *refinements* of Turing's analysis in (i), or the hypothetical concept above are in the same class. But it cannot be denied that there is a certain false glamour (even about Turing's analysis) which comes from the quite unfounded idea that *correct* conceptual analyses are rare and *ipso facto* remarkable. For a sane, that is balanced view of the subject of recursion theory itself its use for (stating) Turing's conceptual analysis of computations is not paramount. For example in actual mathematical practice this use plays a very small role indeed. Perhaps the principal truly practical use of Turing's analysis is that it gives moral certainty to *routine appeals* to Church's thesis mentioned in footnote 6, which saves us the trouble of writing out recursion equations. Also it is a fact that, mathematically. (recursive) undecidability results are often not only simple corollaries of *representation* or *definability* theorems, as very clearly stressed by Tarski, but simply less useful than the definability theorems; cf. also Higman's results [17] which imply the undecidability of the word problem of a particular group. Amusingly, the single most useful discovery in this whole area corresponds to the 'easy' part of Church's thesis: we verify that masses of useful functions can be defined by recursion equations, and the latter can be realized by computing engines. This has changed modern life.

(iii) To develop the project adumbrated in (ii) it is probably best to study it in a specific context, without playing with mathematical generalities. So I shall look at some known facts and some problems about *operations on functions of finite type* (over ω). These are not only of interest in themselves, but they enter indirectly into the *finite description* of objects whose full analysis is infinite. (For example, the objects of metarecursion theory are finitely described by means of Π_1^1-formulae, containing a higher type, that is function quantifier, with its principal interpretation.)

Our main tool is Kleene's g.r.t. [21] on functions of finite type, based on his infinitistic schemata S1–S9. As in any other applied mathematics we do not expect to take over all details, neither of the formal work nor of the interpretation. Specifically, Kleene insists on letting his variables range over the *full* hierarchy, that is if N_σ and N_τ are the ranges of variables of types σ and τ respectively, $N_{\sigma\to\tau}$ consists of *all* functions: $N_\sigma \to N_\tau$. (In this way, he gets hierarchy theorems and can manipulate quantifiers, an enterprise which requires strong closure conditions.) But the definition principles S1–S9 make prefectly good sense for *thinner* hierarchies denoted by a dash where, at each stage, only certain functions: $N'_\sigma \to N'_\tau$, are used. There is no need then to extend N'_σ first to N_σ and apply the schemata afterwards. In contrast to Kleene, we shall not consider N_σ here, simply because it is hard to see how anything *realistic* can be said about possible presentations of our knowledge of all these objects (for $\sigma \neq 0$). But this would be a prerequisite for a genuine analysis of the concept of (ii), concerning operations *fitting* (or: implicit in) the concept of function of finite type.

For pure mathematics, where the concept of (ii) plays a considerable, but purely heuristic role the full hierarchy may well be most useful just because it is 'fully' extensional (suppressing questions of knowledge or even of description). But for our present purpose a thinner hierarchy seems much more promising, namely the *constructive* functions of finite type described by Gödel in the first half of [14], not to be confused with the particular definition schemata codified in the system T at the end of [14]. The intended notion admits *rules* which are not necessarily formal but may refer to meaning or other abstract constructive objects (already mentioned in (ii) in particular, footnote 6 on intuitionistic rules) and *proofs*, by any constructive means, of being a well defined rule. We consider first a beginner's view of the matter, illustrated by my pertinent publications in the late fifties.

Case 1. Gödel's notion is *replaced* by more comfortable variants which, as would be expected, have finitistic presentations or approximations. The two principal variants are the *effective operations*, say N_σ^E (for finite types σ) and socalled *continuous operations*, say $N_\sigma^{\mathcal{G}}$ with *neighbourhood functions* required to belong to $\mathcal{G}$. For 'extreme' choices they are *extensionally* equivalent to Kleene's *countable* functionals (defined by use of the *full* hierarchy) and to the effective operations respectively; the latter if all neighbourhood functions are recursive.

The relations to S1–S9 are now completely understood. Both variants are *closed* under S1–S9; for $\sigma = 1$ or 2, where $1 = 0 \to 0$ and $2 = 1 \to 0$, the *recursive* elements in N_σ' are actually *generated* by S1–S9 applied to N_σ', $\tau \leqslant \sigma$; this was shown early on; see e.g. Part II of [21]. But for the next type $(2 \to 0)$, as shown by Gandy and Tait respectively (in unpublished work, though the problems had been published!), *neither* the effective *nor* the countable operations are generated by S1–S9. It is open whether it would be sufficient to add a few simple additional basic operations to S1–S9 (to generate the two variants). But, whatever the answer, the plain fact is this. *If we really knew enough about the possibilities of forming constructive rules and proofs* (involved in Gödel's notion) *to assert that either variant is extensionally equivalent to the notion of constructive function then* S1–S9 *would* **not** *be adequate.*

Case 2. Naturally, after this conclusion, we look at our notion from a diametrically opposite point of view and ask:

> If we use only what we really know about the notion of constructive function of finite type, have we reason to assert closure under S1–S9 at all?

Clearly we cannot *establish* a negative answer without actually proving that the notion of [14] is distinct from either variant considered in Case 1. But a positive answer is quite implausible for general reasons considered in reviews of [21], and also for a reason stressed by Moschovakis [37] in connection with *prime* and *search* computability; cf. 2c. (The value of his distinction is of course not affected by certain weaknesses in his informal discussion which are corrected in the review of [37] in MR.) The point is this. When used for defining functions of type > 2, the inductive clauses in S1–S9, *prima facie* involve *quantification* over the whole domain of some type; and by Kleene's explicit form of the predicate 'is defined' in [21], this predicate demonstrably involves such quantification (provided the notion of [14] satisfies the assumptions made in Kleene's proof). Now, even if we understand individual infinitistic (higher type) notions, this does not *ipso facto* make quantification over the whole domain, even of lower type, into a fitting operation; it would correspond to number quantification in the *ordinary* theory of recursive functionals. The special role of the existential number quantifier (applied to decidable predicates) is, in terms of [37], that it is realized by a genuine *search*; in contrast, for higher types, S1–S9 generally involve a 'search' through domains of higher type which are given without a well ordering; the assumption of a search is here simply unrealistic. (It is true that, in some of his writings on the thinking subject, Brouwer postulates an ω-ordering for proofs and definitions; but even if this were not inconsistent with his much more persuasive assertion that certain proofs of well foundedness, when fully analyzed, consist of a *trans*finite sequence of steps, the ω-ordering postulated is not itself implicit in, say, the notion of constructive function of finite type.)

Evidently our problem cannot be decided by pure mathematics in its modern *abstract* sense, as used for instance in Bourbaki, where the 'nature' of the objects is not considered; more precisely, where objects are supposed to be given together with their analysis, consisting of a list of their properties relevant to a particular circle of problems; these properties are 'put into' (the relations and functions that make up) the abstract structures of pure mathematics. But pure mathematical, formal work can certainly *help* in a conceptual analysis. Specifically, the notions of prime and search computability or S1–S9 are sufficiently plausible to pursue the following project.

> Describe in familiar terms the extensions of these notions applied to, say, the full hierarchy (together with suitable combinators) and compare them. Try to establish the conclusions without the use of axioms which, though valid for the full structure, are problematic for the notion of [14].

> Do not merely attempt to prove equivalences, but allow your attention to be drawn to discrepancies and reflect on them.

After all, it was precisely this kind of experimentation which brought to light the defects, mentioned in 2b(ii), of the structure used in [42] when *uncountable* segments of the ordinals are considered. (I assume here that the notion of *correct* structure, or structure *reflecting the build-up* of the ordinals is not significantly clearer than the notion of constructive function of [14].)

It cannot be denied that this kind of experimentation is not exactly what is immediately conveyed by the phrase, used in (ii), of 'an *a priori* answer'. But here it is to be remembered that similar intellectual processes were employed in the past to give unambiguous answers to questions which must have looked like our question (ii) applied to the notion of [14]; for example questions of establishing correct definitions. (From my own experience with the non-mathematical public, the possibility of a correct definition of *convex* is found to be very impressive.) As to a justification of the choice of the words '*a priori*', is it not true that, after experimentation has drawn our attention to the principles needed for our conclusions, then these principles are found to be convincing as they stand? in fact they are less dubious than any *explicit* analysis we could give for the relation between the case studies and the conclusion. And doesn't the *a priori* character of these *particular* conclusions consist in the fact (of intellectual experience) that our understanding is not helped by reference to *prior* development? in contrast, for instance, to the case of formal notions which have been *consciously* 'abstracted' from cases.

Evidently, to repeat the theme of this whole article, and particularly of the digression in (ii), conceptual analysis is certainly *not* the only use of g.r.t., and the analysis of a particular concept such as the one considered here in (iii) cannot be expected to have importance similar to, say, Turing's analysis.

(iv) Let me conclude with a comment on the literature. Strong, on p. 475 of [55], wants to use his axiomatic theory for an analysis of the structure of computations. This project seems quite implausible. How can one even begin to formulate anything about the *structure* of computations in the language considered! One really cannot expect to have it both ways: to have an *intrinsic* formulation of recursion theory, also emphasized by Wagner and Strong and already discussed in 4b(i), that is independent of algorithmic details, *and* to use it for analyzing the structure of computations. (In contrast, the usual numbering of partial recursive functions via a numbering of defining equations does have something to do with computations, at least if one means that particular *deterministic* process of computation which is involved in Kleene's T-predicate [20].) I say 'implausible' and not 'impossible' because, as we all know, for instance from set theory, a limited vocabulary can have quite unexpected expressive power, when used in a sufficiently sophisticated way.

Of course we know already genuine mathematical use of this axiomatization by 4b, and we shall consider uses *connected* with computations in (ii) of the next, and final, subsection.

(d) *Incidental uses of axiomatic recursion theory*; specifically for pedagogic purposes ~~described in 1d~~

While 4a – 4c describe purposes for which various g.r.t. were *actually* developed, here I only wish to draw attention to some by-products of this work. The first example is typical of the following kind of exercise: We have some fairly elaborate mathematical construction, for instance of a model of certain axioms $A \wedge B$, and inspection shows that the 'hard' part is to satisfy A. Well, it sometimes pays to be more explicit and to verify formally that B can be satisfied 'trivially' (on the basis of existing knowledge), that is by writing down a model of B using only familiar standard material. The second example is typical of another kind of exercise, not wholly unlike 4b: Inspection of some branch of mathematics gives the impression that certain theorems, say T_1 and T_2 'belong' together and others, say T_3 and T_4 do not, and it turns out that the impression can be analyzed in terms of derivability and non-derivability from suitable axioms; thus in the natural formalizations of the proofs of T_1 and T_2 the same axioms, in those of the proofs of T_3 and T_4 two logically independent sets of axioms are used. The examples remain exercises as long as the intrinsic significance of the various axioms has not been established.

(i) Self application. This topic gets its interest from the paradoxes which have been variously regarded as mere oversights (e.g. by Russell, according to his autobiography, after discovering the most famous one) and as a kind of supreme challenge. Later on, and purely to avoid misunderstanding, I shall make an autobiographical remark too. But first, I wish to use g.r.t. and the axiomatic theory of Wagner and Strong to answer the following question.

> Can we set up some axiomatic theory which permits self application (on an obvious reading), is consistent, but has considerable proof theoretic strength?

One answer is quite clear. We consider the language of the axiomatic theory BRFT (as presented e.g. by Friedman [11]), interpret the variables to range over (numbers of) definitions of partial recursive functions and interpret $a \cdot b = c$ to mean

> the function (defined by) *a applied* to *b* has a value and this value is *c*.

For example, if *a* is the number of a total function, we have $\exists c(a \cdot a = c)$ and thus self application in a natural sense. To get an extension in the language of BRFT which has proof-theoretic strength equivalent to some set theory Σ, we use the following facts from Section 2.

First of all, the axioms of BRFT itself are satisfied if we take any model M of Σ, use its universe also as the range of the variables of BRFT, and interpret $a \cdot b = c$ by means of the g.r.t. on collections of sets, e.g. in 2b. For relevant technical details when the universe M is or is not well ordered by a generalized recursive relation, see e.g. [11].

Secondly, the $\in$-relation in M is itself 'recursive' for the g.r.t. in question. We can therefore consistently add to BRFT a constant, say $e_\in$, which will be interpreted as the (definition of the) characteristic function of the 'class' $\in$, and transcribe the axioms of Σ into the language of BRFT together with $e_\in$. (If Σ has only finitely many axioms, $e_\in$ can be eliminated by existential quantification.) The new system, say BRFT(Σ), is consistent because it is so set up that a g.r.t. on M is a model for it. Conversely, Σ can be interpreted in BRFT(Σ) and hence is, proof theoretically, of equal strength.

As fas as our *present* question is concerned, there are two standard ways of treating *extensionality*. One is standard by using known work on the relative proof theoretic strength of (the usual) Σ with and without extensionality. Another is to appeal to an *enumeration without repetition*; but I have not stopped to verify the obviously essential point whether the axioms of BRFT are in fact satisfied for such an enumeration without repetition; (cf. also [11]; however, 'finite' r.e. sets must not be neglected because, in the reduction above, the sets of M are mapped into generalized finite partial functions).

It is perhaps not superfluous to remark that, from a similarly crude point of view, we can easily set up a *proof theoretically equivalent axiomatic theory of* **total** *functions*. The intended model is quite clear:

> We simply transcribe all assertions about partial functions in terms of the primitive recursive functions which enumerate their graphs (more precisely, the canonical primitive recursive definitions corresponding to the definitions of the partial functions).

To add just a little substance to this recipe for constructing models, we may recall Kleene's recursion theorem. This is a remarkably strong *function existence principle* which includes a version of the principle of λ-conversion but goes far beyond it. We consider any term t (in the language of the axiomatic theory) containing possibly the free variable e; the recursion theorem states the existence of e_t such that

$$e_t \cdot n = e_t \cdot n \rightarrow e_t \cdot n = t\,[e/e_t]$$

with free variable n, where we use the convention that $t\,[e/e_t]$ denotes the result of substituting e_t for e in t. The reason for our particular version, that is for adding the hypothesis $e_t \cdot n = e_t \cdot n$ in the sense of $\exists c\,[e_t \cdot n = c]$, is that $=$ is interpreted as above and not like Kleene's $\simeq$ [20]. And the reason for preferring $=$ over $\simeq$ is that it is realized by an r.e. relation in the intended model and $\simeq$ is not.

There is an obvious transcription of Kleene's recursion theorem into the language of *total* functions; 'obvious' provided one looks for it. We consider here Kleene's original formulation in [19] in connection with r.e. sets in place of partial functions. For any description $A(e, n, n_1, ..., n_k)$ of a primitive recursive (p.r.) relation he obtains, p.r. in A, a number e_A such that, in his notation

$$\exists m T(e_A, n, m) \leftrightarrow \exists n_1 \ldots \exists n_k A(e_A, n, n_1, ..., n_k)\,. \qquad (*)$$

Now, *if* one is interested in a quantifier-free p.r. formulation, one looks at the proof of $(*)$ and finds p.r. functions $f_1, ..., f_k$ of m, g of $n, n_1, ..., n_k$ such that

$$T(e_A, n, m) \rightarrow A\,[e_A, n, f_1(m), ..., f_k(m)]$$

and

$$A(e_A, n, n_1, ..., n_k) \rightarrow T\,[e_A, n, g(n, n_1, ..., n_k)]$$

where f_i needs only one argument (and not also the additional argument n) since

$$[T(e_A, n, m) \wedge T(e_A, n', m)] \rightarrow n = n'\,.$$

As it happens we can make a *modest use* of the material above. In [60], Scott gives mathematically very attractive lattice theoretic models which satisfy all the axioms of the λ-calculus. I am not competent to analyze explicitly any notion which the founders of the λ-calculus were trying to formulate and, *a fortiori*, I am not competent to analyze the interest of finding a model for the particular calculus at which they arrived. But, by comparing Scott's model with the models indicated above we can certainly point to a *distinguishing feature which makes his model more mathematical*. The mindless models above pick out a class of functions by *definability conditions*, which, indeed, is the essence of any application of g.r.t. In contrast, having once decided to work with continuous functions on his domain D_∞, Scott is able to show that *all* continuous mappings: $D_\infty \to D_\infty$ are canonically represented by elements of D_∞, not only some subclass distinguished by definability properties. (Amusingly this feature which, at least to me, is most striking, is not at all required in traditional presentations of the λ-calculus.) It does not seem farfetched to suppose that definability requirements are alien to what strikes us as mathematical [7], where, in accordance with 1a, *finite generation* is not thought of as a definability condition.

Autobiographical remark on the (functional) paradoxes. Inevitably, personal taste and judgement affect one's view of the paradoxes and the 'dramatic' uses made of them: drama easily becomes pathetic unless it is both genuine and restrained, and sound practical sense is needed if drama is to be introduced for effect. Specifically, the pedagogic value of dramatic paradoxes is dubious if they only lead us to ask: well, if we can't trust our intuitive impressions at all, what can we trust? Speaking for myself, I simply do not find the paradoxes dramatic: halfway through the argument, that is well before any hint of a paradox appears, my attention begins to wander as in free association. One product of such free association is this.

In the paradoxes there appears a phrase, having the grammatical form of a sentence or of a term, *which cannot, by any stretch of the imagination, be said to have a well determined value*; truth value in the case of sentences, individual object in the case of terms; we shall have an example of the latter in a moment. The paradoxes then establish the by no means obvious fact that this phrase *cannot be assigned any value* at all, consistent with certain tacitly understood general laws.

[7] It is very easy to find coherent 'explanations'; practically, definability conditions would force us to verify constantly exactly how the predicates we are talking about are defined; logically, no axiom *schema* could ever be immediately evident if its validity depended on the restriction to substitution instances defined by given means; for instance the axiom schema of induction in arithmetic is not evident *because* we confine ourselves to first order formulae built up from + and ×! My own impression, for what it is worth, is that these 'explanations' are rationalizations rather than true reasons.

The familiar comparison of this situation with the case of division by zero (preserving the laws: $x \cdot x^{-1} = 1$ and $0 \cdot y = 0$) seems apt enough, but does not go far enough. We have the truly amazing fact that many operations originally defined for a limited domain, say the natural numbers, can be extended, for example to include 0; moreover, our 'natural' choice of the extension usually preserves many familiar laws [8]. For example, if $m \cdot n$ and m^n are originally defined by 'add m to itself n times', resp. by 'multiply m by itself n times', these definitions simply do not make sense when $n = 0$. But persistance or permanence of very *few* laws requires

$$m \cdot 0 = 0 \quad \text{and} \quad m^0 = 1$$

and these laws in turn imply the permanence of very many laws [9]. Equally remarkable is the extension of number theoretic functions such as the factorial to the whole complex plane by means of the Γ-function. Occasionally we have essential singularities or natural boundaries (to analytic continuation).

It is hard to imagine that any 'abstract nonsense' about the definite article and assigning conventional values to λ-terms could replace delicate considerations of the kind described which use the specific mathematical content of the subject matter. (The same applies to some extent even to the now familiar extension of the ∈-relation which puts $a \in b$ to be *false* when, on the old doctrine of types, $a \in b$ would have been meaningless: by confining ourselves to the *specific* sets obtained by iterating the cumulative power set operation only, we can see the implications of making the convention above for atomic formulae $a \in b$ and then extending it to compound formulae in the language of set theory in the usual way [10]. But here it may be well to remember the com-

[8] If the laws in question are quantifier-free, a *consistent* extension automatically preserves, in fact implies the law for the extended domain.

[9] Gödel has drawn a very striking conclusion from $0^0 = 1$, which makes 1 weakly, but not strongly inaccessible; cf. the postscript in [15].

[10] The restriction to this language prevents the temptation of trying to extend the convention *unreasonably*, as can be seen from the 'paradoxes' connected with truth definitions. Let α denote strings of symbols, of the 'object' language, $\neg_F$ formal negation and $\neg$ negation, of the metalanguage. As is well known the requirement on the 'truth predicate' T

$$T(\neg_F \alpha) \leftrightarrow \neg T(\alpha)$$

cannot be satisfied *if* one insists that each side must have a truth value. Clearly the device of giving a conventional value to atomic formulae which have no sense *per se*, cannot be used because (i) both $T(\alpha)$ and $T(\neg_F \alpha)$ are atomic and (ii) for the familiar diagonal formula, say α_D, which produces a paradox, neither $T(\alpha_D)$ nor $T(\neg_F \alpha_D)$ makes, *prime facie*, any sense. (In short, the device which works for the ϵ-relation above, does not work here.)

putational mathematician who worked out patiently an explicit inverse of a matrix A with det $A \neq 0$, and showed it to hold both for left and for right multiplication; perhaps the relevance of abstract group theory may have been difficult to imagine too.

Concluding then these free associations produced by the paradoxes, the latter may fairly be said to present a situation which is familiar enough in our mathematical experience. Naturally this does not deny that, *at a certain stage of development*, problems of extending an operation consistent with given laws may be very real indeed; particularly *if the laws in questions are the only laws available*. (This was the situation with 'unlimited' comprehension or λ-abstraction when they were the *principal* (existential) axioms.)

Here then is the example, promised earlier on, of a standard 'functional' paradox. We consider the following language: variables for monadic functions or 'rules'; = for equality; juxtaposition for function evaluation; constants x_0 and y_0 for two distinct functions; a constant d_0 for the function determined by the rule: for variable f

$$d_0 f = x_0 \quad \text{if } f \neq x_0;$$

$$d_0 f = y_0 \quad \text{if } f = x_0 \qquad \text{(definition by cases).}$$

For the argument below it is not necessary to decide whether extensional or definitional equality is meant.

The paradox comes about as follows (granted the rule d_0 with the property stated above).

Since

$$x_0 \neq y_0,\ d_0 f \neq f \text{ for variable } f. \tag{0}$$

The principal *additional existential assertion* is: there is a rule, call it c_0 whose action is given by

$$f \mapsto d_0(ff) \tag{1}$$

(which involves the *application* of f to *itself*). But, by the basic meaning of juxtaposition, c_0 also has the action

$$f \mapsto c_0 f\,. \tag{2}$$

Putting c_0 itself for f, we have, by (1) and (2)

$$c_0 \mapsto d_0(c_0 c_0), \quad \text{and} \quad c_0 \mapsto c_0 c_0;$$

for consistency we must have $c_0 c_0 = d_0(c_0 c_0)$ which conflicts with (0) if $c_0 c_0$ is put for f in (0).

Another way of putting this is to use the λ-notation; rewriting c_0 as $\lambda f d_0(ff)$. Then (0) conflicts with the *existential principle expressed by λ-conversion* (that is, the existence of operations satisfying this principle); namely suppose t is a term containing the variable f: then we suppose there is an operation, denoted by $\lambda f t$ such that

$$(\lambda f t)g = t[t/g]$$

in other words, $\lambda f t$ has the action

$$f \mapsto t\,.$$

Evidently, there is no paradox here if we think of *partial* functions; $c_0 c_0$ is *undefined*. Of course the mere absence of a paradox provides no guarantee that we know what we are talking about! But the model above, by use (of numbers) of partial recursive functions, does in fact give an interpretation for the *whole* argument (provided $a = b$ is interpreted as $a \simeq b$ or, better still, as $(a = a \wedge b = b) \to a = b$).

More interestingly, at least to me, let us look at the steps of the argumen by the light of nature. When we assert (1) we suppose, tacitly, that *before* applying the rule c_0, a value is to be assigned to ff. So in particular, a value is to be assigned to $c_0 c_0$. Certainly *neither* (1) *nor* (2) *give even the remotest hint of how* $c_0 c_0$ *is to be determined*: and (1) and (2) are *all* that we have said about c_0! It is at this point that, as mentioned above, one's attention begins to wander: why should c_0 itself be defined at c_0, merely because it is defined at each f which is defined at f?

(ii) As our second logical exercise, we consider the *axiomatic analysis of our knowledge* about mechanically computable number theoretic functions; axiomatic analysis in the tradition of geometry; more concerned with relationships between propositions than basic questions of *applicability* treated in 4c. We can be quite specific here.

The theory of mechanically computable partial functions corresponds to 3-dimensional geometry; and the partial recursive functions perhaps with a specific numbering correspond to E_3, again perhaps with a specific co-ordinate system. (Also the relational 'type' of the Wagner-Strong language corresponds to the relational type of some standard language of Euclidean geometry.) In contrast to 4c, we do not analyze the reasons for identifying mechanically computable and partial recursive functions, nor even the exact meaning of this assertion. This corresponds to the tradition of axiomatic analysis in geometry, in contrast to studies of the *Raumproblem* in the sense of Riemann or Helmholtz, on the hypotheses and facts behind (the application of) geometry.

There is an example at the very beginning of (an old-fashioned education in) geometry which gives the flavour of the kind of exercise I have in mind. (*Exempla trahunt* for those with an old-fashioned education.) We can *state* Pythagoras' theorem in the elementary language of geometry. We learn two proofs early on, a very simple one using similar triangles, a more sophisticated one, preferred by Euclid. *What is the difference*? Now the first uses the congruence of a triangle and its mirror image, the second does not. (The difference applies of course to the *proofs*; the theorem does not need the stronger assumption, as Euclid's proof shows!) *What is the interest of the distinction*? Well, the obvious interest cannot be stated within the language of *elementary* geometry, but concerns *motions* of the triangles, and the full interest cannot be stated within *planar* geometry because there we should have no 'particular' reason for supposing that a triangle and its mirror image are congruentatall.

Let me stress at once that both the difference and its interest are quite intelligible without referring to objects other than E_3. (We do not need the current view of axiomatic analysis in terms of classes of structures for which the axioms are valid [11]). The natural way here is simply to think of the knowledge about E_3 which we use. Also, of course, the analysis, in particular the elimination of an unnecessary hypothesis by Euclid's version, has nothing to do with *increasing* our *conviction*, cf. 4c(ii); our interest does not depend on doubting the existence of a third dimension.

Now let me look at the theory of mechanically computable functions in

[11]) Perhaps this is the main reason why our logical exercise may have the pedagogic value considered in 1d; as a corrective to the model theoretic analysis of axioms (by their class of models). The latter is not directly concerned with the *discovery* of axioms though the analysis may shows us where we go wrong, for instance by means of a contradiction; cf. 4a(i). One possible help, particularly for the discovery of existential axioms, is practice in *seeing* the general in the particular; to become sensitive to delicate aspects of familiar material; cf. the delicate problems of 2b(ii) and 4c(iii). (This interest in exercises conflicts with the robust principle of 'modern' mathematics: Don't waste time on mere illustrations, but attack important problems directly. There is no evidence that this principle is useful for the kind of phenomenological analysis considered here where 19th century mathematics was more successful.)

the light of the example above; specifically at propositions formulated in the language of Wagner-Strong. *In what terms can we expect to analyze our knowledge, that is existing proofs*? or, more elaborately, how should we *choose* the additional axioms by reference to which this analysis is made? (logical dependence on or independence from the axioms). According to our example, we do not expect to explain the choice of axioms also in the language of Wagner-Strong. (To adapt a *bon mot* from the introduction of Dirac's book, you'd hardly speak of 'analysis' if it is explained in the same terms as the ideas to be analyzed.)

As far as I can see there is no one outstanding problem here nor any particular principle for the choice of axioms which recommends itself above any others; in short, as mentioned in footnote 11, the situation is *delicate*; at the *present* time the use of g.r.t. here considered, provides a luxury rather than a necessity (to be avoided by those who do not have the taste, in both senses of the word, required for enjoying luxuries or even feel guilty about them). Let me consider two extremes.

One extreme (for a possible choice of axioms) is nothing but a *suggestion*; it goes back to what was called 'superthesis' in 4c(i). Are there theorems, in the language of BRFT applied to ordinary recursion theory, whose *proof* uses special properties of the enumeration of partial recursive functions? Put differently, if the whole proof were to be valid for mechanically computable functions we should be using not only Church's thesis, but the superthesis. Correspondingly, in formal terms, such a proof would be expected to use axioms radically different from those of BRFT. (Evidently this possibility is not excluded by Friedman's 'relative' isomorphism theorems in [11].)

As an example of the other extreme we have a perfectly standard distinction, which has nothing to do with Turing's analysis, but with whether we operate on *r.e. sets* or on: r.e. sets given *together* with an enumeration (cf. the use of total functions in (i) above). For short we shall speak of *extensionality* in the former case; trivially, whenever we have made a discovery about an intensional object, *post factum* we can generalize the discovery in extensional form, by considering not the 'extension' of the object itself, but the object together with the additional information we have used about it.

The example below was emphasized in Lacombe's thesis [31], but it was not (intended to be) used for the present purpose.

Let F_R ('R' for Rosser) be the familiar operation which associates to each pair (e_1, e_2) of numbers of r.e. sets X_1, X_2 a pair (e_1', e_2') of X_1', X_2' such that

$$X_1 \cup X_2 = X_1' \cup X_2',\ X_1' \subset X_1,\ X_2' \subset X_2,\ X_1' \cap X_2' = \emptyset .$$

Note that the condition *on* F_R, that is the relation between (X_1, X_2) and (X'_1, X'_2), is extensional. Also the *condition* can of course be expressed in the language of Wagner-Strong.

Now, as pointed out by Lacombe very clearly, there is something 'peculiar' about F_R. There is no extensional *recursive* operation, that is 'effective' in the sense of Myhill-Shepherdson [40], satisfying the (extensional) conditions on F_R. For any effective operation is continuous (in X_1, X_2 with product topology on 2^ω) and quite evidently no continuous operation, even if not recursive, will do.

If we look at the obvious 'Rosser' definition of F_R (when $n \in X_1 \cap X_2$, put n into X'_1 if it isn't put into X_2 'before' it is put into X_1) we really appeal to the *order* of enumeration.

As already mentioned the example has nothing to do with 'refinements' of Church's Thesis. If we consider mechanically enumerable sets given *together* with a mechanical enumeration the 'Rosser' definition makes perfectly good sense for these notions and yields the required mapping; indeed, we do not at all use the *numbers* (or definitions) e_1, e_2 of our enumerations, but the extensional enumeration functions.

It seems pretty evident that the existence of F_R does not follows from the Wagner-Strong axioms which tell us nothing about an order of the universe. But the example belongs to 'intrinsic' recursion theory because we do not use 'algorithmic' details of the computation *process*, of *how* we step from the argument of a function to its value. Nevertheless the example may be sufficient to *illustrate* the point about overemphasis on the intrinsic character of recursion theory made in 4b(i). It is fair to say that the kind of reasons which persuade one of the need for an intrinsic treatment are similar to those that recommend a purely 'extensional' treatment of r.e.-sets.

Remark. Apart from the sources cited in the bibliography below I have profited very substantially from detailed comments by my colleagues S. Feferman and D. Prawitz. W.A. Howard who (in unpublished work) has greatly extended Curry's ideas mentioned in 4c(i) connecting formal deductions and terms, explained this material to me; only the, admittedly speculative, suggestion *loc. cit.* to use the convertibility relation between terms for an analysis of *identity between proofs* is my own responsibility.

Historical note. Recursion theory on sets and, in particular, the theory of admissible sets refine a tradition in set theory which was not taken into account at all in [25]. Specifically, one important aim of von Neumann's exposition [65], was to use the replacement schema instead of the power set axiom for the development of basic set theory. This aim may well have been connected with (to me) quite unconvincing polemics against the power set operation. Be that as it may, the theory used has many interesting and manageable models which do not satisfy the power set axiom. In particular, let σ be a regular cardinal and $\mathfrak{P}^\sigma$ the operation mapping the set x into the collection of those subsets of x which have cardinal $< \sigma$. Then the 'thin' hierarchy

$$C_0^\sigma = E,\ C_\alpha^\sigma = \bigcup_{\beta<\alpha} C_\beta^\sigma \cup \mathfrak{P}^\sigma(C_\beta^\sigma) \text{ for } \alpha \neq 0$$

where E is a collection of individuals, satisfies the axioms considered for all $\alpha > \sigma$ (or equivalently, for $\alpha = \sigma$ since $C_\alpha^\sigma = C_\sigma^\sigma$ for $\alpha > \sigma$); in other words, the collection of sets of hereditary cardinality $< \sigma$, built up from E satisfies the axioms, even if the second-order version of the replacement principle is meant. But Zermelo's own axioms are satisfied by these C_σ^σ only for relatively large σ, e.g. for the ωth beth number where $\beth_0 = \aleph_0$ and $\beth_\alpha = \bigcup_{\beta<\alpha} 2^{\beth_\beta}$. Of course; the replacement axiom is not satisfield for $\sigma = \beth_\omega$; for strongly inaccessible σ, Zermelo's own axioms *and* the replacement axiom, that is the usual axioms of Zermelo-Fraenkel, are satisfield (even if taken in their second-order form).

Work on admissible sets constitutes a *refinement* because it concerns consequences of only special (socalled Σ_1) instances of the replacement scheme. Here it is to be mentioned that precisely these instances were involved in early work, especially by Fraenkel (who used the notation of functions rather than sets, particularly for definition by transfinite recursion for which the Σ_1 instances suffice). The current replacement schema, usually attributed to Fraenkel, was put down by Skolem [67].

Evidently, the significance of this refinement could hardly have been stated convincingly before the theory of recursive ordinals was developed, since the only really familiar description (even today) of an admissible set which does not satisfy the usual replacement schema is just the collection of the hereditarily hyperarithmetic sets. Since it turns out that this object is well-suited to the study of infinitary languages, we have here an *application of classical recursion theory* too.

References

[1] J.W. Addison, Current problems in descriptive set theory, Proc. Symp. Pure Math. Am. Math. Soc. 13 (1970).

[2] J. Barwise, Infinitary logic and admissible sets, J. Symbolic Logic 34 (1969) 226–252.
[3] J. Barwise, Applications of strict Π_1^1 predicates to infinitary logic, J. Symbolic Logic 34 (1969) 409–423.
[4] J. Barwise, R.O. Gandy and Y.N. Moschovakis, The next admissible set, J. Symbolic Logic, to appear.
[5] G.C. Driscoll Jr., Metarecursively enumerable sets and their meta degrees, J. Symbolic Logic 33 (1968) 389–411.
[6] S. Feferman, Persistent and invariant formulas for outer extensions, Compositio Math. 20 (1968) 29–52.
[7] S. Feferman, Autonomous transfinite progressions and the extent of predicative mathematics, in: Logic, methodology and philosophy of science III (1968) 121–136.
[8] S. Feferman, Uniform inductive definitions and generalized recursion theory, ASL meeting, Cleveland, Ohio, April 30, 1969.
[9] R. Fraissé, Une notion de récursivité relative, in: Infinitistic methods (1961) pp. 323–328.
[10] R.M. Friedberg, Three theorems on recursive enumeration, J. Symbolic Logic 23 (1958) 309–316.
[11] H. Friedman, Axiomatic recursive function theory, this volume, p.
[12] H. Friedman and R. Jensen, Note on admissible ordinals, Lecture Notes in Mathematics 72 (1968) 77–79.
[13] R.O. Gandy and G.E. Sacks, A minimal hyperdegree, Fundamenta Math. (1967) 215–223.
[14] K. Gödel, Über eine bisher noch nicht benutzte Erweiterung des finiten Standpunktes, Dialectica 12 (1958) 280–287.
[15] K. Gödel, What is Cantor's continuum problem, in: Philosophy of mathematics, eds. P. Benacerraf and H. Putnam (1964) 258–273.
[16] J. Gregory, Dissertation, University of Maryland (1969).
[17] G. Higman, Subgroups of finitely presented groups, Proc. Roy. Soc. A262 (1961) 455–474.
[18] C. Karp, this volume, p.
[19] S.C. Kleene, On the form of the predicates in the theory of constructive ordinals, Am. J. Math. 66 (1944) 41–58.
[20] S.C. Kleene, Introduction to metamathematics, Princeton (1952).
[21] S.C. Kleene, Recursive functionals and quantifiers of finite types, Trans. Am. Math. Soc. 91 (1959) 1–52; 108 (1963) 106–142, reviewed in Zentralblatt 88, p. 13 and 121, p. 13.
[22] S.C. Kleene, Herbrand-Gödel-Style recursive functionals of finite type, in: Proc. Symp. Pure Math. Am. Math. Soc. 5 (1962) 49–75.
[23] G. Kreisel, Model theoretic invariants; applications to recursive and hyperarithmetic operations, in: The theory of models (1965) pp. 190–205.
[24] G. Kreisel, Relative recursiveness in metarecursion theory, J. Symbolic Logic 33 (1967) 442.
[25] G. Kreisel, Choice of infinitary languages by means of definability criteria; generalized recursion theory, Lecture Notes in Mathematics 72 (1968) 139–151.
[26] G. Kreisel and G.E. Sacks, Metarecursive sets, J. Symbolic Logic 28 (1963) 304, 305; 30 (1965) 318–338.

[27] G. Kreisel and W.W. Tait, Finite definability of number theoretic functions and parametric completeness of equation calculi, Z. math. Logik und Grundlagen 7 (1961) 28–38.
[28] S. Kripke, Transfinite recursion on admissible ordinals I, II, J. Symbolic Logic 29 (1964) 161, 162.
[29] K. Kunen, Implicit definability and infinitary languages, J. Symbolic Logic 33 (1968) 446–451.
[30] A.H. Lachlan, The elementary theory of recursively enumerable sets, Duke Math. J. 35 (1968) 123–146.
[31] D. Lacombe, Dissertation, University of Paris (1965).
[32] D. Lacome, Deux généralisations de la notion de récursivité, Compt. Rend. Sci. Paris 258 (1964) 3141–3143 and 3410–3413.
[33] E.G.K. Lopez-Escobar, Remarks on an infinitary language with constructive formulas, J. Symbolic Logic 32 (1967) 305–318.
[34] J.M. MacIntyre, Dissertation, M.I.T. (1968).
[35] A.I. Malcev, Constructive algebra I, Usp. Math. Nauk. 166 (1961) 3–60.
[36] R.M. Montague, Recursion theory as a branch of model theory, in: Logic, methodology and philosophy of science III (1968) 63–86.
[37] Y.N. Moschovakis, Abstract first order computability, Trans. Am. Math. Soc. 30 (1969) 427–504.
[38] Y.N. Moschovakis, this volume, p.
[39] A. Mostowski, Representability of sets in formal systems, Proc. Symp. Pure Math. Am. Math. Soc. 5 (1962) 29–48.
[40] J.R. Myhill and J.C. Shepherdson, Effective operations on partial recursive functions, Z. math. Logik und Grundlagen 1 (1955) 310–317.
[41] J.C. Owings Jr., Π_1^1-sets, ω-sets and metacompleteness, J. Symbolic Logic 34 (1969) 194–204; reviewed MR 39 (1970) 970–971.
[42] R. Platek, Dissertation, Stanford University (1966).
[43] M.B. Pour-El, Independent axiomatization and its relation to the hypersimple set, Z. math. Logik und Grundlagen 14 (1968) 449–456.
[44] Julia Robinson, Recursive functions of one variable, Proc. Am. Math. Soc. 19 (1968) 815–820.
[45] H. Rogers Jr., Gödel numberings of partial recursive functions, J. Symbolic Logic 23 (1958) 331–341.
[46] G.E. Sacks, Degrees of unsolvability, Ann. Math. Studies, Princeton (1963).
[47] G.E. Sacks, The recursively enumerable sets are dense, Ann. Math. 80 (1964) 300–312.
[48] G.E. Sacks, Post's problem, admissible ordinals and regurality, Trans. Am. Math. Soc. 124 (1966) 1–23.
[49] G.E. Sacks, Metarecursion theory, sets, models and recursion theory (1967) pp. 243–263.
[50] J.R. Shoenfield, Undecidable and creative theories, Fund. Math. 49 (1961) 171–179.
[51] C.L. Siegel, Indefinite quadratische Formen und Funktionentheorie, Math. Ann. 124 (1952) 17–54 and 364–387.
[52] E.P. Specker, this volume, p.
[53] C. Spector, On degrees of recursive unsolvability, Ann. Math. 64 (1956) 581–592.

[54] C. Spector, Recursive well-ordering, J. Symbolic Logic 20 (1955) 151–163.
[55] H.R. Strong, Algebraically generalized recursive function theory, IBM J. Research and Development 12 (1968) 465–475.
[56] W.W. Tait, Infinitely long terms of transfinite type, in: Formal systems and recursive functions (1965) pp. 176–185.
[57] E.G. Wagner, Uniformly reflexive structures: On the nature of gödelizations and relative computability, Trans. A.M.S. 144 (1969) 1–42.
[58] A. Weil, Sur la formule de Siegel dans la théorie des groupes classiques, Acta Math. 113 (1967) 1–87.
[59] C.E.M. Yates, Three theorems on the degrees of recursively enumerable sets, Duke Math. J. 32 (1965) 461.
[60] D. Scott, Lattice-theoretic models for the λ-calcules (to appear).
[61] H.B. Curry and R. Feys, Combinatory logic (Amsterdam, 1958).
[62] W. Ackermann, Begründung des "Tertium non datur", Math. Ann. 93 (1924) 1–36.
[63] C. Jockusch, Ramsey's theorem and recursion theory, Notices A.M.S. 17 (1970) 672, 673.
[64] G. Kreisel, Mathematical logic, Lectures on modern mathematics, Vol. III, ed. Saaty (1965) 95–195.
[65] J. von Neumann, Eine Axiomatisierung des Mengenlehre, Journ. f. Math. 154 (1925) 219–240.
[66] ———, Zur Hilbert'schen Beweistheorie, Math. Zeitschrift 26 (1927) 1–46.
[67] T. Skolem, Einige Bemerkungen zur axiomatischen Begrundung der Mengenlehre, Ve Congr. des math scand. 1922 (1923) 217–232.

AXIOMS FOR COMPUTATION THEORIES – FIRST DRAFT

Yiannis N. MOSCHOVAKIS [1)]
University of California, Los Angeles, Calif., USA

There have been many recent attempts to generalize recursion theory to domains other than the integers. The most extensively studied concepts are Kleene's *recursive functionals of finite type* and *metarecursion on an admissible ordinal*, usually associated with the names Kreisel-Sacks, Kripke and Platek. A list of several other such concepts was given in the beginning paragraph of Moschovakis (1969a), and since then still more definitions have been given, e.g. Kunen (1968).

One of the difficulties in trying to compare and classify these theories has been the lack of a definition of "recursion theory". When can one claim to have "given" a "recursion theory"? Is it enough to specify a class of functions on a set satisfying certain closure properties or must one give specific definitions for these functions and study the properties of these definitions? Perhaps one should go even further and specify "machines" and the way in which these machines "compute" functions, before it is granted that a recursion theory has been given.

In this paper we propose a definition of what we shall call *computation theories* and we study these structures enough to get at least the first step in classifying the known theories.

The basic relation in recursion theory, whether on the integers or some abstract set A is usually denoted by

$$(0\text{–}1) \quad \{e\}(x_1, ..., x_n) \simeq y \, ,$$

which reads

[1)] Most of the work for this paper was done during the academic year 1968–1969, when the author was a Guggenheim Fellow, on sabbatical from UCLA. The author received additional financial support during the summers of 1968, 1969 from a National Science Foundation Grant.

machine e with input $x_1, \ldots, x_n$ *has output* y.

Sometimes we have *definitions* or *instructions* rather than machines, but the effect is the same: partial (or even multiple-valued) functions on A to A are defined via objects which can be given *names* in A, so that relation (0–1) is itself defined on A – and in most cases is a *semicomputable (recursively enumerable)* relation. We tried hard to find axioms which will separate those relations of the form (0–1) which can truly claim to generate a recursion theory. The best we could do is what we now call *precomputation theories* in Section 1. They are practically identical with the *uniformly reflexive structures* of Wagner (1969), which in fact are presented much more elegantly and in the algebraic spirit that this approach demands. But we became convinced that the most interesting aspects of recursion theory cannot be axiomatized in this framework. We are particularly thinking here of the *priority* constructions in ordinary recursion theory and the *boundedness* arguments in hyperarithmetic theory, which seem to utilize extensively *computations* of some type.

Computations are very elusive objects and in some theories they are not even made explicit. But they always have *length* – and if we decide to leave analog computations out of the picture, this length is an ordinal, finite or infinite. It turns out that a very decent axiomatization of recursion theory can be given if we take as primitive concepts a relation of the form (0–1) together with an assignment of an ordinal $|e, x_1, \ldots, x_n, y|$ to each $e, x_1, \ldots, x_n, y$ such that $\{e\}(x_1, \ldots, x_n) \simeq y$. Our axioms are fairly strong, but we shall attempt to justify them as natural. They are given in Section 2 in terms of the key concept of a *computable functional*.

In Sections 3–7 we give a brief outline of the basic properties of computation theories. This includes proving some of the basic results of ordinary recursion theory in this context, e.g. the *first recursion theorem* in Section 5, as well as identifying some of the known theories as computation theories, e.g. recursive functionals of finite types in Section 7 and the *prime* and *search computability* of Moschovakis (1969a, b) in Sections 3 and 6.

The important concept of *finiteness relative to a theory* is introduced and studied briefly in Section 8. Our approach here differs from the "recursive and bounded" definition of metarecursion theory, since we do not start with a given wellordering on the basic domain. A set then is finite relative to a theory if the functional representing quantification over that set is computable in the theory. This is surely a natural approach and allows for direct generalization of the fundamental properties of finite sets.

A theory is *normal* (Section 9) if it satisfies certain regularity conditions (e.g. the relation of equality is computable) and if "computations are finite" in the following sense: whenever $\{e\}(x_1, ..., x_n) \simeq y$,

$$\{(e', x'_1, ..., x'_n, y') : \{e'\}(x'_1, ..., x'_n) \simeq y' \;\&\; |e', x'_1, ..., x'_n, y'| \leqslant |e, x_1, ..., x_n, y|\}$$

is finite relative to the theory, uniformly in $e, x_1, ..., x_n, y$. One of the following *alternatives* must hold for normal theories: either the basic domain is finite (relative to the theory) or it is not. In Sections 10 and 11 we show that in the second case the theory is much like metacursion on an admissible ordinal and in the first case the theory is much (though not quite as much) like hyperarithmetic theory. The main mathematical result of the paper is the axiomatic characterization of "metarecursion theory on an admissible prewellordering" given in Section 10. (If the prewellordering in question is actually a wellordering, this is precisely the Kripke-Platek theory.)

The Manchester meeting came at an awkward stage in the development of the theory presented here: the main results had been proved, but few of the details had been worked out and many important problems had not been attacked at all. One possibility would be to write up a brief note giving the axiomatization and announcing some of the results, in particular the material in Sections 4, 5, 10 and 11. Experimentation in this direction resulted in a note that was none too brief and was incomprehensible to all save the handful of experts in abstract recursion theory. We then decided to simply put down what we know now about computation theories and to include enough discussion of the definitions and enough examples so that the paper will be accessible to the non-expert. We have avoided most proofs and all formal details and technicalities. Many of the results are mentioned in the text without special billing, sometimes italicized for emphasis. If we have succeeded in our aim, the non-specialist should be able to get an idea of what we are trying to do and at the same time meet some of the most interesting examples of abstract recursion theory. On the other hand, a reader familiar with recursion on higher types and metarecursion theory should be able to supply all needed details, especially if he has perused our Moschovakis (1969a, b, c).

We should warn the specialist that some of the most important aspects of abstract recursion theory are left untouched here, not because we consider them uninteresting but simply because we have not yet answered the questions they pose in this context. Chief among these is *relative computability*.

We hope to say something about this in a future publication on computation theories [2].

1. Precomputation theories

In a typical theory of computations, there is a set $\mathcal{M}$ of machines and a domain A of objects. Each machine $M \in \mathcal{M}$ will accept as input any finite sequence σ of elements of A and after "computing" for some time may (or may not) yield as output some $z \in A$. Often the machines are not *deterministic*, i.e. the computation and the output z are not determined uniquely by M and σ. In any case, the basic relation of such a theory is

(1–1) $(M, \sigma, z) \in \Theta \Longleftrightarrow$ *machine* M *with input* σ *has* z *as one of its outputs.*

The mathematics in such a situation usually begin by substituting for (1–1) a relation with all its arguments in A. One *codes* (or *indexes*) the machines using elements of some $C \subseteq A$ as names, and the basic relation becomes

(1–2) $(a, \sigma, z) \in \Theta \Longleftrightarrow a \in C$ *and machine* (*with code*) a *with input* σ *has* z *as one of its outputs* .

In the case of Turing machine theory, with $A = \omega =$ the integers, one takes $C = \omega$, i.e. each element of A codes a Turing machine. In the case of recursive functional theory on $\omega \cup {}^{\omega}\omega$, one again takes $C = \omega$, but now C is a proper subset of A. In more abstract theories we find cases where C is a proper subset of A but contains (an isomorphic copy of) ω as a proper subset. In all cases however, the set of codes C has some rudimentary structure on it – at least a copy of ω and a pairing function with its inverses.

Definition 1. A *computation domain* (or simply *domain*) is a structure

[2] For some time now we have been planning to write up an expository note on abstract computability and in fact we listed such a note "in preparation" in the bibliographies of Moschovakis (1969a, b). We will not publish this note, since much of the material we intended to put there has been included in this paper in a form that makes it accessible to the non-expert.

$$\mathfrak{A} = \langle A, C, N, s, M, K, L \rangle$$

with the following properties.

(1–3) $N \subseteq C \subseteq A$.

(1–4) $\langle N, s \restriction N \rangle$ *is isomorphic to the structure of the (non-negative) integers with the successor operation* .

(1–5) *M is a binary function which is a pair on C, i.e.*

$$a, b \in C \Longleftrightarrow M(a, b) \in C ,$$

$$M(a, b) = M(a', b') \in C \Rightarrow a = a' \;\&\; b = b' .$$

(1–6) *The unary functions K, L map C into C and are inverses to M, i.e.*

$$c = M(a, b) \in C \Rightarrow a = K(c) \;\&\; b = L(c) .$$

If $\mathfrak{A}$ is a domain fixed for the discussion, we customarily denote elements of A by $x, y, z, \ldots$, elements of C by $a, b, c, \ldots$, elements of N (integers) by $i, j, k, \ldots$ (and the numerals 0, 1, 2, ...) and finite sequences (tuples) from A (including the empty tuple) by $\sigma, \tau, \ldots$. The *concatenation* of σ and τ is denoted by σ, τ or (σ, τ) and $lh(\sigma)$ is the number of elements in the sequence σ.

Let Θ be a collection of tuples from A all of which have length $\geqslant 2$, suppose we interpret membership in Θ by (1–2). To conform with standard notation of recursion theory, it is natural to put for each n and each $a \in C$,

$$\{a\}^n_\Theta(\sigma) \to z \Longleftrightarrow lh(\sigma) = n \;\&\; (a, \sigma, z) \in \Theta$$

and to associate with each $a \in C$ and each n the n-ary *partial-multiple-valued* (p.m.v.) function

$$\{a\}^n_\Theta(\sigma) = \{z : \{a\}^n_\Theta(\sigma) \to z\} .$$

(When Θ is fixed and n understood or irrelevant, we shall be kind to the printer and skip the embelishments.)

We must allow for multiple-output machines in order to include the important example of search computability. As in Moschovakis (1969a 1.7) we think of p.m.v. functions as functions whose values are sets (including $\emptyset$) and we put

$$(1-7)\quad f(\sigma)\to z \Longleftrightarrow z\in f(\sigma)$$

$$\Longleftrightarrow z \textit{ is one of the values of } f \textit{ at } \sigma\ ,$$

$$(1-8)\quad f(\sigma)=g(\tau)\Longleftrightarrow (\forall z)\,[f(\sigma)\to z \Longleftrightarrow g(\tau)\to z]\ ,$$

$$(1-9)\quad f(\sigma)=z\Longleftrightarrow f(\sigma)=\textit{singleton } \{z\}$$

$$\Longleftrightarrow f(\sigma)\to z \ \&\ (\forall u)\,[f(\sigma)\to u\Rightarrow u=z]\ ,$$

$$(1-10)\quad f\subseteq g\Longleftrightarrow (\forall\sigma)(\forall z)\,[f(\sigma)\to z\Rightarrow g(\sigma)\Rightarrow z]\ .$$

We identify a single-valued (perhaps partial) function $f(\sigma)$ with the p.m.v. function whose value at each σ is the singleton $\{f(\sigma)\}$, so that convention (1–9) leads to no confusion. For brevity, we shall call totally defined single-valued functions *mappings.*

Definition 2. If $\mathfrak{A}$ is a domain, Θ a set of tuples of length $\geqslant 2$ and first elements in C and f an n-ary p.m.v. function on A, then f is *Θ-precomputable* if for some $\hat{f}\in C$,

$$(1-11)\quad f=\{\hat{f}\}^n_\Theta\ ,$$

i.e. for all σ, z,

$$f(\sigma)\to z\Longleftrightarrow \{\hat{f}\}^n_\Theta(\sigma)\to z\ .$$

Whenever (1–11) holds, we call $\hat{f}$ a *Θ-precode* of f.

In addition to p.m.v. functions on A, we must discuss p.m.v. *functionals* on A. These are functions $\varphi(\mathbf{f},\sigma)=\varphi(f_1,\ldots,f_l,x_1,\ldots,x_n)$ with arguments p.m.v. functions on A (of any number of arguments, say f_i is n_i-ary) and elements of A, and subsets of A as values, including $\emptyset$. (We use the same conventions (1–7)–(1–10) for these p.m.v. functionals as for p.m.v. functions on A.) The functionals we deal with are all *continuous*, i.e.

$$f_1\subseteq g_1\ \&\ \ldots\ \&\ f_l\subseteq g_l\ \&\ \varphi(\mathbf{f},\sigma)\to z\Rightarrow \varphi(\mathbf{g},\sigma)\to z\ .$$

Definition 3. If $\mathfrak{A}$ is a domain, Θ a set of tuples of length $\geqslant 2$ and first elements in C and $\varphi(f_1, ..., f_l, \sigma)$ a continuous p.m.v. functional on A (with f_i ranging over n_i-ary p.m.v. functions), then φ is *Θ-precomputable* if for some $\hat{\varphi} \in C$ and all $e_1, ..., e_l \in C, \sigma, z \in A$,

$$(1\text{–}12) \quad \varphi(\{e_1\}_{\Theta}^{n_1}, ..., \{e_l\}_{\Theta}^{n_l}, \sigma) \to z \Longleftrightarrow \{\hat{\varphi}\}_{\Theta}^{l+n}(e_1, ..., e_l, \sigma) \to z\,.$$

Whenever (1–12) holds we call $\hat{\varphi}$ a *Θ-precode* of φ.

Thus a continuous p.m.v. functional φ is Θ-precomputable if the restriction of φ to Θ-precomputable p.m.v. functions is Θ-precomputable in terms of the Θ-precodes. This generalizes the *effective operations* of ordinary recursion theory. Later on we shall talk about *Θ-computable functionals*, which generalize the concept of *partial recursive functional.*

Roughly speaking, a *precomputation theory* will be a set of tuples Θ such that certain functions and functionals are Θ-precomputable. There are infinitely many functions and functionals involved, but they fall into thirteen schemes. Each scheme asserts that a certain family of functions or functionals is *uniformly* Θ-precomputable, in the sense that a Θ-precode for the x'th object of the sequence can be computed via a Θ-precomputable mapping.

Definition 4. A set Θ of tuples on A of length $\geqslant 2$ and with first elements in C is a *precomputation theory* on $\mathfrak{A}$, if there exist Θ-precomputable mappings $p_1, ..., p_{13}$, so that the functions and functionals in I–XII below are Θ-precomputable with Θ-precodes the elements given on the right and the Weak Iteration Property XIII holds. (Here σ varies over n-tuples and τ over m-tuples on A.)

I. $f(\sigma) = y$ — $p_1(n, y)$ ($y \in C$; *constants*)

II. $f(y, \sigma) = \begin{cases} y & \text{if } y \in C \\ 0 & \text{if } y \notin C \end{cases}$ — $p_2(n)$ (*identity*)

III. $f(y, \sigma) = s(y)$ — $p_3(n)$ (*successor*)

IV. $f(y, \sigma) = \begin{cases} 0 & \text{if } y \in C \\ 1 & \text{if } y \notin C \end{cases}$ — $p_4(n)$ (*characteristic function of C*)

V. $f(y, \sigma) = \begin{cases} 0 \text{ if } y \in N \\ 1 \text{ if } y \notin N \end{cases}$ $\qquad p_5(n)$ *(characteristic function of* N*)*

VI. $f(x, y, \sigma) = M(x, y)$ $\qquad p_6(n)$ *(pair)*

VII. $f(y, \sigma) = K(y)$ $\qquad p_7(n)$ *(first projection)*

VIII. $f(y, \sigma) = L(y)$ $\qquad p_8(n)$ *(second projection)*

IX. $\varphi(f, g, \sigma) = f(g(\sigma), \sigma)$ $\qquad p_9(n)$ *(composition)*

X. $\varphi(f, g, y, \sigma) = \begin{cases} 0 \quad \text{if } y \notin N \\ g(\sigma) \text{ if } y = 0 \\ f(\varphi(f, g, x, \sigma), x, \sigma) \\ \qquad \text{if } y = s(x) \in N \end{cases}$ $\qquad p_{10}(n)$ *(primitive recursion)*

XI. $\varphi(f, x_1, \ldots, x_n) = f(x_{j+1}, x_1, \ldots, x_j, x_{j+2}, \ldots, x_n)$

$p_{11}(n, j)$ $(1 \leqslant j \leqslant n$; *interchange*)

XII. $\varphi(f, \sigma, \tau) = f(\sigma)$ $\qquad p_{12}(n, m)$ *(point evaluation)*

XIII. *Weak Iteration Property. For each* n, m, $p_{13}(n, m)$ *is a* Θ*-precode of a mapping* $S^n_m(a, x_1, \ldots, x_n)$, *such that for all* $a, x_1, \ldots, x_n \in C, y_1, \ldots, y_m \in A$,

$$\{a\}^{n+m}_\Theta(x_1, \ldots, x_n, y_1, \ldots, y_m) = \{S^n_m(a, x_1, \ldots, x_n)\}^m_\Theta(y_1, \ldots, y_m) .$$

There is nothing terribly original about these axioms; we have simply carried to an arbitrary domain the clauses in the definition of partial recursive functional of Kleene (1959), just as we defined prime computability in Moschovakis (1969a). Notice that XII is the abstract version of the *enumeration theorem*, Kleene's S7 – in the abstract case, when C may be uncountable, a more appropriate name is *parametrization property.* A restatement of XII that makes no reference to functionals reads: *for each n, m and all* $a \in C, \sigma$, $\tau \in A$,

$$\{p_{12}(n, m)\}_\Theta^{n+m+1}(a, \sigma, \tau) = \{a\}_\Theta^n(\sigma) .$$

We have splurged in listing the axioms, e.g. I follows trivially from II and XIII. Our aim has not been elegance or economy. Rather we are interested in ease of application and immediate recognition of the richness of the theory. The *uniformly reflexive structures* of Wagner (1969) are very similar to precomputation theories, in fact they only differ in technical detail. Wagner gives for these a very elegant set of only three axioms. In a polished exposition of this material one would presumably utilize some of Wagner's combinatory logic methods and reduce our axioms to a small, manageable set.

Ordinary recursion theory can be identified as a precomputation theory in several ways. First we turn ω into a domain by taking $A = C = N = \omega$, $s =$ the successor function and choosing some primitive recursive pair with primitive recursive inverses. We can then put

(1–13) $(a, \sigma, z) \in \Theta \Leftrightarrow$ *the Turing machine with Gödel number a*

and input σ *has z as output* ,

and easily verify the axioms.

There are several other precomputation theories on ω that arise naturally, e.g. one can define Θ so that the Θ-precomputable p.m.v. functions are the Σ_n^0, or the Σ_n^1, or the Π_n^1 ($n \geqslant 1$) relations on ω. A more interesting example is the theory of *continuous partial functions* defined in Moschovakis (1970), which had been part of the folklore of the subject long before that paper. The domain of this theory is $\omega \cup {}^\omega\omega$ with $C = \omega \cup {}^\omega\omega$ and in order to fit the concept given here we must alter a bit the definition in Moschovakis (1970) to allow for integer codes in addition to real codes and choose some continuous M, K, L.

The most important result about precomputation theories is the

Second recursion theorem. Let Θ *be a precomputation theory on* $\mathfrak{A}$. *For each* $n+1$*-ary* Θ*-precomputable p.m.v. function* $f(x, \sigma)$ *there is some* $a \in C$ *so that for all* σ,

$$f(a, \sigma) = \{a\}_\Theta^n(\sigma) .$$

This can be proved by the two line argument given for the corresponding result in Kleene (1959), but it is very useful in applications. The key to the results in Moschovakis (1970) was precisely the fact that the continuous partial functions form a precomputation theory and satisfy this theorem.

2. Computation theories

We tried to find natural conditions on a precomputation theory with interesting, non-trivial consequences, but we failed. Now we are convinced that in order to axiomatize the more advanced parts of recursion theory we must enlarge our stock of primitive concepts – we must in fact say something about *computations*. Since it is not clear what computations are, we have chosen to refer to them by talking about their *lengths*, which we take to be ordinals. Our concept of *computation theory* comes then from that of precomputation theory by adding the new primitive *length* and strengthening the axioms accordingly.

Definition 5. A *computation set on a domain* $\mathfrak{A}$ is a set Θ of tuples on A of length $\geqslant 2$ and first elements in C, together with a function $|\ \ |_\Theta$ mapping Θ into the ordinals. If $(a, \sigma, z) \in \Theta$, we call $|a, \sigma, z|_\Theta$ *the length* (of the *computation* or *assertion*) $(a, \sigma, z) \in \Theta$.

We shall often say "the computation set Θ", where pedantry would call for "the computation set $\langle \Theta, |\ \ |_\Theta \rangle$".

As fas as computation of functions goes, we forget about lengths and copy Definition 2.

Definition 6. If $\mathfrak{A}$ is a domain, Θ a computation set on $\mathfrak{A}$ and f an n-ary p.m.v. function on A, then f is Θ-*computable* (with Θ-*code* $\hat{f}$) if and only if f is Θ-precomputable (with Θ-*precode* $\hat{f}$).

For functionals we strengthen the concept of Θ-precomputation by bringing in lengths and insisting that "computations" be somewhow "natural".

Definition 7. If $\mathfrak{A}$ is a domain, Θ a computation set on $\mathfrak{A}$ and $\varphi(f_1, ..., f_l, x_1, ..., x_n)$ a continuous p.m.v. functional on A (with f_i ranging over n_i-ary p.m.v. functions on A), then φ is Θ-*computable* if for some $\hat{\varphi} \in C$ and all $e_1, ..., e_l \in C, x_1, ..., x_n, z \in A$ the following two conditions hold.

(a) $\varphi(\{e_1\}_\Theta^{n_1}, ..., \{e_l\}_\Theta^{n_l}, x_1, ..., x_n) \to z \iff \{\hat{\varphi}\}_\Theta^{l+n}(e_1, ..., e_l, x_1, ..., x_n) \to z.$

(b) If $\varphi(\{e_1\}_\Theta^{n_1}, ..., \{e_l\}_\Theta^{n_l}, x_1, ..., x_n) \to z$,

then there exist p.m.v. functions $g_1, ..., g_l$ such that

$$(2\text{–}1) \quad g_1 \subseteq \{e_1\}_\Theta^{n_1}, ..., g_l \subseteq \{e_l\}_\Theta^{n_l}, \varphi(g_1, ..., g_l, x_1, ..., x_n) \to z ,$$

$$(2\text{–}1) \quad |\hat{\varphi}, e_1, ..., e_l, x_1, ..., x_n, z|_\Theta \geqslant maximum(\eta_1, ..., \eta_l) ,$$

where for $1 \leqslant i \leqslant l$,

$$(2\text{–}3) \quad \eta_i = supremum\{ |e_i, t_1, ..., t_{n_i}, u|_\Theta + 1 : g_i(t_1, ..., t_{n_i}) \to u\} .$$

If these conditions are satisfied, we call $\hat{\varphi}$ a Θ-code of φ.

Given φ and arguments $f_1, ..., f_l, x_1, ..., x_n$, what is the natural way to compute $\varphi(f_1, ..., f_l, x_1, ..., x_n)$? In the imagery of Kleene (1959), we think of φ as represented by an oracle, which will give us the value $\varphi(f_1, ..., f_l, x_1, ..., x_n)$ if we give it enough values of the functions $f_1, ..., f_l$. The oracle asks for "the value $f_i(t_1, ..., t_{n_i})$", "the value $f_j(s_1, ..., s_{n_j})$", etc., perhaps leading us into a transfinite computation if after ω steps we still have not give it enough information to determine $\varphi(f_1, ..., f_l, x_1, ..., x_n)$. (Of course the oracle is free to ask for values in whatever order he pleases.) Sooner or later the oracle says "enough", and gives us a value z; if the values we have given at that time define subfunctions $g_1, ..., g_l$ of $f_1, ..., f_l$ respectively, it must be that $\varphi(g_1, ..., g_l, x_1, ..., x_n) \to z$, so that $\varphi(f_1, ..., f_l, x_1, ..., x_n) \to z$ by the continuity of φ.

If this is the natural way to compute $\varphi(f_1, ..., f_l, x_1, ..., x_n)$, how long has this computation been? This of course depends on how long we took to compute each $f_i(t_1, ..., t_{n_i})$ that the oracle requested. Suppose $f_1, ..., f_l$ are Θ-computable with Θ-codes $e_1, ..., e_l$ and that we use these codes to make the required computations. Then before the oracle gives a value we must establish $(e_i, t_1, ..., t_{n_i}, u) \in \Theta$ for each $t_1, ..., t_{n_i}, u$ such that $g_i(t_1, ..., t_{n_i}) \to u$, $i = 1, ..., l$. Thus the length of the total computation is *greater* than $|e_i, t_1, ..., t_{n_i}, u|_\Theta$ for all $t_1, ..., t_{n_i}, u$ such that $g_i(t_1, ..., t_{n_i}) \to u$, $i = 1, ..., l$, which is precisely what condition (b) of Definition 7 demands.

We would have liked to insist in condition (b) that there is a machine in the theory Θ which computes $\varphi(f_1, ..., f_l, x_1, ..., x_n)$ in the way described above. However there are no "machines" and no "computations" in our setup. The best we can do is to insist that the assignment of lengths to computations is consistent with this model for computing the values of functionals.

The introduction of length as a primitive and this definition of Θ-computable functional are the key elements of our axiomatization. In this section we use functionals only for convenience, to make the definition of a computation theory short and comprehensible. But many of the more advanced concepts of recursion theory can be defined only in terms of functionals; the most important of these is the concept of *Θ-finiteness* with which we shall deal in Section 8. It is not crystal clear that we have the right concept – the reader must judge this from the development below. Of course we have checked that in all examples of computation theories where there is a natural concept of computable functional, it coincides with the one given here.

Definition 8. A computation set Θ is a *computation theory* on the domain $\mathfrak{A}$, if there exist Θ-computable functions $p_1, ..., p_{13}$ so that the functions and functionals in I–XII of Definition 4 are Θ-computable with Θ-codes the elements given on the right and the *(Strong) Iteration Property* below holds.

Iteration property. For each $n, m, p_{13}(n, m)$ is a Θ-code of a mapping $S^n_m(a, x_1, ..., x_n)$, such that for all $a, x_1, ..., x_n \in C, y_1, ..., y_m \in A$,

$$(2\text{–}4)\qquad \{a\}^{n+m}_\Theta(x_1, ..., x_n, y_1, ..., y_m) = \{S^n_m(a, x_1, ..., x_n)\}^m_\Theta(y_1, ..., y_m)$$

and

$$(2\text{–}5)\qquad \{a\}^{n+m}_\Theta(x_1, ..., x_n, y_1, ..., y_m) \to z \Rightarrow |a, x_1, ..., x_n, y_1, ..., y_m, z|_\Theta$$
$$< |S^n_m(a, x_1, ..., x_n), y_1, ..., y_m, z)|_\Theta \,.$$

The strengthening of the iteration property that brings in the ordinals goes hand-in-hand with the definition of Θ-computable functional. Our justification for putting it in is again in terms of the *natural* way in which machine $S^n_m(a, x_1, ..., x_n)$ would compute – given $y_1, ..., y_m$ it copies the computation establishing $(a, x_1, ..., x_n, y_1, ..., y_m, z) \in \Theta$ and then it gives z as output, so that (2–5) holds.

We can easily identify Turing machine theory as a computation theory by taking Θ as in (1–13) and putting

$$(2\text{–}6)\qquad |a, \sigma, z|_\Theta = \textit{least } m\ [m \textit{ is the Gödel number of some computation establishing } (a, \sigma, z) \in \Theta]\ ,$$

where the Gödel-numbering is, say, that of Kleene (1952). The standard proofs that the functions and functionals I–XII are Turing computable and that the Weak Iteration Property holds, actually prove the stronger length properties that we imposed, with this (or any other "natural") definition of length. A careful examination of these familiar constructions gives evidence that the restrictions we imposed in the axioms on the behavior of the length function are indeed natural.

3. Prime computability

Suppose $\mathfrak{A} = \langle A, C, N, s, M, K, L\rangle$ is a domain,

$$\mathbf{f} = f_1, \dots, f_l$$

a sequence of p.m.v. functions on A, f_i being n_i-ary. Is there a computation theory Θ on $\mathfrak{A}$ such that each f_i is Θ-computable and such that Θ is *least* with this property? We define here the *prime computation theory PR* $[\mathbf{f}]$ and then explain in the next section in what sense it is *least*.

We shall need n-tuple mappings, for coding n-tuples of elements of C by single elements of C. Define then for each $n \geqslant 0$ a mapping $\langle x_1, \dots, x_n\rangle$ by the induction

$$(3\text{–}1) \quad \langle\ \rangle = 1$$

$$(3\text{–}2) \quad \langle x_1, \dots, x_{n+1}\rangle = M(x_1, \langle x_1, \dots, x_{n+1}\rangle)\ ;$$

now

$$\langle x_1, \dots, x_n\rangle \in C \Longleftrightarrow x_1, \dots, x_n \in C\ ,$$

$$\langle x_1, \dots, x_n\rangle = \langle y_1, \dots, y_n\rangle \in C \Rightarrow x_1 = y_1\ \&\ \dots\ \&\ x_n = y_n\ .$$

We also define by induction on $i \in N$ the binary function $(x)_i$,

$$(3\text{–}3) \quad \begin{cases} (x)_1 = K(x)\ , & \\ (x)_{s(i)} = (L(x))_i\ , & \textit{if} \quad i \in N,\ i \geqslant 1\ , \\ (x)_i = 0 \quad \textit{if} \quad i = 0 \quad \textit{or} \quad i \notin N\ ; & \end{cases}$$

now

$$x \in C \Rightarrow (x)_i \in C\,,$$

$$\langle x_1, ..., x_n \rangle \in C \;\&\; 1 \leqslant i \leqslant n \Rightarrow (\langle x_1, ..., x_n \rangle)_i = x_i\,.$$

The mappings $\langle x_1, ..., x_n \rangle$ are clearly Θ-computable for every computation theory Θ on . The same is true for the function $(x)_i$, since the nested recursion (3–3) can be reduced to a primitive recursion along familar lines.

The idea for defining *PR* [f] is to conceive of the axioms for a computation theory as the clauses of an inductive definition, following Kleene 1959. This is basically a *coding problem*, since for each function or functional that we wish to put in the theory we must invent a name. We give this one induction in full, sordid detail here, and we promise the reader that it is for the last time.

For each set Θ of tuples on A, let $\Gamma(\Theta)$ be the set of tuples defined as follows.

I *If* $y \in C$, *then* $(\langle 1, y \rangle, \sigma, y) \in \Gamma(\Theta)$.

II *If* $y \in C$, *then* $(\langle 2, 0 \rangle, y, \sigma, y) \in \Gamma(\Theta)$.

If $y \notin C$, *then* $(\langle 2, 0 \rangle, y, \sigma, 0) \in \Gamma(\Theta)$.

III $(\langle 3, 0 \rangle, y, \sigma, s(y)) \in \Gamma(\Theta)$.

IV *If* $y \in C$, *then* $(\langle 4, 0 \rangle, y, \sigma, 0) \in \Gamma(\Theta)$.

If $y \notin C$, *then* $(\langle 4, 0 \rangle, y, \sigma, 1) \in \Gamma(\Theta)$.

V *If* $y \in N$, *then* $(\langle 5, 0 \rangle, y, \sigma, 0) \in \Gamma(\Theta)$.

If $y \notin N$, *then* $(\langle 5, 0 \rangle, y, \sigma, 1) \in \Gamma(\Theta)$.

VI $(\langle 6, 0 \rangle, x, y, \sigma, M(x, y)) \in \Gamma(\Theta)$.

VII $(\langle 7, 0 \rangle, y, \sigma, K(y)) \in \Gamma(\Theta)$.

VIII $(\langle 8, 0 \rangle, y, \sigma, L(y)) \in \Gamma(\Theta)$.

IX $\quad$ *If* $(\exists u)\,[(\hat{g}, \sigma, u) \in \Theta \;\&\; (\hat{f}, u, \sigma, z) \in \Theta]$,

then $\quad (\langle 9, 0\rangle, \hat{f}, \hat{g}, \sigma, z) \in \Gamma(\Theta))$.

X $\quad$ *If* $y \notin N$, $\quad$ *then* $\quad (\langle 10, 0\rangle, \hat{f}, \hat{g}, y, \sigma, 0) \in \Gamma(\Theta))$.

If $y = 0 \;\&\; (\hat{g}, \sigma, z) \in \Theta$, $\quad$ *then* $\quad (\langle 10, 0\rangle, \hat{f}, \hat{g}, y, \sigma, z) \in \Gamma(\Theta))$.

If $y = s(x) \in N \;\&\; (\exists u)\,[(\langle 10, 0\rangle, \hat{f}, \hat{g}, x, \sigma, u) \in \Theta$

$\&\; (\hat{f}, u, x, \sigma, z) \in \Theta]$, $\quad$ *then* $\quad (\langle 10, 0\rangle, \hat{f}, \hat{g}, y, \sigma, z) \in \Gamma(\Theta)$.

XI $\quad$ *If* $(\hat{f}, x_{j+1}, x_1, \ldots, x_j, x_{j+2}, \ldots, x_n, z) \in \Theta$

then $\quad (\langle 11, j\rangle, \hat{f}, x_1, \ldots, x_n, z) \in \Gamma(\Theta) \quad (1 \leqslant j \leqslant n)$.

XII $\quad$ *If* $(\hat{f}, \sigma, z) \in \Theta$, $\quad$ *then* $\quad (\langle 12, n\rangle, \hat{f}, \sigma, \tau, z) \in \Gamma(\Theta)$.

XIII $\quad$ *If* $a, x_1, \ldots, x_n \in C \;\&\; (a, x_1, \ldots, x_n, y_1, \ldots, y_m, z) \in \Theta$,

then $\quad (\langle 13, a, x_1, \ldots, x_n\rangle, y_1, \ldots, y_m, z) \in \Gamma(\Theta)$.

XIV $\quad$ *If* $f_i(t_1, \ldots, t_{n_i}) \to z$,

then $\quad (\langle 14, i\rangle, t_1, \ldots, t_{n_i}, \sigma, z) \in \Gamma(\Theta) \quad (1 \leqslant i \leqslant l)$ -

(Here $x, y, u, z, \hat{f}, \hat{g}$ vary over A, σ over n-tuples in A, τ over m-tuples in A.)

The clauses of this definition are such that

$$\Theta \subseteq \Theta' \Rightarrow \Gamma(\Theta) \subseteq \Gamma(\Theta') ,$$

i.e. Γ is a *monotone* operator. Hence Γ has a *least fixed point,* i.e. there is a set of tuples Θ^* determined uniquely by the conditions

$$\Gamma(\Theta^*) = \Theta^*$$

$$\Gamma(\Theta) = \Theta \Rightarrow \Theta^* \subseteq \Theta .$$

Moreover

$$\Theta^* = \bigcup_{\xi} \Theta^{\xi} ,$$

where the sets Θ^{ξ} are defined by the induction,

$$(3\text{–}4) \qquad \Theta^{\xi} = \Gamma(\bigcup_{\eta<\xi} \Theta^{\eta}) .$$

(Actually for this Γ it is trivial that $\Theta^* = \Theta^{\omega}$.) Put

$$(3\text{–}5) \qquad PR\,[\mathbf{f}] = \Theta^*$$

and for $(a, \sigma, z) \in PR\,[\mathbf{f}]$, put

$$(3\text{–}6) \qquad |a, \sigma, z|_{RF[\mathbf{f}]} = \mathit{least}\ \xi\,[(a, \sigma, z) \in \Theta^{\xi}] .$$

It is a routine matter to prove that the set $PR\,[\mathbf{f}]$ with the length function (3–6) is a computation theory and that each f_i is $PR\,[\mathbf{f}]$-computable. E.g., to verify the latter, it is enough to show

$$f_i(t_1, \ldots, t_{n_i}) \to z \Longleftrightarrow (\langle 14, i\rangle, t_1, \ldots, t_{n_i}, z) \in PR\,[\mathbf{f}] .$$

The left-to-right implication follows directly from clause XIV in the definition of $PR\,[\mathbf{f}]$ and the right-to-left implication can be shown by a trivial induction on $|\langle 14, i\rangle, t_1, \ldots, t_{n_i}, z|_{PR\,[\mathbf{f}]}$. The proofs that the functions and functionals I–XII are $PR\,[\mathbf{f}]$-computable and that the Iteration Property holds are very similar to this, e.g. a $PR\,[\mathbf{f}]$-code for the function $M(x, y)$ is $\langle 6, 0\rangle$, a $PR\,[\mathbf{f}]$-code for the composition functional is $\langle 9, 0\rangle$ and we can easily define $p_{13}(n, m)$ so that it is a Θ-code of

$$S^n_m(a, x_1, \ldots, x_n) = \langle 13, a, x_1, \ldots, x_n\rangle .$$

Let $\varphi(f)$ be a unary functional on unary functions which is Θ-computable, for some computation theory Θ, let $g(x, y)$ be a fixed Θ-computable function, consider the function

$$h(x) = \varphi(\lambda y g(x, y)) .$$

Is this Θ-computable? If $\hat{\varphi}, \hat{g}$ are Θ-codes for φ and g respectively, then *for each* $x \in C$,

$$h(x) = \{\hat{\varphi}\}_{\Theta}\,(S^1_1(\hat{g}, x)) ,$$

so that if $C = A$, h is indeed Θ-computable. It is the "normal" case that $C = A$, but there are interesting theories with $C \subsetneqq A$. What we need is a stronger concept of computability for functionals which allows us to compute $\varphi(f)$ not only when f is Θ-computable, but also when f is obtained by "substitution of variables" in a Θ-computable function.

Definition 9. Let $\varphi(f)$ be a continuous unary functional on unary p.m.v.'s on A, for each n let φ^n be defined by

$$(3\text{--}7) \qquad \varphi^n(f, \sigma) = \varphi(\lambda y f(\sigma, y))$$

on $n + 1$-ary p.m.v.'s and n-ary tuples on A. We call φ *uniformly Θ-computable* if there is a Θ-computable mapping $p(n)$ such that for each n, φ^n is Θ-computable with Θ-code $p(n)$.

A similar definition of uniform computability can be given for functionals with more complicated sequences of arguments, but we shall not commit it to print.

Suppose now that we are given not only p.m.v. functions $\mathbf{f} = f_1, \ldots, f_l$, but also p.m.v. functionals $\varphi = \varphi_1, \ldots, \varphi_k$ on A. We wish to define $PR\,[\mathbf{f}, \varphi]$, so that $f_1, \ldots, f_l$ are $PR\,[\mathbf{f}, \varphi]$-computable and $\varphi_1, \ldots, \varphi_k$ are uniformly $PR\,[\mathbf{f}, \varphi]$-computable. The procedure is as before, except we add k more clauses to the definition of $\Gamma(\Theta)$, the ith clause guaranteeing that φ_i will be uniformly computable. Let us write only one such clause, for the case that φ_i is a unary functional on unary p.m.v. functions on A – the other cases are similar only a bit more complicated.

XV *If there is some p.m.v. function g on A such that*

$$(\forall u)(\forall v)\,[g(u) \to v \Rightarrow (g^{\#}, \sigma, u, v) \in \Theta] \ \&\ \varphi(g) \to z,$$

then $\quad (\langle 14, i\rangle, g^{\#}, \sigma, z) \in \Gamma(\Theta)\,.$

The sets Θ^{ξ} are defined for this new Γ by (3–4) (the induction may be transfinite now) and the length function is given by (3–6), using these Θ^{ξ}. Proof that $PR\,[\mathbf{f}, \varphi]$ is a computation theory and that each φ_i is uniformly $PR\,[\mathbf{f}, \varphi]$-computable is again routine. (If φ_i is unary on unary p.m.v.'s, then the constant function $p(n) = \langle 15, i\rangle$ shows that φ_i is uniformly $PR\,[\mathbf{f}, \varphi]$-computable.)

It is clear how we could simplify clause XV so that the φ_i's become computable but not necessarily uniformly computable in the resulting theory. We chose this definition of $PR\,[\mathbf{f}, \varphi]$, because it seems to us that uniform computabiligy is a more natural concept than computability. Recall that in the "normal" case, when $C = A$, the two concepts coincide.

Instead of $PR\,[\mathbf{f}, \varphi]$-computable, we say *prime computable in* $\mathbf{f}$, φ. (The concept makes sense when either $\mathbf{f}$ or φ are empty sequences of course.) In the next section we shall argue that prime computability in p.m.v. functions $\mathbf{f}$ as defined here is equivalent to the namesake concept studied in Moschovakis (1969a). One interesting result proved in that paper for functions, but true also of functionals by a similar proof, is the

Transitivity property. If G, H are either p.m.v. functions or p.m.v. functionals on A, if G is prime computable in H, $\mathbf{f}$, φ *and H is prime computable in* $\mathbf{f}$, φ, *then G is prime computable in* $\mathbf{f}$, φ.

Finally consider the case when we are given not only $\mathfrak{A}, \mathbf{f}, \varphi$ (where again either $\mathbf{f}$ or φ may be an empty seuqnece) but also a computation theory H on $\mathfrak{A}$. We wish to define a theory $\Theta = \mathrm{H}\,[\mathbf{f}, \varphi]$ on $\mathfrak{A}$ which will extend H, such that each f_i will be Θ-computable and such that Θ will be *least* with these properties. Perhaps the simplest way to do this is to use the same operator Γ, defined by clauses I–XV, but define now Θ^ξ by the induction

$$(3\text{–}8)\qquad \Theta^\xi = \Gamma(\bigcup_{\eta<\xi} \Theta^\eta) \cup \{(\langle 16, a\rangle, \sigma, z) : (a, \sigma, z) \in \mathrm{H}\ \&\ |a, \sigma, z|_{\mathrm{H}} \leqslant \xi\}.$$

If we now put

$$(3\text{–}9)\qquad \mathrm{H}\,[\mathbf{f}, \varphi] = \Theta^* = \bigcup_\xi \Theta^\xi ,$$

$$(3\text{–}10)\qquad |a, \sigma, z|_{\mathrm{H}\,[\mathbf{f}, \varphi]} = \mathit{least}\ \xi\,[(a, \sigma, z) \in \Theta^\xi] ,$$

then it is again simple to verify that $\mathrm{H}\,[\mathbf{f}, \varphi]$ has the required properties. Moreover

$$(a, \sigma, z) \in \mathrm{H} \Longleftrightarrow (\langle 16, a\rangle, \sigma, z) \in \mathrm{H}\,[\mathbf{f}, \varphi] ,$$

which implies that each H-computable function on A is $\mathrm{H}\,[\mathbf{f}, \varphi]$-computable.

We should perhaps observe that the set of tuples $\mathrm{H}\,[\mathbf{f}, \varphi]$ is the least fixed point of the operator Γ defined by clauses I–XV and

XVI If $(a, \sigma, z) \in \Theta$, *then* $(\langle 16, a\rangle, \sigma, z) \in \Gamma(\Theta)$.

However definition (3–9) makes it simpler to define the length function (3–10) – and why we want that length function rather than the one naturally associated with this operator Γ will be made clear in the next section.

4. Extensions

Suppose Θ and H are computation theories on the same domain $\mathfrak{A}$. One natural way to define "H extends Θ", is "each Θ-computable p.m.v. function on $\mathfrak{A}$ is H-computable". We study instead a much stronger reflexive, transitive relation on computation theories which involves the concept of length.

Definition 10. Let Θ, H be computive theories on the same domain $\mathfrak{A}$. We say that H *extends* Θ, in symbols

$$\Theta \leqslant \mathrm{H} ,$$

if there is an H-computable mapping $p(a, n)$ taking C into C and such that for each n-tuple $\sigma, a \in C$, $z \in A$,

$$(4\text{–}1) \quad (a, \sigma, z) \in \Theta \Longleftrightarrow (p(a, n), \sigma, z) \in \mathrm{H} ,$$

$$(4\text{–}2) \quad (a, \sigma, z) \in \Theta \Rightarrow |a, \sigma, z|_{\Theta} \leqslant |p(a, n), \sigma, z|_{\mathrm{H}} .$$

If $\Theta \leqslant \mathrm{H}$ and $\mathrm{H} \leqslant \Theta$, we call Θ and H *equivalent* and write $\Theta \sim \mathrm{H}$

It is immediate that $\leqslant$ is reflexive and transitive.

Perhaps a pragmatic justification of why we study this extension relation and this equivalence notion would be enough: clearly if we can prove $\Theta \sim \mathrm{H}$ we know much more than if we only show the equivalence of Θ-computability with H-computability and in practice, whenever we can prove the second we always seem to be able to prove the first. The standard arguments that Turing computability on the integers coincides with general recursiveness defined via Herbrand-Gödel-Kleene systems of equations, e.g. those given in Kleene (1952), also show the equivalence of the computation theories involved. What lies behind this happy situation is that reduction proofs of this type tend to be *constructive*. E.g. we show that each Turing machine computation can be *imitated* by a computation from a system of equations. Since the proof is constructive, it yields a recursive function that assigns to each Turing machine the appropriate system of equations; since the latter imitates the Turing machine computation *step by step*, and for each Turing machine step it must execute at least one step to imitate it, its computations are at least as long as the Turing machine computations.

We formulated Turing machine theory as a computation theory at the end of Section 2, using a particular Gödel numbering of Turing machines and computations. For much the same reasons as above, any other "reasonable" Gödel numbering will lead to an equivalent theory. We can characterize Turing machine theory up to this notion of equivalence as the prime computation theory on the integers.

Using this partial ordering on theories we can make precise the sense in which $\Theta[\mathbf{f}, \varphi]$ is the *least* extension of Θ in which $\mathbf{f}, \varphi$ are computable.

(i) *Let* Θ *be a computation theory on some domain* $\mathfrak{A}$, $\mathbf{f}, \varphi$ *a sequence of p.m.v. functions and continuous functionals on A. Then*

$$(4\text{–}3) \qquad \Theta \leqslant \Theta[\mathbf{f}, \varphi]$$

and if H *is any computation theory in* $\mathfrak{A}$,

$$(4\text{–}4) \qquad \Theta \leqslant \mathrm{H}\ \&\ [\mathbf{f}\ \textit{are}\ \mathrm{H}\textit{-computable}]\ \&\ [\varphi\ \textit{are uniformly}\ \mathrm{H}\textit{-computable}]$$

$$\Rightarrow \Theta[\mathbf{f}, \varphi] \leqslant \mathrm{H}\,.$$

That $\Theta \leqslant \Theta[\mathbf{f}, \varphi]$ is trivial, taking $f(x, n) = \langle 16, x \rangle$ and using the definition of Θ^{ξ} in (3–8) and the length function on $\Theta[\mathbf{f}, \varphi]$ given by (3–10) – it is precisely for this result that we chose the length function as we did.

The proof of the second assertion of (i) is technically non-trivial, but proceeds along paths opened by Kleene (1955, and earlier Kleene papers) and utilized extensively since then by many people, including ourselves in Moschovakis (1967, 1969a, b). The required function p is defined by cases, using the second recursion theorem for the theory H, so that it is H-computable. Then (4–2) and direction $\Rightarrow$ of (4–1) (with $\Theta[\mathbf{f}, \varphi]$ for Θ of course) are proved by induction on $|a, \sigma, z|_{\Theta[\mathbf{f}, \varphi]}$, while direction $\Leftarrow$ of (4–1) is proved by induction on $|p(a, n), \sigma, z|_{\mathrm{H}}$.

Properties (4–3) and (4–4) characterize $\Theta[\mathbf{f}, \varphi]$ uniquely up to equivalence. Of course the same arguments show that for any theory H on $\mathfrak{A}$,

(4–5) [**f** *are* H-*computable*] & [φ *are uniformly* H-*computable*]

$$\Rightarrow PR\,[\mathbf{f}, \varphi] \leqslant \mathrm{H}\,,$$

which gives a charcaterization of the prime computation theory $PR\,[\mathbf{f}, \varphi]$.

From (i) we immediately get the following simple property.

(ii) *If* Θ, H *are computation theories on* $\mathfrak{A}$, **f**, φ *a sequence of p.m.v. functions and continuous functionals on* A *and* $\Theta \leqslant \mathrm{H}$, *then* $\Theta\,[\mathbf{f}, \varphi] \leqslant \mathrm{H}\,[\mathbf{f}, \varphi]$.

The relation $\leqslant$ gives a simple characterization of uniform Θ-computability for a continuous p.m.v. functional.

(iii) *Let* Θ *be a computation theory on* $\mathfrak{A}$, φ *a continuous p.m.v. functional on* A. *Then* φ *is uniformly* Θ-*computable if and only if* $\Theta\,[\varphi] \leqslant \Theta$.

That $\Theta\,[\varphi] \leqslant \Theta$ if φ is uniformly Θ-computable follows again by (i). For the converse, suppose $\Theta\,[\varphi] \leqslant \Theta$ via a Θ-computable mapping p, for simplicity assume that φ is unary, with unary p.m.v.'s as arguments. If f is any p.m.v. function with Θ-code $\hat{f}$, then f has $\Theta\,[\varphi]$-code $\langle 16, \hat{f}\rangle$ and since φ is $\Theta\,[\varphi]$-computable with $\Theta[\varphi]$-code $\langle 15, 1\rangle$, we have

$$\varphi(f) \to z \Longleftrightarrow (\langle 15, 1\rangle, \langle 16, \hat{f}\rangle, z) \in \Theta\,[\varphi]$$

$$\Longleftrightarrow (p(\langle 15, 1\rangle, 1), \langle 16, \hat{f}\rangle, z) \in \Theta\,.$$

Now choose a fixed $a^* \in C$ such that

(4–6) $\{a^*\}_\Theta(x) = \{p(\langle 15, 1\rangle, 1)\}_\Theta(\langle 16, x\rangle)\,,$

(4–7) $\{p(\langle 15, 1\rangle, 1)\}_\Theta(\langle 16, x\rangle) \to z$

$$\Rightarrow |a^*, x, z|_\Theta \geqslant |p(\langle 15, 1\rangle, 1), \langle 16, x\rangle, z|_\Theta\,.$$

(For those that enjoy computation, we may take

$$a^* = S_1^2(p_9(1), S_1^1(p_{12}(1, 1), p(\langle 15, 1\rangle, 1)), c)\,,$$

where c is some Θ-code of the function $\langle 16, x\rangle$ and the functions S_m^n, p_9, p_{12} are those guaranteed by the axioms for Θ.) Then

$$\varphi(f) \to z \Longleftrightarrow (a^*, \hat{f}, z) \in \Theta ,$$

and it is easy to verify that a^* is indeed a Θ-code of the functional φ – here of course it is essential that we use the ordinal increasing property (4–2) of p.

This argument shows only that φ is Θ-computable, but an easy notational variance proves the assertion for all φ^n defined by (3–7), thus showing that φ is uniformly Θ-computable.

(iv) *If* Θ, H *are computation theories on* $\mathfrak{A}$, $\Theta \sim$ H *and* φ *is a uniformly* Θ-*computable p.m.v. functional, then* φ *is uniformly* H-*computable.*

This is a trivial consequence of (ii) and (iii), Because of (ii), since $\mathrm{H} \leqslant \Theta$ we have $\mathrm{H}[\varphi] \leqslant \Theta[\varphi]$, by (iii) we have $\Theta[\varphi] \leqslant \Theta$, so that $\mathrm{H}[\varphi] \leqslant \Theta$ and since by hypothesis $\Theta \leqslant \mathrm{H}$, we have $\mathrm{H}[\varphi] \leqslant \mathrm{H}$ which again by (iii) implies that φ is uniformly H-computable.

Thus equivalent computation theories have the same uniformly computable functionals. (A similar proof shows that they have the same computable functionals.) We do not know any reasonable definition of equivalence for precomputation theories for which this result holds, as we do not know an analog of (i) for precomputation theories.

Suppose B is some set. Let 0 be some object not in B, (x, y) some pairing function such that no element of $B \cup \{0\}$ is a pair and as in Moschovakis (1969a), let B^* be the closure $B \cup \{0\}$ under (x, y). The natural domain that we associate with B is

$$(4\text{–}8) \qquad \mathfrak{B}^* = \mathfrak{B}^*_B = \langle B^*, B^*, \omega, s, (\ ,\ ), \pi, \delta \rangle ,$$

where $\omega = \{0, (0, 0), ((0, 0), 0), ...\}$, $s(x) = (x, 0)$ and π, δ are the functions inverse to (x, y) defined by (1.4) of Moschovakis (1969a). If $\mathbf{f} = f_1, ..., f_l$ is a sequence of p.m.v. functions on B^*, let $PC(B, \mathbf{f})$ be $\{(e, \sigma, z) : \{e\}_p(\sigma) \to z\}$ in the sense of Section 5 of Moschovakis (1969a) with the length function $|\ \ |_p$ defined on these tuples in the same section of that paper. It is routine to prove this theory equivalent to $PR(\mathbf{f})$ as defined here.

If $C \subseteq B^*$, C closed under (x, y), $\omega \subseteq C$, put

$$\mathfrak{B}^*_C = \langle B^*, C, \omega, s, (\ ,\ ), \pi, \delta \rangle$$

as above. Again the theory $PC(C, \mathbf{f})$ of functions prime computable in $\mathbf{f}$ *from* C as defined in Moschovakis (1969a) is equivalent to $PR(\mathbf{f})$ for the domain $\mathfrak{B}^*_C$, *provided C is prime computable in* $\mathbf{f}$ *from C* in the terminology of

Moschovakis (1969a). The simplest way to insure this is to include the characteristic function of C in **f**, but sometimes it comes for free, as when $C = \{0\}^*$ (*absolute* prime computability). This restriction, that the set of codes be a computable set seems to be necessary for several results. In Moschovakis (1969a, b) where we did not include it as a blanket assumption, we were forced to insert it as a special hypothesis in many places.

5. The first recursion theorem

Ordinary recursion theory on the integers derives much of its interest from an extramathematical contention, *Church's Thesis*: it is claimed that the precise concept of *recursive function* coincides with the intuitive notion of *effective procedure*. It is hard to attach meaning to Church's Thesis for a computation theory on some abstract domain, where we lack a clear notion of effective procedure. But a computation theory should satisfy the purely mathematical properties implied by Church's Thesis, and in particular one should not be able to define Θ-computable functions by combining Θ-computable and intuitively effective operations. The most significant closure property of the class of general recursive functions is embodied in Kleene's First Recursion Theorem, which we now prove for an abstract theory.

First recursion theorem. Let Θ be a computation theory on $\mathfrak{A}$, let $\varphi(f, x)$ be a Θ-computable continuous p.m.v. functional over A, let f^* be the smallest solution of the equivalence

$$(5\text{--}1) \quad (\forall x \in A)\,[\varphi(f, x) = f(x)] \; .$$

Then f^* is Θ-computable.

Proof. That f^* exists is easily seen, in fact

$$f^*(x) \to z \Longleftrightarrow (\forall g)\{(\forall x)\,[\varphi(g, x) = g(x)] \to g(x) \to z\} \, .$$

We can also define f^* by induction,

$$(5\text{--}2) \quad f^\xi(x) = \varphi(\bigcup_{\eta<\xi} f^\eta, x) \, ,$$

$$(5\text{--}3) \quad f^* = \bigcup_{\xi} f^\xi \, ,$$

where for any collection $\{g^\eta\}$ of p.m.v. functions,

$$(\bigcup_\eta g_\eta)(x) \to z \iff (\exists\eta)\,[g_\eta(x) \to z]\ .$$

Choose a Θ-computable function $g(a, b, x)$ with Θ-code $\hat{g}$, such that for all $a, b \in C$, $x \in A$,

$$(5\text{–}4)\qquad g(a, b, x) = \{a\}_\Theta(\{b\}_\Theta(0), x)\ ,$$

$$(5\text{–}5)\qquad g(a, b, x) \to z \Rightarrow |a, \{b\}_\Theta(0), x, z|_\Theta < |\hat{g}, a, b, x, z|_\Theta\ ;$$

that such a g exists follows easily from the axioms. Let $\hat{\varphi}$ be a Θ-code of the functional $\varphi(f, x)$ and by the second recursion theorem choose $\hat{p}$ so that for all t,

$$\{\hat{p}\}_\Theta(t) = S^2_1(\hat{g}, \hat{\varphi}, \hat{p})\ .$$

Finally put

$$c = \{\hat{p}\}(0) = S^2_1(\hat{g}, \hat{\varphi}, \hat{p})\ .$$

We now claim that

$$(5\text{–}6)\qquad f^*(x) \to z \iff \{c\}_\Theta(x) \to z\ .$$

We prove direction $\Rightarrow$ of (5–6) by transfinite induction on ξ. If $f^\xi(x) \to z$, then

$$(5\text{–}7)\qquad \varphi(h^\xi, x) \to z\ ,$$

where $h^\xi = \bigcup_{\eta<\xi} f^\eta$, and by induction hypothesis,

$$h^\xi(u) \to v \Rightarrow \{c\}_\Theta(u) \to v\ .$$

Since $\hat{\varphi}$ is a Θ-code of φ, we have from (5–7)

$$\{\hat{\varphi}\}_\Theta(c, x) \to z\ ,$$

and then successively

$$g(\hat{\varphi}, \hat{p}, x) \to z \ ,$$

$$\{S_1^2(\hat{g}, \hat{\varphi}, \hat{p})\}_\Theta(x) \to z \ ,$$

$$\{c\}_\Theta(x) \to z \ ,$$

which is what we needed to show.

The converse direction of (5–6) is proved by induction on $|c, x, z|_\Theta$. Assuming $\{c\}_\Theta(x) \to z$ and using (5–5) and the ordinal-increasing property of S_1^2, we get

$$g(\hat{\varphi}, \hat{p}, x) \to z \ , \qquad |\hat{g}, \hat{\varphi}, \hat{p}, x, z|_\Theta < |S_1^2(\hat{g}, \hat{\varphi}, \hat{p}), x, z|_\Theta \ ,$$

and hence

$$\{\hat{\varphi}\}_\Theta(c, x) \to z \ , \qquad |\hat{\varphi}, c, x, z|_\Theta < |\hat{g}, \hat{\varphi}, \hat{p}, x, z|_\Theta \ .$$

By the definition of Θ-computable functional, there is some p.m.v. function h, such that

$$\text{(5–8)} \qquad \varphi(h, x) \to z \ ,$$

$$h(u) \to v \Rightarrow [\{c\}_\Theta(u) \to v \ \& \ |c, u, v|_\Theta < |\hat{\varphi}, c, x, z|_\Theta] \ ;$$

hence by induction hypothesis,

$$h(u) \to v \Rightarrow f^*(u) \to v \ ,$$

and by (5–8) and the continuity of φ,

$$\varphi(f^*, x) \to z$$

which implies immediately $f^*(x) \to z$.

The theorem is significant as evidence that our concept of a computation theory is natural. One of the main reasons we abandoned the effort to axiomatize recursion theory by strengthening the axioms for precomputation theories without adding new primitives was our inability to prove any results of this type in that context.

We mentioned in Section 1 that there is a precomputation theory Θ on the integers, such that the Θ-precomputable p.m.v. functions are precisely the

Σ_1^1 relations. Can we insist that Θ be a computation theory? The answer is trivially positive – simply take $PR[f]$, where the relation $f(n, m) \to k$ is Σ_1^1 and enumerates (in n) all Σ_1^1 binary relations on the integers.

Suppose now that $\varphi(f, x)$ is some *arithmetical* continuous p.m.v. functional on the integers. Can we find a computation (precomputation) theory Θ, such that the Θ-computable (Θ-precomputable) relations are precisely the Σ_1^1 relations and such that $\varphi(f, x)$ is Θ-computable (Θ-precomputable)? Now the answer is positive for the case of a precomposition theory. but it may be negative of we insist on a computation theory, e.g. if the minimum fixed point f^* is a Π_1^1 relation which is not Σ_1^1. In particular, *there is no computation theory Θ on the integers such that Θ-computable p.m.v. functions are the Σ_1^1 relations and such that each arithmetical continuous functional is Θ-computable.*

6. Selection operators

Let Θ be a computation theory on some domain $\mathfrak{A}$. A relation $R(\sigma)$ on A is *Θ-semicomputable* if there is some Θ-computable p.m.v. function $f(\sigma)$ such that

$$(6\text{–}1) \qquad R(\sigma) \Longleftrightarrow f(\sigma) \to 0\ ;$$

we call $R(\sigma)$ *Θ-computable*, if there is a Θ-computable mapping $f(\sigma)$ such that

$$(6\text{–}2) \qquad R(\sigma) \Longleftrightarrow f(\sigma) = 0\ .$$

As usually we classify subsets of A via their representing relations.

Much of the theory of recursively enumerable and recursive sets generalizes to the abstract case for these Θ-semicomputable and Θ-computable sets, e.g. there exists Θ-semicomputable non Θ-computable sets, there exist Θ-semicomputable disjoint sets which cannot be separated by a Θ-computable set, etc. However the basic results that a set is recursive if and only if both it and its complement are recursively enumerable does not generalize without some additional hypothesis.

For a p.m.v. function $f(\sigma)$ put

$$f(\sigma)\downarrow \Longleftrightarrow (\exists z)\,[f(\sigma) \to z]\ .$$

Definition 11. Let Θ be a computation theory on a domain $\mathfrak{A}$. An *n-ary selection operator* for Θ is any $n+1$-ary Θ-computable function $q(a, \sigma)$ (perhaps multiple-valued on partial) with Θ-code $\hat{q}$, such that

$$(6\text{–}3)\quad (\exists y)\,[\{a\}_\Theta(y, \sigma) \to 0] \Rightarrow q(a, \sigma)\downarrow \;\&\; \{a\}_\Theta(q(a, \sigma), \sigma) \to 0\,,$$

$$(6\text{–}4)\quad \{a\}_\Theta(y, \sigma) \to 0 \;\&\; q(a, \sigma) \to y \Rightarrow |a, y, \sigma, 0|_\Theta < |\hat{q}, a, \sigma, y|_\Theta\,.$$

A *(uniform) selection operator* for Θ is any Θ-computable mapping $p(n)$, such that for each n, $p(n)$ is the Θ-code of some n-ary selection operator for Θ.

Perhaps this time we should give up trying to motivate the length condition (6–4) and simply admit that we need it for technical reasons, e.g. in Section 8. However the vague remarks on "naturalness" that we employed in trying to motivate the length condition in the definition of computable functionals apply here too. Because any "naturally defined" n-ary selection operator will presumable verify that $\{a\}_\Theta(y, \sigma) \to 0$ before giving output 0, so that the computation establishing $q(a, \sigma) \to y$, will be longer than that establishing $\{a\}_\Theta(y, \sigma) \to 0$.

As in the case of functionals, the uniformity hypothesis is only needed for theories with $C \subsetneqq A$. It is easy to verify that when $C = A$, then Θ has a (uniform) selection operator if and only if it has a 0-ary selection operator.

It is fairly simple to show that *if Θ has a selection operator, then the class of Θ-semicomputable relations is closed under disjunction and existential quantification over A and a relation is Θ-computable if and only if both it and its negation are Θ-semicomputable.* However it is still possible for such theories that equality may not be Θ-computable and that not all constant functions (with values in A) are Θ-computable.

We allowed selection operators to be multiple-valued in order to include the important example of *search computability*. Let $\nu(f)$ be the functional defined on unary p.m.v.'s and elements of A by

$$(6\text{–}5)\quad \nu(f) \to z \Longleftrightarrow f(z) \to 0\,.$$

If this functional ν is Θ-computable with Θ-code $\hat{\nu}$, then the function $\{\hat{\nu}\}_\Theta(\hat{f})$ is clearly a 0-selection operator for Θ – and if ν is uniformly Θ-computable, then on easily checks that Θ has a selection operator. This is the only natural selection operator that comes to mind in the absence of a natural wellordering of A. For a given sequence $\mathbf{f}, \varphi$ of functions and functionals on A, we call

$$SC(\mathbf{f}, \varphi) = PR(\mathbf{f}, \varphi, \nu)$$

the *search computation theory* determined by $\mathbf{f}, \varphi$ on $\mathfrak{A}$. On a domain $\mathfrak{B}^*$ defined by (4–8) and when there are no functionals, $SC(\mathbf{f})$ is easily proved equivalent to the theory of search computability studied in Moschovakis (1969a).

It is simple to verify that *if* Θ *has a selection operator, then* Θ *is weakly equivalent to* $\Theta[\nu]$, *in the following sense.*

Definition 11. Let Θ, H be computation theories on a domain $\mathfrak{A}$. We say that H *weakly extends* Θ, in symbols

$$\Theta \leqslant_w \mathrm{H},$$

if there is a Θ-computable mapping $p(a, n)$ such that for each n-tuple σ, each $a \in C$, $z \in A$,

$$(6\text{–}6) \quad (a, \sigma, z) \in \Theta \Leftrightarrow (p(a, n), \sigma, z, 0) \in \mathrm{H},$$

$$(6\text{–}7) \quad (a, \sigma, z) \in \Theta \Rightarrow |a, \sigma, z|_\Theta \leqslant |p(a, n), \sigma, z, 0|_\mathrm{H}.$$

If $\Theta \leqslant_w \mathrm{H}$ and $\mathrm{H} \leqslant_w \Theta$, we call Θ and H *weakly equivalent* and write

$$\Theta \sim_w \mathrm{H}.$$

If $\Theta \sim_w \mathrm{H}$, then Θ and H have the same semicomputable relations. If in addition Θ and H have selection operators, then they have the same computable relations and the same computable *single-valued* partial functions and functionals. In studying a theory Θ with a selection operator, it is often convenient to consider $\Theta[\nu]$ which looks much like Θ but allows more freedom in defining p.m.v. functions.

If $\mathbf{f}$ is a sequence of *mappings* on A, i.e. *totally-defined, single-valued* functions, then the search computation theory $SC(\mathbf{f})$ has many interesting structure properties. There is a *normal form theorem*, i.e.

$$\{a\}_\Theta(\sigma) = U(a, \sigma, \nu(\lambda y t(a, \sigma, y) = 0)$$

for all a, σ, for fixed Θ-computable mappings U and t. Also a *projective hierarchy* of classes of relation on A can be defined which generalizes the familiar arithmetical hierarchy on the integers. These and other results for $SC(\mathbf{f})$ are proved in Moschovakis (1969a). We also proposed an abstract form of *Church's Thesis* in that paper. In the terminology we use here, this suggests identifying the $SC(\mathbf{f})$-computable functions on a domain

$$\mathfrak{A} = \langle A, A, N, s, M, K, L\rangle$$

(i.e. with $C = A$) with the functions on A that are *effectively computable given all the constants in* A. The latter intuitive concept is not as clearly understood as the notion of computable functions on the integers. Thus we only content that the most *natural* way to define "effectively computable" on an abstract domain is via search computability. Evidence for such a contention is given in Moschovakis (1969c), Gordon (1968), which prove the equivalence of search computability with practically every other notion of (first-order) abstract computability known, at least when the characteristic function of equality occurs in the sequence $\mathbf{f}$.

7. Computability in higher types

Let $\mathfrak{A} = \langle A, C, N, s, M, K, L\rangle$ be a domain, put

$$A^{(0)} = A\ ,$$

$$A^{(j+1)} = {}^{(A^{(j)})}A = \textit{all unary functions on } A^{(j)} \text{ to } A\ .$$

For each $m < \omega$ consider the domain

$$^{(m)} = \langle \bigcup_{j\leqslant m} A^{(j)}, C, N, s^{(m)}, M^{(m)}, K^{(m)}, L^{(m)}\rangle\ ,$$

where $s^{(m)}, M^{(m)}, K^{(m)}, L^{(m)}$ extend s, M, K, L and give value 0 when one of their arguments is not in $A^{(0)}$.

In discussing the domains $\mathfrak{A}^{(m)}$ it is convenient to use variables α^j ranging over each $A^{(j)}, j \leqslant m$. One naturally associates with $\mathfrak{A}^{(m)}$ the following functions and functionals.

$$c_j(x) = \begin{cases} 0 & \textit{if } x \in A^{(j)} \\ 1 & \textit{if } x \notin A^{(j)} \end{cases} \qquad (0 \leqslant j \leqslant m)$$

$$e_j(x, y) = \begin{cases} x(y) & \text{if } x \in A^{(j+1)} \ \& \ y \in A^{(j)} \\ 0 & \text{otherwise} \end{cases} \qquad (0 \leqslant j < m)$$

$$\lambda_j(f, x) = \begin{cases} x(\lambda\alpha^{j-1} f(\alpha^{j-1})) & \text{if } x \in A^{(j+1)} \\ 0 & \text{otherwise} \end{cases} \qquad (0 < j < m)$$

Here of course $\lambda\alpha^{j-1} f(\alpha^{j-1})$ is the restriction of an arbitrary f on $A^{(0)} \cup \ldots \cup A^{(m)}$ to $A^{(j)}$ and if $x \in A^{(j+1)}$, then

$$x(\lambda\alpha^{j-1} f(\alpha^{j-1})) \to z \Longleftrightarrow (\exists\alpha^j)\{(\forall\alpha^{j-1})[f(\alpha^{j-1}) \to \alpha^j(\alpha^{j-1})]$$
$$\& \ x(\alpha^j) = z\}.$$

Several interesting computation theories can be defined in terms of the sequences

$$\mathbf{c}^{(m)} = c_0, \ldots, c_m, \qquad \mathbf{e}^{(m)} = e_0, \ldots, e_{m-1},$$
$$\boldsymbol{\lambda}^{(m)} = \lambda_1, \ldots, \lambda_{m-1},$$

in particular $PR[\mathbf{c}^{(m)}, \mathbf{e}^{(m)}, \boldsymbol{\lambda}^{(m)}]$ and $PR[\mathbf{c}^{(m)}, \mathbf{e}^{(m)}, \boldsymbol{\lambda}^{(m)}, \mathbf{f}, \boldsymbol{\varphi}]$, where $\mathbf{f}, \boldsymbol{\varphi}$ are given sequences of functions and functionals on $\mathfrak{A}^m$.

If $\mathfrak{A}$ is the domain of the integers (with some choice of recursive M, K, L), then the elements of $\mathfrak{A}^{(m)}$ are the objects of type $\leqslant m$ over ω. Kleene defined and studied a very natural computation theory on these objects. Since our axioms are nothing but an abstract version of the Kleene schemata, the Kleene theory should be equivalent to $PR[\mathbf{c}^{(m)}, \mathbf{e}^{(m)}, \boldsymbol{\lambda}^{(m)}]$. This is in fact the case, but only after we reformulate slightly the Kleene definition so that it yields a computation theory in our sense.

Let us fix m and consider functions on the domain $\mathfrak{A}^{(m)}$ with $\mathfrak{A}$ the integers. Kleene does not study arbitrary n-ary functions on $\mathfrak{A}^{(m)}$ to ω, but only functions whose domain is some cartesian product of the form $\omega^{k_0} \times (\omega^{(1)})^{k_1} \times \ldots \times (\omega^{(m)})^{k_m}$. His basic relation is

$$\{e\}(\sigma) \simeq z,$$

where e, z vary over ω and σ varies over some $\omega^{k_0} \times (\omega^{(1)})^{k_1} \times \ldots \times (\omega^{(m)})^{k_m}$, where in fact the sequence $k_0, k_1, \ldots, k_m$ is recursively retrievable from the index e. This is not a computation theory in our sense, since e.g. the constant function $f(x) = 0$ on $\mathfrak{A}^{(m)}$ is not computable, simply because its domain is not a Cartesian product of this form. However each n-ary function on $\mathfrak{A}^{(m)}$ decomposes into a finite system of functions on such Cartesian products and it is natural to call it Kleene-computable if all these components are recursive in the sense of Kleene.

If $\sigma = \alpha_1^{t_1}, \ldots, \alpha_n^{t_n}$, put

$$ch(\sigma) = \langle t_1, \ldots, t_n \rangle ,$$

$$height(\sigma) = maximum\, \{t_1, \ldots, t_n\},$$

and let

$$(e, \sigma, z) \in \mathbf{K}^{(m)} \Longleftrightarrow height(\sigma) \leqslant m \;\&\; \{(e)_{ch(\sigma)}\}(\sigma) \simeq z ,$$

where $\{e\}(\sigma) \simeq z$ is the Kleene relation. Now this relation was defined by induction, so that it naturally associates an ordinal $|e, \sigma, z|_{\mathbf{K}}$ with each e, σ, z such that $\{e\}(\sigma) \simeq z$; put then for $(e, \sigma, z) \in \mathbf{K}^{(m)}$,

$$|e, \sigma, z|_{\mathbf{K}^{(m)}} = |(e)_{ch(\sigma)}, \sigma, z|_{\mathbf{K}} .$$

It is a routine matter to check that $\mathbf{K}^{(m)}$ *is a computation theory, equivalent with* $PR\,[\mathbf{c}^{(m)}, \mathbf{e}^{(m)}, \lambda^{(m)}]$ *in the sense of Definition* 10.

Suppose Θ is a theory on the domain $\mathfrak{A}^{(m)}$ and α^m is some fixed element of $A^{(m)}$. It is not hard to verify that

$$(7\text{–}1) \qquad \mathrm{H} = \{(a, \sigma, z) : \sigma \textit{ is a sequence on } \mathfrak{A} \;\&\; (a, \alpha^m, \sigma, z) \in \Theta\}$$

together with the length function it inherits from Θ is a computation theory on $\mathfrak{A}$. Particularly interesting theories are obtained when we choose $\alpha^m = {}^m\mathrm{E}$, the functional representing quantification over $A^{(m-2)}$,

$${}^m\mathrm{E}(x) = \begin{cases} 0 & \textit{if } x \in A^{(m-1)} \;\&\; (\exists \alpha^{m-2})\,[x(\alpha^{m-2}) = 0] , \\ 1 & \textit{otherwise} ; \end{cases}$$

e.g. with $\mathfrak{A}$ the integers, $\Theta = PR\,[\mathbf{c}^{(m)}, \mathbf{e}^{(m)}, \lambda^{(m)}]$ and $\alpha^m = {}^m\mathrm{E}, m = 2, 3$, we have respectively the theories of *hyperarithmetic* and *hyperanalytic* functions on ω, see Kleene (1959), Moschovakis (1967).

With $\mathfrak{A}$ some arbitrary domain it is more natural to take $\Theta = PR[\mathbf{c}^{(m)}, \mathbf{e}^{(m)}, \lambda^{(m)}, \mathbf{f}, \nu^{(0)}]$, where $\mathbf{f}$ is a sequence of functions on $A^{(0)}$ and $\nu^{(0)}$ is the search functional on $A^{(0)}$,

$$\nu^{(0)}(f) \to z \Longleftrightarrow z \in A^{(0)} \;\&\; f(z) \to 0\,.$$

With this Θ and $\alpha^2 = {}^2\mathrm{E}$, the theory H defined by (7–1) is "equivalent" to the theory of *hyperprojective* functions on $\mathfrak{A}$ – where again we need the quotations around "equivalent", since the hyperprojective theory was defined only for $\mathfrak{A} = \mathfrak{B}^*$ and even then only for Cartesian products as domains, so that it must be doctored up to make the equivalence precise. (Hyperprojective theory gives a decent generalization of hyperarithmetic theory to domains other than the integers; see Moschovakis (1969a, b, c); Barwise-Gandy-Moschovakis 1970.)

There are other approaches to computability in higher types over a domain, e.g. the theory on *hereditarily consistent functions* in Platek (1966). Some of the most interesting recent results in abstract recursion deal with such theories, e.g. results about selection operators in Gandy (1962), Moschovakis (1967), Platek (1966), Grilliot (1967). It is our hope that the axiomatization given here will help clarify and unify these results and perhaps lead to some new ones. However in this first draft we confine ourselves to the simple remarks above whose sole purpose was to identify the Kleene theory of recursive functionals with our prime computation theory on the appropriate domain.

8. Finiteness relative to a theory

Some of the most interesting constructions of ordinary recursion theory deal with finite sets and utilize their very special properties. Because of this, a good abstract recursion theory must generalize not only the concept of "recursive" but also that of "finite". This observation is due to Kreisel who must be given credit for driving it home with frequent and forceful repetition, e.g. see Kreisel-Sacks (1965), Kreisel (1965). Kreisel approaches abstract recursion theory in model-theoretic terms, via *invariant definability*, and his *finiteness* (relative to a given class of structure) is *absolute* invariant definability. Another definition of relative finiteness is in *metarecursion theory*, where a set is *metafinite* if it is recursive and bounded, see Kreisel and Sacks (1965).

Our approach is via computable functionals: roughly speaking, a set will be "finite" if we can computably *quantify* over it. This will not be a natural definition of finiteness, unless the relevant theory is *regular* in the following sense.

Definition 13. Let Θ be a computation theory on a domain $\mathfrak{A} = \langle A, C, N, s, M, K, L\rangle$. We call Θ *regular* (on $\mathfrak{A}$) if the following conditions are satisfied.

(8–1) $C = A$.

(8–2) *The relation* $x = y$ *on* A *is* Θ*-computable*.

(8–3) Θ *has a selection operator*.

Condition (8–1) insures that for each $y \in A$, the constant function λxy is Θ-computable with Θ-code a Θ-computable function of y. As was mentioned in Section 6, this conditions also insures that if Θ has a 0-ary selection operator, then it has a uniform selection operator. Letting $q(a)$ be such a 0-ary selection operator for Θ and $q^n(a, \sigma)$ the n-ary selection operator associated with this q for each n, put

(8–4) $\nu z\,[\{a\}_\Theta(z, \sigma) \to 0] = q^n(a, \sigma)$.

(Caution: this is a different use of ν than that of Moschovakis (1969a), e.g. if $q^n(a, \sigma)$ is a single-valued function.) We shall often write expressions of the form

$$\nu z R(z, \sigma)$$

whenever $R(z, \sigma)$ is a Θ-semicomputable relation, keeping in mind that the values of $\nu z R(z, \sigma)$ depend not only on the relation $R(z, \sigma)$ but also on a particular *code* of $R(z, \sigma)$, i.e. any code $\hat{f}$ of a function f such that

$$R(z, \sigma) \Longleftrightarrow f(z, \sigma) \to 0 \;.$$

Definition 14. Let Θ be a regular theory on $\mathfrak{A}$, suppose $B \subseteq A$. The *B-quantifier* is the continuous p.m.v. functional $\mathbf{E}_B(f)$ on unary p.m.v.'s defined by

(8–5) $$\mathbf{E}_B(f) \to \begin{cases} 0 & \textit{if}\ (\exists x \in B)\,[f(x) \to 0]\ , \\ 1 & \textit{if}\ (\forall x \in B)\,[f(x) \to 1]\ . \end{cases}$$

We call B *Θ-finite* with *canonical Θ-code e*, if the B-quantifier $\mathbf{E}_B$ is Θ-computable with Θ-code e.

For the results (i)–(v) below assume that Θ is a regular theory on some domain $\mathfrak{A}$.

(i) *If B is a finite subset of A, then B is Θ-finite.*

Proof. For simplicity, let $B = \{y_1, y_2\}$ be a doubleton and notice that

$$\mathbf{E}_B(f) = \nu z\{[[f(y_1) \to 0 \vee f(y_2) \to 0] \ \& \ z = 0]$$

$$\vee \ [f(y_1) \to 1 \ \& \ f(y_2) \to 1 \ \& \ z = 1]\} .$$

To see that $\mathbf{E}_B$ is Θ-computable we must use all three properties (8–1) – (8–3) of the regular theory Θ. (The length condition (6–4) in the definition of a selection is needed in order to prove that functionals defined this way are Θ-computable.)

(ii) *If B is Θ-finite, then B is Θ-computable.*

Proof. For each $x \in A$, let

$$h_x(u) = \begin{cases} 0 & \text{if } u = x \ , \\ 1 & \text{if } u \neq x \ ; \end{cases}$$

the regularity of Θ implies that for some Θ-computable $g(x)$,

$$(8\text{–}6) \quad h_x = \{g(x)\}_\Theta \ .$$

Now the function

$$f(x) = \mathbf{E}_B(h_x)$$

is the characteristic function of B, and it is easily seen to be Θ-computable, from (8–6) and the definition of a Θ-computable functional.

The next three properties can be shown by similar constructions, only a bit more complicated in the case of (v).

(iii) *If B is Θ-finite, $D \subseteq B$ and D is Θ-computable, then D is Θ-finite.*

(iv) *If B is Θ-finite and f a Θ-computable mapping, then $f[B]$ is Θ-finite.*

(v) *If B is Θ-finite, f a Θ-computable mapping and for each $x \in B$, $f(x)$ is a Θ-canonical code of some Θ-finite set B_x, then $\bigcup_{x \in B} B_x$, $\bigcap_{x \in B} B_x$ are Θ-finite.*

It is clear that (i)–(v) are direct generalizations of the fundamental properties of finite sets. Of course one property of true finiteness that does nor generalize is that *every* subset of a finite set is finite; we can only assert this about computable subsets.

9. Normal theories

One of the fundamental properties of Turing machine computability is that computations are finite. By analogy, we would like to call a computation theory Θ *normal*, if computations in Θ are Θ-finite. Since "computations" are not among our primitive or defined concepts, we must replace this by a finiteness condition which makes sense in our context and which says basically the same thing. Our choice: the set of computations of length smaller than any given computation is Θ-finite.

If B is a set of n-tuples in A, let

$$\langle B \rangle = \{\langle x_1, ..., x_n \rangle : x_1, ..., x_n \in B\},$$

where $\langle x_1, ..., x_n \rangle$ is defined by (3–1), (3–2). Call *B Θ-finite with Θ-canonical code e*, if $\langle B \rangle$ is Θ-finite with Θ-canonical code e.

Definition 15. A computation theory Θ on a domain $\mathfrak{A}$ is *normal* if Θ is regular and there is a Θ-computable mapping $p(n)$ such that for each $(a, \sigma, z) \in \Theta$ with $lh(\sigma) = n$,

$$\{(a', \sigma', z') : (a', \sigma', z') \in \Theta \;\&\; |a', \sigma', z'|_\Theta \leqslant |a, \sigma, z|_\Theta\}$$

is Θ-finite with Θ-canonical code $\{p(n)\}_\Theta(a, \sigma, z)$.

Thus Θ is normal if it is regular and the *initial segments* of Θ in the prewellordering induced by $|\quad|_\Theta$ are *uniformly* Θ-finite.

One can extend to normal theories several nice properties of ordinary recursion theory which do not hold for all computation theories. Here are two such properties which have fairly simple proofs.

(i) *Let Θ be normal on $\mathfrak{A}$. If B, D are Θ-semicomputable subsets of A, then there exist Θ-semicomputable sets B_1, D_1 such that $B_1 \subseteq B$, $D_1 \subseteq D$, $B \cup D = B_1 \cup D_1$, $B_1 \cap D_1 = \emptyset$. (The reduction lemma.)*

It is known that the class of Π^1_1 subsets of ω satisfies the reduction property but that the class of Σ^1_1 subsets of ω does not. Hence by (i), *there is no normal theory on the integers whose semicomputable sets are precisely the Σ^1_1 sets*. On the other hand, *there is a normal theory on the integers whose semicomputable sets are precisely the Π^1_1 sets*, namely the theory of hyperarithmetic sets described in Section 1. Normality is very essential here, since *for each $n \geqslant 1$* there is a *computation theory whose semicomputable sets are the Σ^1_n* (or the Π^1_n) *sets*; such a theory is $PR[f]$, with f a single-valued partial function with range $\{0\}$ and such that $\lambda xyf(x, y) = 0$ is a *complete* Σ^1_n (or Π^1_n) relation.

(ii) *Let Θ be normal on $\mathfrak{A}$, suppose f is a totally defined Θ-computable multiple-valued function on A to A. There exists a totally defined Θ-computable multiple-valued function $g \subseteq f$, such that the relation $g(x) \to y$ is Θ-computable. (The normal branch lemma.)*

Of course this last property is of no significance for single-valued theories, since if f is single-valued, total and Θ-computable then the relation $f(x) = y$ is automatically Θ-computable. The lemma is of some interest in the case of multiple-valued theories, e.g. see Moschovakis (1969b), Lemmas 53, 55 and Theorem 9.

Normality is not preserved under the equivalence relation of Definition 10; e.g. *Turing machine theory as defined in* Section 2 *is normal, and it is equivalent to the prime computation theory over the integers, which is not normal.* Perhaps we need a stronger equivalence relation which will preserve normality – this will certainly make our theory smoother. On the other hand the two "presentations" of ordinary recursion theory in the counterexample are certainly "equivalent" in some natural interpretation of this word. The situation here is quite similar to that of metric space theory, where *completeness* is a very natural metric property which is not preserved by the equally natural equivalence relation of *homeomorphism*.

In some vague sense, a normal theory Θ should be "determined" by the

Θ-finite sets. Of course the simplest question one may ask about these sets is whether the whole space A is Θ-finite. The answer to just this one question gives us quite a bit of information about Θ, so we shall introduce terminology suggestive of the expected results.

Definition 16. Let Θ be a normal theory on a domain $\mathfrak{A}$. If the whole space A is not Θ-finite, we call Θ a *Friedberg theory*. If A is Θ-finite, we call Θ a *Spector theory*.

As we shall see in the next two secctions, Friedberg theories are much like ordinary recursion theory and its Kreisel-Sacks-Kripke-Platek generalization to admissible ordinals. Spector theories share many properties with the theory of hyperarithmetic sets and its generalization to arbitrary structures, hyperprojective theory.

10. The first alternative: Friedberg theories

Suppose ξ is an ordinal, $\mathbf{R} = R_1, ..., R_k$ are relations on ξ and ξ is *admissible* relative to $\mathbf{R}$ on the sense of Kripke. We associate with ξ the domain $\langle \xi, \xi, \omega, s, M, K, L \rangle$, where $s(\beta) = \beta + 1$ and M, K, L are any metarecursive pair with its inverses. It is not hard to verify that metarecursion theory relative to $\mathbf{R}$ on this domain, say as formulated by Kripke, is a Friedberg theory in our sense.

In this section we shall prove a structure theorem for Friedberg theories which is a kind of converse to this observation. With each Friedberg theory Θ on some domain $\mathfrak{A}$ we shall associate a *prewellordering* $\lesssim_\Theta$ of A and certain relations $\mathbf{R}_\Theta$ and p.m.v. functions $\mathbf{f}_\Theta$ on A. The pair $\langle \mathfrak{A}, \lesssim_\Theta \rangle$ will be *admissible* with respect to $\mathbf{R}_\Theta$ in a sense which is similar to Kripke's and which reduces to Kripke's if $\lesssim_\Theta$ happens to be a wellordering. Moreover the functions $\mathbf{f}_\Theta$ will satisfy certain "nicety" conditions also. We will then give a general procedure for constructing from any $\mathfrak{A}, \lesssim, \mathbf{R}, \mathbf{f}$ which satisfies all the above conditions a Friedberg theory Θ: of course the key result will be that if we start from Θ, get $\lesssim_\Theta, \mathbf{R}_\Theta, \mathbf{f}_\Theta$ and then construct Θ' from these, Θ will be equivalent to Θ'. Thus the construction will yield (up to equivalence) all Friedberg theories.

Suppose Θ is normal on $\mathfrak{A}$, choose $i \in A$ to be a Θ-code of the identity function, i.e. for all $x \in A$,

$$\{i\}_\Theta(x) = x \;.$$

The relation

$$(10\text{–}1)\quad x \precsim_\Theta y \Leftrightarrow |i, x, x|_\Theta \leqslant |i, y, y|_\Theta$$

is a prewellordering on A, i.e. it has all the properties of a wellordering except (perhaps) antisymmetry. (We shall write $x \prec_\Theta y$ for $x \precsim_\Theta y \;\&\; \neg(y \precsim_\Theta x)$.) The key property of a normal theory implies easily that *$x \precsim_\Theta y$ is a Θ-computable relation and for each y, $\{x : x \precsim_\Theta y\}$ is Θ-finite*. In fact it is not hard to show the following stronger result.

(i) *If Θ is normal, then the functional*

$$(10\text{–}2)\quad \mathrm{E}^{\precsim_\Theta}(f, x) \to \begin{cases} 0 & \text{if } (\exists y \prec_\Theta x)\,[f(y) \to 0]\ , \\ 1 & \text{if } (\forall y \prec_\Theta x)\,[f(y) \to 1] \end{cases}$$

is Θ-computable.

This implies immediately that *each Θ-computable $\precsim_\Theta$-bounded set is Θ-finite*. For Friedberg theories the converse also holds.

(ii) *If Θ is a Friedberg theory on $\mathfrak{A}$ and B is Θ-finite, then B is $\precsim_\Theta$-bounded, i.e. for some $y \in A$,*

$$B \subseteq \{x : x \prec_\Theta y\}\,.$$

Proof. If not, then

$$A = \bigcup_{x \in B} B_x\ ,$$

where

$$B_x = \{y : y \prec_\Theta x\}\,.$$

It is easy to define a Θ-computable mapping f such that for each $x \in A$, $f(x)$ is a Θ-canonical code of B_x. Then A is Θ-finite by (v) of Section 8, contradicting the key property of Friedberg theories.

(iii) *Let Θ be a Friedberg theory on $\mathfrak{A}$. For each n there is a Θ-computable relation $R_n(a, \sigma, y, z)$, such that for all σ with $lh(\sigma) = n$,*

$$\{a\}_\Theta(\sigma) \to y \Longleftrightarrow (\exists z) R_n(a, \sigma, n, z) .$$

(*Normal form theorem.*)

Proof. Take the case $n = 1$, the others being very similar. If $(a, x, y) \in \Theta$, then there must be some z such that

$$|a, x, y|_\Theta \leqslant |i, z, z|_\Theta ,$$

since otherwise $A = \{z : |i, z, z|_\Theta \leqslant |a, x, y|_\Theta\}$ and this set is easily seen to be Θ-finite. Hence

$$\{a\}_\Theta(x) \to y \Longleftrightarrow (\exists z)\, \{|a, x, y|_\Theta \leqslant |i, z, z|_\Theta\}$$

and we can put

$$R_1(a, x, y, z) \Longleftrightarrow |a, x, y|_\Theta \leqslant |i, z, z|_\Theta$$

which is easily seen to be Θ-computable.

Suppose $\precsim$ is any prewellordering on some domain $\mathfrak{A} = \langle A, A, N, s, M, K, L\rangle$, $\mathbf{R} = R_1, ..., R_k$ is a sequence of relations on A. The language $\mathcal{L} = \mathcal{L}(\mathfrak{A}, \precsim, \mathbf{R})$ has relation and function symbols $R_1, ..., R_k, \precsim, =, N, s, M, K, L$ which are given the obvious interpretations, variables over A, logical symbols $\Rightarrow$, &, $\vee$, $\neg, \exists, \forall$, a relation symbol $\leqslant_N$ which is interpreted by the natural ordering on N and a function symbol $p(x, i)$ which is interpreted by the function $(x)_i$ defined by (3–3). (The last two will allow us to talk about tuples from A in $\mathcal{L}$.) $\Sigma_0(\precsim, \mathbf{R})$ *formulas* are built up from the prime formulas by substitutions of terms and the operations $\Rightarrow$, &, $\vee$, $\neg$, $(\exists x \precsim y)$, $(\forall x \precsim y)$, $(\exists i \leqslant_N j)$, $(\forall i \leqslant_N j)$, $\Sigma_1(\precsim, \mathbf{R})$ *formulas* are those of the form $(\exists x)\theta$ with θ $\Sigma_0(\precsim, \mathbf{R})$, $\Pi_1(\precsim, \mathbf{R})$ formulas are negations of $\Sigma_1(\precsim, \mathbf{R})$ formulas. Relations on A are called $\Sigma_0(\precsim, \mathbf{R})$, $\Sigma_1(\precsim, \mathbf{R})$, $\Pi_1(\precsim, \mathbf{R})$ if they can be defined by formulas in these classes, allowing parameters from A; relations are $\Delta_1(\precsim, \mathbf{R})$ if they are both $\Sigma_1(\precsim, \mathbf{R})$ and $\Pi_1(\precsim, \mathbf{R})$. Of course we will often skip the embelishments and write, $\Sigma_0(\mathbf{R})$, Σ_1, etc. if $\precsim$ or $\mathbf{R}$ are clearly understood by the context.

(iv) *Let Θ be a Friedberg theory on $\mathfrak{A}$. There exists Θ-computable relations $\mathbf{R}_\Theta = R_1, R_2$ such that the Θ-semicomputable relations are precisely the $\Sigma_1(\precsim_\Theta, \mathbf{R}_\Theta)$ relations.*

Proof. Take

$$\text{(10–3)}\quad R_1(a, x, z, w) \Longleftrightarrow |a, x, z|_\Theta \leqslant |i, w, w| ,$$

$$\text{(10–4)}\quad R_2(a, b, c) \Longleftrightarrow a = S^1_1(b, c) ,$$

$$\text{(10–5)}\quad \mathbf{R}_\Theta = R_1, R_2 .$$

We prove by induction on $n \geqslant 1$ that each relation

$$\{a\}_\Theta(\sigma) \to z \Longleftrightarrow (a, \sigma, z) \in \Theta \qquad (lh(\sigma) = n)$$

is $\Sigma_1(\mathbf{R}_\Theta)$. For $n = 1$,

$$\{a\}_\Theta(x) \to z \Longleftrightarrow (\exists w) R_1(a, x, z, w)$$

and then inductively,

$$\{a\}_\Theta(x, \sigma) \to z \Longleftrightarrow \{S^1_1(a, x)\}_\Theta(\sigma) \to z$$

$$\Longleftrightarrow (\exists b)\, [R_2(b, a, x) \,\&\, \{b\}_\Theta(\sigma) \to z] ,$$

from which the result follows using some easily shown closure properties of Σ_1 relations. For $n = 0$, notice that by axiom XII there is a fixed b so that

$$\{a\}_\Theta(\ ) \to z \Longleftrightarrow \{b\}_\Theta(a) \to z ,$$

from which the result follows by using the case for $n = 1$. That all $\Sigma_0(\mathbf{R}_\Theta)$ relations are Θ-computable is an easy consequence of (i). Finally that all $\Sigma_1(\mathbf{R}_\Theta)$ relations are Θ-semicomputable follows from the fact that Θ is regular, so that the Θ-semicomputable relations are closed under existential quantification.

Suppose now that a domain $\mathfrak{A}$, a prewellordering $\lesssim$ on A and relations $\mathbf{R}$ on A are given. Can we find some Friedberg theory Θ on $\mathfrak{A}$ whose canonical prewellordering is $\lesssim$ and whose semicomputable relations are the $\Sigma_1(\lesssim, \mathbf{R})$ relations? Of course we do not expect this to be true for all $\mathfrak{A}, \lesssim, \mathbf{R}$, but only those which are *admissible* in a sense which we shall presently make precise.

For any $n \geqslant 1$, let X vary over sets of n-tuples from A. $\Sigma_0(\lesssim, X, \mathbf{R})$,

$\Sigma_1(\precsim, X, \mathbf{R})$ formulas are defined exactly like $\Sigma_0(\precsim, \mathbf{R})$, $\Sigma_1(\precsim, \mathbf{R})$ formulas, except that we allow *positive* occurrences of prime formulas

$$\sigma \in X \qquad (lh(\sigma) = n) .$$

If $\theta(\sigma, X)$ is $\Sigma_1(\precsim, X, \mathbf{R})$ with exactly n free variables (recall that we allow parameters from A), it determines a *monotone operator*

$$(10\text{–}6) \quad \Gamma_\theta(X) = \{\sigma : \theta(\sigma, X)\}$$

on sets of n-tuples. As usual we put

$$(10\text{–}7) \quad X^\xi = \Gamma_\theta(\bigcup_{\eta<\xi} X^\eta) ,$$

$$(10\text{–}8) \quad X^* = \bigcup_{\xi} X^\xi ,$$

so that X^* is the least fixed point of Γ_θ. We call X^* *the set defined inductively by* $\theta(\sigma, X)$.

Definition 17. Let $\mathfrak{A}$ be a domain, $\precsim$ a prewellordering on A, $\mathbf{R}$ a sequence of relations on A. The pair $\langle \mathfrak{A}, \precsim \rangle$ is $\mathbf{R}$-*admissible* if for each $\Sigma_1(\precsim, X, \mathbf{R})$ formula $\theta(\sigma, X)$, the set X^* defined inductively by $\theta(\sigma, X)$ is $\Sigma_1(\precsim, \mathbf{R})$.

(v) *Let* Θ *be a Friedberg theory on* $\mathfrak{A}, \precsim_\Theta, \mathbf{R}_\Theta$ *defined by* (10–1), (10–5), *let* $\mathbf{R}$ *be any sequence of* Θ-*computable relations which extends* $\mathbf{R}_\Theta$. *Then* $\langle \mathfrak{A}, \precsim_\Theta \rangle$ *is* $\mathbf{R}$-*admissible.*

Proof. For each $\Sigma_1(\precsim_\Theta, X, \mathbf{R})$ formula $\theta(\tau, X)$, consider the functional $\varphi(f, \tau)$ on n-ary p.m.v.'s f and m-tuples τ defined by

$$\varphi(f, \tau) \to 0 \Leftrightarrow \theta(\tau, \{\sigma : f(\sigma) \to 0\}) .$$

An elementary induction on the construction of $\theta(\tau, X)$ shows that φ is Θ-computable. From this the result follows by the first recursion theorem and (iv).

We shall eventually show that for each $\mathbf{R}$-admissible $\langle \mathfrak{A}, \precsim \rangle$ there is a Friedberg theory Θ such that $\precsim_\Theta = \precsim$ and whose Θ-semicomputable relations are precisely the $\Sigma_1(\precsim, \mathbf{R})$ relations. However in order to make the construction fine enough to yield all Friedberg theories up to equivalence, we must first develop some of the theory of admissible prewellorderings.

(vi) *Assume* $\langle \mathfrak{A}, \precsim \rangle$ *is* **R**-*admissible. There is a regular theory* Θ *on* $\mathfrak{A}$ *whose semicomputable relations are precisely the* $\Sigma_1(\precsim, \mathbf{R})$ *relations. Moreover the functional* $\mathbf{E}^{\precsim}$ *defined by*

$$(10\text{–}9) \quad \mathbf{E}^{\precsim}(f, x) \rightarrow \begin{cases} 0 & \text{if } (\exists y \prec x)\,[f(y) \rightarrow 0]\ , \\ 1 & \text{if } (\forall y \prec x)\,[f(y) \rightarrow 1] \end{cases}$$

is Θ-*computable and the whole domain* A *is not* Θ-*finite.*

Proof. Take $\Theta = PR[\precsim, R, =, \mathbf{E}^{\precsim}, \nu]$, where to be precise we should substitute for the relations within the brackets their characteristic functions. This is surely regular and all $\Sigma_1(\precsim, \mathbf{R})$ relations are Θ-semicomputable. To prove the converse, consider the relation

$$(m, x) \in B \Leftrightarrow m \in N \;\&\; m \geqslant 2 \;\&\; ((x)_1, \ldots, (x)_m) \in \Theta\ .$$

The inductive definition of Θ yields a $\Sigma_1(\precsim, X, \mathbf{R})$ formula $\theta(m, x, X)$ which defines B inductively, so that B is $\Sigma_1(\precsim, \mathbf{R})$, and hence for each n, the relation

$$(a, \sigma, z) \in \Theta \qquad (lh(\sigma) = n)$$

is $\Sigma_1(\precsim, \mathbf{R})$. Of course $\mathbf{E}^{\precsim}$ is Θ-computable, by the definition of prime computability. If A were Θ-finite, then the class of Θ-computable relations would be closed under quantification over A and in particular each $\Sigma_1(\precsim, \mathbf{R})$ relation would be Θ-computable; but then each Θ-semicomputable relation would be Θ-computable, which is absurd.

(vii) *Assume* $\langle \mathfrak{A}, \precsim \rangle$ *is* **R**-*admissible, let* $\theta(x)$ *be a* $\Sigma_0(\precsim, \mathbf{R})$ *formula. Then*

$$(10\text{–}10)\ (\forall u \prec v)(\exists x)\theta(x) \Rightarrow (\exists w)(\forall u \prec v)(\exists x \prec w)\theta(x)\ ,$$

$$(10\text{–}11)\ (\forall i <_N j)(\exists x)\theta(x) \Rightarrow (\exists w)(\forall i <_N j)(\exists x \prec w)\theta(x)\ .$$

Proof. Let f be a mapping with values 0, 1 such that

$$f(u,x)=0 \Leftrightarrow [\neg(u \prec v) \;\&\; x=0]$$
$$\vee\ [u \prec v \;\&\; \theta(x) \;\&\; (\forall y \prec x)[\neg\theta(y)]]\ ,$$

put

$$g(u)=\nu(\lambda x f(u,x))\ .$$

Now g is Θ-computable in the theory defined in (vi) above. If the antecedent of (10–10) holds and the consequent fails, then g is total and $\precsim$-unbounded; it is then not hard to show that A is Θ-finite, contradicting (vi). The proof of (10–11) is similar.

One of the corollaries of (vii) is the closure of $\Sigma_1(\precsim, \mathbf{R})$ under bounded quantification of both sorts (over $\precsim$ on A and $\leqslant_N$ on N) and both kinds ($\forall$ and $\exists$).

If $\theta(\sigma, X)$ is a $\Sigma_1(\precsim, X, \mathbf{R})$ formula, then for some ordinal ξ,

(10–12) $X^* = X^\xi$

in the notation established by (10–7), (10–8). We call the least ξ for which (10–12) holds *the closure ordinal* of Γ_θ (or θ).

Let

(10–13) $\rho : A \to |\precsim|$

be the unique mapping that takes A *onto* some (uniquely defined) ordinal $|\precsim|$ and carries $\precsim$ onto the ordering of $|\precsim|$. We call $|\precsim|$ *the length of the prewellordering* $\precsim$.

(viii) *Assume* $\langle \mathfrak{A}, \precsim \rangle$ *is* $\mathbf{R}$*-admissible, let* $\theta(\sigma, x)$ *be a* $\Sigma_1(\precsim, X, \mathbf{R})$ *formula that defines a monotone operator* Γ_θ *by* (10–6). *Then the relation*

(10–14) $P(\sigma, x) \Leftrightarrow \sigma \in X^{\rho(x)}$

is $\Sigma_1(\precsim, \mathbf{R})$. *If* $\theta(\sigma, X)$ *is* $\Sigma_0(\precsim, X, \mathbf{R})$, *then* $P(\sigma, x)$ *is* $\Delta_1(\precsim, \mathbf{R})$.

Proof. Here we must use the theory Θ defined in the proof of (vi) in a more essential way. The idea is to use the second recursion theorem for this Θ and find some $\hat{p} \in A$ such that

$$P(\sigma, x) \Longleftrightarrow \{\hat{p}\}_{\Theta}(\sigma, x) \rightarrow 0 \ ,$$

(for the second assertion we find $\hat{p}$ such that

$$P(\sigma, x) \Longleftrightarrow \{(\hat{p})_1\}_{\Theta}(\sigma, x) \rightarrow 0 \ ,$$

$$\neg P(\sigma, x) \Longleftrightarrow \{(\hat{p})_2\}_{\Theta}(\sigma, x) \rightarrow 0) \ .$$

We omit the details.

(ix) *Assume* $\langle \mathfrak{A}, \precsim \rangle$ *is* **R**-*admissible, let* $\theta(\sigma, X)$ *be a* $\Sigma_1(\precsim, X, \mathbf{R})$ *formula that defines a monotone operator* Γ_θ *by* (10–6). *Then the closure ordinal of* Γ_θ *is no larger than* $|\precsim|$.

Proof. It is enough to show that

$$\Gamma_\theta(X^{|\precsim|}) = X^{|\precsim|} \ ,$$

i.e.

$$\theta(\sigma, \{\sigma' : (\exists x)P(\sigma', x)\}) \Rightarrow (\exists x)P(\sigma, x) \ ,$$

where $P(\sigma, x)$ is the relation defined by (10–14). For this again it is enough to show

$$(10\text{–}15)\ \theta(\sigma, \{\sigma' : (\exists x)P(\sigma', x)\}) \Rightarrow (\exists w)\theta(\sigma, \{\sigma' : P(\sigma', w)\}) \ ,$$

since

$$\theta(\sigma, \{\sigma' : P(\sigma', w)\}) \Rightarrow P(\sigma, w')$$

whenever $w \prec w'$. We show (10–15) by rpoving that *for every* $\Sigma_1(\precsim, X, \mathbf{R})$ formula $\theta'(X)$ (allowing parameters from A),

$$\theta'(\{\sigma : (\exists x)P(\sigma, x)\}) \Rightarrow (\exists x)\theta'(\{\sigma : P(\sigma, x)\}) \ .$$

This is seen easily by induction on the construction of $\theta'(X)$, using (vii) and (viii).

In Kripke-Platek style metarecursion theory, admissible ordinals can be characterized by (ix): *ξ is admissible if for each Σ_1 θ, Γ_θ has a closure ordinal no larger than ξ*. We do not know if this is enough in our more general setting of prewellorderings. We also do not know if $\langle \mathfrak{A}, \lesssim \rangle$ must be $\mathbf{R}$-admissible if it satisfies the conclusions of both (vii) and (ix).

Assume $\langle \mathfrak{A}, \lesssim \rangle$ is $\mathbf{R}$-admissible, let Θ be the regular theory constructed on $\mathfrak{A}$ in (vi), put

$$(10\text{–}16)\; G = \{(a, x) : \{a\}_\Theta(x) \to 0\}\,.$$

Now G is $\Sigma_1(\lesssim, \mathbf{R})$*-universal*, i.e. it is a $\Sigma_1(\lesssim, \mathbf{R})$ set of pairs and for every $\Sigma_1(\lesssim, \mathbf{R})$ set B there is some a such that

$$(10\text{–}17)\; B = G_a = \{x : (a, x) \in G\}\,.$$

When (10–17) holds we call a a $\Sigma_1(\lesssim, \mathbf{R})$*-code* of B. A $\Sigma_1(\lesssim, \mathbf{R})$ *selection function* is any p.m.v. function $p(a)$ such that the following two conditions hold.

$$(10\text{–}18)\; \textit{The relation } p(a) \to x \textit{ is } \Sigma_1(\lesssim, \mathbf{R})\,.$$

$$(10\text{–}19)\; G_a \neq \emptyset \Rightarrow [p(a)\downarrow \;\&\; (\forall x)\,[p(a) \to x \Rightarrow x \in G_a]]\,.$$

We now proceed to the main construction of this section. Assume given a domain $\mathfrak{A}$, a prewellordering $\lesssim$ on A, relations $\mathbf{R} = R_1, \ldots, R_k$ on A and p.m.v. functions $\mathbf{f} = f_1, \ldots, f_l$, p on A, such that the following conditions hold.

(a) *$\langle \mathfrak{A}, \lesssim \rangle$ is $\mathbf{R}$-admissible.*

(b) *For each i, the relation $f_i(\sigma_i) \to z$ is $\Sigma_1(\lesssim, \mathbf{R})$.*

(c) *p is a $\Sigma_1(\lesssim, \mathbf{R})$ selection function.*

The idea is to tamper with the definition of $PR[\lesssim, \mathbf{R}, \mathbf{f}, p_1, =, \mathbf{E}^{\lesssim}]$ so that we get a normal theory.

Since $\mathbf{f}, p$ have $\Sigma_1(\lesssim, \mathbf{R})$ graphs, we can find $\Sigma_0(\lesssim, \mathbf{R})$ relations $P_1, \ldots, P_l, P_p$ such that

$$(10\text{–}20)\; f_i(\sigma_i) \to z \Longleftrightarrow (\exists w) P_i(\sigma_i, w)\,,$$

$$(10\text{–}21)\quad p(a) \to x \Longleftrightarrow (\exists w) P_p(a, x, w) .$$

As usually, the desired theory will be the least fixed point Θ^* of some operator $\Gamma(\Theta)$ on sets of tuples from A with length $\geqslant 2$. For each such set Θ, let

$$\Theta^\dagger = \{x : (\forall u \prec x)(\langle 2, 0\rangle, u, u) \in \Theta\} .$$

We define Γ by clauses I$'$–XV$'$, which are variations on the clauses defining $PR[\precsim, \mathbf{R}, \mathbf{f}, p, =, \mathbf{E}^{\precsim}]$. *Clauses* I$'$–XIII$'$ *are obtained by restricting all arguments and quantifiers in clauses* I–XIII *of* Section 3 *to* $\Theta^\dagger$, except for clause II$'$ which gets special treatment, because we allow $\langle 2, 0\rangle$ as an argument, even though it need not be in $\Theta^\dagger$. For example, recalling that $\mathfrak{A} = \langle A, A, N, s, M, K, L\rangle$, i.e. $C = A$, here are sample clauses:

I$'$ *If* $\langle 1, y\rangle, \sigma, y \in \Theta^\dagger$, *then* $(\langle 1, y\rangle, \sigma, y) \in \Gamma(\Theta)$.

II$'$ *If* $y, \sigma \in \Theta^\dagger$, *then* $(\langle 2, 0\rangle, y, \sigma, y) \in \Gamma(\Theta)$.

IX$'$ *If* $\langle 9, 0\rangle, \hat{f}, \hat{g}, \sigma, z \in \Theta^\dagger$ *and for some* $u \in \Theta^\dagger$, $(\hat{g}, \sigma, u) \in \Theta$, $(\hat{f}, u, \sigma, z) \in \Theta$, *then* $(\langle 9, 0\rangle, \hat{f}, \hat{g}, \sigma, z) \in \Gamma(\Theta)$.

Clause XIV$'$ introduces the characteristic functions of $\precsim, R_1, \ldots, R_k, =$ and the functions $f_1, \ldots, f_l, p$. For the former we have clauses:

If $R_i(t_1, \ldots, t_{n_i})$ *and* $\langle 14, s(i)\rangle, t_1, \ldots, t_{n_i}, 0 \in \Theta^\dagger$,

then $(\langle 14, s(i)\rangle, t_1, \ldots, t_{n_i}, 0) \in \Gamma(\Theta)$;

if $\neg R_i(t_1, \ldots, t_{n_i})$ *and* $\langle 14, s(i)\rangle, t_1, \ldots, t_{n_i}, 1 \in \Theta^\dagger$,

then $(\langle 14, s(i)\rangle, t_1, \ldots, t_{n_i}, 1) \in \Gamma(\Theta)$.

(Here $\precsim$ is considered the first relation, getting code $\langle 14, 1\rangle$, and $=$ is the $k + 2$'nd relation, getting code $\langle 14, k + 2\rangle$.) The functions $f_1, \ldots, f_l, p$ will get codes $\langle 14, k + 3\rangle, \ldots, \langle 14, k + l + 3\rangle$ respectively, but we use the relations $P_1, \ldots, P_l, P_p$ to introduce them. A typical clause is:

If $\langle 14, k+2+i\rangle, t_1, ..., t_{n'_i}, z \in \Theta^\dagger$

and for some $w \in \Theta^\dagger, P_i(t_1, ..., t_{n'_i}, z, w)$,

then $(\langle 14, k+2+i\rangle, t_1, ..., t_{n'_i}, z) \in \Gamma(\Theta)$.

Finally clause XV′ introduces the functional $\mathbf{E}^{\precsim}$.

XV′ If $\langle 15, 1\rangle, \hat{f}, x, 0 \in \Theta^\dagger$ and $(\exists y \prec x)\,[(\hat{f}, y, 0) \in \Theta]$,

then $(\langle 15, 1\rangle, \hat{f}, x, 0) \in \Gamma(\Theta)$.

If $\langle 15, 1\rangle, \hat{f}, x, 1 \in \Theta^\dagger$ and $(\forall y \prec x)\,[(\hat{f}, y, 1) \in \Theta]$,

then $(\langle 15, 1\rangle, \hat{f}, x, 1) \in \Gamma(\Theta)$.

In the usual way we now put

(10–22) $\Theta^\xi = \Gamma(\bigcup_{\eta<\xi} \Theta^\eta)$,

(10–23) $PR(\precsim, \mathbf{R}, \mathbf{f}, p) = \Theta^* = \bigcup_\xi \Theta^\xi$,

(10–24) $|a, \sigma, z|_{PR(\precsim, \mathbf{R}, \mathbf{f}, p)} = |a, \sigma, z| = \text{least } \xi\,[(a, \sigma, z) \in \Theta^\xi]$.

(*Notice that the definition of* $PR(\precsim, \mathbf{R}, \mathbf{f}, p)$ *depends not only on* $\mathbf{f}, p$ *but also on the particular* $\Sigma_0(\precsim, \mathbf{R})$ *relations* $P_1, ..., P_l, P_p$ *chosen to define* $\mathbf{f}, p$.)

Trivial transfinite inductions show that

(x) $(\langle 2, 0\rangle, y, y) \in \Theta^\xi \Leftrightarrow \rho(y) \leqslant \xi$,

(xi) $(a, x_1, ..., x_n, z) \in \Theta^\xi \;\&\; a \neq \langle 2, 0\rangle$

$\Rightarrow \rho(a), \rho(x_1), ..., \rho(x_n), \rho(z) \leqslant \xi$,

where $\rho(x)$ is defined by (10–13). Using (x) is easy to verify that at least we have a theory.

(xii) *The set* $PR(\precsim, \mathbf{R}, \mathbf{f}, p)$ *with the length function* (10–24) *is a computa-*

tion theory on $\mathfrak{A}$, *such that the relations, functions and functional* $\precsim, \mathbf{R}, \mathbf{f}, p, =, \mathbf{E}^{\precsim}$ *are computable.*

As in the proof of (vi), put

$$(m, x) \in B \Leftrightarrow m \in N \;\&\; m \geqslant 2 \;\&\; ((x)_1, \ldots, (x)_m) \in PR(\precsim, \mathbf{R}, \mathbf{f}, p)\,.$$

It is easy to verify that B is inductively defined by some $\Sigma_1(\precsim, X, \mathbf{R})$ formula, and since $\langle \mathfrak{A}, \precsim \rangle$ is $\mathbf{R}$-admissible we have

(xiii) *all semicomputable relations in the theory* $PR(\precsim, \mathbf{R}, \mathbf{f}, p)$ *are* $\Sigma_1(\precsim, \mathbf{R})$.

The converse of (xiii) will follow from the existence of a selection operator for $PR(\precsim, \mathbf{R}, \mathbf{f}, p)$. However at this point we can only show that $PR(\precsim, \mathbf{R}, \mathbf{f}, p)$ has a *weak selection operator*, which nonetheless implies the converse of (xiii).

(xiv) *In the theory* $PR(\precsim, \mathbf{R}, \mathbf{f}, p)$ *there is a computable p.m.v. function* $q(a)$, *such that*

$$(10\text{–}25)\; (\exists x)\,[\{a\}(x) \to 0] \Rightarrow q(a)\downarrow \;\&\; (\forall x)\,[q(a) \to x \Rightarrow \{a\}(x) \to 0]\;.$$

Proof. Let Θ be the regular theory on $\mathfrak{A}$ constructed in the proof of (vi). Since the basic relation $\{a\}(x) \to 0$ of the theory $PR(\precsim, \mathbf{R}, \mathbf{f}, p)$ is $\Sigma_1(\precsim, \mathbf{R})$ by (xiii), there is a fixed $b \in A$ such that

$$\{a\}(x) \to 0 \Leftrightarrow \{b\}_\Theta(a, x) \to 0 \Leftrightarrow \{S^1_1(b, a)\}_\Theta(x) \to 0\,.$$

Since $p(a)$ is a $\Sigma_1(\precsim, \mathbf{R})$ selection operator, the function

$$q(a) = p(S^1_1(b, a))$$

satisfies (10–25). To prove that q is computable in $PR(\precsim, \mathbf{R}, \mathbf{f}, p)$ we must recall the definition of Θ as a prime computation theory and in particular that $S^1_1(b, a) = \langle 13, b, a \rangle$, which is certainly $PR(\precsim, \mathbf{R}, \mathbf{f}, p)$-computable.

(xv) *The* $PR(\precsim, \mathbf{R}, \mathbf{f}, p)$*-semicomputable relations are precisely the* $\Sigma_1(\precsim, \mathbf{R})$ *relations and the* $PR(\precsim, \mathbf{R}, \mathbf{f}, p)$*-computable relations are precisely the* $\Delta_1(\precsim, \mathbf{R})$ *relations.*

Proof. The first assertion is shown by induction on the construction of the $\Sigma_1(\precsim, \mathbf{R})$ formula defining a given relation, using the computability of $\mathbf{E}^{\precsim}$ and (xiv). The second assertion follows easily from (xiii) and (xiv).

Before attempting to show that $PR(\precsim, \mathbf{R}, \mathbf{f}, p)$ has a selection operator, we prove that it satisfies the key property of a normal theory. We define a new operator $\Gamma(w, \Theta)$, where w is a parameter ranging over A, by restricting all arguments and quantifiers in the definition of $\Gamma(\Theta)$ to $\precsim_w$, again excepting the term $\langle 2, 0\rangle$ in II$'$; i.e. each clause $x \in \Theta^\dagger$ is replaced by $x \in \Theta^\dagger$ & $x \precsim w$. Let Θ_w^ξ be the sets of tuples defined inductively by this operator for each fixed w, Θ_w^* the least fixed point for each w.

(xvi) *If* $\xi \leqslant \rho(w)$, *then* $\Theta^\xi = \Theta_w^\xi$.

This is immediate using (x).

The operators $\Gamma(\Theta)$, $\Gamma(w, \Theta)$ are defined on sets of tuples from A, of arbitrary length. In the proof of (vi) we showed how to "reduce" such operators to operators on sets of pairs. Here we must look into this reduction for $\Gamma(w, \Theta)$ with a little more care. For each Θ, put

$$X(\Theta) = \{(m, x) : m \in N \;\&\; m \geqslant 2 \;\&\; ((x)_1, ..., (x)_m) \in \Theta\}$$

and for each set X of pairs whose first members are integers $\geqslant 2$ put

$$(10\text{–}26)\quad \Theta(X) = \{((x)_1, ..., (x)_m) : (m, x) \in X\} .$$

Consider the operator $\overline{\Gamma}(w, X)$ on sets of pairs defined by

$$\overline{\Gamma}(w, X) = X(\Gamma(w, \Theta(X))) .$$

If the sets of pairs defined inductively by this operator for each fixed w are X_w^ξ, it is immediate that

$$(10\text{–}27)\quad \begin{cases} X_w^\xi = X(\Theta_w^\xi) , \\ \Theta_w^\xi = \Theta(X_w^\xi) . \end{cases}$$

Now $\overline{\Gamma}(w, X)$ is defined by a $\Sigma_0(\precsim, X, \mathbf{R})$ formula. Hence by (viii), the relation

$$(m, x) \in X_w^{\rho(y)}$$

is $\Delta_1(\precsim, \mathbf{R})$. We compute,

$$\text{(10–28)}\ (a, \sigma, z) \in \Theta^{\rho(y)} \Longleftrightarrow (\exists w)\,[y \precsim w \;\&\; (a, \sigma, z) \in \Theta_w^{\rho(y)}]$$

(by (xvi)

$$\Longleftrightarrow (\exists w)\,[y \precsim w \;\&\; (n+2, \langle a, \sigma, z\rangle) \in X_w^{\rho(y)}]$$

(by (10–27))

which implies that the relation $(a, \sigma, z) \in \Theta^{\rho(y)}$ is $\Sigma_1(\precsim, \mathbf{R})$. Similarly,

$$\text{(10–29)}\ (a, \sigma, z) \in \Theta^{\rho(y)} \Longleftrightarrow (\forall w)\,[y \precsim w \Rightarrow (a, \sigma, z) \in \Theta_w^{\rho(y)}]$$

$$\Longleftrightarrow (\forall w)\,[y \precsim w \Rightarrow (n+2, \langle a, \sigma, z\rangle) \in X^{\rho(y)}]\ ,$$

so that the relation $(a, \sigma, z) \in \Theta^{\rho(y)}$ is $\Delta_1(\precsim, \mathbf{R})$.

(xvi), (10–27) and the fact that $\overline{\Gamma}(w, X)$ is definable by a $\Sigma_0(\precsim, X, \mathbf{R})$ formula imply by (ix) that the closure ordinal of Γ is at most $|\precsim|$, i.e. whenever $\{a\}(\sigma) \to z$, there is some y so that $|a, \sigma, z| = \rho(y)$. Hence

$$|a, \sigma, z| = \rho(y) \Longleftrightarrow (a, \sigma, z) \in \Theta^{\rho(y)} \;\&\; (\forall u \prec y) \neg (a, \sigma, z) \in \Theta^{\rho(u)}\ ,$$

$$|a', \sigma', z'| \leqslant |a, \sigma, z| \Longleftrightarrow (\exists y')(\exists y)\,[|a', \sigma', z'| = \rho(y')$$

$$\&\ |a, \sigma, z| = \rho(y) \;\&\; y' \precsim y]$$

$$\Longleftrightarrow (\forall y')(\forall y)\,[|a', \sigma', z'| = \rho(y')$$

$$\&\ |a, \sigma, z| = \rho(y) \Rightarrow y' \precsim y],$$

so that $\{(a', \sigma', z') : |a', \sigma', z'| \leqslant |a, \sigma, z|\}$ is $\Delta_1(\precsim, \mathbf{R})$ and hence computable. Moreover the computations above, starting with the fixed computable relation $(m, x) \in X_w^{\rho(y)}$ and proceeding with substitutions and applications of computable operations, can be carried out computably in the codes. so that the following holds: *for some computable mapping* $q(n)$, *whenever* $\{a\}(\sigma) \to z$, $\lambda a'\sigma'z' \{q(n)\}(a, \sigma, z, a', \sigma', z')$ *is a mapping with values* 0, 1 *and*

$$|a', \sigma', z'| \leqslant |a, \sigma, z| \Longleftrightarrow \{q(n)\}(a, \sigma, z, a', \sigma', z') = 0 .$$

We want to show that $\{(a', \sigma', z') : |a', \sigma', z'| \leqslant |a, \sigma, z|\}$ is finite in the theory. Consider the functional

$$\varphi(f, a, \sigma, z, y) \to \begin{cases} 0 & \textit{if } \{a\}(\sigma) \to z \;\&\; (\exists a', \sigma', z')\,[\langle a', \sigma', z'\rangle \precsim y \\ & \;\&\; |a', \sigma', z'| \leqslant |a, \sigma, z| \;\&\; f(a', \sigma', z') \to 0], \\ 1 & \textit{if } \{a\}(\sigma) \to z \;\&\; (\forall a', \sigma', z')\,[\langle a', \sigma', z'\rangle \precsim y \\ & \;\&\; |a', \sigma', z'| \leqslant |a, \sigma, z| \Rightarrow f(a', \sigma', z') \to 1]; \end{cases}$$

it is not hard to show that this is computable, using the computability of $\mathbf{E}^{\precsim}$. Whenever $\{a\}(\sigma) \to z$ and $|a, \sigma, z| \leqslant \rho(y)$, $\lambda f \varphi(f, a, \sigma, z, y)$ is precisely the functional representing quantification over $\{\langle a', \sigma', z'\rangle : |a', \sigma', z'| \leqslant |a, \sigma, z|\}$. Since the relation $|a, \sigma, z| = \rho(y)$ is computable, it is easy to use (xiv) and get some computable $r(a, \sigma, z)$ such that

$$\{a\}(\sigma) \to z \Rightarrow [r(a, \sigma, z)\downarrow \;\&\; (\forall y)\,[r(a, \sigma, z) \to y \Rightarrow |a, \sigma, z| = \rho(y)]] .$$

Although $r(a, \sigma, z)$ may be multiple-valued, the functional $\varphi(f, a, \sigma, z, r(a, \sigma, z))$ is clearly single-valued for single-valued f and whenever $\{a\}(\sigma) \to z$, it represents as a functional on f quantification over $\{\langle a', \sigma', z'\rangle : |a', \sigma', z'| \leqslant |a, \sigma, z|\}$. From this it is only a matter of tracing the uniformity of the definitions and proving a couple of properties of relatively finite sets of tuples in order to complete the proof of the required result.

(xvii) *The theory* $PR(\precsim, \mathbf{R}, \mathbf{f}, p)$ *satisfies the key property of a normal theory.*

(xviii) *The theory* $PR(\precsim, \mathbf{R}, \mathbf{f}, p)$ *is normal and the prewellordering associated with it by* (10–1) (*taking* $i = \langle 2, 0\rangle$) *is* $\precsim$.

Proof. The second assertion follows immediately from (x). To complete the proof of the first assertion we only need show that Θ has a 0-ary selection operator, i.e. a computable $q(a)$ with code $\hat{q}$ such that

$$(10\text{–}30)\; (\exists x)\, \{a\}(x) \to 0 \Rightarrow q(a)\downarrow \;\&\; (\forall x)\,[q(a) \to x \Rightarrow \{a\}(x) \to 0] ,$$

$$(10\text{–}31)\ \{a\}(x) \to 0 \ \&\ q(a) \to x \Rightarrow |a, x, 0| < |\hat{q}, a, x| .$$

In (xiv) we showed that there is a $q(a)$ satisfying (10–30). Choose $\hat{f}$ so that

$$\{\hat{f}\}(a, w) \to 0 \Longleftrightarrow |a, (w)_0, 0| \leqslant |\langle 2, 0\rangle, (w)_1, (w)_1|$$

$$\&\ |\hat{q}, a, (w)_0| \leqslant |\langle 2, 0\rangle, (w)_1, (w)_1| ,$$

let

$$g(u, v) = u$$

and put

$$q^*(a) = g((q(S^1_1(f, a)))_0, (q(S^1_1(\hat{f}, a)))_1) .$$

It is easy to verify that $q^*(a)$ satisfies (10–30). If $\hat{q}^*$ is the code assigned to $q^*(a)$ canonically by a sequence of applications of the composition axiom, $\{a\}(x) \to 0$ and $q^*(a) \to x$, then for some z,

$$|\hat{q}^*, a, x| > |b, a, z| ,$$

where b is some element of A such that

$$\{b\}(a) = (q(S^1_1(f, a)))_1 .$$

Now the definition of $\{\hat{f}\}(a, w)$ implies that

$$|a, x, 0| \leqslant |\langle 2, 0\rangle, z, z|$$

and (xi) implies

$$|b, a, z| \geqslant |\langle 2, 0\rangle, z, z|$$

so that $q^1(a)$ with the code $\hat{q}^*$ satisfies (10–31).

We can now reap the corollaries of the construction.

(xix) *Assume* $\langle \mathfrak{A}, \precsim\rangle$ *is* **R**-*admissible. Then there is some normal theory* Θ *on* $\mathfrak{A}$, *with* $\precsim_\Theta = \precsim$ *and whose semicomputable relations are precisely the* $\Sigma_1(\precsim, \mathbf{R})$ *relations.*

Proof. It is enough to show that there exists some $\Sigma_1(\precsim, \mathbf{R})$ selection function, and we can take

$$p(a) \to x \Longleftrightarrow \{a\}(x) \to 0\ ,$$

where $\{a\}(x) \to z$ is relative to the regular theory on $\mathfrak{A}$ constructed in (vi).

In the proof of the next theorem we shall need the following technical result: *let Θ be a Friedberg theory on $\mathfrak{A}$, suppose $f(x_1, ..., x_y)$ is a Θ-computable function with Θ-code $\hat{f}$. For each t, there is a Θ-code $f^{\#}$ of f such that*

$$(10\text{–}32)\ f(x_1, ..., x_n) \to z \Rightarrow \rho(t), \rho(x_1), ..., \rho(x_y), \rho(z), |\hat{f}, x_1, ..., x_n, z|_\Theta$$

$$\leqslant |f^{\#}, x_1, ..., x_n, z|_\Theta\ ,$$

where ρ is associated with $\precsim_\Theta$ via (10–3). This is easily proved by choosing some $h(z, t, x_1, ..., x_n)$ with Θ-code $\hat{h}$ satisfying

$$h(z, t, x_1, ..., x_n) = z\ ,$$

$$|\hat{h}, z, t, x_1, ..., x_n, z|_\Theta \geqslant \rho(z), \rho(t), \rho(x_1), ..., \rho(x_n)\ ,$$

noticing that

$$f(x_1, ..., x_n) = \{h\}_\Theta(\{f\}_\Theta(x_1, ..., x_n), t, x_1, ..., x_n)$$

and taking $f^{\#}$ to be the code assigned to f by axioms IX, XIII, when we view f as a composition in this way.

(xx) *Suppose Θ is a Friedberg theory on $\mathfrak{A}$. There exist Θ-computable $\langle \mathfrak{A}, \precsim_\Theta \rangle$-admissible relations $\mathbf{R}_\Theta$, Θ-computable p.m.v. functions $\mathbf{f}_\Theta$ and a Θ-computable $\Sigma_1(\precsim_\Theta, \mathbf{R}_\Theta)$ selection function p such that Θ is equivalent to $PR(\precsim_\Theta, \mathbf{R}_\Theta, \mathbf{f}_\Theta, p)$.*

Proof. We take $\mathbf{R}_\Theta$ as defined by (10–5) and for $\mathbf{f}_\Theta$ we pick

$$f_1(a, x) = \{a\}_\Theta(x)\ ,$$

$$f_2(b, c) = S_1^1(b, c)\ .$$

The fact that all $\Sigma_1(\precsim_\Theta, \mathbf{R}_\Theta)$ sets are Θ-semicomputable and that Θ is regular implies easily that some $\Sigma_1(\precsim_\Theta, \mathbf{R}_\Theta)$ selection function p is Θ-computable and anyone will do.

As was remarked after the construction of $PR(\precsim, \mathbf{R}, \mathbf{f}, p)$, it is important *how* we pick P_1, P_2, P_p so that (10–20), (10–21) hold. Calling $p = f_3$ for convenience, we choose a Θ-code $f_j^\#$ for each f_j which satisfies (10–32) with $t = \langle 14, 4 + j\rangle$. For f_1 we also make sure that $\hat{f}_1$ is chosen so that

$$f_1(a, x) \to z \Rightarrow |\hat{f}_1, a, x, z|_\Theta \geqslant |a, x, z|_\Theta \, .$$

Now for $j = 1, 2$

$$f_j(a, x) \to z \Longleftrightarrow \{S_1^1(f_j^\#, a)\}(x) \to z$$

$$\Longleftrightarrow (\exists w)\, [S_1^1(f_j, a), x, z|_\Theta \leqslant |i, w, w|_\Theta]]$$

and we take

$$P_j(a, x, z, w) \Longleftrightarrow |S_1^1(f_j^\#, a), x, z|_\Theta \leqslant |i, w, w|_\Theta \, .$$

Similarly,

$$P_3(a, x, w) \Longleftrightarrow |f_3^\#, a, x|_\Theta \leqslant |i, w, w|_\Theta \, .$$

With this choice of P_1, P_2, P_3, it is very easy to verify $\Theta \leqslant PR(\precsim_\Theta, \mathbf{R}_\Theta, \mathbf{f}_\Theta, p)$. For the converse reducibility we must define a Θ-computable mapping $q(a, n)$ such that

$$\{a\}_{\mathrm{H}}(\sigma) \to z \Longleftrightarrow \{q(a, n)\}_\Theta(\sigma) \to z \, ,$$

$$\{a\}_{\mathrm{H}}(\sigma) \to z \Rightarrow |a, \sigma, z|_{\mathrm{H}} \leqslant |q(a, n), \sigma, z|_\Theta \, ,$$

with $\mathrm{H} = PR(\precsim_\Theta, \mathbf{R}_\Theta, \mathbf{f}_\Theta, p)$. The definition is in terms of a Θ-code $\hat{q}$, by another standard application of the second recursion theorem (in Θ) and we shall omit the details. (The only difficulty comes in defining $q(a, n)$ when a is a code in H of one of the given functions, i.e. the characteristic function of $\precsim_\Theta, \mathbf{R}_\Theta, =$ or f_j, for some j, e.g. if $a = \langle 14, 5\rangle$, when $\{\langle 14, 5\rangle\}_{\mathrm{H}} = f_1$. For $\langle 14, 5\rangle$ specifically, $p(\langle 14, 5\rangle, 2) = f_1^\#$ works, because of the special choice of $f_1^\#$ above, and a similar procedure can be follows in treating the cases for $\precsim_\Theta, \mathbf{R}_\Theta, =$.)

If $\mathfrak{A} = \langle \xi, \xi, \omega, s, M, K, L \rangle$ is the domain associated with an admissible ordinal in the beginning of this section and Θ is metarecursion theory on ξ, then Kripke's work gives a very satisfactory representation of Θ, in terms of systems of equations. One would hope for a similar representation of every Friedberg theory, one that would justify our contention that we have here "computation" theories. It is not clear that trivial modifications of the Kripke construction will yield such a strong structure theorem.

Our choice of the term "Friedberg theory" is a little presumptuous, since at this point we do not even have a satisfactory general definition of relative computability. However preliminary investigations have convinced us that with the proper definitions the priority argument can be carried out for any Friedberg theory and we hope to prove this in a future paper.

11. The second alternative: Spector theories

If Θ is a Friedberg theory, then the prewellordering $\precsim_\Theta$ defined by (10–1) is nontrivial and gives us quite a bit of information about Θ. If Θ is a *Spector theory* in the sense of Definition 16, then $\precsim_\Theta$ may well be completely trivial, e.g. it may have length 1. However there is another prewellordering which again tells us a lot about Θ.

For any theory Θ on $\mathfrak{A}$, put

$$(11\text{–}1) \quad G = \{(a, x) : \{a\}_\Theta(x) \to 0\}$$

and for $(a, x), (b, y) \in G$ put

$$(11\text{–}2) \quad (a, x) \precsim^\Theta (b, y) \Longleftrightarrow |a, x, 0|_\Theta \leqslant |b, y, 0|_\Theta \,.$$

If Θ is normal, then the initial segments of $\precsim^\Theta$ are (uniformly) Θ-finite, in particular Θ-computable. For Spector theories we can use $\precsim^\Theta$ to give a simple test of Θ-computability for Θ-semicomputable sets.

Boundedness theorem. Assume Θ is a Spector theory on a domain $\mathfrak{A}$. Then G is Θ-semicomputable and for each Θ-semicomputable set B, there is some $a \in A$ such that

$$(11\text{–}3) \quad B = G_a = \{x : \{a\}_\Theta(x) \to 0\} \,.$$

Moreover, whenever (11–3) *holds, then B is Θ-computable if and only if* *there is some* $(b, y) \in G$ *such that*

$$(11\text{–}4) \quad x \in B \Rightarrow (a, x) \lesssim^{\Theta} (b, y) .$$

Here only the last statement is non-trivial and it has a well-known proof which we shall omit. The idea of using a result of this type as a tool to study the structure of particular Spector theories goes back to Spector (1955). It has been particularly effective in the case of hyperarithmetic, hyperanalytic and hyperprojective theory, see Moschovakis (1967, 1969a, b, c). Many of the results in these papers can be generalized to arbitrary Spector theories; in fact the generalization is so direct that we see no point in listing the theorems here. Perhaps, as a sample, we might mention one computation which implies that $\lesssim^{\Theta}$ is a non-trivial prewellordering:

$$|\lesssim^{\Theta}| = \textit{supremum}\,\{|\lesssim| : \lesssim \textit{is a } \Theta\textit{-computable prewellordering on } A\} .$$

The known examples of Spector theories include the hyperarithmetic and hyperanalytic theories (on ω) and hyperprojective theory (on any $\mathfrak{A}$) – these were defined in Section 7. Since there are important structural differences between the first two of these theories (see Moschovakis, 1967), we cannot expect a general representation theorem for all Spector theories, as concrete as that for Friedberg theories given in Section 10. However one would hope that interesting structural conditions can be found which will separate and classify significant classes of Spector theories, e.g. those arising naturally from computation theories on higher types. Some theorems of this type should follow from a generalization of the methods and results of Barwise et al. [1]. Such a generalization appears feasible, but not entirely trivial.

References

[1] J. Barwise, R.O. Gandy and Y.N. Moschovakis, The next admissible set, to appear in J. Symbolic Logic.

[2] R.O. Gandy, General recursive functionals of finite type and hierarchies of functions, in: Proceedings of the Symposium on Mathematical Logic held at the University of Clermont Ferrand (1962).

[3] C. Gordon, A comparison of abstract computability theories, Ph. D. Thesis, University of California, Los Angeles (1968).

[4] T.J. Grilliot, Recursive functions of finite higher types, Ph. D. Thesis, Duke University (1967).

[5] S.C. Kleene, Introduction to metamathematics (Van Nostrand, New York, 1952).

[6] S.C. Kleene, On the forms of predicates in the theory of constructive ordinals, II, Am. J. Math. 77 (1955) 405–428.

[7] S.C. Kleene, Recursive functionals and quantifiers of finite types, I. Trans. Am. Math. Soc. 91 (1959) 1–52.

[8] G. Kreisel, Model theoretic invariants: Applications to recursive and hyperarithmetic operations, The theory of models (North-Holland Publ. Co., Amsterdam, 1965) pp. 190–205.

[9] G. Kreisel and G.E. Sacks, Metarecursive sets, J. Symbolic Logic 30 (1965) 318–337.

[10] K. Kunen, Implicit definability and infinitary languages, J. Symbolic Logic 33 (1968) 446–451.

[11] Y.N. Moschovakis, Hyperanalytic predicates, Trans. Am. Math. Soc. 129 (1967) 249–282.

[12] Y.N. Moschovakis, Abstract first order computability I, Trans. Am. Math. Soc. 138 (1969a) 427–464.

[13] Y.N. Moschovakis, Abstract first order computability, II, Trans. Am. Math. Soc. 138 (1969b) 465–504.

[14] Y.N. Moschovakis, Abstract computability and invariant definability, J. Symbolic Logic 34 (1969c) 605–633.

[15] Y.N. Moschovakis, Determinacy and prewellorderings of the continuum, to appear in Proceedings of a colloquium in Jerusalem (1970).

[16] R.A. Platek, Foundations of recursion theory, Ph. D. Thesis, Stanford Univ., Stanford, Calif. (1966).

[17] C. Spector, Recursive well-orderings, J. Symbolic Logic, 20 (1955) 151–163.

[18] E.G. Wagner, Uniformly reflexive structures: on the nature of Gödelizations and relative computability, to appear in Trans. Am. Math. Soc. (1969).

A COUNTABLE HIERARCHY FOR THE SUPERJUMP

Richard A. PLATEK [1)]
Cornell University, Ithaca, N.Y., USA

In [4] Kleene introduced his notion of a recursive functional of finite type (for an informal account see [3] ; also our [6] or better the forthcoming [7]) and illustrated the usefulness of this concept by showing that the hyperarithmetical sets of integers are exactly the sets recursive in the type 2 object ${}^2E$ where

$${}^2E(\alpha^1) = \begin{cases} 0 & \text{if } \exists x[\alpha(x) = 0] \\ 1 & \text{otherwise .} \end{cases}$$

${}^2E$ can be thought of as the number quantifier. Even if one has doubts about the "correctness" of Kleene's notion of recursiveness for objects of arbitrary finite type it seems to be the general opinion of logicians working in definability theory that this concept of a set of integers being recursive in a fixed type 2 object is extremely fruitful particularly when the type 2 object is thought of as a "jump" (for a definition of an abstract level 2 jump see [1]) for then Kleene's result mentioned above can be rephrased "the hyperarithmetical sets of integers are exactly the sets recursive in the ordinary jump

$$A \to \{x \mid \{x\}^A(x) \quad \text{is defined}\}\text{"}$$

(because the ordinary jump is of the same degree as ${}^2E$; i.e. each is recursive in the other). This immediately suggests asking for a more detailed analysis of the sets recursive in other jumps, e.g. the hyperjump

$$A \to \{x \mid \forall \alpha^1 \exists y T^A(x, x, \bar{\alpha}(y))\}$$

1) The research for this paper was supported by NSF GP 14363.

which is of the same degree as the Souslin-quantifier E_1 where

$$E_1(\alpha) = \begin{cases} 0 & \text{if } \exists\beta^1 \forall x [\alpha(\bar{\beta}(x)) = 0] \\ 1 & \text{otherwise} \end{cases}$$

which can also be thought of as the operation A of descriptive set theory. This problem was solved by Shoenfield in [13] [2] (see Section 1 below) where to any type 2 object F at least as strong as ${}^2E$ a system of ordinal notations $O^F \subseteq N$ is generated together with sets of integers H_a^F for each $a \in O^F$ and it is shown that

(i) the Turing degree of H_a^F depends only on $|a|^F$ the latter being the ordinal named by a;

(ii) a set of integers is recursive in F if and only if it is recursive in some H_a^F;

(iii) the least ordinal not named by an a in O^F is ω_1^F the latter being the least ordinal not represented by a well ordering of the integers which is recursive in F.

By results (i) and (iii) we can introduce the Turing degrees d_α^F, $\alpha < \omega_1^F$. The idea behind the proof of (ii) is that a set of integers is recursive in F and the computation always terminates in $< \alpha$ steps if and only if it is of degree $< d_\alpha^F$. The set H_a^F codes up all possible computations from F which terminate in $< \alpha = |a|^F$ steps. Since in the actual definition of H_a^F no mention is made of the notion of being recursive in F (cf. Section 1 below) we have a reduction of this notion to the ordinary notion of relative recursiveness between type 1 objects.

It is natural to ask whether a similar reduction can be accomplished for the notion of recursiveness in a type 3 object. For example in [4] Kleene defined "hyperanalytical" to mean recursive in the type 3 object ${}^3E$

$${}^3E(\alpha^2) = \begin{cases} 0 & \text{if } \exists\beta^1 [\alpha(\beta) = 0] \\ 1 & \text{otherwise} \end{cases}$$

which can be thought of as the function quantifier. He also proposed a countable hierarchy which he hoped would exhaust this class. In [5] Moschovakis showed that there are hyperanalytical sets of integers which are not enumerated by Kleene's proposed hierarchy and in fact no countable

[2] Also by P.G. Hinman in his doctoral dissertation.

hierarchy (i.e. a hierarchy indexed by numbers) can work the essential point being that a computation from a type 3 object of even a set of integers is in general an uncountable object whereas in the type 2 case computations are infinite but countable. The Moschovakis observation is illustrated in Section 4 below. Hence there can be no countable hierarchy which generates the class of sets of integers recursive in a type 3 object G which is at least as strong as ${}^3E$ (although uncountable hierarchies can be found, this is the result of [5]).

In [2] Gandy introduced the superjump S which is a level 3 object which maps jumps J into their diagonal $S(J)$ where

$$S(J)(A) = \{x \mid \{x\}^J(x, A) \quad \text{is defined}\} .$$

$S(J)$ is a more powerful jump than J, e.g. the hyperjump is (of the same degree as) S of the ordinary jump and the latter is S of the identity map which is recursive. S is of lower degree than ${}^3E$ although it is stronger than the ordinary jump, the hyperjump, the hyper-hyperjump, etc. It also is essentially of type 3, i.e. it is not of the same degree as any type 2 object just as the ordinary jump is not of the same degree as any type 1 object. From a recursion theoretic point of view S appears to be the proper type 3 analogue of ${}^2E$ = ordinary jump and not ${}^3E$. The purpose of this paper is to present a countable hierarchy which generates the sets of integers recursive in S. This is possible because uncountable computations from S can be replaced by countable ones, an insight of Gandy [2] (cf. footnote 4).

We now describe the contents of this paper in more detail. In Section 1 we review Shoenfield's construction and sketch why it works. In Section 2 we describe the superjump S in more detail and give the modified countable computation which makes use of Shoenfield's hierarchy. In Section 3 we follow the Shoenfield scheme of Section 1 and present a countable hierarchy of type 2 jumps J_α, $\alpha < \sigma$, recursive in S and show that the sets of integers recursive in S are exactly the sets recursive in some J_α and furthermore the ordinal $\sigma = \omega_1^S$. On the other hand the jumps J_α do not exhaust all type 2 objects recursive in S. This is shown in Section 4 and the example given there shows the existence of ordinals $\sigma^* < \omega_1^S$ which are "recursively regular with respect to the superjump". In Section 5 we use our constructed hierarchy to derive some results about recursiveness in S. In Section 6 we discuss how the basic construction of the paper can be extended to other higher jumps, e.g. the diagonal of the superjump.

1. Let F be an arbitrary type 2 object i.e. a map from number-theoretic functions into integers. With F we associate the "jump" j^F where for any set of integers A

$$j^F(A) = \{\langle x, y\rangle \mid \{x\}^A \text{ is a total type 1 function } \phi \text{ and } y = F(\phi)\}\ .$$

Following [13] we inductively define a set O^F of ordinal notations and for each $a \in O^F$ an ordinal $|a|^F$ and a set of integers H_a^F by

(1) $1 \in O^F$; $|1|^F = 0$; $H_1^F = N$;

(2) $a \in O^F \Rightarrow 2^a \in O^F$; $|2^a|^F = |a|^F + 1$; $H_{2^a}^F = j^F(H_a^F)$;

(3) Suppose that $a \in O^F$; that $\{e\}^{H_a}$ is a total function ϕ; that $\phi(n) \in O^F$ and $|\phi(n)|^F < |\phi(n{+}1)|^F$ for all n; and that $\phi(0) = a$. Then $3^a5^e \in O^F$; $|3^a5^e|^F = \sup_n |\phi(n)|^F$; and $H_{3^a5^e}(x) \Longleftrightarrow H_{\phi((x)_0)}((x)_1)$.

(The above definition is slightly different from the one in [13] since we have avoided defining a relation $<_0$ which is used in the original 03 instead of $|\phi(n)|^F < |\phi(n{+}1)|^F$. This change does not affect the proofs of [13].)

Shoenfield shows in [13] that the Turing degree of H_a^F depends only on $|a|^F$ and that if the number-theoretic function ϕ is recursive in F then it's recursive in some H_a^F, $a \in O^F$. The latter statement is proved by induction on the length of computation of ϕ. Since the only way F can be used in such a computation is in an application of Kleene's scheme S0

$$(*) \qquad \phi(-) \simeq F(\lambda y \psi(y, -))$$

where ψ is a number-theoretic function recursive in F we see that we never have to consider functions whose arguments have type higher than 0.

Conversely, if ${}^2E$ is recursive in F then so is each H_a^F (${}^2E$ is needed since there are number quantifiers built into the definition of j^F when one says "is total") hence for such F being recursive in it is the same for number-theoretic functions as being recursive in some H_a^F. For a more symmetric statement of the result we can say that for an arbitrary type 2 F a number-theoretic ϕ is recursive in F and ${}^2E$ if and only if it is recursive in some H_a^F with $a \in O^F$. To see that this is true let F^* be the recursive join of F and ${}^2E$. Then since ${}^2E$ is recursive in F^* we have the Shoenfield result that ϕ is recursive in F^* iff it is recursive in $H_a^{F^*}$ for some $a \in O^{F^*}$. But ϕ is recursive in F, ${}^2E$ iff it is recursive in F^*. Furthermore there is a fixed e such that for all A

$$j^{F^*}(A) \leqslant_e j^F(A)$$

as is easily seen so that the Turing degrees of the $H_a^{F^*}$'s are exactly the degrees of the H_a^F's. For more information on the Shoenfield hierarchy see the forthcoming [8].

For any set of integers A we can define the relativized objects $O^{F,A}$, $H_a^{F,A}$ which are generated just like the unrelativized ones except that $H_1^{F,A}$ is A.

We note that if F is a partial map from type 1 to type 0 then we can still start generating O^F and the H_a^F's. It may happen that for some $a \in O^F$ and some e, $\{e\}^{H_a^F}$ is total but not in F's domain. In this case $j^F(H_a^F)$ cannot be defined and the procedure breaks down. On the other hand it could be the case that the generation of O^F and H_a^F's can be completed even though F is partial because everytime we have to apply F the function involved is in its domain.

For any such partial F we can adopt the Kleene schemes and still talk about functions being recursive in F. This is easily done since the only way F is used is in the scheme (∗) above and then $\phi(-)$ is said to be undefined if either $\lambda y \psi(y, -)$ is partial or it's total but not in F's domain. We call a partial F acceptable if it is defined for all α^1 which are recursive in F, ${}^2E$. Then it is easy to show that for such acceptable F the hierarchy H_a^F, $a \in O^F$ can be generated and the result that α^1 is recursive in F if and only if it is recursive in some H_a^F can be proved as in [13].

2. To any jump J we can associate its diagonal [3] $S(J)$ defined by

$$A \to \{\langle x, y\rangle \mid x \in O^{J,A} \wedge y \in H_x^{J,A}\}.$$

$S(J)$ is more powerful than J (i.e. has higher degree as a level 2 object). For example if J is the ordinary jump then $S(J)$ is essentially (i.e. has the same degree as) the hyperjump. The map S is called the superjump and the ordinary jump, the hyperjump, the hyper-hyperjump, etc. are all recursive in S. S is of level 3 in the fill simple theory of types of predicates and functions but in order to study recursions relative to S it is useful to have a type 3 object of the same degree as S. A useful object is $\mathcal{E}$ defined by

[3] The diagonal $S(J)$ as defined in the introduction is different than the one described here but they are of the same degree; this can be proved using the result of Section 1 and the material in [2]; cf. Sections 5 and 6.

$$\mathcal{E}(F) = \begin{cases} 0 & \text{if } \exists \alpha^1 [\alpha \text{ is recursive in } F, {}^2E \text{ and } F(\alpha) = 0] \\ 1 & \text{otherwise .} \end{cases}$$

$\mathcal{E}$ is used in recursions through the scheme

$$(**) \qquad \phi(-) \simeq \mathcal{E}(\lambda \alpha^1 \psi(\alpha^1, -))$$

where ψ is a function already known to be recursive in $\mathcal{E}$. We note that even if ϕ is number-theoretic ψ isn't since α^1 is of type one. Hence in analyzing the number-tehoretic functions recursive in $\mathcal{E}$ we have to also consider the level 2 functionals recursive in $\mathcal{E}$. Also we note that the computation of ϕ in $(**)$ is uncountable since to compute $\phi(-)$ we must first compute $\psi(\alpha^1, -)$ for all type 1 α^1, see what type 2 object F is thereby defined and then evaluate $\mathcal{E}$ at F. But we note from the definition that the value $\mathcal{E}(F)$ depends only on the value of F at countable many α^1 namely those recursive in F, ${}^2E$. This isn't much help since we have to know what F is as a total object in order to determine which α^1 are recursive in it. On the other hand by the remark made in Section 1 we can start generating the Shoenfield hierarchy H_a^F even when $F = \lambda\alpha\psi(\alpha, -)$ is partial. If F is total (or acceptable) then this procedure will not break down and will give all the functions recursive in F, ${}^2E$ and we can then test F at all these functions in order to determine $\mathcal{E}(F)$. Thus the computation to be performed by scheme $(**)$ is countable.

More precisely we inductively define the predicate

$$(***) \qquad \{e\}^{\mathcal{E}}(x_0, \ldots, x_{n_0-1}, \alpha_0^1, \ldots, \alpha_{n_1-1}^1) \simeq_w y$$

which is to be read "the eth functional partial recursive in $\mathcal{E}$ is weakly defined at $\vec{x}$, $\vec{\alpha}$ with value y". The definition is just like the inductive definition of "is defined" (which we will call "strongly defined") given in [4] except when $(e)_0 = 0$ i.e.

$$\{e\}^{\mathcal{E}}(\vec{x}, \vec{\alpha}) \simeq \mathcal{E}(\lambda\beta\{(e)_3\}^{\mathcal{E}}(\vec{x}, \beta, \vec{\alpha}))$$

in Kleene's notation (this is just scheme $(**)$). Let F be the partial type 2 object defined by

$$F(\beta) \simeq z \iff \{(e)_3\}^{\mathcal{E}}(\vec{x}, \beta, \vec{\alpha}) \simeq_w z \ ,$$

where the weak computation is used on the right hand side. Then $(***)$ will be true if the following holds. Begin generating O^F; H_a^F, $a \in O^F$. If at any time this procedure breaks down because F is undefined at some object which it must be tested at then (∗∗∗) is false for all y and the weak computation is said to be undefined. If the procedure does not break down and there is some $\beta^1 = \{f\}^{H_a^F}$ for which $F(\beta) \simeq 0$ then (∗∗∗) is true for $y = 0$ but if $F(\beta)$ is defined for all such β and $\neq 0$ then (∗∗∗) is true for $y = 1$. In these two cases the weak computation is said to terminate.

Shoenfield's hierarchy result guarantees that if $\{e\}^{\mathcal{E}}(\vec{x}, \vec{\alpha})$ is strongly defined with value y then it's weakly defined with value y. But we can have weak computations when there are no strong ones. If a total number theoretic ϕ is (strongly) recursive in $\mathcal{E}$ then by the above remark it is weakly recursive in $\mathcal{E}$. The converse (for total ϕ!) follows from our hierarchy result [4].

3. We now modify the hierarchy of Section 1 to get a hierarchy of level 2 jumps J_a^S. We inductively define O^S, $|a|^S$ and J_a^S by

(01) $1 \in O^S$; $|1|^S = 0$, J_1^S = ordinary jump;

(02) $a \in O^S \Rightarrow 2^a \in O^S$; $|2^a| = |a| + 1$; $J_{2^a}^S = S(J_a^S)$;

(03) Suppose that $a \in O^S$; that $\{e\}^{J_a^S}$ is a total number-theoretic function ϕ; that $\phi(n) \in O^S$ and $|\phi(n)|^S < |\phi(n+1)|^S$ for all n; and that $\phi(0) = a$. Then $3^a 5^e \in O^S$; $|3^a 5^e|^S = \sup_n |\phi(n)|^S$; and $J_{3^a 5^e}^S$ is the jump

$$A \to \{x \mid (x)_1 \in J_{\phi((x)_0)}^S(A)\}.$$

By adopting the Spector style uniqueness theorem of [13] we have

Uniqueness. There are partial recursive functions L and M such that if $a, b \in O^S$ and $|a|^S \leqslant |b|^S$ then $L(a, b)$, $M(b)$ are defined and J_a^S is recursive in J_b^S with index $L(a, b)$ and $\{y \,\|\, y|^S < |b|^S\}$ is recursive in $J_{2^b}^S$ with index $M(b)$.

Hence the degree of the jump J_a^S depends only on $|a|^S = \alpha$ and we can use the notation J_α^S. Intuitively J_α^S is the jump that is gotten by applying the superjump to the ordinary jump α times. We can call J_α^S the α-hyper-jump and can think of S as the hyper-operator. We do not give the proof of the

[4] The idea of replacing uncountable computations from $\mathcal{E}$ by countable ones is due to Gandy [2] but a careful reading of his proposed way of doing it shows that it is circular and hence not well defined. Our interpolation of a Shoenfield hierarchy construction at the crucial points does just what Gandy wished to do but unfortunately this construction was not available to him when [2] was being prepared.

uniqueness theorem since it follows so closely the construction in [13].

We now come to our principal result namely

Theorem 1. A set of integers is recursive in S if and only if it is recursive in some J_a^S, $a \in O^S$.

This gives a countable hierarchy for such sets of integers since a set will be recursive in J_a^S if and only if it is recursive in some $H_b^{J_a^S}$ with $b \in O^{J_a^S}$ and the notion of being recursive in S is reduced to ordinary relative recursiveness.

For the if part we use the recursion theorem to define a partial recursive Z such that if $a \in O^S$ then $Z(a)$ is defined and is an index J_a^S from S. This is straightforward.

For the converse we prove that if a total number-theoretic ϕ is computable from $\mathcal{E}$ using weak computability then ϕ is recursive in some J_a^S with $a \in O^S$ and indeed a and an index of ϕ from J_a^S can be found recursively from an index of ϕ from $\mathcal{E}$. Since strong (i.e. ordinary) computability implies weak computability we get the only if part and the

Corollary II. The same total number-theoric ϕ are weakly computable from $\mathcal{E}$ as are strongly computable from $\mathcal{E}$.

Again our proof will follow the proof in [13] except that we have extra cases. If $\phi = \{e\}^{J_a^S}$ is a total 1 place number-theoretic function we will call the ordered pair $\langle a, e\rangle$ a J-number for ϕ and let $\mathcal{R}$ be the collection of ϕ's which have J numbers. Our result shows that $\mathcal{R}$ is just the 1-section of S. We first have

Boundedness lemma. If $\phi \in \mathcal{R}$ and $\phi(n) \in O^S$ for all n then there is a $b \in O^S$ with $|\phi(n)|^S < |b|^S$. Furthermore such a b can be found recursively from a J number of ϕ.

The proof is exactly as in [13] namely a partial recursive ordinal addition function $+_0$ is defined and then if $\langle a, e\rangle$ is a J number for ϕ then $b = 3^a 5^f$ where

$$\{f\}^{J_a^S}(0) = a; \qquad \{f\}^{J_a^S}(n+1) = \{f\}^{J_a^S}(n) +_0 \phi(n)\,.$$

To finish the proof we define partial recursive functions G and P and prove that if

$$\{e\}^{\mathcal{E}}[x,\alpha^1] \simeq_w y$$

where $\alpha^1 \in \mathcal{R}$ and the bracket notation is from [4] and if j is a J number for α^1 then $G(e,x,j)$ is defined and is some $a \in O^S$ and $\{P(e,x,j)\}^{J_a^S} \simeq y$. To see that this suffices assume that $\{e\}^{\mathcal{E}}$ is a total number-theoretic function using weak computations. Then by the lemma there is a $b \in O^S$ with $|G(e,x)|^S < |b|^S$ for all x and using the L of the uniqueness theorem $\{e\}^{\mathcal{E}}$ is recursive in J_b^S.

P and G are defined simultaneously by the recursion theorem (as well as some auxiliary functions mentioned below) and the result is proved by induction on the predicate "is weakly defined". The construction proceeds just as in [13] the only new cases are $(e)_0 = 0$ or 7. The latter is trivial for then

$$\{e\}^{\mathcal{E}}[x,\alpha^1] = (\alpha^1)_0((x)_0) = y$$

and we can take $G(e,x,\langle a,f\rangle) = a$ and from f and x we can compute a $P(e,x,j)$ with

$$\{(P(e,x,j))\}^{J_a^S} = y\ .$$

The case $(e)_0 = 0$ is where the weak computation is used. Let us suppose that $\{e\}^{\mathcal{E}}[x,\alpha^1]$ is weakly defined and let F be the partial type 2 object which for any β^1 is equal to z if and only if

$$\{e'\}^{\mathcal{E}}[x,\beta^1\frown\alpha^1] \simeq_w z$$

where $e' = (e)_3$ and $\beta\frown\alpha$ is the concatenation of β and α. Since $\{e\}^{\mathcal{E}}[x,\alpha^1]$ is weakly defined we know that $F(\beta^1)$ is defined for all β^1 in the Shoenfield hierarchy for F. Using the recursion theorem we can define a partial recursive H with property that

$$a \in O^F \Rightarrow H(a) \text{ is a } J \text{ number for } H_a^F\ .$$

The only non-trivial step is defining H at successor notations. But knowing a J number for H_a^F we can find J numbers for each total ϕ recursive in H_a^F as well as telling for which c, $\{c\}^{H_a^F}$ is total (since the ordinary jump is recursive in each J_b^S). Then using G and P we can calculate $F(\phi)$ for all such ϕ and finally find a J number for $H_{2^a}^F$. The use of G and P is justified by the induction assumption $H(a) = H(e,x,j,a)$ where j is a J number for α^1. By the lemma there is a $d \in O^S$ (computable from a Gödel number of H) such that every ϕ in F's hierarchy is recursive in J_d^S. Using G and P we can find an $F^{\#}$ recursive in $S(J_d)$ which is total and agrees with F on all ϕ generated in the Shoenfield hierarchy for F. But by the construction of the hierarchy $O^F = O^{F^{\#}}$ and $H_a^F = H_a^{F^{\#}}$. Hence the value of $\mathcal{E}(F)$ is $\mathcal{E}(F^{\#})$ and this can be calculated from $S(S(J_d))$. Hence $G(e, x, j) = 2^{2^d}$.

To clean up loose ends we describe how $F^{\#}$ is computed from $S(J_d)$. Given any β^1 we ask whether it is recursive in J_d. If not we let $F^{\#}(\beta^1) = 0$. If it is we ask whether there is a J number for it of the form $\langle d, t\rangle = j'$ such that $G(e', x, j'')$ is defined with value in $\{y \| y | < | d |\}$ where $e' = (e)_3$ and j'' is a J number for β^1 concatenated with α^1 and α^1 has J number j. If no, we set $F^{\#}(\beta^1) = 0$. If yes, we ask whether $\{P(e', x, j'')\}^{J_{G(e',x,j'')}}$ is defined. If no again, we set $F^{\#}(\beta^1) = 0$ and if yes, we set $F^{\#}(\beta^1)$ equal to the value. Now by choice of d and the induction assumption for G and P, $F^{\#}$ will agree with F for all β^1 in the hierarchy for F and $F^{\#}$ will be recursive in $S(J_d)$.

By an application of the boundedness lemma together with the M of the uniqueness theorem we get that the least ordinal not named in O^S is $\omega_1^{\mathcal{E}}$.

We also note that a relativized hierarchy $O^{S,F}$, $J^{S,F}$, etc. can be generated where now $J_1^{S,F} = J_F$ and we can relativize the proof to show that we get a hierarchy for all sets recursive in $\mathcal{E}, F$ where F is any total type 2 object or a partial one that is "acceptable" now in a new sense.

4. In this section we look at what information our hierarchy gives about type 2 objects (partial and total) which are recursive in $\mathcal{E}$. By the construction in Section 3 we have

Theorem III. Let $F = \lambda\beta^1\phi(\beta^1, \mathcal{E})$ be partial recursive in $\mathcal{E}$ and acceptable. Then there is a total $F^{\#}$ recursive in some J_d, $d \in O$ that generates the same hierarchy as F. In particular F has the same 1-section [5] as $F^{\#}$.

Corollary IV. Suppose that the total type 2 object F is recursive in $\mathcal{E}$ then there is an $F^{\#}$ recursive in some J_d, $d \in O$ which has the same 1-section as F and which agrees with F on this 1-section.

We also get

Corollary V. Suppose the partial F is partial recursive in S and for some

[5] The k-section of a higher type object T is all the function of type $\leqslant k$-recursive in T.

$\alpha < \omega_1^S F$ is defined (weakly or strongly) for each β^1 recursive in J_α. Then there is a total type 2 object α^2 recursive in some J_β which agrees with F on J_α's 1-section. A J number for α^2 can be found recursively from an index of F from $\mathcal{E}$.

In contrast to these results we have

Theorem VI. There is a jump T recursive in $\mathcal{E}$ which is not recursive in any J_α, $\alpha < \omega_1$. Indeed there isn't even a jump recursive in some J_α that agrees with T on $\mathcal{E}$'s 1-section.

Proof. The idea is to diagonalize through the hierarchy. T will be a jump with two arguments A, B. We describe how to compute $T(A, B)$ from $\mathcal{E}$: For any A construct a well ordering of length ω_1^A recursively from A and the hyperjump. Then iterate the superjump up this well ordering to generate a sequence of jumps J_β, $\beta < \omega_1^A$ and combine them all into a single jump $J_{\omega_1^A}$ and let $T(A, B) = J_{\omega_1^A}(B)$. T is recursive in $\mathcal{E}$. Now suppose that there were a T^* recursive in some J_α^S in the hierarchy which agrees with T on $\mathcal{E}$'s 1-section. Using the M of the uniqueness theorem we have that $A = \{\langle x, y\rangle \mid |x|^S < |y|^S < \alpha\}$ is recursive in $J_{\alpha+1}^S$ and so is T^*. But $\alpha+2 < \omega_1^A$ and $\lambda B T^*(A, B)$ is recursive in $J_{\alpha+1}^S$ and the former is equal to $\lambda B T(A, B)$ for all B in $\mathcal{E}$'s 1-section. But this makes $\{e \mid \{e\}^{J_{\alpha+1}}$ is defined$\}$ recursive in $J_{\alpha+1}$ and that's a contradiction.

So our hierarchy does not exhaust the 2-section of $\mathcal{E}$. A given type 2 F will be recursive in J_α if and only if it is recursive in $\mathcal{E}$ and for all A

$$\omega_1^{F,A} \leqslant \omega_1^{J_\alpha, A} .$$

What is shown in the above theorem is that for any $\alpha < \omega_1^{\mathcal{E}}$ there is an A with

$$\omega_1^{T,A} > \omega_1^{J_\alpha, A}$$

and indeed A can be taken in $\mathcal{E}$'s 1-section.

It is interesting to ask which J_α are recursive in T. If $\alpha < \omega_1^T < \omega_1^S$ then J_α is recursive in T since J_α can be recovered from $\lambda B T(A, B)$ where A is some T-recursive set with $\omega_1^A > \alpha$. But we claim that $J_{\omega_1^T}$ is not T-recursive because every α is $< \omega_1^{J_\alpha}$ (by induction on α) and if $J_{\omega_1^T}$ were T-recursive we would have $\omega_1^T < \omega_1^{J_{\omega_1^T}} \leqslant \omega_1^T$. Hence T and $J_{\omega_1^T}$ are incomparable type 2 objects whose degrees provide upper bounds for $\{J_\alpha \mid \alpha < \omega_1^T\}$.

Let σ^* be the ordinal ω_1^T. Then $\sigma^* < \omega_1^S$ and it has the following property

(t) $\alpha < \sigma^* \Rightarrow \omega_1^{J_\alpha} < \sigma^*$

(because $\alpha < \sigma^*$ implies $J_{\alpha+1}$ is T-recursive so $\omega_1^{J_\alpha} < \omega_1^{J_{\alpha+1}} \leqslant \omega_1^T = \sigma^*$). We can interpret (t) in the following way: suppose we only consider those sets of integers which are recursive in the superjump and have (weak) computations of length $< \sigma^*$. Then any well ordering in that set will have ordinal $< \sigma^*$. We call such σ^* "recursively regular with respect to the superjump". There are ω_1^S of them and they provide natural stopping places for a more detailed analysis of $\mathcal{E}$'s 1-section.

5. In this section we apply our hierarchy and the weak countable computations to derive some results about recursivity in $\mathcal{E}$.

Theorem VII (selection operators). There is a function $\phi(e, \alpha^1, \alpha^2)$ partial recursive in $\mathcal{E}$ such that if for some x, $\{e\}^{\mathcal{E}}[x, \alpha^1, \alpha^2]$ is weakly defined then so is $\phi(e, \alpha^1, \alpha^2]$ and its values is an x for the above.

The proof uses the countable computations and proceeds just like the general proofs of [6, 7]. The proof given there for type 3 (also in [5]) does not apply in this case since ${}^3E$ is not recursive in S. We do not go through the consequences of this theorem but refer the reader to [6, 7] where there is a general discussion of what happens when a recursion theory possess selection operators.

Call a set $\mathcal{E}$-semi-recursive if it is the domain of a function partial recursive in $\mathcal{E}$ using weak computations.

Theorem VIII (Gandy-style normal forms). There is a predicate $T(e, x, \alpha^1, \beta^1, y)$ primitive recursive in $\mathcal{E}$ such that for all $\alpha^1 \in \mathcal{R}$ (= 1-section of $\mathcal{E}$)

$$\{e\}^{\mathcal{E}}[x, \alpha^1] \simeq_w y \Longleftrightarrow \exists \beta^1 \in \mathcal{R}\, T(e, x, \alpha^1, \beta^1, y) .$$

In particular

$$\exists \beta^1 \in \mathcal{R} M(x, \beta^1)$$

with M primitive recursive in $\mathcal{E}$ is the general form for $\mathcal{E}$-semi-recursive sets of integers.

The proof follows the description in [2]. $\beta^1 \in \mathcal{R}$ is the computation considerede as a complete whole. A relativized theorem holds for any type 2 object α^2 but then we have

$$\exists \beta^1 [\beta^1 \text{ is recursive in } \mathcal{E}, \alpha^2]$$

for the prefix.

We mention that the set

$$K^{\mathcal{E}} = \{x \mid (x)_0 \in O^S, \ (x)_1 \in O^{J^S_{(x)_0}}, \ (x)_2 \in H^{J^S_{(x)_0}}_{(x)_1}\}$$

is complete for $\mathcal{E}$-semi-recursive sets of integers.

As far as the ordinal $\sigma = \omega_1^{\mathcal{E}}$ is concerned we have

Theorem IX. σ is recursively regular. The σ-recursive subsets of ω are exactly the $\mathcal{E}$-recursive ones; the σ-r.e. subsets of ω are exactly the $\mathcal{E}$-semi-recursive ones. σ is projectible into ω (the set $K^{\mathcal{E}}$ above can be taken as the image of σ). σ is recursively inaccessible, the limit of recursively inaccessibles, and is α-hyper-recursively inaccessible for all $\alpha < \sigma$, etc. σ is less than the first non projectible and hence $< \beta_0$ (the former ordinal gives the ordinal of the minimal β-model for Σ^1_2-comprehension).

For terminology for the above see [10]. The theorem implies [10] that the sets of integers recursive in $\mathcal{E}$ are exactly those constructible before σ. For proofs compare [11]; the main technique is to use O^S to define a $\mathcal{E}$-semi-recursive well ordering of $K^{\mathcal{E}}$ of length σ all of whose proper segments are $\mathcal{E}$-recursive (uniformly).

We note that $\mathcal{R}$ is a β-model of Δ^1_2 comprehension but not of Σ^1_2 comprehension even though $\mathcal{R}$ has strong closure properties from a recursion-theoretic point of view. On the other hand those closure properties are reflected in the extreme recursive inaccessibility of σ.

6. In this section we describe how our work can be extended. To any type 3 object α^3 we can define its diagonal $D(\alpha^3)$ which is the type 3 object

$$D(\alpha^3)(\beta^2) = \begin{cases} 0 & \text{if } \{(\beta)_0^0\}[(\beta)_1^0, (\beta)_2^1, (\beta)_2^2, \alpha^3]\ ^{6)} \\ & \text{is defined} \\ 1 & \text{otherwise .} \end{cases}$$

Let $\mathcal{E}^{\#}$ be the diagonal of $\mathcal{E}$. $\mathcal{E}^{\#}$ is of higher degree than $\mathcal{E}$. By Theorem VIII of the last section $\mathcal{E}^{\#}$ has the same degree as

$$\mathcal{E}^*(\alpha^2) = \begin{cases} 0 & \text{if } \exists\beta^1\,[\beta^1 \text{ recursive in } \mathcal{E}, \alpha^2 \text{ and } \alpha(\beta) = 0] \\ 1 & \text{otherwise .} \end{cases}$$

Now computation from $\mathcal{E}^*$ can be replaced by weak countable computations where

$$\mathcal{E}^*(\lambda\beta^1\psi(\beta^1, -))$$

is computed by generating the hierarchy of Section 4 for $\mathcal{E}, F$ where

$$F(\beta) \simeq_w \psi(\beta, -)$$

and testing whether there was an appropriate β^1 with $F(\beta^1) \simeq 0$. We can thus derive the Gandy style normal form for $\mathcal{E}^*$-semi-recursive sets and use this to construct countable computations for the diagonal of $\mathcal{E}^{\#}$, etc. Finally let us consider the level 4 diagonal operator D itself. It has the property that if the procedure described in this paper works for α^3 then it works for $D(\alpha^3)$. This fact can be exploited to yield weak countable computations from D and a subsequent countable hierarchy. We can then go to the diagonal operator at level 5, etc. All this is explored in [12] where we construct countable computations for all higher type diagonals.

6) $(\beta)_j^i$ is the jth type i function coded into β.

References

[1] H.B. Enderton, Hierarchies in recursive function theory, Trans. Am. Math. Soc. 111 (1964) 457–471.

[2] R.O. Gandy, General recursive functionals of finite type and hierarchies of functions (A paper given at the Symposium on Mathematical Logic held at the University of Clermont Ferrand, June 1962).
[3] R.O. Gandy, Computable functionals of finite type I, in: Sets, models and recursion theory (North-Holland Publ. Co., Amsterdam, 1967) pp. 202–242.
[4] S.C. Kleene, Recursive functionals and quantifiers of finite types, Trans. Am. Math. Soc. 91 (1959) 1–52; 108 (1963) 106–142.
[5] Y.N. Moschovakis, Hyperanalytical predicates, Trans. Am. Math. Soc. 129, 249–282.
[6] R.A. Platek, Foundations of recursion theory, Ph. D. thesis, Stanford University (1966).
[7] R.A. Platek, Recursive definability, to appear in Ann. Math. Logic.
[8] R.A. Platek, On the Shoenfield hierarchy, to appear.
[9] R.A. Platek, Recursive functionals of finite type based on partial functions, to appear.
[10] R.A. Platek, The theory of recursively regular ordinals and admissible sets, to appear.
[11] R.A. Platek, Recursively regular ordinals and higher type recursion theory, to appear.
[12] R.A. Platek, Countable computations for higher type objects, to appear.
[13] J.R. Shoenfield, A hierarchy based on a type 2 object, Trans. Am. Math. Soc. 134 (1968) 103–108.

RECURSIVELY MAHLO ORDINALS AND INDUCTIVE DEFINITIONS

Wayne RICHTER [1)]
University of Minnesota, USA

1. Introduction

Gandy has characterized ω_1, the first non-recursive ordinal, as the maximum of the closure ordinals of Π_1^0 transfinite inductive definitions. We provide here a characterization of the first recursively Mahlo (i.e. ρ_0) ordinal, the first recursively hyper-Mahlo ordinal, etc. It turns out that these rather large countable ordinals are obtained as the closure ordinals of Π_2^0 inductive definitions.

Inductive definitions play a central role in hierarchy theory. A classic example is the theory of recursive ordinals. The usual systems of notations for the recursive ordinals are inductively defined by very simple (arithmetic) operations. More recently, two developments have led to generalizations of the theory of recursive ordinals. One is the Kleene [5] theory of recursive functionals of finite types; the second is the Kripke [7, 8], Platek [13] theory of admissible sets. Inductive definitions play an important role in both these areas.

Such inductive definitions may be described as follows. Let Γ be an operation on $P(X)$, the power set of an infinite set X. Γ is *inclusive* if $Y \subseteq \Gamma(Y)$ for all $Y \subseteq X$. All operations considered below are assumed to be inclusive. Γ is *monotone* if $Y \subseteq Z$ implies $\Gamma(Y) \subseteq \Gamma(Z)$ for all $Y, Z \subseteq X$. Γ defines a transfinite sequence $\langle A_\xi : \xi \in On \rangle$ of subsets of X, where $A_0 = \phi, ..., A_{\alpha+1} = \Gamma(A_\alpha)$, and $A_\lambda = \cup \{A_\xi : \xi < \lambda\}$ for λ a limit ordinal. Since Γ is inclusive, $\alpha \leqslant \beta$ implies $A_\alpha \subseteq A_\beta$. Let $|\Gamma|$, the *closure ordinal* of Γ, be the smallest ordinal α such that $A_{\alpha+1} = A_\alpha$, and let $A = \cup A_\xi$. It is clear that $A = A_{|\Gamma|}$ and

[1)] The results in this paper are an extension of those reported at the 1969 Logic Colloquium in Manchester. We are grateful for the opportunity at Manchester to discuss these topics with numerous logicians; in particular, we especially benefitted from conversations with P. Aczel, R. Gandy, and H. Putnam.

$|\Gamma| < \overline{\overline{X}}^{+}$. A is said to be *inductively defined by* Γ. For $a \in A$, let $|a| = \mu\alpha \,.\, a \in A_{\alpha+1}$. Thus A may be regarded as a system of notations for the ordinals less than $|\Gamma|$.

Given a set $\mathcal{C}$ of operations of a certain form on $P(X)$, we would like to be able to characterize $|\mathcal{C}| = \sup\{|\Gamma| : \Gamma \in \mathcal{C}\}$. We shall obtain such characterizations in certain cases in terms of recursive analogues of large cardinals.

The closure ordinals of monotone operations on $P(\omega)$ have been discussed by Spector [19]. Recently, interesting results on the closure ordinals of special kinds of monotone operations have been obtained in the more abstract setting of admissible sets and generalized recursion theory (cf. Barwise et al. [2], Barwise [1], and Moschovakis [12]).

Examples of non-monotone operations have also appeared in print. Systems of notations for recursive analogues of the first inaccessible ordinal and beyond, have been given by Kreider and Rogers [6], Putnam [14], and Richter [16, 17]. As we observe in Section 3, these systems are defined by non-monotone operations [2]. Another example of the use of non-monotone operations is the Grilliot [4] hierarchy of objects recursive in a given type n object.

Very little is known about the closure ordinals of non-monotone operations. The main results for operations on $P(\omega)$ are due to Putnam [15] and Gandy but they apply only to very special cases. Our results show that characterizing the closure ordinals of even very simple kinds of arithmetic operations on $P(\omega)$ requires non-trivial techniques and leads to surprisingly large ordinals.

The following theorem of Spector points out the distinction between the monotone and non-monotone case. All operations are henceforth operations on $P(\omega)$. ω_1 is the first ordinal which is not recursive. We say that Γ is Π^n_m if the binary relation $n \in \Gamma(X)$ is Π^n_m.

Theorem 1.1 (i). If Γ is monotone and Π^1_1 then $|\Gamma| \leqslant \omega_1$; *(ii)* There is a monotone and Π^0_1 Γ such that $|\Gamma| = \omega_1$.

Note the contrast between (i) and (ii). As far as monotone operations are concerned, Π^0_1 operations are just as strong as Π^1_1 operations. As we now observe (i) is trivially false if the condition of monotonicity is dropped.

[2] The earlier confusion about extensions of the constructive ordinals is due precisely to the fact that the operations are non-monotone. A discussion of this confusion appears in [6].

Definition. Let $\mathscr{A}$ be the set of operations Γ of the form

$$\Gamma(X) = \begin{cases} \Gamma_0(X) & \text{if } \Gamma_0(X) \neq X , \\ \Gamma_1(X) & \text{otherwise} , \end{cases}$$

where Γ_0 and Γ_1 are Π_1^0 operations.

Thus in constructing the transfinite sequence $\langle A_\xi : \xi \in On \rangle$ for $\Gamma \in \mathscr{A}$ we apply a Π_1^0 operation Γ_0 until we reach its closure ordinal, i.e. until $\Gamma_0(A_\alpha) = A_\alpha$, and then apply a Π_1^0 operation Γ_1, etc. $\mathscr{A}$ is perhaps the simplest natural extension of Π_1^0. Since Γ_0 and Γ_1 are inclusive,

$$n \in \Gamma(X) \equiv n \in \Gamma_0(X) \vee [\Gamma_0(X) = X \;\&\; n \in \Gamma_1(X)] \; .$$

Since the predicate $\Gamma_0(X) = X$ is Π_2^0 for $\Gamma_0 \in \Pi_1^0$, $\mathscr{A} \subseteq \Pi_2^0$. Choose $\Gamma_0 \in \Pi_1^0$ so that $|\Gamma_0| = \omega_1$. By choosing a suitable $\Gamma_1 \in \Pi_1^0$ it is clear that the corresponding Γ will have closure ordinal greater than ω_1. In view of 1.1 (i) such a Γ is not monotone. It is conceivable that in constructing the A_α's, at ω_1 we apply Γ_1 to get A_{ω_1+1} and then apply Γ_0 repeatedly until the ordinal ω_1^O (Kleene's O is reached, then Γ_1 is applied once, then Γ_0 is applied repeatedly up to $\omega_1^{O^O}$, etc. Thus it is plausible that $|\mathscr{A}|$ is quite large. Our main result is the rather surprising:

Theorem 1.2. $|\mathscr{A}|$ is the first recursively Mahlo ordinal.

Let $\mathscr{A}_k$ be the set of operations Γ of the form:

$$\Gamma(X) = \begin{cases} \Gamma_0(X) & \text{if } \Gamma_0(X) \neq X , \\ \vdots & \\ \Gamma_i(X) & \text{if } (\forall j < i)\,[\Gamma_j(X) = X] \;\&\; \Gamma_i(X) \neq X , \quad i < k , \\ \vdots & \\ \Gamma_k(X) & \text{if } \Gamma_j(X) = X \text{ for } j < k , \end{cases}$$

where Γ_i is Π_1^0, $i \leqslant k$. An easy extension of our work shows that $|\mathscr{A}_2|$ is the first recursively hyper-Mahlo ordinal, $|\mathscr{A}_3|$ is the first recursively hyper-hyper-Mahlo ordinal, etc. Furthermore, $\mathscr{A}_k \subseteq \Pi_2^0$ for all $k < \omega$. Thus $|\Pi_2^0|$ is very large in terms of recursively Mahlo ordinals [3]. It seems likely that this

[3] These results provide a counterexample to statements announced by Gandy in 1966 in a mimeographed abstract titled Inductive definitions. The Gandy abstract has been referred to in [17] and [4].

work can be carried over to a treatment of "the next Mahlo set" in the manner of [2] but we have not yet done this.

The proof that $|\mathscr{A}|$ is the first recursively Mahlo ordinal occupies Sections 2–5. In Section 5 a proof of Theorem 4.1 is outlined. This theorem is powerful enough to show that many familiar ordinals are admissible. In particular an easy consequence is the fact that $\omega_1^{\mathbf{F}}$ is admissible for $\mathbf{F}$, a type 2 functional in which $\mathbf{E}$ (see Kleene [5]) is recursive. Its main value, however, is in showing that certain ordinals are admissible, even though not obviously of the form $\omega_1^{\mathbf{F}}$.

We assume familiarity with ordinary recursion theory and the basic properties of the theory of admissible sets. In Section 5 some knowledge is required of the theory of recursive functionals of type 2. In particular a version of the Gandy theorem [3] on the existence of selection operators is the basic tool.

2. $|\mathscr{A}| \leqslant \rho$

We begin with a proof of Gandy's theorem on the closure ordinals of Π_1^0 operations which is apparently quite different from his original unpublished proof.

Theorem 2.1. $|\Pi_1^0| = \omega_1$.

In view of 1.1 (ii) it suffices to show $|\Pi_1^0| \leqslant \omega_1$. For $X \subseteq \omega$, let $\bar{X}$ be the set of sequence numbers with values in X. For example, $2^{3+1}3^{2+1}5^{6+1} \in \bar{X}$ iff $3, 2, 6 \in X$. Let $\bar{\phi} = \{1\}$. We need the following lemma. The proof is essentially given in the proof of Lemma 7.2 of [16].

Lemma 2.2. Let Q be a Σ_1^0 predicate on $\omega \times P(\omega)$. There is a recursive predicate R on ω^3 such that for all $X \subseteq \omega$,

$$Q(n, X) \equiv (\exists x \in \bar{X})(\exists y \in \overline{\sim X}) R(n, x, y) .$$

Proof of Theorem 2.1. Let Γ be a Π_1^0 operation. By 2.2 there is a recursive R such that

$$(1) \qquad n \in \Gamma(X) \equiv (\forall x \in \bar{X})(\forall y \in \overline{\sim X}) \neg R(n, x, y)$$

$$\equiv (\forall x \in \bar{X}) \forall y [R(n, x, y) \Rightarrow y \in \sim\overline{\sim X}] .$$

Note that both $\overline{X}$ and $\overline{\sim\sim X}$ are monotone in X. Let $\langle A_\xi : \xi \in On \rangle$ be the sequence defined by Γ. Suppose $n \in A_{\omega_1+1} = \Gamma(A_{\omega_1})$. We show $n \in A_{\omega_1}$. By (1),

$$n \in \Gamma(A_{\omega_1}) \equiv (\forall x \in \overline{A}_{\omega_1}) \forall y \, [R(n, x, y) \Rightarrow y \in \overline{\sim\sim A}_{\omega_1}] \ .$$

We first show there is an ω_1-recursive function f which is normal on ω_1, such that for $\alpha < \omega_1$,

$$(2) \qquad (\forall x \in \overline{A}_\alpha) \forall y \, [R(n, x, y) \Rightarrow y \in \overline{\sim\sim A}_{f(\alpha)}] \ .$$

Note that the functions $\langle \overline{A}_\alpha : \alpha < \omega_1 \rangle$ and $\langle \overline{\sim\sim A}_\alpha : \alpha < \omega_1 \rangle$ are Σ_1 on L_{ω_1}, the set of constructible sets of order less than ω_1. For $x \in \overline{A}_{\omega_1}$ let:

$$g(x, y) = \begin{cases} \mu\alpha \, . \, y \in \overline{\sim\sim A}_\alpha & \text{if } R(n, x, y) \ , \\ 0 & \text{otherwise} \ . \end{cases}$$

Then $g : \overline{A}_{\omega_1} \times \omega \to \omega_1$. Since $\overline{A}_{\omega_1}$ is Σ_1 on L_{ω_1} so is g. For $\alpha < \omega_1$ let $h(\alpha) = \sup \{ g(x, y) : x \in \overline{A}_\alpha \ \& \ y \in \omega \}$. Let f be defined by recursion so that $f(0) = h(0)$; $f(\alpha+1) = f(\alpha) + h(\alpha) + 1$; $f(\lambda) = \sup \{ f(\xi) : \xi < \lambda \}$ for λ a limit ordinal. Then f is normal ω_1-recursive, and satisfies (2). Thus f has a fixed point $\beta < \omega_1$. Hence

$$(\forall x \in \overline{A}_\beta) \forall y \, [R(n, x, y) \Rightarrow y \in \overline{\sim\sim A}_\beta] \ ;$$

hence by (1), $n \in \Gamma(A_\beta) \subseteq A_{\omega_1}$.

By analogy with the classical situation one can define large admissible ordinals. Following Kripke [9] (and Mahlo [10]) let $\lambda\alpha . \tau^0_\alpha$ be the monotone enumeration of the admissible ordinals; and let $\lambda\alpha . \tau^\beta_\alpha$ be the monotone enumeration of the admissible ordinals α such that for all $\xi < \beta$, $\alpha = \tau^\xi_\alpha$. Thus τ^1_0 is the first recursively inaccessible, τ^2_0 is the first recursively inaccessible which is a fixed point of recursively inaccessibles, etc. An ordinal α is *recursively Mahlo* (i.e. ρ_0) if α is admissible and every α-recursive function which is normal on α has a fixed point which is admissible. This definition is equivalent to that of Kripke. Let ρ be the first recursively Mahlo ordinal. As in the classical case, ρ is a very large τ-number.

Remark. The τ numbers are closely related to ordinals associated with recursive functionals of type 2. For any type 2 functional $\mathbf{F}$ let $d(\mathbf{F})$ be the type 2 functional equivalent to the representing functional of the predicate $[\lambda x\alpha . \{x\}(\alpha, \mathbf{F})$ is defined$]$. Let $\mathbf{E}_0 = \mathbf{E}$; for $n < \omega$ let $\mathbf{E}_{n+1} = d(\mathbf{E}_n)$. Let $\omega_1^{\mathbf{F}}$ be the first ordinal not recursive in $\mathbf{F}$. Kripke and Platek have observed that $\omega_1^{\mathbf{E}_1} = \tau_0^1$. A generalization of this shows that for each $0 < n < \omega$, $\omega_1^{\mathbf{E}_n} = \tau_0^n$. Gandy [3] discusses a powerful functional of type 3 called the *superjump* which is equivalent to the operation d. Platek has announced a hierarchy for the sets of integers recursive in the superjump. From his description of this hierarchy it appears to us that the first ordinal not recursive in the superjump is equal to the smallest ordinal α such that $\alpha = \tau_0^\alpha$, which is much smaller than the first recursively Mahlo ordinal [4].

Theorem 2.3. $|\mathcal{A}| \leqslant \rho$.

Proof. Let $n \in \Gamma(X) \equiv n \in \Gamma_0(X) \vee [\Gamma_0(X) = X \;\&\; n \in \Gamma_1(X)]$, where Γ_0 and Γ_1 are Π_1^0. Let $\langle A_\xi : \xi \in On \rangle$ be the sequence defined by Γ. The proof of the following lemma is a simple modification of the proof of 2.1.

Lemma 2.4. If $\alpha > \omega$ is admissible then $\Gamma_0(A_\alpha) = A_\alpha$.

It follows that if α is admissible then $\Gamma_1(A_\alpha) = \Gamma(A_\alpha)$. Now suppose $n \in \Gamma(A_\rho)$. We show $n \in A_\rho$. Since ρ is admissible, $n \in \Gamma_1(A_\rho)$. Let f be the ρ-recursive normal function constructed in the proof of 2.1 (replacing there Γ by Γ_1). Since ρ is recursively Mahlo, f has a fixed point $\beta < \rho$ which is admissible. Hence, as in the proof of 2.1,

$$n \in \Gamma_1(A_\beta) = \Gamma(A_\beta) \subseteq A_\rho \;.$$

3. A notational analogue of the first Mahlo ordinal

We wish to show that $|\mathcal{A}| = \rho$. The proof is in two stages. First we find a $\Gamma \in \mathcal{A}$ which inductively defines a set which, regarded as a system of notations for ordinals, provides a *notational* analogue of the first Mahlo ordinal. Next we show that the notational analogue really is ρ.

Previous methods (cf. [6, 14, 18]) of defining systems of notations for ordinals greater than ω_1 consisted of introducing names for recursive versions of transfinite number classes. A typical system would be defined by an inductive definition consisting of several cases, depending on whether the ordinal reached at a given stage was zero, a successor, 'notationally' singular,

4) [Added in print.] Aczel has shown that the ordinal of the superjump in fact is equal to the first recursively Mahlo ordinal.

'notationally' regular, etc. For illustration, consider a system of notations for the ordinals less than the first 'notationally' inaccessible ordinal. Combining the 'zero', 'successor' and 'singular' cases into one, the definition will consist of two cases; the 'non-regular' and 'regular' case. These two cases correspond roughly to operations Γ_0 and Γ_1. The desired set A of notations is then obtained as the union of a transfinite sequence $\langle A_\xi : \xi \in On \rangle$ obtained by applying the operation Γ_0 until closure is reached, i.e. until a 'notationally' regular ordinal is reached, and then applying the operation Γ_1. Thus A is the set inductively defined by Γ where:

$$\Gamma(X) = \begin{cases} \Gamma_0(X) & \text{if } \Gamma_0(X) \neq X\,, \\ \Gamma_1(X) & \text{otherwise}\,. \end{cases}$$

It is not clear that different notational versions of the first inaccessible will actually be the same. If things are done properly, however, it is possible to prove that the ordinal reached really is the first recursively inaccessible. With a little ingenuity one can choose Γ_0 and Γ_1 to be Π_1^0 operations. One can even continue beyond the first inaccessible in this way by introducing names for inaccessible number classes. Just how far one can go by this ad hoc procedure of naming number classes and using only Γ_0 and Γ_1 which are Π_1^0 seems limited only by one's ingenuity in finding names for number classes. Whenever a new kind of fixed point is reached one must discover a new type of naming procedure. It is possible to show that this procedure can continue beyond the first inaccessible, fixed point of inaccessibles, etc. It is clear, however, that this ad hoc procedure of naming number classes, will never reach the first recursively Mahlo ordinal.

To obtain a notational analogue of the first (weakly) Mahlo ordinal in set theory we make the following observation. Let α be a regular ordinal. Then α is not Mahlo iff there is some function $f : \alpha \to \alpha$ such that:

(*) for every regular $\beta < \alpha, \beta < \sup\{f(\xi) : \xi < \beta\}$.

Our system of notations is therefore defined so that when we have constructed a set M_α of notations for the ordinals less than α, α a notational analogue of a regular ordinal, we let the notations for α consist of (roughly speaking) all those names for (primitive recursive) functions f such that:

(i) $f : M_\alpha \to M_\alpha$;

(ii) names for f have not already been put in M_α.

If α is less than the first 'notationally' Mahlo ordinal we expect by $(*)$ that there is such an f. More precisely, we proceed as follows: Let $\lambda y.\{z\}_p(y, X)$ be the function primitive recursive in $X \subseteq \omega$ with index z. Similarly without the X. In the following inductive definition of M, $|x| = \mu\alpha . x \in M_{\alpha+1}$, for $x \in M$. The definition of M makes use of some simplifications due to P. Aczel.

Definition of M

(i) $M_0 = \phi$;

(ii) $M_{\alpha+1} = M_\alpha \cup \{1\} \cup \{2^x : (x)_0 \in M_\alpha$

$\& \forall y\, [\{(x)_1\}_p(y, M_{|(x)_0|}) \in M_\alpha]\}$

$\cup \{3^{\langle x,y \rangle} : x \in M_\alpha \vee y \in M_\alpha\}$,

if the right side is not equal to M_α;

(iii) $M_{\alpha+1} = M_\alpha \cup \{5^x : (\forall y \in M_\alpha)\, [\{x\}_p(y) \in M_\alpha]\}$, otherwise;

(iv) $M_\lambda = \cup \{M_\xi : \xi < \lambda\}$ for λ a limit ordinal.

Let $M = \cup M_\xi$ and $|M| = \sup\{|x| : x \in M\}$. For $x \notin M$, let $|x| = |M|$.

Remarks. (1) Case (ii) covers the case where α is either a successor ordinal or a notational analogue of a singular limit ordinal. For if in case (ii) the right side is equal to M_α, then M_α is 'closed' under mappings primitive recursive in M_β, for all $\beta < \alpha$. Note that if $2^x \in M$, then $|2^x|$ is the larger of $|(x)_0| + 1$ and $\sup\{|\{(x)_1\}_p(y, M_{|(x)_0|})| + 1 : y \in \omega\}$.

(2) the use of relative *primitive* recursiveness instead of partial relative recursiveness in (ii) avoids the introduction of an extra quantifier to express the predicate '$\{z\}(y, X)$ is defined'.

(3) From (ii) we have:

(a) $x \in M \vee y \in M \Leftrightarrow 3^{\langle x,y \rangle} \in M$,

(b) $x \in M \vee y \in M \Rightarrow \inf\{|x|, |y|\} < |3^{\langle x,y \rangle}|$.

(a) and (b) seem to be needed for technical reasons in Section 5. We do not know if the clause introducing $3^{\langle x,y\rangle}$ is essential.

Note that in (ii) and (iii) there is just one quantifier which is universal. (ii) and (iii) correspond roughly to operations Γ_0 and Γ_1, respectively. Because of the occurrence of $|\ |$, however, this is not directly expressible as an inductive definition of the proper form. To avoid this we give an inductive definition of $\langle M'_\alpha : \alpha \in On\rangle$ where $M'_\alpha = \{\langle x, y\rangle : x, y \in M_\alpha \ \&\ |x| \leqslant |y|\}$. It is a routine, but tedious, chore to define Π^0_1 inclusive operations Γ_0 and Γ_1 so that the corresponding $\Gamma \in \mathscr{A}$ inductively defines $\langle M'_\alpha : \alpha \in On\rangle$. Hence $|\Gamma| = |M|$. Thus in order to show that $|\mathscr{A}| = \rho$ it suffices to show that $|M| = \rho$.

4. $|M| = \rho$

A pair $(A, ||)$ is called an *ordinal system* if A is a non-empty subset of ω and $||$ maps A onto an initial segment of ordinals. (This is a minor change from the definition in [16].) $||$ need not be one-to-one. Let $|A| = \sup\{|a| + 1 : a \in A\}$ and $A_\alpha = \{a \in A : |a| < \alpha\}$. Thus $A = A_{|A|}$. If $a \notin A$, we set $|a| = |A|$.

Let $B \subseteq \omega$. $(A, ||)$ is said to be *B-restricted productive* if there is a recursive function p such that:

$$(1)\qquad \forall x\,[\{n\}(x, B) \in A] \Rightarrow [p(n) \in A \ \&\ |p(n)| \geqslant \sup\{|\{n\}(x, B)| + 1 : x \in \omega\}]\ ;$$

$$(2)\qquad p(n) \in A \Rightarrow \forall x\,[\{n\}(x, B) \in A]\ .$$

p is called a *B-restricted productive function for* $(A, ||)$.

Definition. The ordinal system $(A, ||)$ is *acceptable* if there are recursive functions g, p, and Ⓥ such that:

(i) $g : A \to A$ and for $a \in A$, if $|g(a)| \leqslant \alpha$ then $j(A_{|a|}) \leqslant_t A_\alpha$ uniformly in a (where $j(X) = \{z : \exists y T^X(z, z, y)\}$);

(ii) If $a \in A$ then $\lambda n.p(a, n)$ is an $A_{|a|}$-restricted productive function for $(A, ||)$;

(iii) $x \in A \vee y \in A \Rightarrow x$ Ⓥ $y \in A \ \&\ \inf\{|x|, |y|\} \leqslant |x$ Ⓥ $y|$.

Theorem 4.1. Let $(A, ||)$ be acceptable. Then:

(i) $|A|$ is admissible;

(ii) If $f: |A| \to |A|$ is $|A|$-recursive then there is a recursive function h such that:

$$\text{(a)} \qquad x \in A \Rightarrow h(x) \in A ,$$

$$\text{(b)} \qquad f(|x|) \leqslant |h(x)| \qquad \text{for } x \in A .$$

We do not know whether clause (iii) in the definition of acceptable is necessary for the proof of Theorem 4.1.(iii) is used only in the proof of Lemma 5.4 below.

The proof of Theorem 4.1 will be outlined in Section 5. We now show how to apply Theorem 4.1 to prove that $|M| = \rho$. In Lemmas 4.2 and 4.3 below we assume λ is either $|M|$ or of the form $|5^x|$ where $5^x \in M$. We shall show that $(M_\lambda, ||_\lambda)$ is acceptable, where $||_\lambda$ is the restriction of $||$ to M_λ.

Lemma 4.2. Let $\beta < |M|$. Then $j(M_\beta) \leqslant_t M_\alpha$ for all $\alpha > \beta + 1$ uniformly in a notation for β.

Proof. Let $|a| + 1 = \beta + 1 < \alpha$. Let e be a recursive function such that

$$\{e(a, x)\}_p(t, M_\beta) = \begin{cases} a & \text{if } \neg\, T^{M_\beta}(x, x, t) , \\ 0 & \text{otherwise} . \end{cases}$$

Then $x \notin j(M_\beta) \equiv 2^{\langle a, e(a,x)\rangle} \in M_\alpha$.

Lemma 4.3. For $\beta < \lambda$, $(M_\lambda, ||_\lambda)$ is M_β-restricted productive, uniformly in a notation for β.

Proof. Let $p(a, n) = 2^{\langle a,n \rangle}$.

From Lemmas 4.2 and 4.3 and the definition of $(M, ||)$ it follows that $(M_\lambda, ||_\lambda)$ is acceptable. Hence by Theorem 4.1,

Lemma 4.4. λ is admissible.

Theorem 4.5. $|M| = \rho$.

Proof. We have already shown that $|M| \leqslant p$. Let $f: |M| \to |M|$ be normal and

$|M|$-recursive. It suffices to show that f has a fixed point which is admissible. Let h be a recursive function such that $f(|x|) \leqslant |h(x)|$. Let e_0 be a Gödel number of h, $|a| = \omega$, and g be a primitive recursive function such that

$$\{g(x)\}_p(t, M_\omega) = \begin{cases} (t)_0 & \text{if } T^{M_\omega}(e_0, x, t), \\ a & \text{otherwise}. \end{cases}$$

Let $\{e\}_p(x) = 2^{\langle a, g(x) \rangle}$. Then $\{e\}_p$ is primitive recursive and $x \in M \Rightarrow |h(x)| < |\{e\}_p(x)|$. Thus $\{e\}_p : M \to M$ and $\sup\{|\{e\}_p(t)| : t \in M\} = |M|$. Hence, from the definition of $(M, \|)$, $5^e \in M$. Since $|5^e|$ is admissable it suffices to show that $f(|5^e|) = |5^e|$. Let $|u| < |5^e|$. Then,

$$|u| \leqslant f(|u|) \leqslant |\{e\}_p(u)| < |5^e|.$$

Hence, since f is normal,

$$f(|5^e|) = \sup\{f(|u|) : |u| < |5^e|\} = |5^e|.$$

5. Acceptable ordinal systems

In this section we sketch a proof of Theorem 4.1. Let $\mathfrak{A} = (A, \|)$ be acceptable with associated functions p, g, and ⓥ. Let $|u_0| = 0$ and $u_1 \notin A$.

Lemma 5.1. There is a recursive function $+_A$ such that

(i) $x \in A \;\&\; y \in A \Rightarrow x +_A y \in A \;\&\; |x +_A y| \geqslant \max\{|x|, |y|\}$;

(ii) $x +_A y \in A \Rightarrow x \in A \;\&\; y \in A$.

Proof. Let e be a recursive function such that

$$\{e(x, y)\}(t, \phi) = \begin{cases} x & \text{if } t = u_0, \\ y & \text{otherwise}. \end{cases}$$

Let $x +_A y = p(u_0, e(x, y))$.

We now define the class of $\mathfrak{A}$-partial recursive number-theoretic functions.

The predicate $\{z\}^{\mathfrak{A}}(\vec{x}) \simeq y$ is defined inductively as in Kleene [5], but the starting functions in the case S0 are given by

$$\text{S0.}a \qquad \{\langle 0, a\rangle\}^{\mathfrak{A}}(x) = \begin{cases} 0 & \text{if } x \in A_{|a|}, \\ 1 & \text{otherwise}. \end{cases} \qquad \text{for } a \in A,$$

and S8 is

$$\text{S8} \qquad \{\langle 8, n, h\rangle\}^{\mathfrak{A}}(\vec{x}) \simeq \mathbf{E}(\lambda t . \{h\}^{\mathfrak{A}}(\vec{x}, t)),$$

where $\mathbf{E}$ is the type 2 functional describing the jump operator j (see [5]). Since $\vec{x}$ will always be a finite sequence of non-negative integers, there is no S7 clause. $\{z\}^{\mathfrak{A}}$ is called the *$\mathfrak{A}$-partial recursive function with index z*. $\{z\}^{\mathfrak{A}}$ is *$\mathfrak{A}$-recursive* if it is everywhere defined. For each z, a, such that $\{z\}^{\mathfrak{A}}[a]$ (defined in [5]) is defined we associate the ordinal $|\langle z, a\rangle|^{\mathfrak{A}}$ of the computation of the value $\{z\}^{\mathfrak{A}}[a]$. Let

$$D = \{\langle z, a\rangle : \{z\}^{\mathfrak{A}}[a] \text{ is defined}\}.$$

Lemma 5.2. There are recursive functions θ and φ such that:

(i) $\langle z, a\rangle \in D \equiv \varphi(z, a) \in A$;

(ii) If $\langle z, a\rangle \in D$ then for all x,

$$\{z\}^{\mathfrak{A}}[a] = \{\theta(z, a)\}(x, A_{|\varphi(z,a)|}).$$

The functions θ and φ are defined by the recursion theorem. (ii) and (i) in the direction $\Rightarrow$ are then proven by induction on $|\langle z, a\rangle|^{\mathfrak{A}}$. (i) in the direction $\Leftarrow$ is proven by induction on $|\varphi(z, a)|$. The function g is used to define φ in cases S4 and S8. We omit the proof which is similar to that of Lemma 5.3 of [18].

$X \subseteq \omega$ is said to be *$\mathfrak{A}$-r.e.* if X is the domain of an $\mathfrak{A}$-partial recursive function; X is *$\mathfrak{A}$-recursive* if the representing function of X is $\mathfrak{A}$-recursive. Using Lemma 5.2 we have as in [18]:

Lemma 5.3. Let $X \subseteq \omega$.

(i) X is $\mathfrak{A}$-r.e. iff $X \leqslant_1 A$;

(ii) X is $\mathfrak{A}$-recursive iff $X \leqslant_t A_\alpha$ for some $\alpha < |A|$;

(iii) if $f : \omega \to A$ and f is $\mathfrak{A}$-recursive, then

$$\sup\{|f(t)| : t \in \omega\} < |A|.$$

Using (V) and the fact that every A_α is $\mathfrak{A}$-recursive, uniformly in a notation for α, given $x \in A \vee y \in A$ it is possible to decide $\mathfrak{A}$-recursively whether $|x| \leqslant |y|$ or $|y| < |x|$. This leads to the following lemma.

Lemma 5.4. There is an $\mathfrak{A}$-partial recursive function δ such that:

(i) $\quad x \in A \;\&\; |x| \leqslant |y| \Rightarrow \delta(x, y) = 0$;

(ii) $\quad |y| < |x| \Rightarrow \delta(x, y) = 1$.

An important theorem of Gandy [3] establishes the existence of selection functions for partial recursive functionals of type 2. Since we have an infinite number of starting functions in case S0 of our definition of $\mathfrak{A}$-partial recursive, the existence of a selection function is not an immediate consequence of Gandy's theorem. However, a crucial part of his proof (which unfortunately has never been published) is, in our setting, Lemma 5.4. By paraphrasing the Gandy argument (cf. Moschovakis [11], p 268, where the argument is used in a similar context) we obtain the following version of Gandy's theorem.

Lemma 5.5. There is an $\mathfrak{A}$-partial recursive function ν such that if $\{e\}$ is total then

$$\exists t[\{e\}(t) \in A] \Rightarrow \{e\}(\nu(e)) \in A .$$

Let $Q = \{x : x \in A \;\&\; (\forall y < x)\,[|y| \neq |x|]\}$. Q is $\mathfrak{A}$-r.e. and for $\alpha < |A|$, $Q_\alpha = \{y \in Q : |y| < \alpha\}$ is $\mathfrak{A}$-recursive (uniformly in a notation for α). Consider the Kripke [7] equation calculus for the $|A|$-partial recursive functions where the numeral denoting the ordinal α is the unique member a of Q such that $|a| = \alpha$. Let $\langle S^E_\alpha : \alpha \in On\rangle$ be the transfinite sequence defined by the finite set E of equations. It is easy to Gödel number the language so that the set of Gödel numbers of equations is $\mathfrak{A}$-r.e. and hence one-one reducible to A. Furthermore, by 5.3 (i) there is a partial recursive function σ such that for $b \in A$, $\lambda n . \sigma(n, b)$ is total and

$$n \text{ is the Gödel number of an equation belonging to } S^E_{|b|}$$

iff $\sigma(n, b) \in A$.

Proof of 4.1. (i). Let $\alpha < |A|$ and let f be recursive. From inspection of the Kripke rules of inference and the property of σ, it suffices to show that (1) implies (2) below.

(1) $$(\forall x \in Q_\alpha)(\exists y \in A)\,[f(x, y) \in A]\ .$$

(2) $$(\exists \beta < |A|)(\forall x \in Q_\alpha)(\exists y \in A_\beta)\,[f(x, y) \in A]\ .$$

Suppose (1) holds. Let e be a recursive function such that

$$\{e(x)\}(y) = y +_A f(x, y)\ .$$

(1) is equivalent to:

$$(\forall x \in Q_\alpha)\exists y\,[\{e(x)\}(y) \in A]\ .$$

Hence by 5.5,

$$(\forall x \in Q_\alpha)\,[\{e(x)\}(\nu(e(x))) \in A]\ .$$

Let,

$$h(x) = \begin{cases} \{e(x)\}(\nu(e(x))) & \text{if } x \in Q_\alpha\,, \\ u_0 & \text{otherwise}\ . \end{cases}$$

Then $h : \omega \to A$ is $\mathfrak{A}$-recursive. Let $\beta = \sup\{|h(t)| : t \in \omega\}$. $\beta < |A|$ by 5.3 and hence this β satisfies (2).

Lemma 5.6. There is an $\mathfrak{A}$-partial recursive ψ such that:

$$x \in A \Rightarrow \psi(x) \in Q \;\&\; |x| = |\psi(x)|\ .$$

Proof. $B = \{\langle x, y\rangle : x \in A \;\&\; y \in Q \;\&\; |x| = |y|\}$ is $\mathfrak{A}$-r.e. by 5.4. Hence there is a recursive e such that:

$$\langle x, y\rangle \in B \equiv \{e(x)\}(y) \in A\ .$$

Since $(\forall x \in A)\exists y\,[\langle x, y\rangle \in B]$, we have by 5.5,

$$(\forall x \in A)\,[\{e(x)\}(\nu(e(x))) \in A]\ .$$

Thus let $\psi(x) \simeq \nu(e(x))$.

Proof of 4.1 (ii). Let $f: |A| \to |A|$ be $|A|$-recursive. We begin by showing there is an $\mathfrak{A}$-partial recursive function $\{b\}^{\mathfrak{A}}$ such that for $x \in A$, $f(|x|) = |\{b\}^{\mathfrak{A}}[x]|$. Let E be a finite set of equations and **f** a function symbol such that for $x, y \in Q$,

$$f(|x|) = |y| \equiv (\exists a \in A)\,[\text{`}\mathbf{f}(x) = y\text{'} \in S^E_{|a|}]\ .$$

Using the recursive function σ we can find a recursive function τ such that:

$$f(|x|) = |y| \equiv (\exists a \in A)\,[\tau(x, y, a) \in A]\ .$$

Let q be a recursive function such that $q : Q \leqslant_1 A$. Let e be a recursive function such that for $x \in Q$,

$$(\exists y \in Q)(\exists a \in A)\,[\tau(x, y, a) \in A]$$

$$\equiv \exists z\,[q((z)_0) +_A ((z)_1 +_A \tau(x, (z)_0, (z)_1)) \in A]$$

$$\equiv \exists z\,[\{e(x)\}(z) \in A]\ .$$

Let $x \in Q$. Then, $\{e(x)\}(\nu(e(x))) \in A$ and $f(|x|) = |(\nu(e(x)))_0|$. By 5.6, for $x \in A$, $\psi(x) \in Q$ and $f(|x|) = f(|\psi(x)|) = |(\nu e\psi(x))_0|$. Let $\{b\}^{\mathfrak{A}}[x] \simeq (\nu e\psi(x))_0$. By 5.2, for $x \in A$ and all t:

$$\{b\}^{\mathfrak{A}}[x] = \{\theta(b, x)\}(t, A_{|\varphi(b,x)|})\ .$$

Let $h(x) = p(\varphi(b, x), \theta(b, x))$. Then h is recursive and for $x \in A$, $\varphi(b, x) \in A$ and

$$|h(x)| \geqslant \sup\{|\{\theta(b, x)\}(t, A_{|\varphi(b,x)|})| + 1 : t \in \omega\}$$

$$> f(|x|)\ .$$

References

[1] J. Barwise, Applications of strict Π^1_1 predicates to infinitary logic, J. Symbolic Logic 34 (1969) 409–423.

[2] J. Barwise, R.O. Gandy and Y.N. Moschovakis, The next admissible set, to appear.

[3] R.O. Gandy, General recursive functionals of finite type and hierarchies of functions, A paper given at the Symposium on Mathematical Logic held at the University of Clermont Ferrand in June 1962, mimeographed.
[4] T.J. Grilliot, Hierarchies based on objects of finite type, J. Symbolic Logic 34 (1969) 177–182.
[5] S.C. Kleene, Recursive functionals and quantifiers of finite types, I, Trans. Am. Math. Soc. 91 (1959) 1–52.
[6] D.L. Kreider and H. Rogers Jr., Constructive versions of ordinal number classes, Trans. Am. Math. Soc. 100 (1961) 325–369.
[7] S. Kripke, Transfinite recursions on admissible ordinals, I, II, J. Symbolic Logic 29 (1964) 161–162 (abstracts).
[8] S. Kripke, Admissible ordinals and the analytic hierarchy, J. Symbolic Logic 29 (1964) 162 (abstract).
[9] S. Kripke, Transfinite recursion, constructible sets, and analogues of cardinals, in: Lecture notes prepared in connection with the Summer Institute on Axiomatic Set Theory held at University of California, Los Angeles, Calif., July–August, 1967.
[10] P. Mahlo, Über lineare transfinite Mengen, Berichte über die Verhandlungen der Königlich-Sächsischen Gesellfschaft der Wissenschaften zu Leipzig, Mathematisch-Physische Klasse 63 (1911) 187–225.
[11] Y.N. Moschovakis, Hyperanalytic predicates, Trans. Am. Math. Soc. 129 (1967) 249–282.
[12] Y.N. Moschovakis, The Suslin-Kleene theorem for countable structures, to appear.
[13] R.A. Platek, Foundations of recursion theory, Ph. D. Thesis, Stanford University (1966).
[14] H. Putnam, Uniqueness ordinals in higher constructive number classes, in: Essays on the foundations of mathematics, eds. Y. Bar-Hillel, E.I.J. Poznanski, M.O. Rabin and A. Robinson (Magnes Press, Jerusalem, 1961; North-Holland Publ. Co., Amsterdam, 1962).
[15]. H. Putnam. On hierarchies and systems of notations, Proc. Am. Math. Soc. 15 (1964) 44–50.
[16] W. Richter, Extensions of the constructive ordinals, J. Symbolic Logic 30 (1965) 193–211.
[17] W. Richter, Constructive transfinite number classes, Bull. Am. Math. Soc. 73 (1967) 261–265.
[18] W. Richter, Constructively accessible ordinal numbers, J. Symbolic Logic 33 (1968) 43–55.
[19] C. Spector, Inductively defined sets of natural numbers, in: Infinitistic Methods (New York-Oxford-London-Paris and Warsaw, 1961) pp. 97–102.

F-RECURSIVENESS

Gerald E. SACKS [1)]
Massachusetts Institute of Technology and University of Wisconsin, USA

1. Normal F-jumps

What follows is intended to serve as an introduction to [1], a somewhat lengthy paper which generalizes the principal results of [2] from admissible ordinals [3, 4] and hyperdegrees to F-admissible ordinals and F-degrees. The principal technique employed is "forcing with *pointed*, perfect closed sets", a simple example of which was given in the final section of [5]. After proving Corollary 2.2, we learned that R.B. Jensen had obtained an earlier proof by another forcing technique. Jensen's approach has the virtue that it readily provides important extensions of 2.2 of the following type: if $\{\alpha_i \mid i < \omega\}$ is an ascending sequence of F-admissible ordinals, then there exists an $A \subseteq \omega$ such that for all $i < \omega$, α_i is the ith ordinal F-admissible in A. Our approach leads in a natural way to the minimality facts of Section 2 and to the absoluteness properties of Sections 3 and 4. It is interesting to note that H. Friedman [6] independently developed an unorthodox proof of 2.2 for the case of recursively inaccessible ordinals, unorthodox in that it virtually omits forcing in favor of Barwise's compactness theorem [7].

Let F be a set of sentences in the language of Zermelo-Fraenkel set theory. We will identify F with the set of Gödel numbers of members of F. If M is a transitive set, then $M \models F$ means that every member of F is true in M. For each ordinal β and each $T \subseteq \omega$, let L_β (or $L_\beta(T)$) be the set of all sets constructible in the sense of Gödel [8] from the empty set (or from T) via the ordinals less than β. Kripke [3] and Platek [4] call an ordinal α admissible if certain trivial axioms and every Σ_1 instance of the replacement axiom are true in L_α.

We associate some partial functions with F:

1) The preparation of this paper was partially supported by U.S. Army contract DAHCO-4-67-C-0052. The author is grateful to H. Friedman and R. Platek for many stimulating conversations concerning countable, admissible ordinals.

$$\omega_F^T \simeq \mu\beta(L_\beta(T) \models F) \qquad (\mu \text{ means least})$$

$$O_F^T \simeq L_{\omega_F^T}(T)\,.$$

If ω_F^T is defined, then O_F^T is called the *F-jump* of T. If T is empty, then ω_F (O_F respectively) is written in place of ω_F^T (O_F^T respectively). F is said to be *normal* if:

(a) $F \in O_F$;
(b) for all T, ω_F^T is defined and is admissible;
(c) for all T, $O_F^T \models$ every set is countable;
(d) for all T, $L_{\omega_F^T} \models F$;
(e) for all S and T, if $S \in O_F^T$, then $O_F^S \subseteq O_F^T$.

Let F be normal; an ordinal α is said to be *F-admissible* if $L_\alpha \models F$. By (b) and (d), every ordinal of the form ω_F^T is both admissible and F-admissible. Let S be a subset of ω; S is said to be *F-recursive in T* ($S \leqslant_F T$) if $S \in O_F^T$. By (e) the relation $\leqslant_F$ is transitive. Two subsets of ω have the same *F-degree* if each is F-recursive in the other.

R. Gandy calls an ordinal recursively inaccessible if it is both admissible and the limit of admissible ordinals. Let RI be a recursive set of sentences of ZF such that for all α, α is recursively inaccessible if and only if $L_\alpha \models$ RI. For each $n \geqslant 1$, let Σ_n be such that for all α: α is admissible and every Σ_n instance of replacement holds in L_α if and only if $L_\alpha \models \Sigma_n$. Let $\Sigma_\infty = \bigcup_n \Sigma_n$.

Proposition 1.1. RI, Σ_n ($n \geqslant 1$), Σ_∞ are normal.

2. Regular F-jumps

Observe that $\omega_{\Sigma_1}^T = \omega_1^T$, and that the Σ_1-degrees coincide with the hyperdegrees. We regard a normal F-jump as a generalization of the hyperjump of Kleene [9]. Let F be normal; α is said to be F-regular if $\alpha = \omega_F^T$ for some T. F is called *regular* if every countable, F-admissible ordinal is F-regular. Note that clause (d) of the definition of normality requires that every F-regular ordinal be F-admissible. In [2] it was shown that Σ_1 is regular. To be more precise, in [2] the following was shown: let α be a countable admissible ordinal; then there exists a T such that $\omega_1^T = \alpha$ and such that for every B of lower hyperdegree than T, $\omega_1^B < \alpha$.

Theorem 2.1. If F is RI, Σ_n $(n \geqslant 1)$, or Σ_∞, and α is countable and F-admissible, then there exists a T such that $\omega_F^T = \alpha$ and such that

$$(B)\,[B <_F T \rightarrow \omega_F^B < \alpha]\ .$$

Corollary 2.2 (R.B. Jensen). If F is RI, Σ_n $(n > 1)$, or Σ_∞, then F is regular.

Let $\mathcal{K}$ be a countable set of F-degrees. An F-degree d is an upper bound for $\mathcal{K}$ if every member of $\mathcal{K}$ is less than or equal to d; d is a minimal upper bound if no F-degree less than d is an upper bound for $\mathcal{K}$.

Corollary 2.3. If F is RI, Σ_n $(n \geqslant 1)$, or Σ_∞, and β is a countable ordinal, then the set of F-degrees of subsets of ω belonging to L_β has a minimal upper bound.

We sketch a proof of an instructive, special case of 2.1. An interested reader can supply the details missing below by carefully perusing [2]. A complete proof of 2.1 will be given in [1].

Let α be a countable, Σ_∞-admissible ordinal. Assume $L_\alpha \models$ every set is countable. We seek a T such that $\omega_{\Sigma_\infty}^T = \alpha$ and such that

$$(B)\,[B <_{\Sigma_\infty} T \rightarrow \omega_{\Sigma_\infty}^B < \alpha]\ .$$

Let $P, Q, R, \ldots$ denote perfect closed subsets of 2^ω, i.e. uncountable closed subsets of 2^ω without isolated points. P can be encoded by a subset of ω, since P is completely specified by the set of all finite initial segments of members of P. We will abuse notation and let P denote both a perfect closed set and its standard encoding by a subset of ω. P is Σ_∞*-pointed* if

$$(T)\,[T \in P \rightarrow P \in O_{\Sigma_\infty}^T]\ .$$

If P is Σ_∞-pointed, then $\omega_{\Sigma_\infty}^T \geqslant \omega_{\Sigma_\infty}^P$ for every $T \in P$. Thus the use of Σ_∞-pointed P's as forcing conditions makes it possible to "force" $\omega_{\Sigma_\infty}^T$ as high as is needed. Let $\mathcal{L}_\alpha(\mathcal{T})$ be an extension of the language of ZF containing names for all members of $L_\alpha(T)$. $\mathcal{L}_\alpha(\mathcal{T})$ includes: a numeral $\bar{n}$ for each $n < \omega$, unranked variables $x, y, z, \ldots$; and ranked variables x^β, y^β, z^β for each $\beta < \alpha$. It is intended that $\mathcal{T}$ denote T, that x^β range over $L_\beta(T)$, and that x range over $L_\alpha(T)$. A sentence $\mathcal{S}$ is ranked if all its variables are ranked.

We define a forcing relation, $P \Vdash \mathcal{S}$, where P is a Σ_∞-pointed, perfect closed set whose standard encoding as a subset of ω belongs to L_α, and where $\mathcal{G}$ is a sentence of $\mathcal{L}_\alpha(\mathcal{G})$:

(1) $P \Vdash \mathcal{G}$ if $(T)\,[T \in P \to L_\alpha(T) \models \mathcal{G}]$ and $\mathcal{G}$ is a ranked sentence;

(2) $P \Vdash (Ex^\beta)\mathcal{G}(x^\beta)$ if $(Ex^\beta)\mathcal{G}(x^\beta)$ is an unranked sentence and $P \Vdash \mathcal{G}(c)$ for some constant c denoting a member of $L_\beta(T)$;

(3) $P \Vdash (Ex)\mathcal{G}(x)$ if $P \Vdash (Ex^\beta)\mathcal{G}(x^\beta)$ for some $\beta < \alpha$;

(4) $P \Vdash \mathcal{G}\ \&\ \mathcal{G}$ if $P \Vdash \mathcal{G}$ and $P \Vdash \mathcal{G}$;

(5) $P \Vdash \sim \mathcal{G}$ if $\mathcal{F}$ is unranked and $(Q)_{P \supseteq Q} \sim [Q \Vdash \mathcal{G}]$.

T is *generic* if for every sentence $\mathcal{G}$ of $\mathcal{L}_\alpha(\mathcal{G})$ there is a P such that $T \in P$, and either $P \Vdash \mathcal{G}$ or $P \Vdash \sim\mathcal{G}$. Clause (1) of the definition of $\Vdash$ is not orthodox; consequently, the existence of generic T's is not obvious, but instead requires 2.6.

Let Z be a subset of L_α. Z is said to be L_α-definable if there is a formula $\mathcal{G}(x)$ of ZF with parameters in L_α such that

$$Z = \{x \mid x \in L_\alpha \text{ and } L_\alpha \models \bar{\mathcal{G}}(x)\}\ .$$

Kripke [3] and Platek [4] call a function f (from α into L_α) *α-recursive* if its graph is L_α-definable by means of a Σ_1 formula. A set is α-recursively enumerable if it is the range of an α-recursive function. A set is α-finite if it belongs to L_α.

Proposition 2.4 (i). The relation, $P \Vdash \mathcal{G}$, restricted to ranked $\mathcal{G}$'s, is α-recursively enumerable. (ii) for each $n > 0$: the relation $P \Vdash \mathcal{G}$, restricted to $\mathcal{G} \in \Sigma_n$, is Σ_n over L_α.

Proposition 2.4 (i) is, in the main, a consequence of the following fact: if $R(X)$ is a Π^1_1 predicate and α is admissible, then the set $\{X \mid X \in L_\alpha\ \&\ R(X)\}$ is α-recursively enumerable.

Proposition 2.5. $(\gamma)_{\gamma<\alpha}(P)\,(EQ)_{P \supseteq Q}[\omega^Q_{\Sigma_\infty} \geqslant \gamma]$.

Don't forget that 2.5 assumes that α is such that $L_\alpha \models$ every set is countable. The proof of 2.5 is similar to that of 3.2 of [5].

The next lemma, the so-called "sequential" lemma (cf. [10]) or "fusion" lemma, is designed to capture the essential properties of Σ_∞-pointed, perfect closed set forcing. The only worrisome points in the proof arise from the Σ_∞-pointedness requirement. The proof for Σ_1-pointedness [2] is more direct, and if all pointedness requirements are dropped as in [5, 10], the proof is almost immediate.

Lemma 2.6. Let $\{\mathcal{G}_i \mid i < \omega\}$ be an α-finite set of Σ_n sentences of $\mathcal{L}_\alpha(\mathcal{G})$. Let P be such that

$$(i)\,(Q)_{P \supseteq Q}(ER)_{Q \supseteq R}\,[R \Vdash \mathcal{G}_i]\ .$$

Then there exists an $R \subseteq P$ such that $(i)\,[R \Vdash \mathcal{G}_i]$.

Proof. Let $L_\beta \prec_k L_\alpha$ mean that $\beta \leqslant \alpha$ and that any Σ_k sentence of ZF with parameters in L_β is true in L_β if and only if it is true in L_α. Choose k large enough so that formulas of the type

$$(EQ)\,[P \supseteq Q \;\&\; \omega^Q_{\Sigma_\infty} \geqslant \gamma \;\&\; Q \Vdash \mathcal{G}_i] \qquad (i < \omega)$$

are L_α-definable by means of Σ_k formulas.

Let β_0 be the least β such that $L_\beta \prec_k L_\alpha$. Let $R_0 \in L_{\beta_0}$ be such that

$$L_{\beta_0} \models (R_0 \text{ is the "least" } R \subseteq P \text{ such that } R \Vdash \mathcal{G}_0)$$

By "least" we mean least according to the Σ_1 well-ordering of the reals in any admissible initial segment of L. Suppose L_{β_i} and R_i $(i \geqslant 0)$ have been defined. Let β_{i+1} be the least $\beta \geqslant \beta_i$ such that $L_\beta \prec_k L_\alpha$ and

$$L_\beta \models (ER)\,[R_i \supseteq R \;\&\; \omega^R_{\Sigma_\infty} \geqslant \beta_i \;\&\; R \Vdash \mathcal{G}_{i+1}]\ .$$

Choose $R_{i+1} \in L_{\beta_{i+1}}$ to be the "least" R with the above property.

Thus it is possible to define α-finite functions $\lambda_i \mid R_i$ and $\lambda_i \mid \beta_i$ such that $P \supseteq R_i \supseteq R_{i+1}$ and $R_i \in L_{\beta_i} \prec_k L_{\beta_{i+1}}$ for all i. It is also possible to keep the R_i's "fat" in a canonical fashion (as in [5, 10]) so that $\cap\, \{R_i \mid i < \omega\} = R$ is an uncountable, perfect closed set. $R \in L_\alpha$, since $\lambda_i \mid R_i$ is α-finite. Similarly, $\beta_\infty = \lim \beta_i < \alpha$.

But we must check that R is Σ_∞-pointed. Fix $T \in R$. It must be shown that $R \in O^T_{\Sigma_\infty}$. Observe that $L_{\beta_\infty} \subseteq O^T_{\Sigma_\infty}$, since for every i, $T \in R_{i+1}$ and so $\omega^T_{\Sigma_\infty} \geqslant \omega^{R_{i+1}}_{\Sigma_\infty} \geqslant \beta_i$. Let δ be the least Σ_∞-admissible ordinal $\geqslant \beta_\infty$. Then $L_{\beta_\infty} \subseteq L_\delta \subseteq O^T_{\Sigma_\infty}$. And it will suffice to show $\lambda_i \mid \beta_i$ and $\lambda_i \mid R_i$ are δ-finite functions.

Let β'_0 be the least β such that $L_\beta \prec_k L_{\beta_\infty}$. By the Tarski-Vaught elementary chain principle, $L_{\beta_\infty} \prec_k L_\alpha$; so $\beta'_0 = \beta_0$. By iteration of the elementary chain principle $\beta_\infty \leqslant \delta$ and $R \in L_\delta$.

Lemma 2.7. $(\mathcal{G})(P)(EQ)_{P\supseteq Q}[Q \Vdash \mathcal{G} \text{ or } Q \Vdash \sim\mathcal{G}]$.

Proof. It is safe to assume $\mathcal{G}$ is a ranked sentence. Since $L_\alpha \vdash$ every set is countable, it is possible to define an α-recursive notion of "rank" with the property that $\mathcal{G}$ is either atomic or equivalent to a countable (in the sense of L_α) conjunction or disjunction of sentences of lower "rank". Suppose $\{\mathcal{G}_i \mid i < \omega\}$ is an α-finite sequence of sentences of lower "rank" than $\mathcal{G}$ such that

(a) $$(i)(P)(EQ)_{P\supseteq Q}[Q \Vdash \mathcal{G}_i \text{ or } Q \Vdash \sim \mathcal{G}_i] ,$$

(b) $$(T)\,[L_\alpha(T) \models \mathcal{G} \leftrightarrow (Ei)(L_\alpha(T) \models \mathcal{G}_i)] .$$

Fix P. We seek a $Q \subseteq P$ such that $Q \Vdash \mathcal{G}$ or $Q \Vdash \sim\mathcal{G}$. If there is a $Q \subseteq P$ and an $i < \omega$ such that $Q \Vdash \mathcal{G}_i$, then all is well by (b). Suppose no such Q exists. Then by (a),

$$(i)(P)(EQ)_{P\supseteq Q}[Q \Vdash \sim \mathcal{G}_i] .$$

It follows from 2.6 that there is an $R \subseteq P$ such that $(i)\,[R \Vdash \sim \mathcal{G}_i]$. But then $R \Vdash \sim \mathcal{G}$ by (b).

It is immediate from 2.7 that generic T's exist.

Let c be a constant of $\mathcal{L}(\mathcal{G})$ that denotes a subset of ω, and for arbitrary T, let $c(T)$ be the member of $L_\alpha(T)$ denoted by c. It is assumed that the sentence $\bar{n} \in c$ $(n < \omega)$ is ranked.

Lemma 2.8. Let c denote a subset of ω. For each P there is a $Q \subseteq P$ such that either (i) or (ii) holds:

(i) $$(T)\,[T \in Q \to c(T) \in L_\alpha] ;$$

(ii) $$(T)\,[T \in Q \to T \in L_\alpha(c(T))] .$$

Proof. First suppose

$$(EP_0)_{P\supseteq P_0}(n)(Q_0)_{P_0\supseteq Q_0}(Q_1)_{P_0\supseteq Q_1} \sim [Q_0 \Vdash \bar{n} \in c \;\&\; Q_1 \Vdash \bar{n} \notin c] .$$

Then for each n, define

$$\mathcal{G}_n = \begin{cases} \bar{n} \in c & \text{if } (EQ)_{P_0 \supseteq Q}[Q \Vdash \bar{n} \in c] \\ \bar{n} \notin c & \text{if } (EQ)_{P_0 \supseteq Q}[Q \Vdash \bar{n} \notin c] \,. \end{cases}$$

$\lambda_n \mid \mathcal{G}_n$ is α-finite by 2.4 (i); clearly,

$$(n)\,(Q)_{P_0 \supseteq Q}(ER)_{Q \supseteq R}\,[R \Vdash \mathcal{G}_n] \;.$$

It follows from 2.7 that there is an $R \subseteq P$ such that

$$(n)\;[R \Vdash \mathcal{G}_n] \;.$$

Let $D = \{\bar{n} \mid \mathcal{G}_n \text{ is } n \in c\}$; $D \in L_\alpha$, and

$$(T)\,[T \in R \to c(T) = D] \;.$$

Now suppose otherwise:

$$(P_0)_{P \supseteq P_0}(En)\,(EQ_0)_{P_0 \supseteq Q_0}(EQ_1)_{P_0 \supseteq Q_1}[Q_0 \Vdash \bar{n} \in c \;\&\; Q_1 \Vdash \bar{n} \notin c] \;.$$

But then a slight variation on the sequential lemma argument produces a Q with the following properties: $P \supseteq Q$; if T_0, $T_1 \in Q$ and $T_0 \neq T_1$, then

$$(En)\;[n \in c(T_0) \leftrightarrow n \notin c(T_1)] \;.$$

Thus c can be construed as a one-to-one, α-finite map of Q onto $c[Q]$. Let $\delta < \alpha$ be so large that $c : Q \to c[Q]$ belongs to L_δ. Then $T \in L_\delta(c(T))$ for all $T \in Q$.

Lemma 2.9. Let T be generic. Then $\omega^T_{\Sigma_\infty} = \alpha$. If $L_\alpha \models (x)\,(x \subseteq \omega \to \omega^x_{\Sigma_\infty} < \alpha)$, then $(B)\,[B <_{\Sigma_\infty} T \to \omega^B_{\Sigma_\infty} < \alpha]$.

Proof. By 2.5 $\omega^T_{\Sigma_\infty} \geqslant \alpha$. To see that $\omega^T_{\Sigma_\infty} \leqslant \alpha$, suppose $P \Vdash (n)\,(Ey)\,\mathcal{G}(\bar{n}, y)$. It will suffice to find a $Q \subseteq P$ and a $\gamma < \alpha$ such that $Q \Vdash (n)\,(Ey^\gamma)\,\mathcal{G}(\bar{n}, y^\gamma)$. By 2.6 there is a $Q \subseteq P$ such that

$$(n)\;[Q \Vdash (Ey)\mathcal{G}(\bar{n}, y)] \;.$$

By 2.4 there is an α-finite function $\lambda_n \mid \gamma_n$ such that

$$(n)\,[Q \Vdash (Ey^{\gamma_n})\mathcal{G}(\bar{n}, y^{\gamma_n})]\ .$$

Then $\sup\{\gamma_n \mid n < \omega\}$ is the desired γ.

Suppose $B \in L_\alpha(T)$ and $\omega_{\Sigma_\infty}^B = \alpha$. Then $B \notin L_\alpha$. It follows from 2.8 that $T \in L_\alpha(B)$, or in other words, that $T \leqslant_{\Sigma_\infty} B$.

The proof of 2.9 made strong use of the initial assumption that $L_\alpha \models$ every set is countable. We now drop that assumption and sketch the changes required in the proof of 2.9. (Similar changes were required in [2] for the case of Σ_1 replacement.) Let α be a countable, admissible ordinal such that α is Σ_∞-admissible. It is routine to adjoin a generic "class" $K \subseteq L_\alpha$ to L_α so that $L_\alpha(K) \models$ every set is countable & Σ_∞ replacement. By "routine" we mean that K is built up from Lévy-type cardinal collapsing functions via a forcing argument based on finite conditions. In order to obtain a sequential lemma analogous to 2.6, we build K via a forcing argument based on perfect closed conditions. Then we repeat the argument of 2.9 with L_α replaced by $L_\alpha(K)$ and with $L_\alpha(T)$ replaced by $L_\alpha(K)(T)$.

The proof of 2.6 becomes more difficult, since $P \in L_\alpha(K)$ rather than L_α, and $L_\alpha(K)$ lacks the absolute well-ordering of L utilized in 2.6 to show that R is Σ_∞-pointed. The difficulty is overcome by viewing $L_\alpha(K)(T)$ as being obtained in one step, rather than two, from L_α. Let $\mathcal{L}_\alpha(\mathcal{K})$ be an α-recursive language with terms naming all the members of $L_\alpha(K)$. Since $P \in L_\alpha(K)$, there is a term $\mathcal{P}$ such that $P = \mathcal{P}(K)$. Now $\mathcal{P}$, unlike P, is a member of L_α. We call $\mathcal{P}$ a *virtual forcing condition*. Let $J \in L_\alpha$ be a perfect closed condition of the type used to construct K. We call $(J, \mathcal{P})$ a *mixed forcing condition*. Such conditions are typical of iterated forcing arguments.

The proof of 2.6 can be repeated for suitable mixed forcing conditions of L_α and sentences of an α-recursive language $\mathcal{L}_\alpha(\mathcal{K})(\mathcal{T})$. The result is a mixed sequential lemma which has the same bite as 2.6 and which applies to the transition from L_α to $L_\alpha(K)(T)$. It then follows that there is a sequential lemma that covers the transition from $L_\alpha(K)$ to $L_\alpha(K)(T)$, but it is the mixed sequential lemma that is needed to establish the minimality property of T expressed by

$$(B)\,[B <_{\Sigma_\infty} T \to \omega_{\Sigma_\infty}^T < \alpha]\ .$$

3. First-order absoluteness of F

Let F be normal and regular; we say F is *first-order absolute* if for every countable, F-admissible α,

$$L_\alpha = \cap \{O_F^T \mid \omega_F^T = \alpha\} .$$

Theorem 3.1. Let F be Σ_n $(n \geqslant 1)$, Σ_∞, or RI. Then F is first-order absolute.

Let $X \subseteq \omega$ and let α be countable and F-admissible; we say X is *F-recursive in α* if $X \in L_\alpha(T)$ for every T such that $\omega_F^T = \alpha$.

Corollary 3.2. Let F be Σ_n $(n \geqslant 1)$, Σ_∞, or RI. Let $X \subseteq \omega$ and let α be a countable, F-admissible ordinal. Then X is F-recursive in α if and only if $X \in L_\alpha$.

Let $X \subseteq \omega$ and let α be admissible; we say X is *hyperarithmetic in* α if X is hyperarithmetic in every T such that $\omega_1^T = \alpha$.

Corollary 3.3 ([2]). Let $X \subseteq \omega$ and let α be countable and admissible. Then X is hyperarithmetic in α if and only if $X \in L_\alpha$ [2)].

It is possible to improve 3.3 as follows. Let $X \subseteq \omega$ and let α be a countable admissible ordinal; we say X is Π_1^1 *in* α if X is Π_1^1 in every T such that $\omega_1^T = \alpha$.

Theorem 3.4 ([2]). Let $X \subseteq \omega$ and let α be a countable, admissible ordinal. Then X is Π_1^1 in α if and only if X is α-recursively enumerable (cf. Theorem 5.1).

One of the reasons for our interest in 3.4 is the following question. Is it possible to prove a wide class of theorems for all countable, admissible ordinals by first proving them for Church-Kleene ω_1 and then lifting them up to an arbitrary countable, admissible α with the help of 3.4 or some variant of 3.4? Barwise's compactness theorem [7] can be proved in this fashion.

2) Corollary 3.3 strengthens the following result of S. Kripke: Let δ be an arbitrary countable ordinal, and let $X \subseteq \omega$. If X is hyperarithmetic in every T such that $\omega_1^T > \delta$, then X is constructible in the sense of Gödel.

We indicate a proof of 3.2 for the case of Σ_∞. Let $X \subseteq \omega$ and let α be a countable, Σ_∞-admissible ordinal. Suppose $X \notin L_\alpha$. It suffices to find a T such that $\omega_{\Sigma_\infty}^T = \alpha$ and $X \notin L_\alpha(T)$. T is constructed as in Section 2 save for some additional steps designed to omit X from $L_\alpha(T)$. Let c be a term of $\mathcal{L}_\alpha(\mathcal{T})$ which denotes a subset of ω. Fix P. We define a $Q \subseteq P$ such that for every generic $T \in Q$, $X \neq c(T)$.

First suppose

$$(En)(EQ_0)_{P \supseteq Q_0}(EQ_1)_{P \supseteq Q_1}[Q_0 \Vdash \bar{n} \in c \ \& \ Q_1 \Vdash \bar{n} \notin c] \ .$$

Let $Q = Q_0$ if $n \notin X$; otherwise let $Q = Q_1$.

Now suppose otherwise:

$$(n)(Q_0)_{P \supseteq Q_0}(Q_1)_{P \supseteq Q_1} \sim [Q_0 \Vdash \bar{n} \in c \ \& \ Q_1 \Vdash \bar{n} \notin c] \ .$$

Let $Q = P$. If $T \in Q$ and T is generic, then

$$n \in c(T) \leftrightarrow (\mathrm{ER})_{Q \supseteq R}[R \Vdash \bar{n} \in c]$$

$$n \notin c(T) \leftrightarrow (\mathrm{ER})_{Q \supseteq R}[R \Vdash \bar{n} \notin c]$$

for all n; but then $c(T) \in L_\alpha$, since the forcing relation restricted to ranked sentences is α-recursive.

The idea underlying the above argument goes back to Kleene and Post [11].

4. Second-order absoluteness of F

Let $\mathcal{B}$ be a subset of 2^ω, and let α be a countable, admissible ordinal. We say $\mathcal{B}$ is *α-computable* if there exists a ranked sentence $\mathcal{G}$ of $\mathcal{L}_\alpha(\mathcal{S})$ such that

$$\mathcal{B} = \hat{S}(L_\alpha(S) \models \mathcal{G}) \ ^{3)} \ .$$

3) This is not the definition of α-computable subset of 2^ω given in [2], but it is equivalent to it. The definition used in [2] is longer but more to the point, since it is given solely in terms of computations whose heights range over the ordinals less than α.

(The notion of α-computable subset of 2^ω is discussed in more detail in [2].) Since $\mathcal{G}$ is ranked, it has rank equal to some $\delta < \alpha$. Thus the question of whether or not an arbitrary S belongs to $\mathcal{B}$ can be decided by means of a computation that belongs to $L_\alpha(S)$ and whose height is at most δ. Although the computations associated with $\mathcal{B}$ are bounded in height, they are not bounded in complexity, since S is arbitrary. If $\mathcal{B}$ is α-computable, then $\mathcal{B}$ is Borel. It is well-known that $\mathcal{B}$ is lightface Δ_1^1 if and only if $\mathcal{B}$ is ω_1-computable, where ω_1 is the least non-recursive ordinal.

$\mathcal{B}$ is said to be Σ_1 *over* α [2] if there is a Σ_1 sentence $\mathcal{G}$ of $\mathcal{L}_\alpha(\mathcal{S})$ such that

$$\mathcal{B} = \hat{S}(L_\alpha(S) \models \mathcal{G}) .$$

Proposition 4.1 is an immediate consequence of Barwise's compactness theorem [7].

Proposition 4.1. Let $\mathcal{B} \subseteq 2^\omega$ and let α be a countable, admissible ordinal. Then $\mathcal{B}$ is α-computable if and only if both $\mathcal{B}$ and $2^\omega - \mathcal{B}$ are Σ_1 over α.

Let $\mathcal{B}$ is said to be *Borel in T* if there exists Σ_1^1 formulas $P(X, Y)$ and $Q(X, Y)$ such that $\mathcal{B} = \hat{S}P(S, T)$ and $2^\omega - \mathcal{B} = \hat{S}Q(S, T)$. A set is Borel if and only if it is Borel in some T. A set $\mathcal{B}$ is Borel in T if and only if there is a ranked sentence $\mathcal{G}$ of $\mathcal{L}_{\omega_1^T}(\mathcal{T}, \mathcal{S})$ such that

$$\mathcal{B} = \hat{S}(L_{\omega_1^T}(T, S) \models \mathcal{G}) .$$

We say $\mathcal{B}$ is *Borel in* α if B is Borel in every T such that $\omega_1^T = \alpha$. We say $\mathcal{B}$ is *uniformly Borel in* α if there exist Σ_1^1 formulas $P(X, Y)$ and $Q(X, Y)$ such that

$$\mathcal{B} = \hat{S}P(S, T) \quad \text{and } 2^\omega - \mathcal{B} = \hat{S}Q(S, T)$$

for every T such that $\omega_1^T = \alpha$.

Theorem 4.2 ([2]). Let $\mathcal{B} \subseteq 2^\omega$ and let α be a countable, admissible ordinal. Then (i), (ii) and (iii) are equivalent.

(i) $\mathcal{B}$ is Borel in α.

(ii) $\mathcal{B}$ is unformly Borel in α.

(iii) $\mathcal{B}$ is α-computable.

It seems wise to discuss the proof of 4.2 before proceeding to a generalization of 4.2. First we draw the reader's attention to the uniformity expressed by (i) → (ii). We do not know a direct proof of (i) → (ii), although we tend to think there is one; (iii) → (ii) is almost immediate, so we confine our efforts to (i) → (iii).

Suppose $\mathcal{B}$ is not α-computable. We sketch a T such that $\omega_1^T = \alpha$ and $\mathcal{B}$ is not Borel in T. The construction of Section 2 is employed to insure that $\omega_1^T = \alpha$. Some additional steps are woven in so that $\mathcal{B}$ will not be Borel in T. Let $\mathcal{G}$ be a ranked sentence of $\mathcal{L}_\alpha(\mathcal{T}, \mathcal{S})$. Fix P. We show there is a $Q \subseteq P$ such that for every generic $T \in Q$,

$$\mathcal{B} \neq \hat{S}(L_\alpha(T, S) \models \mathcal{G})\ .$$

Fix S and consider the mechanics of adding a T to $L_\alpha(S)$ in the style of Section 2. The α-finite forcing conditions of Section 2 are retained so that ω_1^T will equal α, but all sentences of $\mathcal{L}_\alpha(\mathcal{T}, \mathcal{S})$ are to be forced with $\mathcal{S}$ held fixed at S. The forcing relation $P \Vdash^S \mathcal{G}$, restricted to the α-finite P's of Section 2 and to the ranked sentences of $\mathcal{L}_\alpha(\mathcal{T}, \mathcal{S})$ with $\mathcal{S}$ interpreted as S, is Σ_1 over $L_\alpha(S)$. This last is true uniformly in S; that is, there is a single Σ_1 formula of $\mathcal{L}_\alpha(\mathcal{S})$ that defines the forcing relation in question for all S.

The definition of Q has three cases.

Case 1. $(ES)(EQ)_{P\supseteq Q}[S \notin \mathcal{B}\ \ \&\ Q \Vdash^S \mathcal{G}]$. Choose such an S and such a Q. Devote infinitely many of the steps remaining in the construction of T to making T generic in the sense of $\Vdash^S$. Thus when the construction T is finished, it will be the case that $S \notin \mathcal{B}$ and $L_\alpha(T, S) \models \mathcal{G}$.

Case 2. $(ES)(EQ)_{P\supseteq Q}[S \in \mathcal{B}\ \&\ Q \Vdash^S \sim \mathcal{G}]$. Proceed as in Case 1.

Case 3. Otherwise. It follows that for all S,

$$S \in \mathcal{B} \leftrightarrow (EQ)_{P\supseteq Q}[Q \Vdash^S \mathcal{G}]$$

$$S \notin \mathcal{B} \leftrightarrow (EQ)_{P\supseteq Q}[Q \Vdash^S \sim \mathcal{G}]$$

But then by 4.1, $\mathcal{B}$ is α-computable; so Case 3 never applies.

The overall construction of T has the following form: T is generic in the sense of $\Vdash$ so that $\omega_1^T = \alpha$; T is generic in the sense of $\Vdash^S$ for countably many S's chosen in the course of the overall construction of T so that $\mathcal{B}$ not be Borel in T.

Let F be normal and regular, and let α be a countable, F-admissible ordinal. We say $\mathcal{B}$ is *F-Borel in α* if $\mathcal{B}$ is Borel in every T such that $\omega_F^T = \alpha$.

We say $\mathcal{B}$ is *uniformly F-Borel in* α if there exist Σ_1^1 formulas $P(X, Y)$ and $Q(X, Y)$ such that

$$\mathcal{B} = \hat{S}P(S, T) \quad \text{and} \quad 2^\omega - \mathcal{B} = \hat{S}Q(S, T)$$

for all T such that $\omega_F^T = \alpha$.

Theorem 4.3. Suppose F is Σ_n $(n \geqslant 1)$, Σ_∞, or RI. Let $\mathcal{B} \subseteq 2^\omega$ and let α be a countable F-admissible ordinal. Then (i), (ii), and (iii) are equivalent.

(i) $\mathcal{B}$ is F-Borel in α.
(ii) $\mathcal{B}$ is uniformly F-Borel in α.
(iii) $\mathcal{B}$ is α-computable.

The proof of 4.3 is similar in form to that of 4.2. 4.3 is one mode of generalization of 4.2. Another mode is given by 4.4 for Σ_n $(n \geqslant 1)$.

Let $\mathcal{B} \subseteq 2^\omega$, $n \geqslant 1$, and α be a countable, admissible ordinal. We say $\mathcal{B}$ is Σ_n over L_α if

$$\mathcal{B} = \hat{S}(L_\alpha(S) \models \mathcal{G})$$

for some $\mathcal{G}$ a Σ_n sentence of $\mathcal{L}_\alpha(\mathcal{S})$. We say $\mathcal{B}$ is Σ_n over $L_\alpha(T)$ if

$$\mathcal{B} = \hat{S}(L_\alpha(T, S) \models \mathcal{G})$$

for some $\mathcal{G}$ a Σ_n sentence of $\mathcal{L}_\alpha(\mathcal{T}, \mathcal{S})$. We say $\mathcal{B}$ is Δ_n *over* L_α if $\mathcal{B}$ and $2^\omega - \mathcal{B}$ are Σ_n over L_α. We say $\mathcal{B}$ is Δ_n *over* $L_\alpha(T)$ if $\mathcal{B}$ and $2^\omega - \mathcal{B}$ are Σ_n over $L_\alpha(T)$.

Theorem 4.4. Let $\mathcal{B} \subseteq 2^\omega$ and let α be a countable, Σ_n-admissible ordinal $(n \geqslant 1)$. Then (i) and (ii) are equivalent.

(i) $\mathcal{B}$ is Δ_n over L_α.
(ii) $\mathcal{B}$ is Δ_n over $L_\alpha(T)$ for every T such that $\omega_{\Sigma_n}^T = \alpha$.

The proof of 4.4 is like that of 4.2. The forcing relation $P \Vdash^S \mathcal{G}$ is introduced as before, and then the following observation is made. The relation $P \Vdash^S \mathcal{G}$, restricted to the Σ_n sentences of $\mathcal{L}_\alpha(\mathcal{T}, \mathcal{S})$ with $\mathcal{S}$ interpreted as S, is Σ_n over $L_\alpha(S)$.

The obvious extension of 4.4 to the case of Σ_∞ is true.

5. Further extensions and questions

(Q1) Under what conditions: is a normal F-jump regular?; is a normal, regular F-jump first-order absolute?

(Q2) Is there a natural concept, necessarily broader than first-order absoluteness, which is definable for all normal, regular F-jumps, and which covers the partial improvement of 3.1 expressed by 5.1?

Theorem 5.1. Let $X \subseteq \alpha$ and let α be a countable, Σ_n-admissible ordinal for some $n \geqslant 1$. Then (i) and (ii) are equivalent.

(i) X is Σ_n over L_α.

(ii) X is Σ_n over $L_\alpha(T)$ for every T such that $\omega^T_{\Sigma_n} = \alpha$.

(Q3) Is there a concept called second order absoluteness which is definable for all normal, regular F-jumps, and which covers the results of Section 4?

(Q4) Is there a result about RI which corresponds to 4.4? to 5.1?

(Q5) Is there a proof by the Barwise compactness approach of Friedman and Jensen [12] of 2.1? 2.2?, 4.2?

(Q6) Let F be Σ_n $(n \geqslant 1)$, Σ_∞, or RI. Does every countable set of F-degrees have a minimal upper bound? We conjecture it does, but at the moment we know little more than 2.3.

For each F define F_1 as follows. An ordinal α is F_1-admissible if α is F-admissible and the limit of F-admissible ordinals. Thus RI $= (\Sigma_1)_1$. If $\omega^T_{F_1}$ is defined for all T and F is normal, then F_1 is normal.

(Q7) Suppose F is regular and F_1 is normal. What further conditions on F guarantee that F_1 is regular?

References

[1] G. Sacks, Normal F-jumps, in preparation.

[2] G. Sacks, Countable admissible ordinals and hyperdegrees, to appear.

[3] S. Kripke, Transfinite recursions on admissible ordinals I and II (abstracts). J. Symbolic Logic 29 (1964) 161–162.

[4] R. Platek, Foundations of Recursion Theory, Ph. D. Thesis, Stanford University (1965).

[5] G. Sacks, Forcing with perfect closed sets, in: Proceedings of the 1967 UCLA Set Theory Institute, to appear.

[6] H. Friedman, Killing admissible ordinals (mimeomgraphed notes), Stanford University (1969).

[7] J. Barwise, Implicit definability and compactness in infinitary languages, Lecture Notes in Mathematics 72 (1968) (Springer-Verlag) 1–35.

[8] K. Gödel, The consistency of the axiom of choice and of the generalized continuum hypothesis (Princeton University Press, 1966).

[9] S.C. Kleene, Hierarchies of number-theoretic predicates, Bull. Am. Math. Soc. 61 (1955) 193–213.

[10] R. Gandy and G. Sacks, A minimal hyperdegree, Fund. Math. 61 (1967) 215–223.

[11] S.C. Kleene and E.L. Post, The upper semi-lattice of degrees of recursive unsolvability, Ann. Math. 59 (1954) 379–407.

[12] H. Friedman and R. Jensen, A note on admissible ordinals, Lecture Notes in Mathematics 72 (Springer-Verlag, 1968).

PART IV

INVITED PAPERS ON OTHER TOPICS

A SIMPLIFIED PROOF FOR THE UNSOLVABILITY OF THE DECISION PROBLEM IN THE CASE $\wedge\vee\wedge$

H. HERMES
Freiburg i.Br., Germany

1. Introduction

Using earlier ideas of Wang [1] and Büchi [2] the first proof for the unsolvability of the decision problem in the case $\wedge\vee\wedge$ has been given by Kahr et al. [3]. (The knowledge of this paper is pressuopposed.) Wang, who invented the domino games, proved that the decision problem for the corner game is unsolvable. Büchi inferred that the decision problem for the case $\vee\wedge\vee\wedge\vee$ is insolvable. Kahr et al. treated the diagonal game. The unsolvability of the decision problem for the diagonal game leads to the result that the decision problem for $\wedge\vee\wedge$ is unsolvable.

The essential idea of Kahr et al. was to represent each configuration of a Turing machine M periodically on a diagonal of the first quadrant. The dominoes of the game associated with M have two different pruposes:

(a) to present the machine M,

(b) to generate the periodic patterns on the diagonals.

In order to achieve (a) and (b) the authors used 26 different kinds of dominoes.

In this paper a simplified proof for the unsolvability of the decision problem in the case $\wedge\vee\wedge$ will be indicated, where only 10 different kinds of dominoes occur. The modification of the proof of Kahr et al. may be described as follows:

(1) the dominoes are not used to describe the pattern,

(2) it is not required that each pair of adjoining dominoes match in color,

(3) the pattern is given by a function,

(4) the pattern is used to indicate the edges where the adjoining dominoes should match in color.

A minor change consists in replacing the diagonals by the rows of the first quadrant.

2. Fundamental concepts

Let be Q the set of unit squares of the first quadrant. Let be φ a mapping of Q into $\{0, 1\}$. φ is called a *mosaic* iff

(a) on the lowest row we have alternating values of φ,

(b) if for each quadruple of squares A_1, A_2, A_3, A_4, where A_2 is the right neighbor of A_1 and A_4 is the right neighbor of A_3 and A_3 is immediately above A_1 we have

$$\varphi(A_3) = \varphi(A_4) \quad \text{iff} \quad \varphi(A_1) = 1 \quad \text{and} \quad \varphi(A_2) = 0 .$$

If we associate to the squares of the lowest row the values 0, 1, 0, 1, ..., to the squares of the next now the values 0, 0, 1, 1, 0, 0, 1, 1, ..., to the squares of the next row the values 0, 0, 0, 0, 1, 1, 1, 1, 0, 0, 0, 0, ..., etc., we get an example for a mosaic.

A *domino game* is given by *three* finite sets D, D^0, D^* of dominoes, where $D^0 \subset D$ and $D^* \subset D$.

Let be φ a mosaic and D, D^0, D^* a domino game. A mapping F of Q into D is called a *D, D^0, D^*, φ-covering of Q*, iff

(1) $F(A) \in D^0$ for each A of the lowest row, where $\varphi(A) = 0$,

(2) $F(A) \in D^*$ for each square A, where $\varphi(A) = 1$,

(3) $F(A_1)$ and $F(A_2)$ are matching in color for each pair A_1, A_2 of squares where A_2 is the right neighbor of A_1 and $\varphi(A_1) = \varphi(A_2) = 0$,

(4) $F(A_1)$ and $F(A_2)$ are matching in color for each pair A_1, A_2 of squares where A_2 is immediately above A_1 and $\varphi(A_2) = 0$.

A domino game D, D^0, D^* is called a *good game* iff there is a mosaic φ and a mapping F such that F is a D, D^0, D^*, φ-covering.

In order to show the unsolvability of the decision problem for the case $\bigwedge\bigvee\bigwedge$ it is sufficient to associate in an effective way to every Turing machine M a domino game D_M, D_M^0, D_M^* and to every domino game D, D^0, D^* a formula $\alpha_{DD^0D^*} \in \bigwedge\bigvee\bigwedge$ such that

(*) M does not halt iff D_M, D_M^0, D_M^* is good,

(**) D, D^0, D^* is good iff $\alpha_{DD^0D^*}$ is satisfiable.

The definition of D_M, D_M^0, D_M^* is given in the next section. The definition of $\alpha_{DD^0D^*}$ and the proofs for (*) and (**) can be performed without difficulty, knowing [3]. Cf. also [4].

3. The definition of the domino game D_M, D_M^0, D_M^* associated with a Turing machine M (which is operating on a one-way tape).

M is identified with its table, which may be considered as the set of its rows. We use a as a variable for the letters of the alphabet and q as a variable for the states. 0 is the blank letter and also the initial state. For each pair q, a we have a row $qab\bar{q}$ of M, where $\bar{q}$ is a state and b (the behaviour) is a letter a or one of the symbols r (right), l (left), s (stop).

The *colors* of the dominoes are a, qa, gr, ql, qar, qal, B (beginning), C_1, C_2 (corner), H (horizontal), W (west), S (stop).

We introduce a *color function* ϕ which is defined for each pair q, a, and where ϕ_{qa} is a color, as follows:

$$\phi_{qa} = \begin{cases} \bar{q}\bar{a} & \text{if } qa\bar{a}\bar{q} \in M \\ \bar{q}ar & \text{if } qar\bar{q} \in M \\ \bar{q}al & \text{if } qal\bar{q} \in M \\ S, & \text{if } qas\bar{q} \in M \end{cases}$$

In D_M we have four single dominoes (C), (C'), (B), (W) and six kinds of dominoes (a), (qar), (qal), $[qa]$, $[qar]$, $[qal]$. (C) is the only element of D_M^0 and (0) the only element of D_M^*. The colors of the dominoes are given by the following diagrams:

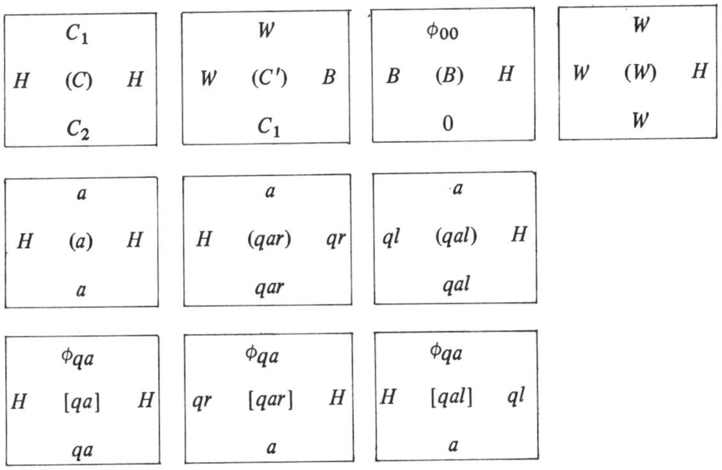

References

[1] H. Wang, Proving theorems by pattern recognition. II, Bell Systems Technical J. 40 (1961) 1–41.
[2] J.R. Büchi, Turing machines and the Entscheidungsproblem, Math. Ann. 148 (1962) 201–213.
[3] A.S. Kahr, E.F. Moore and H. Wang, Entscheidungsproblem reduced to the $\forall\exists\forall$-case, Proc. Natl. Acad. Sci. U.S. 48 (1962) 365–377.
[4] H. Hermes, Entscheidungsproblem und Dominospiele, in: Selecta Mathematica II, hrsg. von K. Jacobs (Springer, 1970) pp. 114–140.

AN INTRINSIC CHARACTERIZATION OF THE HIERARCHY OF CONSTRUCTIBLE SETS OF INTEGERS[1)] 1?

Stephen LEEDS and Hilary PUTNAM
Department of Philosophy, Harvard University, Cambridge, Mass.

Introduction

1. *Notation*

Some of our results will be in second-order number theory, others will be in set theory; we will use the standard (incompatible) notations for each. In second-order number theory, small Roman letters will represent (i.e., be used as constants to designate, as variables to range over) integers; capital Roman letters will represent sets of integers. In set theory, small Roman letters will represent sets. Small Greek letters will always represent ordinals.

We will use Roger's notations [9] in recursive function theory: the e-th A-partial recursive function will be designated by 'ϕ_e^A'; the domain of ϕ_e^A will be designated by 'W_e^A'. J will be a fixed primitive recursive 1–1 onto map from $N \times N$ to N; the inverse maps will be denoted by 'K' and 'L', so that

$$KJ(x, y) = x \qquad LJ(x, y) = y \; .$$

The definition of jumps is as follows: Let A be a set of integers. We define

$$A^{(0)} = A$$

$$A^{(n+1)} = \{x : (\exists y)\, (y = \phi_x^A x)\} (= (A^{(n)})')$$

$$A^{(\omega)} = \{J(x, y) : x \in A^{(y)}\} \, .$$

(We will denote $N^{(1)}$ by 'K'.)

1) This paper is essentially the first author's Ph.D. thesis (of the same title) written under the direction of Professor Putnam and submitted to M.I.T., June, 1969.

Notation in set theory varies little from author to author; ours will be familiar. It should be pointed out that, like Gödel (but unlike, e.g., Cohen), we take $M_\lambda = \bigcup_{\beta<\lambda} M_\beta$, for limit λ. 'L' will denote $\bigcup_\beta M_\beta$, the class of constructible sets. 'ω_1^L' will denote the least L-uncountable ordinal.

2. *Preliminary results and definitions*

If A is a set of integers in L, and $A \in M_{\alpha+1} - M_\alpha$, we will say that A has *order* α. If there is a set of integers of order α, we call α an *index*. Gödel's classical result [7] shows that no index is $\geqslant \omega_1^L$. Putnam has shown [8] that there are an L-uncountable infinity of non-indices; clearly there are an L-uncountable infinity of indices.

(As an example, let M_β be the minimal model for set theory, and let γ be the least uncountable ordinal relative to M_β. Then $\gamma < \beta < \omega_1^L$. Using $V = L \Rightarrow$ all sets of integers are constructed before ω_1, and the absoluteness of the concept 'is an index', we see that there are no sets of integers in $M_{\delta+1} - M$ for $\gamma < \delta < \beta$.)

Let E be a set of integers. It will frequently be useful to consider E as a set of ordered pairs. We call $\{x : (\exists y)(J(x, y) \in E \vee J(y, x) \in E)\}$ the *field* of E (Field E). If there is an ordinal α, and a map f (f not necessarily in L) such that f : Field $E \to M_\alpha$ 1–1 onto, and $J(x, y) \in E \Longleftrightarrow f(x) \in f(y)$, we call E an arithmetical copy of M_α. If, in addition, there is an integer a such that every member of the field of E is of the form $J(x, a)$, i.e.,

$$(\forall x \in E)(LKx = LLx = a)$$

we call E an a-initial arithmetic copy of M_α.

We can now state the fundamental result of Boolos [1]:

Let α be an index $> \omega$. There is an arithmetic copy E_α of M_α, where E_α is of order α. It clearly follows that for any b, there is a b-initial arithmetic copy of M_α, of order α.

If E is a set of integers of order α, we call E (arithmetically) complete (of order α) if every set of integers in $M_{\alpha+1}$ is arithmetical in E. It is easy to see that if E_α is an arithmetical copy of M_α of order α then E_α is a complete set of order α; for any set of integers in $M_{\alpha+1}$ is first-order-definable over M_α, and there is a straightforward way of translating any such definition into a definition over E_α using only number quantifiers.

(Example: Let 'a' name a set in M_α, and let m be the member of Field E_α which, in the natural sense, 'corresponds' to a. Let n be the member of Field E_α which, in the natural sense, 'corresponds' to ω. Then the set of integers in $M_{\alpha+1}$ defined by

$$\{x \in M_\alpha : (\exists y)\,(x \in \omega \;\&\; x \in y \in a)\}$$

is also given by the following expression:

$$\{x \in N : (\exists z \in \text{Field } E_\alpha)\,(\exists y \in \text{Field } E_\alpha)\,(J(z, n) \in E_\alpha \;\&$$

$$J(z, y) \in \text{Field } E_\alpha \;\&\; J(y, m) \in \text{Field } E_\alpha \;\&\; z$$

$$\text{represents } x \text{ in Field } E_\alpha)\}\,.$$

The last clause may be translated as 'z has x 'members'', which has simple translations into expressions arithmetic in E_α.)

It will be noticed that if E is a complete set of order α, so are $E^{(1)}$, $E^{(2)}$, ..., nor is there any obvious way to choose a complete set, or a complete Turing degree, as being somehow unique (e.g., minimal) among the complete sets of a given order. We shall deal with this problem by speaking, not of Turing degrees, but of arithmetical degress, defined as follows:

If A is arithmetical in B, we write '$A \leqslant_a B$'.

If $A \leqslant_a B$ and $B \leqslant_a A$, we write '$A \equiv_a B$'.

We define $\deg_a A = \{B : B \equiv_a A\}$.

Clearly $\equiv_a$ is an equivalence relation; also if $A \in M_\alpha$, $\deg_a \alpha \in M_{\alpha+1}$ [2]. Also, the complete sets of order α constitute a single arithmetical degree. We designate this degree by '$\deg_a(\alpha)$'. The $\deg_a(\alpha)$ have properties quite analogous to those of the $d(K^{(n)})$ in the arithmetical hierarchy:

(1) if $A \in L \cap P(\omega)$, $(\exists \alpha)\,(\deg A \leqslant_a \deg(\alpha))$

(2) if α and β are indices, $\alpha < \beta \Leftrightarrow \deg(\alpha) < \deg(\beta)$.

If we restrict our attention to index α, (1) remains true, and (2) becomes true unconditionally. We find that the sequence of $\deg(\alpha)$ has the comprehensiveness and monotone feature characteristic of the familiar hierarchies. We call the sequence of $\deg(\alpha)$ for index $\alpha > \omega$ the *hierarchy of arithmetical degrees*. This hierarchy is, to this point, apparently less interesting than the

[2] Because $A \leqslant_a B \Leftrightarrow A \leqslant_T B^{(n)}$ for some n, $\Leftrightarrow \exists$ a sequence of n+1 sets $b_0, B_1 \ldots B_n$ such that $B_0 = B$, $B_{i+1} = B_i'$ and $A \leqslant_T B_n$. This last expression is clearly first-order definable over M_α.

familiar hierarchies. What is lacking in its characterization is a feature which we may call 'minimality': we need something analogous to the sense in which $K^{(4)}$, and not $K^{(5)}$, or *any* set r.e. in $K^{(3)}$ is chosen as the successor of $K^{(3)}$ in the arithmetic hierarchy. We would like to define 'next least' in such a way that if α is an index, and β is the least index $> \alpha$, then $\deg(\beta)$ is the next least degree after $\deg(\alpha)$. We are encouraged in thinking that something of this kind is possible, by the following result of Boolos:

The hierarchy of arithmetical degrees is identical with the ramified analytic hierarchy, where the latter is defined; further, it is identical with the hyper-arithmetic hierarchy where that hierarchy is defined – taking into account in the latter case the difference between Turing degrees and arithmetical degrees.

More precisely, Boolos proved

(1) letting $\beta_0 = \mu\alpha(A_\alpha = A_{\alpha+1})$ we have: if $\alpha \leqslant \beta_0 + 1$, $A_\alpha = M_\alpha \cap P(\omega)$

(2) if $\alpha <$ constructive ω_1 (Kleene ω_1) then order $A = \alpha \Rightarrow A \leqslant_a H_b$ for all b such that $|b| = \omega \cdot \alpha$, and if $|c| < \omega \cdot \alpha$, then $A \nleqslant_a H_c$. Also, if $|a| = \omega \cdot \alpha$, then order $H_a = \alpha$.

The sequence of complete degrees is identical with the sequence of $\deg(H_b)$ for $|b|$ a limit ordinal; to the extent to which $\deg(H_b)$ may be seen as a natural 'next' degree after all $\deg(H_a)$, $|a| < |b|$, our question has been answered.

There are two ways in which the answer falls short. First, O may appear somewhat artificial, and the hyperarithmetic hierarchy may seem to inherit this artificiality. Secondly, and more significantly, constructive $\omega_1 < \omega_1^L$, indeed $\beta_0 < \omega_1^L$ – the hierarchy we are considering far outstrips the hyper-arithmetic hierarchy and even the ramified analytic hierarchy.

One might deal with the first difficulty by citing results like those of Enderton [5] on the minimality of O as a notational system. To the extent to which this answer is persuasive, one will also be able to deal with large stretches of our hierarchy, even beyond constructive ω_1. For, as Boolos shows, if G is a complete set of order β, then result (2) above holds, relativized to G. That is, between β and $\beta + \omega_1^G$, the hierarchy of arithmetical degrees is identical with the limit members of the G-hyperarithmetical hierarchy. We will not give details here, as we will make virtually no use of this fact.

The cases that make even the relativized hyperarithmetic hierarchies useless for our purposes are those in which an index β is a limit of non-indices or a successor of a non-index. In such cases, we have the following situation: Let G be a complete set of order $\alpha < \beta$. Then for limit $\delta < \omega_1^G$, there is a member of O^G, say d, such that $\omega \cdot |d| = \delta$. Then if $|b| = \omega \cdot |d|$, H_b^G is of order $\alpha + \delta$. It follows that order $\alpha + \omega_1^G < \beta$, for any complete G, hence any G of order $< \beta$. The consequence is that no relativized hyperarithmetic hierarchy can carry us up to β, or can justify the choice of deg(β) as the 'next least' arithmetic degree.

We conclude the introduction with an important definition that is especially relevant to the above situation:

Definition: 'HYP α' for $(\forall A)(A \in M_\alpha \cap P(\omega) \Rightarrow O^A \in M_\alpha \cap P(\omega))$.

Chapter 1

1. Let E_α be an initial arithmetic copy of M_α, of order α. It is easy to show, as Boolos does, that there is an arithmetic copy of $M_{\alpha+1}$, of order α+1. Let us assume that E_α is an a-initial copy, i.e., $(\forall x)(x \in \text{Field } E_\alpha \Rightarrow Lx = a)$. The idea of the construction is as follows: if y is the Gödel number of a first-order definition over M_α, and that definition gives a set not in M_α, and no definition of the same set has a smaller Gödel number, then we add $J(y, b)$ to Field E_α, using a fixed $b \neq a$. More precisely:

If ϕ is a formula with number quantifiers, containing =, $\in$, and constants from Field E_α, e.g.,

$$(\forall x)(\exists y)(\forall z)(z \in x \Leftrightarrow 5 \in 4)$$

then we may use '$E_\alpha \models \phi$' to express what model theorists would write as

$$\langle \text{Field } E_\alpha, \{\langle x, y\rangle : J(x, y) \in E_\alpha\}, \text{ Field } E_\alpha\rangle \models \phi ,$$

satisfaction being defined by

$$(1)\ E_\alpha \models a \in b \Leftrightarrow J(a, b) \in E_\alpha$$

$$(2)\ E_\alpha \models \forall x \phi x \Leftrightarrow \forall a \in \text{Field } E_\alpha,\ E_\alpha \models \phi_x^a \text{ , etc.}$$

It is apparent that if ϕ contains the constants $a_1 \ldots a_n$, and a_i represents the set x_i in M_α, then

$$E_\alpha \models \phi \Leftrightarrow M_\alpha \models \psi$$

where ψ is like ϕ except in having 'x_i' for 'a_i'.

If ϕ is a formula of this kind, containing a free variable x, then ϕ will correspond to a first-order definition over M_α of a member of $M_{\alpha+1}$. The set defined by ϕ will be in $M_{\alpha+1} - M_\alpha$, just in case

$$E_\alpha \models \sim (\exists z)(\forall w)(\phi w \Leftrightarrow w \in z) .$$

We say in this case that ϕ defines a new set.

If y is a Gödel number of such a formula ϕ, then y is a least Gödel number of ϕ just in case

$$(\forall z)(z < y \;\&\; z \text{ is a Gödel number for } \psi \Rightarrow E_\alpha \models \sim (\forall w)(\phi w \Leftrightarrow \psi w)) .$$

If y is a least Gödel number for ϕ, we let $J(y, b)$ be a name for the set defined by ϕ. That is, we add to E_α

$$\{J(d, J(y, b)) : E_\alpha \models \phi d\} .$$

The resulting set

$$E_\alpha \cup \{J(d, J(y, b)) : y \text{ is a least Gödel number of some formula}$$

$$\phi, \text{ and } y \text{ defines a new set, and } d \in \text{Field } E_\alpha \;\&\; E_\alpha \models \phi d\}$$

is an arithmetic copy of $M_{\alpha+1}$. It is easily seen, if not from our sketch, then from Boolos' more detailed presentation, that the copy is uniformly recursive in $E_\alpha^{(\omega)}$, b. We will designate it by '$(E_\alpha, b)^*$' following Boolos' notation. Because $E_\alpha^{(\omega)} \in M_{\alpha+2}$, so is $(E_\alpha, b)^*$. Also, $(E_\alpha, b)^* \notin M_{\alpha+1}$. To see this, recall that E_α is a complete set of order α; it follows that $(E_\alpha, b)^* \in M_{\alpha+1} \Rightarrow (E_\alpha, b)^* \leqslant_a E_\alpha$. But as we showed in the Introduction, $(E_\alpha, b)^*$, since it is an arithmetical copy of $M_{\alpha+1}$, must be an arithmetical upper bound on $M_{\alpha+1} \cap P(\omega)$ [3]. It follows that $E_\alpha^{(\omega)} \leqslant_a (E_\alpha, b)^* \leqslant_a E_\alpha$, which is impossible. It follows that $(E_\alpha, b)^*$ is of order α+1. The $*$ operation will be our chief tool in the construction of complete sets.

2. Most of our results will have to do with uniform upper bounds, which we define as follows:

[3] We proved that if E_β is an arithmetical copy of M_β, of order β, and $A \in M_{\beta+1} \cap P(\omega)$, then $A \leqslant_a E_\beta$. However, we made no use of the fact that E_β was of order β.

Definition. Let Γ be a set of sets of integers; let K be a set of integers. We say K is a uniform upper bound (uub) on Γ if there is a K-recursive enumeration of K-Gödel numbers for all the sets in Γ. That is

$$(\exists e)\,[(\forall f)\,(f \in W_e^K \Rightarrow \phi_f^K \text{ is a characteristic function})$$

$$\&\,(\forall A)\,(A \in \Gamma \Leftrightarrow (\exists f)\,(f \in W_e^K \;\&\; \phi_f^K \text{ is a characteristic function for } A)]\ .$$

The relevance of uubs to our topic is suggested by the following small theorem: Let $\alpha > \omega$ be an index; let K be a complete set of order α. There is a uub on $M_\alpha \cap P(\omega)$ which is arithmetic in K.

Proof. Let E_α be an arithmetic copy of M_α, of order α. We have $E_\alpha \leqslant_a K$. Let v be the integer in Field E_α that represents ω. Then if $r \in$ Field E_α, r represents a set of integers iff $r \in S$, where $S = \{y : (\forall x)\,(J(x, y) \in E_\alpha \Rightarrow J(x, v) \in E_\alpha)\}$. The set $W = \{J(x, y) : (\exists z)\,(z$ represents x in E_α & $y \in S$ & $J(z, y) \in E_\alpha\}$ is a uub on $M_\alpha \cap P(\omega)$. Also, $W \leqslant_a E_\alpha \leqslant_a K$, giving us the result.

The converse of this result is in one case rather simple.

Theorem. Let $\alpha > \omega$ be an index. Then $\alpha+1$ is also an index, and if $E_{\alpha+1}$ is a complete set of order $\alpha+1$, and K is a uub on $M_{\alpha+1} \cap P(\omega)$, then $E_{\alpha+1}$ is arithmetic in K.

Proof. We know from above that, since α is an index, there is an initial arithmetic copy E_α of M_α, of order α, and hence a set $(E_\alpha, b)^*$, for some b, which is complete of order $\alpha+1$. To prove the second part of the result, it will suffice to show that $(E_\alpha, b)^*$ is arithmetic in K.

We recall that $(E_\alpha, b)^*$ is recursive in $E_\alpha^{(\omega)}$. We will show $E_\alpha^{(\omega)}$ arithmetic in K. We have $E_\alpha \in M_{\alpha+1} \cap P(\omega)$ and $A \in M_{\alpha+1} \cap P(\omega) \Rightarrow A' \in M_{\alpha+1} \cap P(\omega)$. It follows from simple closure properties of $M_{\alpha+1} \cap P(\omega)$ that $E_\alpha^{(\omega)}$ is second-order definable over $M_\alpha \cap P(\omega)$, and the definition can easily be translated into a definition of $E_\alpha^{(\omega)}$ arithmetic in K. We give the definition of $E_\alpha^{(\omega)}$ over $M_\alpha \cap P(\omega)$, and its K-arithmetic translation; we shall later often make tacit use of the same techniques.

$J(x, y) \in E_\alpha^{(\omega)} \Leftrightarrow$ there is a sequence of $y+1$ sets $A_0, A_1 \ldots A_y$ such that $A_0 = E_\alpha$, $A_{i+1} = A_i'$, & $x \in Ay \Leftrightarrow$ there is a sequence of y integers $a_0, a_1 \ldots a_y$ such that $(\forall i)\,(a_i \in W_e^K$ & $W_{a_0}^K = E_\alpha$ & $W_{a_{i+1}}^K = W_{a_i}')$ & $x \in W_{a_y}^K$.

We now have $E_{\alpha+1} \leqslant_a (E_\alpha, b)^* \leqslant_T (E_\alpha)^\omega \leqslant_a K$, which gives the result.

We are interested in showing the relationship in general between complete sets of order α and uubs on $M_\alpha \cap P(\omega)$. The above theorems show that in the case where α is an index and the successor of an index, E_α is a uub on $M_\alpha \cap P(\omega)$, and E_α is arithmetic in any uub on $M_\alpha \cap P(\omega)$. We now show that the same result holds for indices that are limits of indices.

Theorem. Let $\alpha > \omega$ be an index and a limit of indices. Let K be a uub on $M_\alpha \cap P(\omega)$. Then there is a complete set E_α, of order α, arithmetic in K. It follows that every complete set, indeed every set, of order α is arithmatic in K.

Proof. We prove the theorem by considering a number of cases and subcases; different techniques seem appropriate in each case. All the method used can be found in Boolos [1] or Boolos and Putnam [2]; indeed, the most involved part of the proof (Case IIa) is an obvious corollary to the proof of theorem in Boolos and Putnam. The rest of this paper will be clearer, however, if we give the proofs in some detail.

The devision into cases is motivated by the relativized hyperarithmetical hierarchy mentioned earlier: α can lie within such a hierarchy (Case I), beyond all such hierarchies (Case IIa), or at the exact end of such a hierarchy (Case IIb).

Case I. $(\exists\beta < \alpha)$ ($M_\beta \cap P(\omega)$ contains a well-ordering of integers of length α).

We may assume β is an index, for if δ is an index, $\beta < \delta < \alpha$, M_δ will also contain the same well-ordering. Let the given well-ordering be $S \in M_\beta$. Let R be an initial segment of S (perhaps equal to S) of length $\alpha - \beta$. The idea of the construction is as follows: We begin with an initial copy E_β, of order β, and construct an arithmetic copy of E_α by successive applications of $*$. R is used to index the succesive copies; we assume E_β is an 0_R-initial copy.

We define E_α as follows:

$$J(r, s) \in E_\alpha \Leftrightarrow$$

(1) $L_s = 0_R \;\&\; J(r, s) \in E_\beta \;\vee$

$$(2)\ (\exists S)\ [(\forall u)(\forall v)(Lv = 0_R \vee . Lu <_R Lv \leqslant_R Ls \ \&\ Lv \text{ is an } R\text{-successor})$$

$$\&\ (\forall u)(\forall v)(Lv = 0_R \Rightarrow . J(u, v) \in S \Longleftrightarrow J(u, v) \in E_\beta)\ \&$$

$$(\forall u)(\forall v)(Lv \neq 0_R \Rightarrow . J(u, v) \in S \Longleftrightarrow (\exists M)(\exists N)(M =$$

$$\{J(s, t) \in S : Lt <_R Lv\}\ \&\ N = (M, Lv)^*\ \&\ J(u, v) \in N)\ \&$$

$$J(r, s) \in S]\ .$$

The sets that will instantiate S in the above formula will be the successive copies E_δ of M_δ. M and N will be instantiated also by the E_δ, and if M copies M_δ, N will copy $M_{\delta+1}$. A simple induction shows that $E_\delta \in M_{\delta+1}$; in particular, the definition locates E_α in $M_{\alpha+1}$. By a previous argument, order E_α is α. Finally, the above definition is a second-order definition that remains valid when its quantifiers are restricted to $M_\alpha \cap P(\omega)$. It follows that the definition may be given as a definition of E_α, arithmetic in K.

Notice that in this part, we do not use the assumption that α is itself an index. We may draw a

Corollary to the proof. If α is a limit of indices and M_α contains a well-ordering of integers of length α, then α is an index.

Case II. $(\forall \beta < \alpha)$ (there is no well-ordering of integers in M_β of length α)

(In this case it is apparent that $\beta < \alpha \Rightarrow \beta + \beta < \alpha$. For otherwise, let β be an index, $\beta + \beta > \alpha$. Let E_β be an arithmetic copy of M_β, of order β; then $R = \{J(x, y) \in E_\beta : x \text{ and } y \text{ are 'ordinals'}\}$ is a well-ordering of length β, and $R' = \{J(x, y) : Lx \in \text{Field } R \ \&\ Ly \in \text{Field } R \ \&. Kx = 1 \vee Kx = 2 .\&\ Ky = 1 \vee Ky = 2 .\&\ Kx = Ky \Rightarrow LxL_R Ly\}$ is a well-ordering of length $> \alpha$, and $R' \in M_\delta$ for some index δ.)

Case IIa. α is HYP

In this case, it will not be possible to index an iteration of $*$. However, we are given that α is an index; there is therefore a set-theoretic definition of some E_α over M_α. Our goal will be to translate that definition into a definition in second-order number theory over $M_\alpha \cap P(\omega)$, and then into a definition of E_α arithmetical in K.

For concreteness, let us suppose that the definition of E_α over M_α has the following form:

$$x \in E_\alpha \Leftrightarrow (\forall u \in M_\alpha)(\exists v \in M_\alpha) A(u, v, x, c) \quad \text{where } c \in M_\alpha .$$

Bearing in mind that α is a limit of limit ordinals, we may rewrite the definition as

$$x \in E_\alpha \Leftrightarrow (\forall \delta < \alpha, \delta \text{ a limit})(\forall u \in M_\delta)(\exists \delta', \delta < \delta' < \alpha, \delta \text{ a limit})$$

$$(\exists v \in M_{\delta'}) A(u, v, x, c) .$$

As a first step toward a translation of the above formula into second-order number theory, we observe the following fact about M_α:

Lemma. Let μ and ν be limit ordinals, $\mu < \nu < \alpha$. Let E_μ be an arithmetic copy of M_μ; let E_ν be an arithmetic copy of M_ν. Let $E_\mu \in M_\alpha$, $E_\nu \in M_\alpha$. Then there is an arithmetic copy of M_ν, say E'_ν, such that $E'_\nu \in M_\alpha$, and for some r,

$$J(a, b) \in E_\mu \Leftrightarrow J(J(a, r), J(b, r)) \in E'_\nu .$$

We say E'_ν r-extends E_μ.

We postpone the proof of the lemma.

Now let $c \in M_\lambda$ where λ is a limit ordinal $< \alpha$. Let $E \in M_\alpha$ be an arithmetic copy of M_λ [4]. Let p represent c in Field E. Assuming the lemma, we can write

> $x \in E_\alpha \Leftrightarrow (\forall R \in M_\alpha)(\forall r)$ [R is an arithmetic copy of some M_δ, δ a limit, and R r-extends E. $\Rightarrow (\forall z \in \text{Field } R)(\exists S \in M_\alpha)(\exists s)$ (S is an arithmetic copy of some M_δ, δ a limit & S s-extends R & $(\exists w \in \text{Field } S)(\exists n \in \text{Field } S)$ (n represents x in S & $A'(J(z, s), w, n, J(p, r)s)))$]

[4] Such copies are available in M_α for any $\lambda < \alpha$. For let δ be an index, $\lambda < \delta < \alpha$, and let $E_\delta \in M_{\delta+1}$ be an arithmetic copy of M_δ. Let s represent M_λ in M_δ. Then

$$\{J(a, b) \in E_\delta : J(a, s) \in E_\delta \ \& \ J(b, s) \in E_\delta\}$$

is in M_α, and is an arithmetic copy of M_λ.

where A' is like A except in having '$J(u, v) \in S$' where A had '$u \in v$'.

The above definition is close to a definition of E_α over $M_\alpha \cap P(\omega)$ in second-order number theory. To show that a second-order definition can in fact be given, we show that the prose parts of the above formula can be given second-order translations. This is clearly the case for 'X m-extends Y', and because x is always an integer, it is easy to translate 'n represents x'. It remains to show that 'X is an arithmetic copy of some M_δ, δ a limit' can be given a second-order translation.

It is well-known that there is a sentence σ in set theory such that u is well-founded $\Rightarrow \langle u, \epsilon_u \rangle \models \sigma \Longleftrightarrow (\exists\delta)(\delta$ a limit & $\langle u, \epsilon_u \rangle \approx \langle M_\delta, \epsilon_{M_\delta} \rangle$. It follows that there is a sentence τ in second-order number theory such that A is well-founded $\Rightarrow : A \models \tau . \Longleftrightarrow A$ is an arithmetic copy of some M_δ, δ a limit (taking 'A is well-founded' amd '$\models$' in the sense appropriate to the model determined by A). Now it is clear that $A \models \tau . \Rightarrow . A$ is well-founded $\Longleftrightarrow$ the 'ordinals' of A are a well-ordered set. It is therefore sufficient to show that 'X is a well-ordering of integers' is a second-order predicate in $M_\alpha \cap P(\omega)$.

Here we use the fact that α is HYP. From W^R (the set of R-Gödel-numbers of R recursive well-orderings) $\leqslant_T O^R$, we have $R \in M_\alpha \Rightarrow W^R \in M_\alpha$. It follows that for any R in M_α, the standard definition of W^R as an intersection of sets, with R as a constant, is correct when we restrict the quantifiers to range over $M_\alpha \cap P(\omega)$. So that $S = W^R$ is an $M_\alpha \cap P(\omega)$-second-order predicate of S, R. And 'R is a well-ordering' may be written as

$$(\exists S)(\exists n)(S = W^R \ \& \ n \in S \ \& \ \phi_n^R \text{ is a characteristic function for } R)\,.$$

Assembling all our translations, we arrive at a second-order definition of E_α with quantifiers restricted to $M_\alpha \cap P(\omega)$. It follows, as usual, that E_α is arithmetical in K.

It remains to prove the lemma, which we state again:

Let μ and ν be limit indices, $\mu < \nu < \alpha$. Let $E_\mu \in M_\alpha$ and $E_\nu \in M_\alpha$ be arithmetic copies of M_μ and M_ν, respectively. Then there is an arithmetic copy E'_ν of M_ν, $E'_\nu \in M_\alpha$, such that for some r

$$J(a, b) \in E_\mu \Longleftrightarrow J(J(a, r), J(b, r)) \in E'_\nu\,.$$

Proof. We use the familiar fact that there is a predicate in set theory which, when its variables are taken to range over any M_δ, δ a limit, well-orders that M_δ. Let L be the corresponding predicate in arithmetic which, when its variables are restricted to Field E_δ, for some limit δ, well-orders Field E_δ. L will inherit the property of its set-theoretic couterpart – that if E_δ and E_γ are two arithmetic copies, perhaps of different M_η, then the βth element in the E_δ-ordering and the βth element in the E_γ-ordering will, if they both exist, represent the same set.

We observe that the well-orderings induced by L on E_μ, E_ν will be arithmetic in E_μ, E_ν, respectively, and hence will both be in M_α. It follows that they are both of length $< \alpha$ (by the general condition on Case II – though indeed this fact follows from HYP α alone); the following functions are then also in M_α:

$$\begin{aligned} &f(0) = \text{the } L\text{-least member of Field } E_\mu, \\ &f(\beta{+}1) = \text{the } L\text{-successor in Field } E_\mu \text{ of } f(\beta), \\ &f(\lambda) = (\mu x)(x \in \text{Field } E_\mu \ \&\ (\forall\beta)(\beta < \lambda = \\ &\qquad f(\beta) <_L x) \ \&\ (\forall y)(y <_L x \Rightarrow (\exists\beta < \lambda)(y = f(\lambda))) \,, \end{aligned}$$

with the domain of x restricted to the length of E_μ.

g defined similarly, except in having 'E_ν' for 'E_μ'.

Because f and g are in M_α, so is the set

$$z = \{J(x, y) : (\exists\beta)(x = f(\beta) \ \&\ y = g(\beta)\} \,.$$

We can now define E'_ν:

$$\begin{aligned} &\text{Let } n(x) = J(v, r) \quad \text{if } (\exists w)(J(w, x) \in z \ \&\ w = x) \,; \\ &\text{Let } n(x) = J(x, r{+}1) \text{ otherwise} \,. \\ &\text{Let } E'_\nu = \{J(n(x), n(y)) : J(x, y) \in E_\nu\} \,. \end{aligned}$$

E'_ν is arithmetic in z, hence in M_α; it clearly has the required properties. This concludes the proof of Case IIa.

Case IIb. α is not HYP

In this case, we continue to have all well-orderings of integers of length $< \alpha$. It follows that $(\forall A \in M_\alpha)$ (A is a complete set of order $\beta < \alpha \Rightarrow \alpha \geqslant \omega_1^A$). However, we cannot strengthen the inequality, else α would be HYP [5]. It follows that $(\exists A \in M_\alpha)(\alpha = \omega_1^A)$. We now use Boolos' results on relativized hyperarithmetic hierarchies to conclude that $M_\alpha \cap P(\omega)$ consists of precisely the sets of integers hyperarithmetic in A. Clearly all B of order $\leqslant$ order A are

[5] For any R, W^R is easily definable over $M_{\omega_1^R}$.

hyperarithmetic in A. Also, if order $A = \beta <$ order $B < \alpha$, then $B \leqslant_T H^A_{\omega \cdot ((\text{order } B) - \beta)}$. Finally, B hyperarithmetic in $A \Rightarrow B \leqslant_T H^A_{(\omega \cdot \delta) - \beta}$ for some $\delta < \omega_1^A$, and $H^A_{(\omega \cdot \delta) - \beta} \leqslant_a E_\delta \in M_\alpha$. By the Spector hyperarithmetic quantifier theorem [10] W^A is second-order definable over $M_\alpha \cap P(\omega)$. This fact is all we need to imitate the proof of Case IIa, which concludes the proof of the theorem.

3. To this point we have discovered a minimality property for deg(α): we have shown that deg(α) contains uubs on $M_\alpha \cap P(\omega)$, and in certain cases is the minimal degree with that property. The cases we have discussed are those in which α is an index, and either the successor of an index or a limit of indices; in these cases, we have deg(α) a least uub on $M_\alpha \cap P(\omega)$, that is, K a uub on $M_\alpha \cap P(\omega) \Rightarrow \deg(\alpha) \leqslant \deg(K)$. The remaining cases that may arise are

(1) α is an index, but neither a limit of indices nor the successor of an index.
(2) α is not an index.

In these cases, there will be a least upper bound on the indices $< \alpha$, and by (δ an index $\Rightarrow$ there is an E_δ of order $\delta \Rightarrow (E_\delta)^{(\omega)}$ is of order $\delta+1$), that least upper bound cannot itself be an index. It follows that α will lie either at the beginning or middle (Case 2) or at the end (Case 1) of what we may call a 'gap' in the series of indices. The precise definition follows.

Definition. Let α be a non-index ordinal, $\omega < \alpha < \omega_1^L$, and let α be a limit of index ordinals. Let $\alpha+\beta$ be the least index ordinal $> \alpha$. Then we call α a gap ordinal, and we say that there occurs at α a gap of length β.

If α is a gap ordinal, $M_\alpha \cap P(\omega) = M_{\alpha+\delta} \cap P(\omega)$ for $\delta \leqslant \beta$. An arithmetic copy of $M_\alpha \cap P(\omega)$, then, does not appear until $M_{\alpha+\beta+1}$. We shall discover that α and β are characteristics for the properties of deg$(\alpha+\beta)$.

Gap ordinals have several interesting properties which will be of importance to us. $M_\alpha \cap P(\omega)$ is closed under many operations – in particular, it is closed under definition in second-order number theory; $M_\alpha \cap P(\omega)$ is therefore an ω-model for analysis. By the theorems used in Case IIb of the last theorem, we have $A \in M_\alpha \cap P(\omega) \Rightarrow \alpha \geqslant \omega_1^A$, else the relativized hyperarithmetic hierarchy would produce new sets of integers in $M_{\alpha+1}$. Indeed, $\alpha > \omega_1^A$, for if $\alpha = \omega_1^A$ then O^A would be of order α. It follows that α is HYP, and therefore that $M_\alpha \cap P(\omega)$ is a β-model for analysis.

We have shown that if α is an index, then there is a complete set of order α arithmetic in any uub on $M_\alpha \cap P(\omega)$. It is plausible to suppose that if the gap at α is of length 1, and K is a uub on $M_\alpha \cap P(\omega)$, we can retrieve the complete set of order $\alpha+1$ from $K^{(\omega)}$. In general, one might conjecture that if the gap at α is of length β, and K is a uub on $M_\alpha \cap P(\omega)$, then we can retrieve the complete set of order $\alpha+\beta$ from $K^{\omega\cdot\beta}$. This in fact turns out to be the case; however, as the example of the minimal model for set theory shows, β can be very large: indeed, we will need a definition of jumps that allows us to speak of $K^{\omega\cdot\beta}$ for β bounded only by ω_1^L.

We shall use the simplest type of notational systems: any total well-ordering of integers will be available as a system of notations. We define jumps in an appropriately simple way:

Definition. Let R be a total well-ordering of integers; let A be a set of integers. We define

$$A_R^a = A \quad \text{if } |a|_R = 0$$

$$A_R^b = (A_R^c)' \quad \text{if } |b| = |c| + 1$$

$$A_R^b = \{J(r, c) : c <_R b \;\&\; r \in A_R^c\} \quad \text{if } |b|_R \text{ is a limit}.$$

(If $|d|_R = \beta$, we will often write 'A_R^β' for 'A_R^d', or, where there is no ambiguity, 'A^β' or 'A^d'. Ordinary jumps will continue to be written '$A^{(n)}$'.

This definition of jumps has the usual properties of jumps; in particular, $n <_R m \Rightarrow K_R^n \leqslant_T K_R^m$. For let $|m|_R$ be a limit. Then $K^m = \{J(a, b) : b <_R m \;\&\; a \in K^b\}$, so that

$$f : x \to J(x, n)$$

gives a 1–1 reduction of K^n to K^m. Now let $|m|_R = \lambda + r$, λ a limit, r an integer. Then $K^m = (K^\lambda)^{(r)}$, so that if $|n|_R < \lambda$, $K^n \leqslant_T K^\lambda \leqslant_T K^m$, while if $\lambda < |n|_R < \lambda + r$, then $K^n = (K^\lambda)^{(|n|-\lambda)} \leqslant_T K^m$. For $|n|_R = r$ an integer, we have $K_R^n = K^{(r)}$.

We are now able to state the desired

Theorem. Let α be a gap ordinal; let the gap at α be of length β. Let K be a uub on $M_\alpha \cap P(\omega)$, and let K_s^b be a $\omega \cdot \beta$ jump of K. Then there is a complete set of order $\alpha+\beta$, arithmetical in K_s^b. It follows that any complete set of integers of order $\alpha+\beta$, is arithmetical in K_s^b.

Proof. The first step is to construct an arithmetic copy of M_α, arithmetic in K. The copy will be called E_α.

The construction is by now familiar – since α is a limit of indices, there is a complete E_γ of order γ, for some $\gamma < \alpha$. We form E_α by iterating $*$ on E_γ $\alpha - \gamma$ $(=\alpha)$ times. To index the successive copies, we need a well-ordering of integers of length α, arithmetic in K.

Because α is a limit of indices, $M_\alpha \cap P(\omega)$ contains well-orderings of integers of length $> \delta$ for any $\delta < \alpha$. For let $\delta < \alpha$, and let $\delta' \geqslant \delta$ be an index. Let $E_{\delta'}$ be an arithmetic copy of M_δ, of order δ. Then

$$\{J(x, y) \in E_{\delta'} : \text{Ord}\, x \text{ and Ord}\, y\}$$

is a well-ordering of integers in M_α, of length $\delta' \geqslant \delta$.

Let W_e^K be the set of K-Gödel numbers of sets in $M_\alpha \cap P(\omega)$. α is HYP, hence "X is a well-ordering of integers in $M_\alpha \cap P(\omega)$" is expressible by a second-order predicate with quantifiers restricted to $M_\alpha \cap P(\omega)$. It follows that there is a predicate $Q(x)$, arithmetic in K, such that "$x \in W_e^K$ & $Q(x)$" expresses "$x \in W_e^K$ & W_x^K is a well-ordering". We can construct a well-ordering of integers, of length α, by 'laying end to end' the well-orderings in $M_\alpha \cap P(\omega)$, and indexing them by their Gödel numbers, as follows:

$$R = \{J(J(x, a), J(y, b)) : a \in W_e^K \ \&\ b \in W_e^K \ \&\ Q(a) \ \&\ Q(b) \ \&.$$

$$[a < b \ \&\ x \in \text{Field}\ W_a^K \ \&\ y \in \text{Field}\ W_b^K] \vee$$

$$[a = b \ \&\ J(x, y) \in W_a^K]\}.$$

R is clearly arithmetic in K; its length is $> \delta$ for every $\delta < \alpha$. By the construction of R as a joining together of exactly ω well-orderings, any initial segment of R is in $M_\alpha \cap P(\omega)$. It follows that the length of R is exactly α.

We can now use R to construct an arithmetic copy of $M_\alpha \cap P(\omega)$. We are given $E_\gamma \in M_{\gamma+1} \cap P(\omega)$, an arithmetic copy of M_γ. We may assume that E_γ is a 0_R-initial copy of M_γ. Then E_α will be defined by induction over R:

$$E_{0_R} = E_\gamma$$

$$E_a = (E_b, a)^* \quad \text{if } |a|_R = |b|_R + 1$$

$$E_a = \bigcup_{b <_R a} E_b \quad \text{if } |a|_R \text{ is a limit}$$

$$E_\alpha = \bigcup_{a \in \text{Field} R} E_a .$$

E_α is clearly an arithmetic copy of M_α. To prove it arithmetic in K we will first show by induction that $E_a \in M_\alpha \cap P(\omega)$ for $a \in \text{Field } R$.

(1) $E_{0_R} = E_\gamma \in M_\alpha$

(2) if $|a| = |b| + 1$, $E_a \leqslant_T E_b^{(\omega)} \in M_{\delta+1} \subseteq M_\alpha$, if $E_b \in M_\delta$

(3) if $|a|$ is a limit ordinal.

Let $R_a = \{J(x, y) \in R : J(y, a) \in R\}$. R_a is the initial segment of R determined by (not including) a. By a remark above, $R_a \in M_\alpha$. We may define E_a as follows:

$$y \in E_a \Leftrightarrow (\exists S)(\forall x)\{[x \in S \Rightarrow.$$

$$LKx \in \text{Field } R_a \ \&\ LLx \in \text{Field } R_a \ \&\ LKx \text{ not a limit } \&\ LLx \text{ not a limit } \&.$$

$$(LLx = LKx = 0_R) \vee LKx <_R LLy] \ \&\ [LKx = LLx = 0_R \Rightarrow.$$

$$x \in S \Leftrightarrow x \in E_\gamma] \ \&\ [LKx <_{R_a} LLx \Rightarrow.$$

$$(1)\ x \in S \Rightarrow (\exists V)(\exists W)(V = \{z \in S : LLz <_{R_a} LLx\} \ \&\ W = (V, LLx)^* \ \&\ x \in W) \ \&.$$

$$(2)\ (\forall V)(\forall W)(V = \{z \in S : LLz <_{R_a} LLx\} \ \&\ W = (V, LLx)^* \Rightarrow x \in W) \Rightarrow x \in S]$$

$$\&\ y \in S\} .$$

We may also define E_a by

$$y \in E_a \Longleftrightarrow (\forall S)(\forall x)\{[x \in S \Rightarrow.$$

$$LKx \in \text{Field } R_a \ \& \ LLx \in \text{Field } R_a \ \& \ LKx \text{ not a limit } \& \ LLx \text{ not a limit } \&.$$

$$(LLx = LKx = 0_R) \vee (LKx <_{R_a} LLx \leqslant_{R_a} y] \ \& \ [LKx = LLx = 0_R \Rightarrow.$$

$$x \in S \Longleftrightarrow x \in E_\gamma] \ \& \ [LKx <_{R_a} LLx \Rightarrow.$$

$$(1)\, x \in S \Rightarrow (\forall V)(\forall W)(V = \{z \in S : LLz <_{R_a} LLx\} \ \& \ W = (V, LLx)^* \Rightarrow x \in W) \ \&$$

$$(2)\, (\exists V)(\exists W)(V = \{z \in S : LLz <_{R_a} LLx\} \ \& \ W = (V, LLx)^* \ \& \ x \in W) \Rightarrow x \in S]$$

$$\Rightarrow y \in S\}.$$

Now in these two definitions $R_a \in M_\alpha \cap P(\omega)$. The expression $(X, z)^*$ may be replaced by an expression Π_1^1 or Σ_1^1 in X, as the case requires, for $(X, z)^*$ is uniformly recursive in $(X^{(\omega)}, z)$, which is uniformly Π_1^1 and uniformly Σ_1^1 over X, z. Hence, shifting and contracting quantifiers, we get the result that E_a is Δ_1^1 in R_a, that is, $\leqslant_T O^{R_a}$. But $O^{R_a} \in M_\alpha$, hence $E_a \in M_\alpha$.

It is an easy step now to show that E_α is arithmetic in K. E_α is defined just like E_a, with 'R' for R_a; the appropriate V and W are precisely the E_a we showed to be elements of M_α. It follows that the second-order definition of E_α is valid when the quantifiers are restricted to $M_\alpha \cap P(\omega)$, and hence that E_α is arithmetic in K.

Let us review the situation. We are given an $\omega \cdot \beta$ jump of K, K_S^b, and we want to find an arithmetic copy $E_{\alpha+\beta}$ of $M_{\alpha+\beta}$, arithmetic in K_S^b. We have constructed E_α, an arithmetic copy of M_α, in such a way that E_α is arithmetic in K. It follows that $E_\alpha \leqslant_T K^{(n)}$ for some integer n. It follows that $E_\alpha \leqslant_T K_S^a$ for some a, $|a|_S < \omega$, and hence that $E_\alpha \leqslant_T K_S^\omega$. We will construct $E_{\alpha+\beta}$ as usual by reiteration of $*$, this time over the limit ordinals in S, taking care that at each stage $E_{\alpha+\delta} \leqslant_T K_S^{\omega \cdot \delta + \omega}$.

We have S arithmetic in $K^{\omega \cdot \beta}$, for

$$(1)\, x \in \text{Field } S \Longleftrightarrow (\exists r)(J(r, x) \in K^{\omega \cdot \beta}) \vee x = b \text{ where } |b|_S = \omega \cdot \beta$$

(2) $J(r, s) \in S \Longleftrightarrow r \in \text{Field } S \mathbin{\&} s \in \text{Field } S \mathbin{\&} r \neq s$

$$\{J(x, r) \in K^{\omega \cdot \beta}\} \leqslant_T \{J(x, s) \in K^{\omega \cdot \beta}\} .$$

We shall call the sequence of S-limits S'.

$$S' = \{J(x, y) \in S : \lim y \mathbin{\&} (\lim x \vee x = 0_S)\} .$$

Clearly $S' \leqslant_a S \leqslant_a K^{\omega \cdot \beta}$.

The definition of $E_{\alpha+\beta}$ is as follows:

$$E_{O_{S'}} = E_\alpha \text{ (we assume } E_\alpha \text{ to be } O_S = O_{S'}\text{-initial copy)}$$

$$E_a = (E_b, a)^* \text{ where } |a|_{S'} = |b|_{S'} + 1$$

$$E_a = \bigcup_{b <_{S'} a} E_b \text{ where } |a|_{S'} \text{ is a limit ordinal}$$

$$E_{\alpha+\beta} = E_b \text{ where } |b|_{S'} = \beta .$$

The following problems are uniformly arithmetical in K_S^b for any $b \in \text{Field } S'$:

(1) Determining if $x \in \text{Field } S \mathbin{\&} x <_S b$.

$$x \in \text{Field } S \mathbin{\&} x <_S b \Longleftrightarrow (\exists y)(J(y, x) \in K_S^b$$

(2) Determining if $x <_S y <_S b$.

$$x <_S y <_S b \Longleftrightarrow . x \in \text{Field } S \mathbin{\&} x <_S b \mathbin{\&} y \in \text{Field } S \mathbin{\&} y <_S b \mathbin{\&}$$

$$\{J(r, x) \in K_S^b\} \leqslant_T \{J(r, y) \in K_S^b\} \mathbin{\&} y \neq x$$

(3) Determining if $x \in \text{Field } S' \mathbin{\&} x <_S b$.

$$x \in \text{Field } S' \mathbin{\&} x <_S b \Longleftrightarrow x \in \text{Field } S \mathbin{\&} x <_S b \mathbin{\&}$$

$$\sim (\exists y)(y <_S x <_S b \mathbin{\&} (\forall z)(z <_S x <_S b \Rightarrow . z = y \vee z <_S y <_S b))$$

(4) Determing if $x \in \text{Field } S' \mathbin{\&} x <_{S'} b$ and x is an S'-limit.

$$x \in \text{Field } S' \;\&\; x <_{S'} b \;\&\; x \text{ is an } S'\text{-limit} \Longleftrightarrow x \in \text{Field } S' \;\&\; x <_{S'} b \;\&$$

$$\sim (\exists y)(y \in \text{Field } S' \;\&\; y <_S x <_S b \;\&\; (\forall z)(z <_S x <_S b \;\&\; z \in \text{Field } S'.$$

$$\Rightarrow z = y \vee z <_S y <_S b)$$

(5) Given $x \in \text{Field } S$, $x <_S b$, finding the S-successor of x.

$$x' = (\mu y)(\forall z)(x <_S z <_S b \Rightarrow . z = y \vee y <_S z <_S b : \& x <_S y)$$

(6) Given $x \in \text{Field } S'$ and $x <_S b$, and x is an S'-successor, finding the S' Predecessor of x.

$$\text{pred } x = (\mu y)(y \in \text{Field } S' \;\&\; y <_S x <_S b \;\&\; (\forall z)(z \in \text{Field } S' \;\&$$

$$z <_S b \;\&\; z <_S x <_S b \Rightarrow . z = y \vee z <_S y <_S b)) .$$

Let $E_{O_S} \leqslant_T K^{(n)}$. Let $q > n$ be so large that all of (1)–(6) are uniformly recursively solvable in $(K^b)^{(q)}$, for any $b \in \text{Field } S'$. Notice that we also have solutions to the following problems:

(7) If $|b|_{S'} = |a|_{S'} + 1$, we can find a Gödel number of E_b in $E_a^{(\omega)}$, uniformly in b.

(8) Given a Gödel number of A in $B^{(q)}$, we can recursively find a Gödel number of $A^{(\omega)}$ in $B^{(\omega)}$.

(9) If $|b|_{S'} = |a|_{S'} + 1$, we can uniformly find a Gödel number for $(K_S^a)^{(\omega)}$ in $(K_S^b)^{(q)}$.

(10) If b is a limit in S', we can give, uniformly in e, b, a Gödel number for the characteristic function in $(K^b)^{(q)}$ of

$$\{J(x, y) : x <_S b \;\&\; \phi_e^{(K^x)^{(q)}} y = 0\}, \text{ provided } \phi_e^{(K^x)^{(q)}}$$

$$\text{converges everywhere for } x <_S b .$$

(7) is by now familiar. (8) is an easy consequence of the fact that there is a recursive function ϕ_e such that if ϕ_x^B is the characteristic function of A, then $\phi_{\phi_e(x)}^{B'}$ is the characteristic function of A'. (9) follows from the fact that A is uniformly recursive in $A^{(q)}$; given $(K_S^b)^{(q)}$, we can also find pred b, by (6), and the members of S $a <_S x <_S b$ – this is enough information to determine membership in $(K_S^a)^{(\omega)}$. (10) is direct.

We are now in a position to show $E_a \leqslant_T (K_S^a)^{(q)}$ for all $a \in S'$. We use Roger's recursion lemma:

Let P be a 2-place predicate of numbers, and let R be a well-ordering of integers. Let ϕ_r be such that for all $b \in \text{Field } R$ and for all e,

$$(\mathsf{A}a <_R b)P(\phi_e(a), a) \Rightarrow P(\phi_r(e, b)b) .$$

Then there is a ϕ_s such that

$$(\forall a \in \text{Field } R)P(\phi_s(a), a) .$$

Let R be S'. Let $P(x, y)$ be the predicate 'x is a Gödel number for Ey in $(K^y)^{(q)}$'.

Our theorem will be proved if we can find a ϕ_r such that

$$(\forall b \in \text{Field } S')(\forall e)\ [(\forall a <_{S'} b)P(\phi_e a, a) \Rightarrow P(\phi_r(e, b), b)] .$$

We give the definition of r somewhat anthropomorphically, by showing how to calculate in A a function which we call $g^A(e, b, x)$. We assume that e has the appropriate conditions, $b \in \text{Field } S'$, and A is $(K^b)^{(q)}$. We are not concerned with g for other e, b, A.

(1) $E_{O_{S'}}$ is recursive in $K^{(q)}$. Say $\phi_d^{K^{(q)}}$ is a characteristic function for $E_{O_{S'}}$. We set $g(e, O_{S'}, x)^{K(q)} = \phi_d^{K(q)}x$.

(2) Given e, and $b \neq O_{S'}$, discover whether b is a limit in S'.

(a) If b is not a limit, find a such that $|a|_{S'} + 1 = |b|_{S'}$. Calculate $\phi_e(a)$. $\phi_e(a)$ is a Gödel number of E_a in $(K^a)^{(q)}$. We have $E_b \leqslant_T (E_a)^{(\omega)} \leqslant_T (K^a)^{(\omega)} \leqslant_T K^b \leqslant_T (K^b)^{(q)}$. All of these reducibilities are uniform in b; let ϕ_f be a Gödel number of E_b in $(K^b)^{(q)}$. Give as output $\phi_f^{(K^b)(q)}x$.

(b) If b is a limit, then the characteristic function for E_b is given by $x \in E_b \Leftrightarrow Lx <_{S'} b \ \&\ \phi_{\phi_e Lx}^{(K^{Lx})(q)}x = 0$. By (10), there is a $\phi_{h(e,b)}$ such that $\phi_{h(e,b)}^{(K^b)(q)}$ is, for any $b \in \text{Field } S'$, a characteristic function for E_b. Give as output $\phi_{h(e,b)}^{(K^b)(q)}x$.

By the *s-m-n* theorem, we may set

$$g^A(e, b, x) = \phi_{\phi_r(e,b)}^A x .$$

This gives the desired r. It follows that $E_{\alpha+\beta}$, an arithmetic copy of $M_{\alpha+\beta}$, is recursive in $K_S^{\omega \cdot \beta + q}$, i.e., arithmetic in $K_S^{\omega \cdot \beta}$. It follows that every set of integers of order $\alpha+\beta$ is arithmetical in $K_S^{\omega \cdot \beta}$.

Chapter 2

We begin with a definition:

Definition. Let Γ be a set of sets of integers, and let A be a uub on Γ. We call A a δ-least uub on Γ if, for all K, S, and r, if K is a uub on Γ, S a total well-ordering of integers, and K_S^r a δ-jump of K, $A \leqslant_a K_S^r$.

In the last theorem of Chapter 1, we proved a result which, using the above definition, may be stated rather simply: Let α be a gap ordinal, and let the gap at α be of length β. Then any complete set $E_{\alpha+\beta}$ of order $\alpha+\beta$ is an $\omega \cdot \beta$-least uub on $M_\alpha \cap P(\omega) = M_{\alpha+\beta} \cap P(\omega)$.

Notice that the result generalizes to the trivial case, where $\beta = 0$. In this case, α is itself an index, and any complete set E_α of order α is an $\omega \cdot 0 = 0$-least uub on $M_\alpha \cap P(\omega)$, using the convention where $|a|_S = 0, K_S^a = K^{(0)} = K$. In the trivial case, it is clear that the result is best possible – E_α could not be an n-least uub for any $n < 0$; more consequentially, we cannot show that every member of the complete degree is recursive in any uub K, or in $K^{(n)}$ for a fixed n: the complete degree itself contains uubs whose degrees of unsolvability are distant enough to preclude any such simple relation. It is natural to ask whether in the general case the result we have achieved is best possible, i.e., whether the ordinal $\omega \cdot \beta$ can be 'improved'. In this chapter, we shall show that it is in fact best possible, in the sense of the following theorem:

Theorem. Let α be a gap ordinal, and let the gap at α be of length β. Let $\delta < \omega \cdot \beta$. Then there is no δ-least uub A on $M_\alpha \cap P(\omega)$.

In proving the theorem, we may assume $\delta = \omega \cdot \gamma$ for some $\gamma < \beta$, For if $\delta = \omega \cdot \gamma + n$, and A is a δ-least uub, the A is also an $\omega \cdot \gamma$-least uub.

We shall prove the result in a somewhat more constructive version:

Theorem. Let α be a gap ordinal, and let the gap at α be of length β. Let $\gamma < \beta$. There are two total well-orderings of integers R_1 and R_2, and two sets of integers K_1 and K_2 such that

(1) K_1 and K_2 are uubs on $M_\alpha \cap P(\omega)$.

(2) For any set of integers B,

$$B \leqslant_a K_{1_{R_1}}^{\omega \cdot \gamma} \;\&\; B \leqslant_a K_{2_{R_2}}^{\omega \cdot \gamma} \Rightarrow B \in M_\alpha \cap P(\omega)\,.$$

It follows immediately that there can be no $\omega \cdot\cdot \gamma$-least uub on $M_\alpha \cap P(\omega)$. For if B were such a uub, we would have $B \in M_\alpha \cap P(\omega)$, so that $B' \in M_\alpha \cap P(\omega)$ would be recursive in B.

The proof of the theorem is a generalization of a proof of Richard Boyd's [3] ; it may help the reader through the wilderness if we first review that proof.

Boyd showed the following:

If $\mathfrak{A}$ is a countable ω-model for analysis, then there exist two sets of integers A and B such that

(1) A and B are uubs on $\mathfrak{A}$.

(2) If $K \leqslant_a A$ and $K \leqslant_a B$ then K is second-order definable over the sets of integers in $\mathfrak{A}$.

It followed, as in our theorem, that there could be no arithmetically least uub on $\mathfrak{A}$.

The proof, in outline, went as follows:

Forcing was defined for a language containing names for the sets in $\mathfrak{A}$, integer names, and two number predicates A and B. A forcing condition was a pair of finite sequences of sets in $\mathfrak{A}$, $\langle A_1, A_2 \ldots\rangle$, $\langle B_1, B_2 \ldots\rangle$, the crucial forcing condition being as follows:

$$\langle A_1, A_2 \ldots\rangle \Vdash A(J(r, s)) \Leftrightarrow r \in A_s$$

$$\langle B_1, B_2 \ldots\rangle \Vdash B(J(r, s)) \Leftrightarrow r \in B_s \, .$$

A generic sequence of sets was constructed for the model with the two new predicates A and B; the usual forcing = truth lemma went through for that model, it followed from forcing = truth and a standard generic-set argument that $A = \{x : \mathfrak{A} \models Ax\}$ and $B = \{x : \mathfrak{A} \models Bx\}$ were both uubs on $\mathfrak{A}$. It was also possible to prove Feferman's result, on the irrelevance of A-conditions to sentences not containing the letter A.

If K was a set arithmetically definable over each of A and B, say by $\phi_1(A, x)$ and $\phi_2(B, x)$, then some finite set of conditions P forced $(\forall x)(\phi_1(A, x) \Leftrightarrow \phi_2(B, x))$. It followed that the precise constitution of B was irrelevant to the constitution of K, so long as the B-conditions extended P. For, by the Feferman result, the A-conditions were alone sufficient to force each $\phi_1(A, n)$ or $\sim \phi_1(A, n)$, and any B-conditions that extended P would eventually have to force exactly the same sentences. It followed that K could be defined in terms of the forcing concept alone, and the result followed by the definability of forcing in $\mathfrak{A}$.

We shall imitate Boyd's proof in the following way: Given $\alpha < \alpha+\gamma < \alpha+\beta$, we shall construct two sets of integers, K_1 and K_2, which will be generic, not for the model for analysis $M_\alpha \cap P(\omega)$, but for the (partial) model for set theory M_α. The expanded model will contain K_1 and K_2 as sets – they will appear at the α-th level, and the expanded model will be closed under first-order definability up through the $\alpha+\gamma$-th level.

Our models will therefore bear a good deal of resemblance to the model N in Cohen's proof of the independence of $V = L$, at least in this respect – that they will be considerably richer in sets of integers than the ground model. Unfortunately, the presence of K_1 and K_2 in our model does not itself guarantee that the model will be rich enough: we will need a $K_1^{\omega\cdot\gamma}$ and a $K_2^{\omega\cdot\gamma}$, if not in our model as sets, then definable over it as classes, and for this to be possible, we will need well-orderings of length γ present in, or definable over, our model. Nothing in our construction so far has guaranteed (so far as we know) the possibility of defining such well-orderings. We will therefore define two generic well-orderings W_1 and W_2. W_1 and W_2 will appear as 2-place predicates of ordinals, $W_i n\delta$, $n < \omega$, $\alpha \leqslant \delta < \alpha+\gamma$. $\{n : (\exists\delta)(W_i n\delta\}$ will appear only as a class in the model, but at each level, η, $\{n : (\exists\delta < \eta)(W_i n\delta)\}$ will appear as a set. Thus, each level will be the closure of the preceding level under first-order definability, using the predicate letters W_1, W_2: at each level, longer well-orderings will appear, and it will be possible to define $K_1^{\omega\cdot\gamma}$ and $K_2^{\omega\cdot\gamma}$ as classes.

Let $S = S_{\alpha+\gamma}$ if γ is a limit, $S = S_{\alpha+\gamma-1}$ otherwise. S will be the set of all constants in our language. If $\eta = \mu\delta(c \in S_\delta)$, we say rank $c = \eta$.

The rank of formulas is defined, as in Cohen [4], as follows: If ϕ is a limited statement, let rank $\phi = (\delta, i, r)$ where

(1) δ is the least ordinal such that if $\forall\eta$ occurs in ϕ, then $\eta \leqslant \delta$, and if c of rank η occurs in ϕ, then $\eta \leqslant \delta$.

(2) r is the number of symbols in ϕ [a constant counts as a single symbol].

(3) $i = 0$ if δ is a successor ordinal, say $\delta = \eta + 1$, and $\forall\delta$ does not occur in ϕ, and no term of the form $c \in (\cdot)$, $c = (\cdot)$, or $(\cdot) = c$ occurs in ϕ, where rank $c = \eta$. Otherwise $i = 1$.

We define $(\delta, i, r) < (\delta', i', r')$ if $\delta < \delta'$, or $\delta = \delta'$ and $i < i'$, or $\delta = \delta'$ & $i = i'$ & $r < r'$.

Forcing conditions

As always, what we choose to be our forcing conditions is not as important as what sentences they will force. Later, when we come to the definition of forcing for limited statements as a class in $M_{\alpha+\gamma}$, we will need to do some quite artificial manoeuvering in order to identify forcing conditions with sets of low rank. For the present, however, we define forcing in the most natural manner we can; it will be apparent later that our results carry over intact with the more complicated definition.

We define a forcing condition to be a quadruple of sets k_1, k_2, w_1, w_2, where k_i is a sequence $K_{i1} \dots K_{in}$ of sets of integers in M_α.

The language $\mathcal{L}$

We introduce a language for $M_{\alpha+\gamma}$ that differs from the usual one only in containing two generic [6] constants 'K_1' and 'K_2', and two generic predicate-symbols 'W_1' and 'W_2':

Propositional connectives

$\sim, \&$; '$\vee$' will abbreviate $\sim(\sim - \& \sim \dots)$.

'$\Rightarrow$' abbreviates $\sim - \vee \dots$, and '$\Leftrightarrow$' abbreviates '$\dots \Rightarrow - \& - \Rightarrow \dots$'.

Quantifiers $\forall$, $\forall_\delta$ for $\delta < \alpha + \gamma$

'$\exists$' will abbreviate $\sim\forall\sim$, '$\exists_\delta$' will abbreviate $\sim\forall_\delta\sim$.

Predicate symbols

$\in, =, W_1, W_2$ [the last two are α-place predicate]

We define constants in the usual way by ordinal induction:

$S_\phi = \phi$

$S_{\delta+1} = S_\delta \cup$ the set of all expressions of the form $\{x_\delta : \phi x\}$

where ϕ is limited, and all ordinals in ϕ are $\leqslant \delta$, and ϕ may contain constants in S_δ. If $\delta \geqslant \alpha$, ϕ may also contain the predicate letters 'W_1' and 'W_2' [7].

$$S_\delta \text{ for } \delta \text{ a limit} = \bigcup_{\eta<\delta} S_\eta \cup \{K_1\} \cup \{K_2\} \quad \text{if } \delta = \alpha$$

$$= \bigcup_{\eta<\delta} S_\eta \quad \text{if } \delta \neq \alpha$$

and w_i is a finite set whose members are of the form $\langle n, \delta \rangle$ for ordinals n, δ, $n < \omega, \alpha \leqslant \delta < \alpha + \gamma$, such that no n and no δ occurs twice in either w_i. We denote forcing conditions by the letters 'P', 'Q', etc. and define forcing for limited statements by induction on rank [8]:

(1) $P \Vdash \forall_\delta u \phi u$ if for all $c \in S_\delta$ $P \Vdash \phi c$

(2) $P \Vdash \sim\phi$ if $(\forall P' \supseteq P)(\sim P' \Vdash \phi)$

(3) $P \Vdash \phi \vee \psi$ if $P \Vdash \phi$ or $P \Vdash \psi$

[6] We call a *constant* generic, following Cohen, if the value it takes is in a natural way, directly determined by the forcing conditions; likewise for predicates.

[7] Our S_δ, unlike Cohen's, are cumulative – an inessential difference.

[8] We use weak forcing.

(4) $P \Vdash c_1 \in c_2$ where rank $c_1 = \delta > \delta' =$ rank c_2 if $\exists c_3 \in S_{\delta''}, \delta'' < \delta'$, such that $P \Vdash [(\forall_\delta u)(u \in c_1 \Leftrightarrow u \in c_3) \,\&\, c_3 \in c_2]$

(5) $P \Vdash c_1 = c_2$ where rank $c_1 = \delta \geqslant \delta' =$ rank c_2 if $P \Vdash (\forall_\delta u)(u \in c_1 \Leftrightarrow u \in c_2)$

(6) $P \Vdash c_1 \in c_2$ where rank $c_1 < c_2$ and c_2 is not K_1 if $P \Vdash \psi c_1$ where $c_2 = \{x_\delta : \psi x\}$

(7) $P \Vdash c \in K_i$ if there is an integer n such that $P \Vdash c = \{x_n : \text{Ord}\, x\}$, and if $n = J(r, s)$, then $r \in K_{is}$

(8) $P \Vdash W_i c_1 c_2$ if there exist ordinals $n < \omega, \alpha \leqslant \delta < \alpha + \gamma$ such that $P \Vdash [c_1 = \{x_n : \text{Ord}\, x]$ and $P \Vdash [c_2 = \{x_\delta : \text{Ord}\, x\}]$, and $\langle n, \delta\rangle \in W_i$

(1)–(6) reduce rank. (7) and (8) do not, but it is clear that, with some sacrifice of perspicuity, they could be rewritten to reduce rank. Thus, e.g.,

(7′) $P \Vdash c \in K_i$ if there is an integer n such that $P \Vdash (\forall_\delta x)(x \in c \Leftrightarrow x \in \{x_n : \text{Ord}\, x\})$ where rank $c = \delta + 1$, etc.

Forcing for unlimited statements is defined in the usual way, in terms of forcing for limited statements, and by induction on complexity.

The model N

We assume here the standard results on forcing; in particular, the existence of a complete sequence: this result follows from only the most general facts about forcing, and the countability of $\mathcal{L}$. We pick a particular complete sequence $\{P_n\}$, which will remain fixed for the remainder of the argument.

'P_i', 'P_j' will always denote members of $\{P_n\}$. In N, K_i will denote $\{n : (\exists P_i)(P_i \Vdash n \in K_i)\}$. We will stipulate that in N, $W_i n, \delta$ is true iff $(\exists P_j)(\langle n, \delta\rangle \in W_i$ in $P_j)$.

We define N by ordinal induction, for $\alpha \leqslant \delta \leqslant \alpha + \gamma$:

$$N_\alpha = M_\alpha \cup \{K_1\} \cup \{K_2\}$$

$$N_{\delta+1} = \{x \subseteq N_\delta : x \text{ is first-order definable over } N_\delta \text{ using the predicates } W_1 \,\&\, W_2\}$$

$$N_\delta \text{ for limit } \delta = \bigcup_{\eta<\delta} M_\eta$$

$$N = N_{\alpha+\gamma}\,.$$

In N, constants will have their natural designation: thus

$$\{x_\delta : (\forall_\delta y)(\phi yx\}$$

will designate

$$\{x \in N_\delta : (\forall y \in N_\delta)(\phi yx)\}.$$

We shall use 'δ' to abbreviate '$\{x_\delta : \text{Ord}\, x\}$'.

The following basic result holds true of N: Let ϕ be any sentence in $\mathcal{L}$. Then

$$N \models \phi \Leftrightarrow (\exists P_i)(P_i \Vdash \phi).$$

The proof is sufficiently familiar not to need repetition.

$K_{1R_1}^{\omega\cdot\gamma}$ and $K_{2R_2}^{\omega\cdot\gamma}$

Using the truth=forcing lemma, we can prove the following

Lemma. (1) K_i is a uub on $M_\alpha \cap P(\omega)$.

(2) $R_i = \{J(a, b) : (\exists\delta)(\exists\delta')(W_i a\delta \;\&\; W_i b\delta' \;\&\; \delta < \delta')\}$ is a γ-long well-ordering of integers.

Proof. (1) Let $z \in M_\alpha \cap P(\omega)$.

Let $P \Vdash (\forall y) \sim (z = \{x_\omega : \text{Ord}\, x \;\&\; J(x, y) \in K_1\})$.

Let $P = \langle k_1, k_2, w_1, w_2 \rangle$ where $k_1 = K_{11} \ldots K_{1n}$.

Let $Q = \langle k_1', k_2, w_1, w_2 \rangle$ where $k_1' = K_{11} \ldots K_{1n} z$.

Then $Q \supseteq P$ and $Q \Vdash z = \{x_\omega : \text{Ord}\, x \;\&\; J(x, n{+}1) \in K_1\}$, which is a contradiction. It follows that no P forces

$$(\forall y)(\sim z = \{x_\omega : \text{Ord}\, x \;\&\; J(x, y) \in K_1\},$$

so that $(\exists y)(z = \{x_\omega : \text{Ord}\, x \& J(x, y) \in K_1\}$ is true in N. But $N_\alpha = M_\alpha$, so the above formula is absolute; it follows that K is a uub on $M_\alpha \cap P(\omega)$. Similarly for K_2.

(2) Similarly, for each δ, $\alpha \leqslant \delta < \alpha + \gamma$,

$$(\forall n) \sim W_i n\delta$$

can never be forced; it follows that each δ, $\alpha \leqslant \delta < \alpha + \gamma$ appears paired with exactly one n in the extension of W_i.

N is a transitive set; it follows that $\{x_\delta : \text{Ord}\, x\}$ defines an actual ordinal, in fact, by an easy induction, δ. Bearing in mind that every ordinal δ,

$\alpha \leqslant \delta < \alpha + \gamma$ appears in the right field of W_i, and that $\delta \in N_{\delta+1}$, we see that

$$\{J(a, b) : (\exists_{\alpha+\eta} \delta)(\exists_{\alpha+\eta} \delta')(W_i a\delta \ \&\ W_i b\delta' \ \&\ \delta < \delta')\}$$

defines a well-ordering of integers of length η in $N_{\alpha+\eta+1}$.

We can expand this ordering to an $\omega \cdot \eta$-long well-ordering:

$$R^i_{\omega \cdot \eta} = \{J(J(a, r), J(b, s)) : (\exists_{\alpha+\eta} \delta)(\exists_{\alpha+\eta} \delta')(W_i a\delta \ \&\ W_i b\delta' \ \&\ (\delta = \delta' \ \&\ r < s) \vee \delta < \delta')\}.$$

We always have $R^i_{\omega \cdot \eta} \in N_{\alpha+\eta+1}$. All the $R^i_{\omega \cdot \eta}$ for $0 \leqslant \eta < \gamma$ are compatible, indeed, they are all initial segments of the following well-ordering, which is a class in N:

$$R^i_{\omega \cdot \gamma} = \{J(J(a, r), J(b, s)) : (\exists \delta)(\exists \delta')(W_i a\delta \ \&\ W_i b\delta' \ \&\ (\delta = \delta' \ \&\ r < s) \vee \delta < \delta')\}.$$

We can use $R_i = R^i_{\omega \cdot \gamma}$ to define the $\omega \cdot \gamma$-jump of K_i. Observe that $K_i \in N_\alpha$, all finite jumps of K_i appear in $N_{\alpha+1}$, and $K_i^{(\omega)}$ appears in $N_{\alpha+2}$. It follows that $K^{\omega}_{iR_i}$ is definable over $N_{\alpha+1}$, thus:

$$K^{\omega}_{iR_i} = \{J(x, y) : (\exists z)(x \in K_i^{(z)} \ \&\ z = |y|_{R_{\omega \cdot 1}})\}$$

where each $K_i^{(z)} \in N_{\alpha+1}$, and $R_{\omega \cdot 1}$ may be replaced by its definition in $N_{\alpha+1}$. It follows that $K^{\omega \cdot 1}_{iR_i} \in N_{\alpha+2}$; we may show by induction that $K^{\omega \cdot \eta}_{iR_i} \in N_{\alpha+\eta+1}$, for $\eta < \gamma$: If $K_i^{\omega \cdot \eta} \in$[9] $N_{\alpha+\eta+1}$, we have $R^i_{\omega \cdot \eta + \omega} \in N_{\alpha+\eta+2}$; also all finite jumps of $K_i^{\omega \cdot \eta}$ are in $N_{\alpha+\eta+1}$. Hence,

$$K_i^{\omega \cdot \eta + \omega} = \{J(x, y) : (\exists z)(x \in (K_i^{\omega \cdot \eta})^{(z)} \ \&\ z = |y|_{R^i_{\omega \cdot \eta + \omega}} - |b|_{R^i_{\omega \cdot \eta + \omega}})\}$$

where b is the $\omega \cdot \eta$ member of R_i. Replacing $R^i_{\omega \cdot \eta + \omega}$ by its definition over $N_{\alpha+\eta+2}$, we see that $K_i^{\omega \cdot \eta + \omega} \in N_{\alpha+\eta+2}$.

For η a limit, we have

[9] We drop for convenience the subscript 'R_i'.

$$K_i^{\omega\cdot\eta} = \{J(x, y) : (\exists S)(S \text{ an initial segment of } R^i_{\omega\cdot\eta} \;\&$$

$$y \in \text{Field } S \;\&\; x \in K_{iS}^y)\} \in N_{\alpha+\eta+1}\,.$$

Finally, since all the δ-jumps of K_i, $\delta < \omega\cdot\gamma$, and all initial segments of R, are in N, we may define

$$K_i^{\omega\cdot\gamma} = \{J(x, y) : (\exists S)(S \text{ an initial segment of } R_i \;\&$$

$$y \in \text{Field } S \;\&\; x \in K_{iS}^y)\}\,, \text{ a class in } N\,.$$

The most important fact about these definitions is that the symbols 'K_2' and 'W_2' do not appear in the formula that defines $K_{1R_1}^{\omega\cdot\gamma}$ over N, nor do the symbols 'K_1' and 'W_1' occur in the formula that defines $K_{2R_2}^{\omega\cdot\gamma}$ over N.

To make use of this fact, we now imitate Feferman in developing a small theory of transformations. Accordingly, we now abandon the model N temporarily. Our goal will be to show the following lemma.

Lemma. If $P = \langle k_1, k_2, w_1, w_2\rangle$, and $Q = \langle k_1, k_2', w_1, w_2\rangle$ and ϕ is a statement in $\mathcal{L}$ that is K_2, W_2-free (i.e., does not contain the letters K_2, W_2), then if $P \Vdash \phi$, $Q \Vdash \phi$.

Transformations

Because we are using weak forcing, our main result will be somewhat simpler than Feferman's. The relevant facts are these:

1. $\sim(P \Vdash \phi) \Leftrightarrow (\exists P' \supseteq P)(P' \Vdash \sim\phi)$

Proof. $\Leftarrow$ trivial

$\Rightarrow$ by induction on rank for limited statements, on complexity for unlimited statements. The interesting cases are as follows:

(a) If ϕ is $W_1 n\delta$ [10]

If $\sim(P \Vdash \phi)$, then $\langle n, \delta\rangle \notin w_1$. If $\langle n, \eta\rangle$ or $\langle m, \delta\rangle$ is in w_1, for some η or some m, then $P \Vdash \sim\phi$ [11]. Otherwise, let $r < \omega$ be such that $\langle r, \eta\rangle$ does not

[10] Recall that this expression abbreviates $W_1\{x_n : \text{Ord}\, x\}\{x_\delta : \text{Ord}\, x\}$.

[11] This statement perhaps requires proof: Say $\langle n, \eta\rangle \in w_i$, and $\sim(P \Vdash \sim\phi)$. Let $Q \supseteq P$, $Q \Vdash \phi$. Then $\langle n, \delta\rangle \notin w_1$, by the restriction on forcing conditions that no ordinal may appear twice. Then $Q \Vdash \{x_n : \text{Ord}\, x\} = \{x_{n'} : \text{Ord}\, x\}$, & $\{x_\delta : \text{Ord}\, x\} = \{x_{\delta'} : \text{Ord}\, x\}$ for either $n \neq n'$, $\delta \neq \delta'$. Extending Q to a complete sequence, we arrive at a model N' in which either '$n=n'$' is true, and $n \neq n'$, or '$\delta = \delta'$' is true, and $\delta \neq \delta'$. By the induction we used to show the same result in N, this is not possible.

appear in w_1 for any η. Let $w_1' = w_1 \cup \{\langle r, \delta \rangle\}$. Let P' be like P, except in having w_1' for w_1. Then $P' \supseteq P$ and $P' \Vdash \sim \phi$.

(b) If ϕ is '$n \in K_1$', where $n = J(r, s)$

If $\sim (P \Vdash n \in K_1)$ and $\sim (P \Vdash n \in K_1)$ then k_1 has fewer than s members. Let $K_{11} \ldots K_{1t}$ be the members of k_1. Let $K_{1,t+1} \ldots K_{1s}$ be distinct non-empty members of $M_\alpha \cap P(\omega)$, distinct from the members of k_1, and such that $r \notin K_{1s}$. Let $k_1' = \langle K_{11} \ldots K_{1t}, K_{1,t+1} \ldots K_{1s} \rangle$. Let P' be like P except in having k_1' for k_1. Then $P' \supseteq P$ and $P' \Vdash \sim \phi$.

(c) If ϕ is $\forall_\delta x \, \psi x$

If $\sim (P \Vdash \phi)$, then for some $c \in S_\delta$. $\sim (P \Vdash \psi c)$. Let $P' \supseteq P$ and $P' \Vdash \sim \psi c$. Then $P' \Vdash \sim \forall_\delta x \, \psi x$.

(d) If ϕ is $\sim \psi$

If $\sim (P \Vdash \sim \psi)$, then for some Q, $Q \supseteq P$, $Q \Vdash \psi$. Then $Q \Vdash \sim\sim \psi$ (a standard result), i.e., $Q \Vdash \sim \phi$.

2. $P \Vdash \phi \Leftrightarrow P \Vdash \sim\sim \phi$

Proof $\Rightarrow$ trivial

$\Leftarrow$ if $P \Vdash \sim\sim \phi$ but $\sim (P \Vdash \phi)$, then some $P' \supseteq P$ forces $\sim \phi$. But then $P' \Vdash \sim\sim\sim \phi$; but $P' \supseteq P \Rightarrow P' \Vdash \sim\sim \phi \Rightarrow \Leftarrow$

We now introduce transformations on forcing conditions:

We define a transformation τ to be a set of four functions, and four associated sets in $M_\alpha \cap P(\omega)$, $\tau_1, \tau_2, \rho_1, \rho_2, t_1, t_2, r_1, r_2$, satisfying the following conditions:

$$\tau_i : \omega \to 2$$

$$\rho_i : \omega \to \omega \text{ 1–1 onto, with } \rho_i \text{ a product of disjoint 2-cycles}$$

$$\tau_i(x) = n \Leftrightarrow J(x, n) \in t_i$$

$$\rho_i(x) = n \Leftrightarrow J(x, n) \in r_i \, .$$

If $P = \langle k_1, k_2, w_1, w_2 \rangle$ is a forcing condition,

$$k_i = K_{i1} \ldots K_{in_i} \, , \qquad w_i = \langle m_{i1}, \delta_{i1} \rangle \ldots \langle m_{ir_i}, \delta_{ir_i} \rangle \, ,$$

we define

$$\tau(K_{ij}) = \{J(r, j) : \tau_i(J(r, j)) = 1 \,\&\, r \in K_{ij} \;.\!\vee.\; \tau_i(J(r, j)) = 0 \,\&\, r \notin K_{ij}\}$$

$$\tau\langle m_{ij}, \delta_{ij}\rangle = \langle \rho_i(m_{ij}), \delta_{ij}\rangle \,.$$

Then

$$\tau(P) = \langle \tau(k_1), \tau(k_2), \tau(w_1), \tau(w_2)\rangle =$$

$$\langle\langle \tau(K_{11}) \ldots \tau(K_{1n_1})\rangle, \langle \tau(K_{21}) \ldots \tau(K_{2n_2})\rangle \,,$$

$$\langle \tau\langle m_{11}, \delta_{11}\rangle \ldots \tau\langle m_{1r_1}, \delta_{1r_1}\rangle\rangle \,,$$

$$\langle \tau\langle m_{21}, \delta_{21}\rangle \ldots \tau\langle m_{2r_2}, \delta_{2r_2}\rangle\rangle\rangle \,.$$

Notice that $\tau\tau(P) = P$.

Corresponding to each transformation τ, we define, by simultaneous induction on rank, $\tau(c)$, $\tau(\phi)$ for each constant and formula in $\mathcal{L}$.

For constants and formulas of rank $< \alpha$, we have

$\tau(c) = c$, $\quad \tau(\phi) = \phi$ [we may regard occurrences of W_i in such formulas as ill-formed] .

For constants of rank α, that is, K_1 and K_2, we define

$$\tau(K_i) = \{x_\omega : x \in K_i \,\&\, J(x, 1) \in t_i \;.\!\vee.\; x \notin K_i \,\&\, J(x, 0) \in t_i\}$$

where 't_i' is a name of t_i, of rank $< \alpha$.

For formulas of rank α: $\tau(P)$ = the result of replacing in ϕ all occurrences of K_i by $\tau(K_i)$.

For constants of rank $\delta + 1$: $\tau\{x_\delta : \phi x\} = \{x_\delta : \tau(\phi x)\}$.

For formulas of rank $\delta + 1$, or λ: $\tau(\phi)$ = the result of replacing in ϕ, first, all constants c by $\tau(c)$; next, all occurrences of $W_i c_1 c_2$ by $(\exists_\omega x)(J(c_1, x) \in r_i \,\&\, W_i x c_2)$, and all occurrences of $W y c_2$ where 'y' is a variable, by '$(\exists_\omega x)(J(y, x) \in r_i \,\&\, W_i x c_2)$.

Unlimited formulas are treated in exactly the same way.

We have a simple

Lemma. For any constant c, we have

$$P \Vdash \phi(c) \Leftrightarrow P \Vdash \phi(\tau\tau(c)) .$$

Proof. $\Rightarrow$ Let $Q \supseteq P$, $Q \Vdash \sim \tau\tau(c)$. Extend Q to a complete sequence, and consider the corresponding model N'. In N', we have $\phi(c)$ and $\sim \phi(\tau\tau(c))$, hence $c \neq \tau\tau(c)$ is true in N'. A simple induction shows this to be impossible.

$\Leftarrow$ Similarly.

Lemma. Let $\tau(P)$ be a forcing condition. For any ϕ, we have

$$P \Vdash \phi \Leftrightarrow \tau(P) \Vdash \tau(\phi) .$$

Proof. By induction on rank. Here are the interesting cases:

(1) Let ϕ be $\forall_\delta x \psi x$. Then $P \Vdash \forall_\delta x \psi x \Rightarrow (\forall c \in S_\delta) P \Vdash \psi c$. By inductive assumption, $(\forall c \in S_\delta) \tau(P) \Vdash \tau(\psi x)_x^{\tau(c)}$. Let $c \in S_\delta$. Then $\tau(c) \in S_\delta$, hence $\tau(P) \Vdash \tau(\psi x)_x^{\tau(c)}$; hence, by the lemma above, $\tau(P) \Vdash \tau(\psi x)_x^c$. Hence, $\tau(P) \Vdash \forall_\delta x \psi x$. The converse is similar.

(2) Let ϕ be $\sim \psi$. If $P \Vdash \sim \psi$, assume $\sim \tau(P) \Vdash \sim \tau(\psi))$. Let $Q \supseteq \tau(P)$, $Q \Vdash \sim\sim \tau(\psi)$. By inductive assumption, $\tau(Q) \Vdash \psi$. But $\tau(Q) \supseteq P \Rightarrow \Leftarrow$. The converse is similar.

(3) ϕ is $c \in K_i$. Let c be $\{x_\delta : \psi x_\delta\}$. Let $P \Vdash c \in K_i$. Then for some n, $P \Vdash c = n$. If $\delta < \alpha$, then c is absolute, hence all P force $c = n$, and all P force $\tau(c{=}n)$, i.e., $c = n$. Hence, $\tau(P) \Vdash \tau(c{=}n)$. If $\delta > \alpha$, then

$$P \Vdash c = n \Rightarrow P \Vdash. (\forall_\delta x)(x \in c \Leftrightarrow x \in \{x_n : \operatorname{Ord} x\})$$

$$\Rightarrow \tau(P) \Vdash. (\forall_\delta x)(x \in \tau(c) \Leftrightarrow x \in$$

$$\{x_n : \operatorname{Ord} x\}) \text{ [(inductive step)]}$$

$$\Rightarrow \tau(P) \Vdash. \tau(c) = n, \text{ i.e., } \tau(P) \Vdash \tau(c{=}n) .$$

Now we show that $\tau(P) \Vdash n \in \tau(K_i)$. We have $n = J(r, s)$ and $r \in K_{is}$ in P. Now either $P \Vdash n \in K_i \,\&\, J(n, 1) \in t_i$ or $P \Vdash n \in K_i \,\&\, J(n, 0) \in t_i$. In the first case, we have $r \in K_{is}$ in $\tau(P)$, and $J(n, 1) \in t_i$ is absolute, so $\tau(P) \Vdash n \in K_i \,\&\, j(n, 1) \in t_i$, i.e., $\tau(P) \Vdash \tau(n \in K_i)$. In the second case, we have $r \notin K_{is}$ in $\tau(P)$, and $J(n, 0) \in t_i$ is absolute, so $\tau(P) \Vdash n \notin K_i \,\&\, J(n, 0) \in t_i$, i.e., $\tau(P) \Vdash \tau(n \in K_i)$. In either case, we have $\tau(P) \Vdash \tau(c{=}n) \,\&\, \tau(n \in K_i)$, i.e., $\tau(P) \Vdash \tau(c \in K_i)$.

(4) The case where ϕ is $W_i c_1 c_2$ is quite similar; once again, the transformations on formulas have been set up expressly to allow the desired result. The remaining cases proceed by a straightforward induction.

We can now apply the results on transformations to prove the independence lemma.

Lemma. If $P \Vdash \phi$ and ϕ does not contain 'W_2', 'K_2', then if Q differs from P only in k_2, w_2, $Q \Vdash \phi$.

Proof. We show that no extension of Q forces $\sim \phi$.

We call δ a w_2-ordinal (of a forcing condition, say P_1) if $(\exists n)(\langle n, \delta \rangle \in W_2$ in $P_1)$. We call n a w_2-integer of P_1 if $(\exists \delta)(\langle n, \delta \rangle \in W_2$ in $P_1)$.

Let $Q \supseteq Q$. Choose $P^* \supseteq P$, $P^* = \langle k_1, k_2, w_1, w_2 \rangle$ with the following properties:

k_1 in $P^* = k_1$ in Q'

k_2 in P^* has as many members as does k_2 in Q'

w_1 in $P^* = w_1$ in Q'

the w_2 ordinals of P^* are precisely those which are w_2-ordinals in either P or Q'.

Choose $Q^* \supseteq Q' \supseteq Q$ with the following properties:

k_1 in $Q^* = k_1$ in Q'

k_2 in $Q^* = k_2$ in Q'

w_1 in $Q^* = w_1$ in Q'

the w_2 ordinals of Q^* are precisely those which are w_2-ordinals in P^*.

We now have $P^* \supseteq P$, $Q^* \supseteq Q$, P^* and Q^* have identical k_1 and w_1 sets, k_2 sets of equal length, and w_2 sets with exactly the same ordinals. We can now find a transformation that takes P^* to Q^*: with respect to the k_2 sets, closure properties of $M_\alpha \cap P(\omega)$ clearly allow us to find such a transformation; however, because ρ_2 in any transformation must be a product of disjoint 2-cycles, we proceed as follows:

Let τ be a transformation that leaves k_1 and w_1 fixed, and takes each K_{2i} of P^* to the K_{2i} of Q^*, and takes each w_2 integer of P^* to an integer that is not a w_2 integer of either P^* or Q^*.

Let τ' be a transformation that leaves k_1, w_1, and k_2 fixed, and takes each w_2 integer of $\tau(P^*)$ to a w_2 integer of Q^* in such a way that $\tau'\tau(P^*) = Q^*$.

We have $P \Vdash \phi$, hence $P^* \Vdash \phi$; so $\tau(P^*) \Vdash \tau(\phi)$, hence $\tau'\tau(P^*) \Vdash \tau'\tau(\phi)$. But ϕ contains neither 'K_2' nor 'W_2', so that $\tau'\tau(\phi)$ is just ϕ. Also $\tau'\tau(P^*) = Q^*$. We have, then, $Q^* \Vdash \phi$. But $Q^* \supseteq Q'$, so $\sim (Q' \Vdash \sim \phi)$. But Q' was any extension of Q, so $Q \Vdash \sim\sim \phi$, hence $Q \Vdash \phi$.

By parity of reasoning, the theorem is true with 'K_1', 'W_1' for 'K_2', 'W_2'.

Proof of the theorem

We return now to the model N. We had found two sets of integers $K_{1R_1}^{\omega\cdot\gamma}$ and $K_{2R_2}^{\omega\cdot\gamma}$, both definable over N as classes, and in such a way that the definition of $K_{1R_1}^{\omega\cdot\gamma}$ did not contain the symbols 'K_2' or 'W_2' and the definition of $K_{2R_2}^{\omega\cdot\gamma}$ did not contain the symbols 'K_1' or 'W_1'. Both $K_{1R_1}^{\omega\cdot\gamma}$ and $K_{2R_2}^{\omega\cdot\gamma}$ were $\omega\cdot\gamma$-jumps of uubs on $M_\alpha \cap P(\omega)$.

Let A be a set of integers, $A \leqslant_{\mathrm{a}} K_{1R_1}^{\omega\cdot\gamma}$, $A \leqslant_{\mathrm{a}} K_{2R_2}^{\omega\cdot\gamma}$. Then A is also a class in N, and is given by two definition, of which one is K_2, W_2-free, and the other K_1, W_1-free. It follows that there are two formulas in $\mathcal{L}$, ϕ_1 and ϕ_2 such that

$$x \in A \mathrel{.\Longleftrightarrow.} N \models \phi_1(x, K_1, W_1) \mathrel{.\Longleftrightarrow.} N \models \phi_2(x, K_2, W_2)\,,$$

where ϕ_1 is K_2, W_2-free, and ϕ_2 is K_1, W_1-free. We have

$$N \models . (\forall x)(\phi_1(x, K_1, W_1) \Longleftrightarrow \phi_2(x, K_2, W_2))\,.$$

Let $P_i \in \{P_n\}$ force $(\forall x)(\phi_1(x, K_1, W_1) \Longleftrightarrow \phi_2(x, K_2, W_2))$. Let P_i be $\langle k_1, k_2, w_1, w_2\rangle$. We introduce the symbol '$Q \underset{i}{\supseteq} P$' to mean '$Q \supseteq P$ and k_i and w_i are identical in Q and P'. Using this notation, we have the following lemma:

Lemma. $(\forall n)(n \in A \Longleftrightarrow (\exists Q)(Q \underset{2}{\supseteq} P_i \;\&\; Q \Vdash \phi_1(n, K_1, W_1)$

Proof. $\Rightarrow$ Let $n \in A$. Let $P^* \supseteq P_i$, and $P^* \Vdash \phi_1(n, K_1, W_1)$. Let Q be like P^* in k_1, w_1 and like P_i in k_2, w_2. Then, by the independence lemma, $Q \Vdash \phi_1(n, K_1, W_1)$, also $Q \underset{2}{\supseteq} P_i$.

$\Leftarrow$ Let $Q \underset{2}{\supseteq} P_i$ and $Q \Vdash \phi_1(n, K_1, W_1)$.

Assume $n \notin A$. Let $P_j \supseteq P_i$, $P_j \Vdash \sim\phi_2(n, K_2, W_2)$, $P_j \in \{P_n\}$. Let Q^* be like P_j in k_2, w_2, and like Q in k_1, w_1. Then, again by the independence lemma, $Q^* \Vdash \sim\phi_2(n, K_2, W_2)$; also $Q^* \Vdash \phi_1(n, K_1, W_1)$. But $Q^* \supseteq P_i$, so that $Q^* \Vdash \phi_1(n, K_1, W_1) \Longleftrightarrow \phi_2(n, K_2, W_2)$. This is a contradiction, hence $n \in A$.

We assume the following

Lemma. Under a suitable Gödel-numbering of forcing conditions and sentences, $\{\langle P, \phi\rangle : \phi$ is limited, or, unlimited, of complexity $< n$, and $P \Vdash \phi\}$ is a class in $M_{\alpha+\gamma}$ (i.e., $\in M_{\alpha+\gamma+1}$).

From the preceding two lemmas, we immediately deduce the

Theorem. If $A \leqslant_a K^{\omega \cdot \gamma}_{1R_1}$ and $A \leqslant_a K^{\omega \cdot \gamma}_{2R_2}$, then $A \in M_\alpha$.

Proof. $A = \{n : (\exists Q \underset{2}{\supseteq} P_i)(\psi(n, Q, \phi_1))\}$ where ψ is expressible using quantifiers restricted to $M^{2}_{\alpha+\gamma}$. Hence

$$A \in M_{\alpha+\gamma+1} \cap P(\omega) = M_\alpha \cap P(\omega) .$$

It follows that there is no $\omega \cdot \gamma$-least uub on $M_\alpha \cap P(\omega)$.

Definability of forcing

We turn now to the proof of the lemma assumed above. We shall give only a sketch – to fill in the details would be a formidable, and not particularly useful, exercise.

We shall define forcing, not for th entire language $\mathcal{L}$, but for the K_2, W_2-free part of $\mathcal{L}$ – the definition for all of $\mathcal{L}$ is essentially the same. Accordingly, forcing conditions will be of the form $\langle k, w\rangle$.

We do not specify our Gödel-numbering, except in this respect: the Gödel number of $\langle P, \phi\rangle$, where δ is the largest ordinal appearing, either as a w-ordinal, or as a bound on a quantifier (including quantifiers within constants), will be a member of $M_{\delta+1}$. The reader will easily think of several ways of meeting this requirement.

We redefine ranks of formulas as follows:

The ordinal rank of a formula is δ, where all constants in the formula are in S_δ, and all quantifiers are of the form $\forall_\eta \eta \leqslant \delta$. The complexity of a formula is $s = i + r$, where i and r are as before. The rank of a formula is (δ, s), with $(\delta, s) < (\delta', s')$ iff $(\delta < \delta') \vee (\delta = \delta' \;\&\; s < s')$. The definition of forcing as we gave it at the beginning of the chapter continues to be well-defined by induction on rank.

The following result will be of use:

If $P \Vdash \phi$, ϕ is of rank $\delta > \alpha$, and $Q \subseteq P$, then if the w set of Q contains exactly those pairs $\langle m, \eta\rangle$ in the w set of P such that $\eta < \delta$, and the k set of Q = the k set of Q = the k set of P, then $Q \Vdash P$.

Proof. If Q does not force ϕ, let $Q^* \supseteq Q$, $Q^* \Vdash \sim \phi$. Extend Q^* to a complete sequence; extend P to a complete sequence, and let N_1, N_2 be the associated models. It is apparent that $(N_1)_\delta = (N_2)_\delta$; hence $n_1 \models \phi \Leftrightarrow N_2 \models \phi$. But this is impossible.

Let the rank of a forcing condition be $\eta + 1$, where η is the largest w-ordinal in the condition. In a natural sense, by the result above, only forcing conditions of rank $< \delta$ are relevant to formulas of rank $< \delta$.

We claim that

$$\text{Forcing}_{(\delta,0)} = \{\langle P, \phi\rangle : P \text{ is of rank} \leqslant \delta \text{ and } \phi \text{ is of rank} < (\delta, 0) \text{ and}$$

$$P \Vdash \phi\} \in M_{\delta+1}, \text{ for } \alpha \leqslant \delta \leqslant \alpha + \gamma.$$

To start the induction, observe that

$$\text{Forcing}_{(\alpha,0)} = \{\langle P, \phi\rangle : P \text{ is of rank} \leqslant \alpha \text{ and } \phi \text{ is of rank} < (\alpha, 0) \text{ and}$$

$$P \Vdash \phi\} =$$

$$\{\langle P, \phi\rangle : \text{the } w\text{-set of } P \text{ is empty, and } \phi \text{ is a limited}$$

$$\text{statement in } M_\alpha, \text{ true in } M_\alpha\}.$$

This set is easily seen to be in $M_{\alpha+1}$, by the fact that

$$\{\phi : \phi \text{ is a limited statement in } M_\omega, \text{ true in } M_\omega\} \text{ is recursive,}$$

$$\text{hence} \in M_{\omega+1},$$

and by an induction essentially similar to the one we use in the remainder of the argument.

Assume the result true for δ. To show the result true for $\text{Forcing}_{(\delta+1,0)}$, we first show $\bigcup_n \text{Forcing}_{(\delta,n)} \in M_{\delta+2}$. There is one unpleasant complication: under any usual Gödel-numbering, ordinals will be autonymous – it follows that each $\langle P, \phi\rangle$, for ϕ of rank $< (\delta, n)$, will be of rank $M_{\delta+1}$[12], and hence that $\text{Forcing}_{(\delta,n)}$ will be at best of rank $M_{\delta+2}$. Our way around this problem is somewhat make-shift: for the purposes of the inductive step, we change the

[12] We can do this, by the stipulation on our Gödel-numbering.

Gödel numbering so that 'δ' has a number of rank, say $< \alpha$ [13]).

Such a numbering will allow $\text{Forcing}_{(\delta,n)}$ to appear in $M_{\delta+1}$, for each n, and $\bigcup_n \text{Forcing}_{(\delta,n)}$ to appear in $M_{\delta+2}$; there is then no difficulty in giving δ its rightful name again.

We have

$$\begin{aligned}\text{Forcing}_{(\delta,n+1)} = \{\langle P, \phi\rangle : P \text{ or rank} \leqslant \delta \;\&\; \phi \text{ of rank} < (\delta, n{+}1) \\ \&.\ \langle P, \phi\rangle \in \text{Forcing}_{(\delta,n)} \text{ or } [\phi \text{ is of rank } (\delta, n)\,, \\ \text{and } \phi \text{ is } \forall_\eta x \psi x \text{ and} \\ (\forall c \in S_\eta)(\langle P, \psi c\rangle \in \text{Forcing}_{(\delta,n)})] \text{ or } \ldots\}\end{aligned}$$

where the conditions to be spelled out are those in the definition of forcing. These conditions are finite in number, and both '$c \in S_\eta$' and $\text{Forcing}_{(\delta,n)}$ can be replaced by their definitions in M_δ, so if $\text{Forcing}_{(\delta,n)} \in M_{\delta+1}$, $\text{Forcing}_{(\delta,n+1)} \in M_{\delta+1}$.

We have also

$$\begin{aligned}\bigcup_n \text{Forcing}_{(\delta,n)} = \{\langle P, \phi\rangle : P \text{ of rank} \leqslant \delta \;\&\; (\exists n)(\phi \text{ of rank} < (\delta, n) \\ \&\; (\forall z \in M_{\delta+1})(\text{Forcing}_{(\delta,0)} \subseteq z, \text{ and } z \text{ is} \\ \text{closed under forcing for formulas of rank} \\ \leqslant (\delta, n) \Rightarrow \langle P, \phi\rangle \in z)\}\end{aligned}$$

where by 'z is closed under forcing for formulas of rank $\leqslant (\delta, n)$', we mean that if ϕ is of rank $\leqslant (\delta, n)$, and ϕ is $\forall_\eta x \psi x$ and if $(\forall c \in S_\eta)(\langle P, \psi c\rangle \in z)$, then $\langle P, \phi\rangle \in z$; etc.

It follows that $\bigcup_n \text{Forcing}_{(\delta,n)} \in M_{\delta+2}$, and this formula can be expressed in terms of our fixed Gödel-numbering.

We now have, by the result quoted above,

13) One way to do this is to treat $\forall_\delta$ as unlimited, and replace it by '$\forall$'.

$$\text{Forcing}_{(\delta+1,0)} = \{\langle P, \phi\rangle : P \text{ is of rank} \leqslant \delta + 1, \phi \text{ is of rank}$$

$$< (\delta + 1, 0) \;\&\; (\exists Q)(Q \text{ of rank} \leqslant \delta \;\&\; Q \leqslant P \;\&$$

$$\langle Q, \phi\rangle \in \bigcup_n \text{Forcing}_{(\delta,n)}\} \in M_{\delta+2}$$

At limit ordinals, we have

$$\text{Forcing}_{(\lambda,0)} = \{\langle P, \phi\rangle : P \text{ is of rank} < \lambda, \;\&\; (\exists \delta < \lambda)(\exists n)$$

$$(\text{rank } \phi = (\delta, n) \;\&\; (\forall z \in M_\lambda)(\text{Forcing}_{(\alpha,0)} \subseteq z$$

$$\text{and } z \text{ is closed under forcing for formulas of}$$

$$\text{rank} \leqslant (\delta, n) \Rightarrow (\exists Q)(Q \text{ is of rank } \delta, \text{and } Q \subseteq P$$

$$\text{and } \langle Q, \phi\rangle \in z))\} \in M_{\lambda+1}$$

For $\alpha + \gamma$, we have $\text{Forcing}_{(\alpha+\gamma,0)} \in M_{\alpha+\gamma+1}$. It follows that forcing, for limited ϕ, is definable over $M_{\alpha+\gamma}$. For unlimited statements of complexity $< n$, we do not need to change our numbering: the ordinal $\alpha + \gamma$ does not appear in these statements. We have $\text{Forcing}_{(\alpha+\gamma,n)}$ is a class in $M_{\alpha+\gamma}$, which proves the result.

We conclude with a corollary.

Corollary. There is an $\omega \cdot \beta$-least uub K on $M_\alpha \cap P(\omega)$, such that K and an $\omega \cdot \beta$-jump of K are both of rank $\alpha + \beta$.

Proof: We let $\gamma = \beta$ in the preceding theorem, and restrict our attention to the K_2, W_2-free part of $\mathcal{L}$. Assume once more a reasonable Gödel-numbering for $\mathcal{L}$ in $M_{\alpha+\beta}$[14]. Let $E_{\alpha+\beta}$ be an arithmetic copy of $M_{\alpha+\beta}$, of order $\alpha + \beta + 1$. Then the following relations are arithmetic in $E_{\alpha+\beta}$:

(1) $x \in \text{Field } E_{\alpha+\beta}$ represents a set in $M_{\alpha+\beta}$ that is the Gödel number of a forcing condition. We write 'Cond. x'.

(2) $x \in \text{Field } E_{\alpha+\beta}$ represents a set in $M_{\alpha+\beta}$ that is the Gödel number of a formula. We write 'Form. x'.

[14] That is, a numbering of which the assertions we make will be true. As will appear, we require very little.

(3) $x \in \text{Field}\, E_{\alpha+\beta}$ and $y \in \text{Field}\, E_{\alpha+\beta}$ and x represents the Gödel number of a forcing condition P and y the Gödel number of a forcing condition Q and $P \subseteq Q$. We write 'x pars y'.

(4) $x \in \text{Field}\, E_{\alpha+\beta}$ and $y \in \text{Field}\, E_{\alpha+\beta}$ and x represents the Gödel number of a formula ϕ and y represents the Gödel number of $\sim \phi$. We write '$y = \text{Neg}\, x$'.

(5) $x \in \text{Field}\, E_{\alpha+\beta}$ represents a set in $M_{\alpha+\beta}$ that is the Gödel number of a formula ϕ, ϕ limited, or unlimited, and of complexity $< p$. We write '$\text{Form}_p\, x$'.

Let $m(x, y)$ be a set-theoretic formula with constants in $M_{\alpha+\beta}$, which, when its quantifiers are restricted to $M_{\alpha+\beta}$, defines $\text{Forcing}_{(\alpha+\beta,n)}$ for a certain large n, to be specified later. That is, $m(x, y) \Longleftrightarrow x$ represents the Gödel number of a forcing condition P and y is the Gödel number of a formula ϕ, ϕ limited or unlimited with complexity $< n$, and $P \vdash \phi$.

Let $M(x, y)$ be like $m(x, y)$, except that all quantifiers are arithmetic, and restricted to Field $E_{\alpha+\beta}$[15], and '$x \in y$' is replaced by $J(x, y) \in E_{\alpha+\beta}$.

Then $M(x, y) \Longleftrightarrow x$ represents a set a in $M_{\alpha+\beta}$, and y a set b; and a is the Gödel number of a formula ϕ, limited or of complexity $< n$, and $P \Vdash \phi$.

We say that x is the mth least formula that is limited, or if unlimited, of complexity $< n$ if

$$\text{Form}_n\, x \;\&\; (\exists_m y)(\text{Form}_n\, y \;\&\; y \leqslant x)\,.$$

We now use M and the other predicates arithmetical in $E_{\alpha+\beta}$ to define a 'least' sequence of forcing conditions:

$$
\begin{aligned}
x \in \mathcal{F} \Longleftrightarrow (\exists y)(\exists m)\ \{&\text{Cond.}\, x \;\&\; \text{Cond.}\, y \;\&\; x \text{ pars } y \;\&\\
&\exists \text{ a sequence of integers } x_1 \ldots x_m \text{ such that Cond.}\, x_i,\\
&\&\; x_i \text{ pars } x_{i+1} \;\&\; x_m = y \;\&\\
&(x_1 = (\mu w)(\exists v)(v \text{ is the 1st least formula} \;\&\; \text{Cond.}\, w \;\&.\\
&\qquad M(w, v) \vee M(w, \text{Neg}\, v)) \;\&\\
&((\forall i)_{1 < i \leqslant m}(x_i = (\mu w)(\exists v)(v = \text{the } i\text{th least formula}\\
&\&\; x_{i-1} \text{ pars } w \;\&.\, M(w, v) \vee M(w, \text{Neg}\, v))\}\,.
\end{aligned}
$$

15) We assume m an unlimited formula. If m is limited, or has bounded quantifiers, then $\forall_\delta x$ must be replaced by $(\forall x)\,(x \in M_\delta \Rightarrow \ldots)$.

The restriction to formulas that if unlimited are of complexity $< n$ is merely for the purpose of defining $\mathcal{F}$ arithmetically over $E_{\alpha+\beta}$. $\mathcal{F}$ corresponds naturally to a sequence of forcing conditions that is complete in this sense:

Letting $\{P_i\}$ be the sequence determined by $\mathcal{F}$, $\{P_i\}$ fixes a unique N, and for limited ϕ in $\mathcal{L}$, or unlimited ϕ of complexity $< n$, we have $N \models \phi \Longleftrightarrow (\exists P_i)(P_i \Vdash \phi) \Longleftrightarrow (\exists x \in \mathcal{F})(M(x, b))$ where b represents the Gödel number of ϕ. As before, we have K a uub on $M_\alpha \cap P(\omega)$ and we have a formula $\phi_1(x)$ in $\mathcal{L}$ such that $\{x : N \models \phi_1(x)\}$ is $K_W^{\omega \cdot \beta}$. We choose n to be greater than the complexity of ϕ_1.

Let ρ be a function arithmetic in $E_{\alpha+\beta}$, such that if x represents in $E_{\alpha+\beta}$ the Gödel number in $M_{\alpha+\beta}$ of a constant c in $\mathcal{L}$, then $\rho(x)$ represents in $E_{\alpha+\beta}$ the Gödel number in $M_{\alpha+\beta}$ of the formula $\phi_1(c)$.

Let $L = \{m : (\exists x \in \mathcal{F})(M(x, \rho(m))\}$. Then L is the set of members of $E_{\alpha+\beta}$ which represent Gödel numbers in $M_{\alpha+\beta}$ of members of $K_W^{\omega \cdot \beta}$.

It is easy to define $K^{\omega \cdot \beta}$, and therefore K, arithmetically in L; the definition depends somewhat on the specific Gödel-numbering used. Perhaps the easiest way is to choose a specific set of integer-names in $\mathcal{L}$ – names of the form $\{x_n : \text{Ord}\, x\}$ seem easiest. Then, under a reasonable Gödel-numbering, the set of all Gödel-numbers in $M_{\alpha+\beta}$ of such names is a set in, say $M_{\omega \cdot \alpha}$. Then, if r in Field $E_{\alpha+\rho}$ represents that set, we are interested, first in

$$R = \{x \in \text{Field}\, E_{\alpha+\beta} : J(x, r) \in E_{\alpha+\beta}\},$$

and then in

$$K^{\omega \cdot \beta} = \{y : (\exists x \in k)(x \text{ represents } y)\}.$$

We have

$$K^{\omega \cdot \beta} \leqslant_a R \leqslant_a E_{\alpha+\beta}, \quad \text{also } K \leqslant_a K^{\omega \cdot \beta} \leqslant_a E_{\alpha+\beta}.$$

It follows that all $\omega \cdot \beta$-least uubs on $M_\alpha \cap P(\omega)$ are in $M_{\alpha+\beta+1}$, since they are all $\leqslant_a K^{\omega \cdot \beta} \leqslant_a E_{\alpha+\beta} \in M_{\alpha+\beta+1}$. We have then a fairly satisfactory intrinsic characterization of the hierarchy of constructible sets of integers.

The complete hierarchy may be described as follows:

deg(0) = the degree of arithmetical sets.

$\deg(\alpha+1) = \deg(A^{(\omega)})$ for $A \in \deg(\alpha)$.

$\deg(\lambda)$ = the complete degree of the δ-least uniform upper bounds on the recursive union of $\deg(\alpha)$, $\alpha < \lambda$, where δ is the least ordinal such that there exists such a uniform upper bound.

Alternatively, one may wish to index the degrees by their levels in the *M*-hierarchy, in which case deg(0), above, is best called deg(ω), and deg(λ), deg($\lambda + \eta$), where $\omega \cdot \eta = \delta$. deg($\lambda + \beta$) will then be undefined for $0 < \beta < \eta$.

References

[1] G.S. Boolos, The hierarchy of constructible sets of integers, Ph.D. thesis, Massachusetts Institute of Technology, Department of Humanities, Cambridge, Mass. (1966).
[2] G. Boolos and H. Putnam, Degrees of unsolvability of constructible sets of integers, J. Symbolic Logic 33 (1968) 4.
[3] R. Boyd, Ph.D. thesis, Massachusetts Institute of Technology, Department of Humanities, Cambridge, Mass. (forthcoming).
[4] P. Cohen, Set theory and the continuum hypothesis (W.A. Benjamin, New York, 1966).
[5] H. Enderton, Hierarchies in recursive function theory, Trans. Am. Math. Soc. 111 (1964) 437–471.
[6] S. Feferman, Some applications of the notions of forcing and generic sequences, Fund. Math. 55 (1965) 325–345.
[7] K. Gödel, Consistency-proof for the generalized continuum hypothesis, Proc. Acad. Sci. U.S. 25 (1939) 220–224.
[8] H. Putnam, A note on constructible sets of integers, Notre Dame J. Formal Logic 4 (1963) 270–273.
[9] Hartley Rogers Jr., Recursive functions and effective computability (McGraw-Hill, New York, 1967).
[10] C. Spector, Hyperarithmetic quantifiers, Fund. Math. 48 (1960) 313–319.

PART V

CONTRIBUTED PAPERS

STANDARD AND NON-STANDARD METHODS IN UNIFORM TOPOLOGY

J.E. FENSTAD and A.M. NYBERG

The purpose of this note is to discuss the relationship between standard and non-standard concepts in uniform topology. We have, in particular, been interested in the case where the space carries both an algebraic and uniform structure, e.g. as in the case of a topological group.

We assume that the reader is familiar with the standard theory as presented e.g. in [3] and the non-standard theory as presented in [6].

Recently two contributions to the non-standard approach to uniform topology have been published [4, 5]. There seems to be little overlap with the present discussion. We believe that our emphasise on the notion of a *bounded* point leads to a very clear understanding of the exact relationship between standard and non-standard concepts.

1. Bounded points

Let $(X,\mathcal{U})$ *be a uniform space and let* $^*X$ be a non-standard extension of X. By the *s-topology* on $^*X$ we understand the topology defined by the neighbourhood systems

$$N_x = \{^*U(x) \mid U \in \mathcal{U}\}, \qquad x \in {}^*X .$$

Remark. The uniformity $\mathcal{U}$ can also be defined by an associated family of pseudo-metrics, $\mathcal{D}$. And it is easily shown that the *s*-topology is generated by the following family of open sets in $^*X$,

$$\{S_d(x, r) \mid d \in \mathcal{D}, \qquad x \in {}^*X, \qquad r \in R^+\},$$

where $S_d(x, r) = \{y \mid \mathrm{st}\, d(x, y) < r\}$.

The *monad*, $\mu(x)$, of a point $x \in {}^*X$ is defined to be the set

$$\mu(x) = \bigcap_{U \in \mathcal{U}} {}^*U(x) .$$

Remark. Using the associated family of pseudo-metrics we see that $\mu(x) = \{y \in {}^*X \mid d(x, y) \simeq 0 \text{ for all } d \in \mathcal{U}\}$.

One easily notes that the relation $x \in \mu(y)$ is an equivalence relation on ${}^*X$ which we denote by $x \simeq y$. The space X is Hausdorff iff every monad contains at most one standard point.

We call $x \in {}^*X$ *near-standard* if x belongs to the monad of a standard point.

In the sequel we assume that X is Hausdorff in the associated topology.

Definition. The set B_X of *bounded points* of ${}^*X$ is defined to be the closure of X in ${}^*X$ with respect to the s-topology,

$$B_X = \mathrm{cl}_s X .$$

We shall obtain a characterization of B in terms of Cauchy z-ultrafilters on X.

It is known (see e.g. [2]) that to every $x \in {}^*X$ there is associated a unique z-ultrafilter $\mathcal{F}_x$ on X, and to every z-ultrafilter $\mathcal{F}$ on X there corresponds a point $x \in {}^*X$ such that $\mathcal{F} = \mathcal{F}_x$. ($\mathcal{F}_x$ is the unique z-ultrafilter which extends the prime z-filter $\mathcal{F}'_x = \{F \subseteq X \mid x \in {}^*F \text{ and } F \in Z(X)\}$.)

Proposition. The set of bounded points consists exactly of those $x \in {}^*X$ such that the associated z-ultrafilter is Cauchy, i.e.

$$B_X = \{x \in {}^*X \mid \mathcal{F}_x \text{ is a Cauchy } z\text{-ultrafilter}\} .$$

We sketch the proof. Let $x \in \mathrm{cl}_s X$, we have to show that $\mathcal{F}_x$ is Cauchy. Pick any $U \in \mathcal{U}$ and choose a closed $V \in \mathcal{U}$ such that $V \circ V \subseteq U$ and such that ${}^*V(x) \cap X \neq \emptyset$. Let $p \in {}^*V(x) \cap X$. We now observe that $V(p) \in \mathcal{F}_x$, since $x \in {}^*V(p)$ and we may assume that $V(p) \in Z(X)$. And obviously $V(p) \times V(p) \subseteq U$, which shows that $\mathcal{F}_x$ is Cauchy.

Conversely, let $\mathcal{F}_x$ be Cauchy. Let $U \in \mathcal{U}$ and pick a (symmetric) $U_1 \in \mathcal{U}$ such that $U_1 \circ U_1 \subseteq U$. Since $\mathcal{F}_x$ is Cauchy, there is a $V \in \mathcal{F}_x$ such that $V \times V \subseteq U_1$. It is now possible to pick an $x_0 \in \mu(x)$ such that $x_0 \in {}^*V$. This shows that $(p, x_0) \in {}^*U_1$, hence $p \in {}^*U(x) \cap X$.

Remark. We mention here the following result: X is *complete* iff every bounded point is near-standard. This generalizes the fact that compactness is equivalent to every (non-standard) point being near-standard.

2. The commutative diagram

We noted above that if $x \in {}^*X$, then $\mu(x)$ contains at most one standard point. If p is a standard point in $\mu(x)$, then p is uniquely determined. We call p the *standard part* of x, and write $p = \mathrm{st}(x)$.

Let γX denote the completion of X. Then X is imbedded in both ${}^*X$ and γX and there is a surjection $\pi : B_X \to \gamma X$ such that the following diagram is commutative:

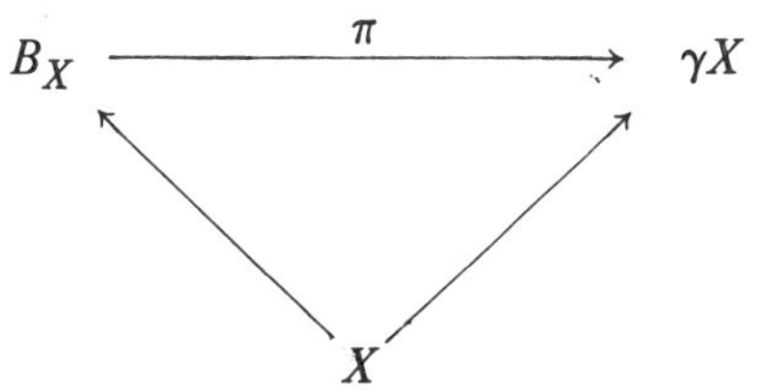

The definition of π is immediate: $\pi(x), x \in B$, is the equivalence class of the Cauchy z-ultrafilter $\mathcal{F}_x$ in γX.

Proposition. Let $(X, \mathcal{U})$ and $(Y, \mathcal{V})$ be uniform spaces, and assume that $(Y, \mathcal{V})$ is complete. Let $f : X \to Y$ be a uniformly continuous map, hence f has an extension to a continious map $\hat{f} : \gamma X \to Y$. The following identity is valid for all $x \in B_X$:

$$\mathrm{st}({}^*f(x)) = \hat{f}(\pi(x)) .$$

As a preliminary remark toward a sketch of the proof, we note that if $f : X \to Y$ is uniformly continuous, then $f(\mu(x)) \subset \mu(f(x))$ for all $x \in {}^*X$. This follows immediately from the definition of uniform continuity, and, in fact, characterizes this notion.

Let $x \in B_X$ be given. Since $\mathcal{F}_x$ is Cauchy and Y is complete, we see that the z-filter $f^{\#}(\mathcal{F}_x) = \{Z \in Z(Y) \mid f^{-1}(Z) \in \mathcal{F}_x\}$ converges toward the point $\hat{f}(\pi(x))$ in Y. Pick a point x_0 such that $x_0 \in {}^*F$. for all $F \in \mathcal{F}_x$. Then $x_0 \in \mu(x)$ and ${}^*f(x_0) \in {}^*Z$, for all $Z \in f^{\#}(\mathcal{F}_x)$. Since $f^{\#}(\mathcal{F}_x)$ converges to

$\hat{f}(\pi(x))$, we see that

$$*f(x_0) \in \bigcap_{Z \in f^{\#}(\mathcal{F}_x)} *Z \subseteq \bigcap_{V \in \mathcal{V}} *V(\hat{f}(\pi(x))) = \mu(\hat{f}(\pi(x))) .$$

The uniform continuity of f now implies that $*f(x) \in \mu(*f(x_0))$. Hence $*f(x) \in \mu(\hat{f}(\pi(x)))$, which exactly means that $\mathrm{st}(*f(x)) = \hat{f}(\pi(x))$.

Remark. This generalizes a result in [2] where we considered the relationship between $*X$ and the Stone-Čech compactification βX of X.

In the first section of this note we showed that the set B_X, which was defined as the s-closure of X in $*X$, is the set of points x such that $\mathcal{F}_x$ is a Cauchy z-ultrafilter on X. From the observations of this section it follows that

$$\mathrm{st}\, d(x, y) = \hat{d}(\pi(x) , \quad \pi(y)) ,$$

for all bounded x and y and all d in the associated family of pseudo-metrics. This identity "explains" why the s-topology as defined by A. Robinson is the appropriate setting for discussing the completion of metric spaces. It also implies that $x \simeq y$ (i.e. $x \in \mu(y)$) iff $\pi(x) = \pi(y)$.

These observations taken together shows that we obtain the completion of X as the set of bounded points modulo monads. Implicit in our observations is also the fact that this non-standard approach is nothing but the "lifting" from βX to $*X$ of a well-known procedure (see e.g. [3]).

But something may be gained. A. Robinson [7] constructed R as Q_f/Q_i, where Q_f is nothing but the bounded points of Q and Q_i are the infinitesimals. The important point here is that the algebraic operations of Q extends to Q_f, i.e. Q_f is a "nice" substructure of $*Q$ which itself is an elementary extension of Q. We will return to this point in the next section.

3. Extending maps from X to γX

Let as above $(X, \mathcal{U})$ and $(Y, \mathcal{V})$ be uniform spaces. and assume that $(Y, \mathcal{V})$ is complete. It is known in the metric case (and fairly straight forward to extend to the uniform case) that a map $f : X \to Y$ is *continuous*, iff f is s-continuous for all standard x, or, equivalently, iff $f(\mu(x)) \subseteq \mu(f(x))$ for all standard x. Here "standard" can be replaced by "near-standard".

And a map $f: X \to Y$ is *uniformly continuous*, iff f is s-continuous for all $x \in {}^*X$, or, equivalently, iff $f(\mu(x)) \subseteq \mu(f(x))$ for all $x \in {}^*X$.

In this section we characterize in a similar way the property of having a continuous extension from X to γX.

Proposition. Let $(X, \mathcal{V})$ *and* $(Y, \mathcal{U})$ *be uniform spaces, and assume that* $(Y, \mathcal{V})$ *is complete. Let f be a map from X to Y. The following three conditions are equivalent.*

(i) *f has a continuous extension to γX.*

(ii) *f is s-continuous for all bounded $x \in {}^*X$.*

(iii) $f(\mu(x)) \subseteq \mu(f(x))$ *for all bounded* $x \in {}^*X$.

We first prove that (i) implies (ii) and (iii). Let $\hat{f}$ be the continuous extension of f to γX. We consider X as a dense subspace of γX and we work with non-standard extensions ${}^*X \subseteq {}^*\gamma X$. The map $\hat{f}$ is continuous from γX to Y. Hence ${}^*\hat{f}$ maps near-standard points of ${}^*\gamma X$ to near-standard points in ${}^*Y$. And ${}^*\hat{f}(\hat{\mu}(x)) \subseteq \mu({}^*\hat{f}(x))$, for all near-standard $x \in {}^*\gamma X$. (Here $\hat{\mu}$ denotes the monad in γX.)

First, note that near-standard is the same as bounded in ${}^*\gamma X$ since γX is complete.

Next, note that X is uniformly imbedded as a dense subset of γX. This means, in particular, that $\mu(x)$, the monad of a point $x \in {}^*X$, is the restriction of the monad $\hat{\mu}(x)$ to ${}^*X$. (This follows from the correspondence between *entourages* in X and γX.) Further, a point $x \in {}^*X$ which is bounded in ${}^*X$ remains bounded in ${}^*\gamma X$.

Putting these things together we see that (i) implies (ii) and (iii). As an example we verify (iii). Thus let x be a bounded point in ${}^*X$. Hence x is bounded, and therefore near-standard in γX. The continuity of $\hat{f}$ then implies that ${}^*\hat{f}(\hat{\mu}(x)) \subseteq \mu({}^*\hat{f}(x))$. Let $y \in \mu(x)$. Then $y \in \hat{\mu}(x)$, hence ${}^*\hat{f}(y) \in \mu({}^*\hat{f}(x))$. But ${}^*f = {}^*\hat{f} \,|\, {}^*X$, hence ${}^*f(y) \in \mu({}^*f(x))$, which was to be proved.

Remark. Since $\hat{f}$ maps near-standard points of ${}^*\gamma X$ to near-standard points in ${}^*Y$, it follows that f maps bounded points in ${}^*X$ to bounded points in ${}^*Y$.

The s-continuity of f on B_X implies immediately that $f(\mu(x)) \subset \mu(f(x))$ for any $x \in B_X$. In fact, let x be bounded and consider an s-open neighbourhood V' of $f(x)$ in ${}^*Y$. There then exists an s-open neighbourhood V of x

such that $f(V \cap B_X) \subseteq V'$. As $\mu(x)$ is the intersection of all s-neighbourhoods of x and $\mu(x) \subseteq B_X$, we see that $f(\mu(x)) \subseteq V'$. Since V' is arbitrary, the result follows.

For the final part of the proof assume that $f(\mu(x)) \subseteq \mu(f(x))$ for all bounded x. Let $\mathcal{F}$ be an arbitrary Cauchy z-ultrafilter on X. We have to show that $f^{\#}\mathcal{F} = \{Z \in Z(Y) \mid f^{-1}(Z) \in \mathcal{F}\}$ is a Cauchy z-filter on Y. Note first that $\mathcal{F} = \mathcal{F}_x$ for some bounded point x.

Consider the family $f(\mathcal{F}_x) = \{f(F) \mid F \in \mathcal{F}_x\}$. One sees that it suffices to show that for all $V \in \mathcal{V}$ there is some $F' \in f(\mathcal{F}_x)$ such that $F' \times F' \subseteq V$.

We prove this by contradiction. Assume not: Then there exists a $V \in \mathcal{V}$ such that the following (standard) sentence is true:

$$(\forall F)\,(F \in \mathcal{F} \to (f \times f)\,(F) \not\subset V)\,.$$

A familiar type non-standard argument now gives an *internal* set $F_0 \in {}^*\mathcal{F}$ such that $F_0 \in {}^*\mathcal{F}$ such that $F_0 \subseteq \mu(x)$ and such that $(f \times f)\,(F_0) \not\subset {}^*V$.

Pick a symmetric $W \in \mathcal{V}$ such that $W \circ W \subseteq V$. It follows that $(f \times f)\,(F_0) \not\subset {}^*W \circ {}^*W$. And since $F_0 \subseteq \mu(x)$, we further obtain that $(f \times f)\,(\mu(x)) \not\subset {}^*W \circ {}^*W$. We may then choose points $u, v \in \mu(x)$ such that $(f(u), f(v)) \notin {}^*W \circ {}^*W$, which easily implies that $f(\mu(x)) \not\subset {}^*W(f(x))$. However, $\mu(f(x)) \subset {}^*W(f(x))$, and the result follows.

Remark. We note the following supplementary characterization of the set of bounded points: a point x is bounded, iff for all uniformly continuous $f: X \to R$, $f(x)$ is bounded, or, equivalently, iff for all $f: X \to R$ which has a continuous extension to γX, $f(x)$ is bounded.

It remains to show that if $x \in {}^*X - B_X$, then there is some uniformly continuous $f: X \to R$ such that $f(x)$ is not bounded. Note that X can be imbedded into a product R^I, where $I = \{f: X \to R \mid f$ is uniformly continous$\}$. If $x \notin B$, then $\mathcal{F}_x$ is a z-ultrafilter which is not Cauchy. Hence there must be some $f \in I$ such that $\mathrm{pr}_f(\mathcal{F}_x)$ is a base for a z-ultrafilter which is not Cauchy. But this means that the z-ultrafilter $\mathcal{F}_{f(x)}$ cannot be Cauchy, hence $f(x)$ is not bounded.

Let X now carry both an algebraic and uniform structure, e.g. let X be a topological group. When is γX an algebraic structure of the same kind as X? The answer must be related to the "degree of continuity" of the algebraic operations. Continuity is known to be too weak and uniform continuity too strong. It turns out that s-continuity of the algebraic operations on the set of bounded points of ${}^*X$ is the right kind of requirement.

For simplicity assume that X as an algebraic structure has certain operations $f_1, ..., f_k$ and that the axioms which X is supposed to satisfy are open, positive sentences (e.g. let X be a group with both group multiplication and inverse operation). We must now consider maps $f_r : X^n \to \gamma X$. (Note that monads and bounded points commute with finite cartesian products, hence our previous results apply.) The *s*-continuity of the operations f_r implies that each f_r maps bounded points to bounded points. Hence the operations f_r, which can be extended to $^*X$ by general model-theoretic considerations, can be restricted to the bounded points. Since *s*-continuity means that $f_r(\mu(b_1), ..., \mu(b_n)) \subseteq \mu(f(b_1, ..., b_n))$, the operations f_r can be further defined on $\gamma X \simeq B/\mu$.

And since γX is a homomorphic image of a subsystem of $^*X$, which itself is an elementary extension of X, the syntactic form of the axioms implies that they are also valid in γX. Thus in this case the *s*-continuity of the algebraic operations ensures that yX is an algebraic structure of the same kind as X, and that the extended operations are continuous in the associated topology. And in this case the condition of *s*-continuity is also necessary.

This includes known results on topological groups and rings. When the axioms are of a more complicated syntactic character, the situation becomes more involved. It is perhaps somewhat doubtful whether a useful general theorem can be stated.

References

[1] J.E. Fenstad, A note on "standard" versus "Non-standard" topology, Indag. Math. 29 (1967) 378–380.

[2] J.E. Fenstad, Non-standard models for arithmetic and analysis, in: Proc. XVth Scandinavian Math. Congress (Springer-Verlag), forthcoming.

[3] L. Gillman and M. Jerison, Rings of continuous functions (Van Nostrand, Princeton, 1960).

[4] W.A.J. Luxemburg, A general theory of monads, in: Applications of model theory, to algebra, analysis, and probability, ed. W.A.J. Luxemburg (Holt, Rinehart and Winston, New York, 1969) pp. 18–86.

[5] M. Machover and J. Hirschfeld, Lectures on non-standard analysis, Springer Lecture Notes (Springer-Verlag, 1969).

[6] A. Robinson, Non-standard analysis (North-Holland Publ. Co., Amsterdam, 1966).

[7] A. Robinson, Non-standard arithmetics, Bull. Am. Math. Soc. 73 (1967) 818–843.

ALGORITHMIC PROCEDURES, GENERALIZED TURING ALGORITHMS, AND ELEMENTARY RECURSION THEORY

Harvey FRIEDMAN
Department of Philosophy, Stanford University, Stanford, Calif., USA

Introduction

Turing's analysis (as improved by Kleene and others) gives a mathematical analysis of configurational computations: one operates only on a fixed finite set of symbols which at each stage are arranged in a finite configuration in which the same symbol may occur in many different positions. (In Turing's analysis the only configurations are linear orderings, a point which is taken up in Section 3.)

Suppose $\{a_1, ..., a_n\}$ are the symbols used in some Turing algorithm. Let $\{a_1, ..., a_n\}^*$ be the set of all finite strings of elements of $A = \{a_1, ..., a_n\}$. Then with this Turing algorithm we may associate an *algorithmic procedure for evaluating* a partial function on $\{a_1, ..., a_n\}^*$, which we now describe.

At any stage in Turing computation there will be a finite number, say m, squares of tape in existence, each of which has an a_i placed on it (a conventional blank symbol will be assumed to be among $\{a_1, ..., a_n\}$), and the reading head splits these i elements into two finite strings α, β whose lengths add up to m. In the associated algorithmic procedure we will have 3 registers r_1, r_2, r_3. In the first register r_1 we will place the α's above that arise; in the second we will place the β's; r_3 will be left empty until we start to give the output. When the procedure terminates, the contents of r_3 will be the output. The instructions for the algorithmic procedure correspond to the Turing instructions: the instruction one is at corresponds to the state one would be in. When no instruction applies the procedure terminates, or equivalently, the instructions are so arranged that the command "terminate" is allowed and some instruction always applies. Each register at any time is either empty or contains exactly one lement of $\{a_1, ..., a_n\}^*$.

The instructions for this algorithmic procedure will be built up of basic *operations* and basic *tests*. For each i, appending a_i at the end, or at the front of a string is a basic operation. Deleting the front symbol, or the end symbol of a string is a basic operation. Testing whether a string is empty is a basic test. For each i, testing whether a_i is the front symbol, or testing whether a_i is the end symbol of a string is a basic test. Hence we associate, to Turing's analysis, relational structures $M = (\{a_1, ..., a_n\}^*, f_1, ..., f_n, g_1, ..., g_n, d_1, d_2, R_1, ..., R_n, S_1, ..., S_n, E)$, where $0 \leqslant n$, $f_i(s) = sa_i$, $g_j(s) = a_j s$, $d_1(a_i s) = d_2(sa_j) = s$, $d_1(\langle\ \rangle) = d_2(\langle\ \rangle) = \langle\ \rangle$, $R_i(x) \equiv (\exists s)\,(x = a_i s)$, $S_j(x) \equiv (\exists s)\,(x = sa_i)$, $E(x) \equiv x = \langle\ \rangle$. The partial functions of many arguments evaluated by some algorithmic procedure applied to M will be the partial recursive functions on $\{a_1, ..., a_n\}^*$ in the sense of ordinary recursion theory.

We have just dexcribed an algorithmic procedure built up from basic operations and tests *applied* to a particular structure, M. In an algorithmic procedure (unapplied) one does not specify the basic operations and tests, or even the domain of application. Thus each algorithmic procedure, when applied to a structure defines a partial function on the domain of that structure.

The difference between configuration computations and algorithmic procedures is twofold. Firstly, in configurational computations the objects are symbols, whereas in algorthimic procedures the objects operated on are unrestricted (or unspecified). Secondly, in configurational computations at each stage one has a finite configuration whose size is not restricted before computation. On the other hand in algorithmic procedures one fixes beforehand a finite number of registers to hold the objects. Thus for some n, at each stage one has at most n objects.

Stepherdson and Sturgis [4], through use of their register machines, analyzed algorithmic procedures applied to (ω, S, P, R), where $S(x) = x + 1$, $P(x) = x - 1$ if $x \neq 0$; 0 if $x = 0$, $R(x) \equiv x = 0$, although they did not think of their work in this connection. Registers which are allowed to hold (and only hold) integers and which can be operated on by use of S, P, R are called *counting registers*. (We use a modification of [4] used in [1].) We consider, in general, algorithmic procedures with counting. These may, when applied to a structure M with domain D, allow auxiliary counting registers in addition to the registers restricted for use in holding elements of D. Of course, this distinction between algorithmic procedures with and without counting is hidden when one applies them to structures corresponding to ordinary recursion theory, but in the general context the distinction is apparent.

Note that in general the equality relation (on the domain of the unspecified structure) is not included as a basic test. For this reason one considers both partial functions and partial relations. A partial relation is something which is, at any appropriate arguments, either undefined or true or false.

Algorithmic procedures for evaluating partial relations differ from those for evaluating partial functions only in the way the output is interpreted. As in evaluation of partial functions, lack of termination corresponds to being undefined. If there is a termination then the output register being empty corresponds to falsity; the output register being nonempty corresponds to truth. Partial functions cannot be considered here as reduced to partial relations since partial characteristic relations of those partial functions obtained by applying algorithmic procedures (with or without counting) need not be obtained by applying algorithmic procedures (with or without counting), since this requires a test for equality. For the same reason, relations cannot be considered here as reduced to partial functions since partial relations whose partial characteristic functions are obtained by applying algorithmic procedures (with or without counting) need not be obtained by applying algorithmic procedures (with or without counting).

In Section 1 we present in detail our mathematical analysis of algorithmic procedures both with and without counting. This is done by describing the action that results from applying our formalized algorithmic procedures to a relational structure at particular arguments (imputs) taken from the domain of the structure, and what constitutes the output if termination occurs.

We then introduce a third concept: the generalization of configurational computation (on symbols) to configurational computations on entities. In Section 1 we formalize this generalized configurational computability via generalized Turing algorithms. (Just as in Turing's analysis, our configurations are linear orderings.)

We then introduce effective definitional schemes. These are, essentially, definitions by infinite cases, where the cases are given by atomic formulae and negations, and the definition has recursively enumerable structure. In Section 1 we show that these effective definitional schemes provide a mathematically perspicuous characterization of the generalized Turing algorithms, since every generalized Turing algorithm is equivalent to some effective definitional scheme and vice versa, in the sense that they define the same partial function (or partial relation) when applied to the same structure.

In Section 1 we show that not all generalized Turing algorithms are equivalent to some algorithmic procedure with counting, and in turn, not every algorithmic procedure with counting is equivalent to some algorithmic procedure without counting.

Given a relational structure one considers the collection of all partial functions (partial relations) given by some effective definitional scheme, or equivalently, some generalized Turing algorithm, modified so that any element of the domain is allowed to be used as a constant. These collections are used in

Section 2 to illustrate important features of elementary recursion theory which often remain hidden (the extra constants are needed in order to have all the constant functions). In particular one must analyze the computational, or better: procedural, aspects of the proofs in elementary recursion theory in order to find meaningful conditions on structures which insure that the various theorems lift. The theorems of elementary recursion theory considered are basically concerned with the equivalence of various notions of recursively enumerable and recursive sets in terms of domains of partial functions and relations and ranges of partial and total functions, and enumeration theorems.

In Section 3 we suggest future research.

In many cases only the intuitive idea of the proof is conveyed. The emphasis here is on presenting basic notions and ideas; this is not the place to look for delicate mathematical arguments.

Section 1

We will be using both partial functions and partial relations. If f, g are n-ary partial functions then $f(x_1, ..., x_n) \simeq g(x_1, ..., x_n)$ and $f(x_1, ..., x_n) = g(x_1, ..., x_n)$ respectively mean that either $f(x_1, ..., x_n)$ and $g(x_1, ..., x_n)$ are both undefined or both defined and equal; $f(x_1, ..., x_n)$ and $g(x_1, ..., x_n)$ are both defined and equal. If R, S are n-ary partial relations then $R(x_1, ..., x_n) \cong S(x_1, ..., x_n)$ and $R(x_1, ..., x_n) \equiv S(x_1, ..., x_n)$ respectively mean that either $R(x_1, ..., x_n)$ and $S(x_1, ..., x_n)$ are both undefined or both are defined and true or both are defined and false; $R(x_1, ..., x_n)$ and $S(x_1, ..., x_n)$ are both defined and true or both defined and false.

Definition 1.1. A relational structure M is a system $(D, c_0, ..., c_k, f_0, ..., f_r, R_0, ..., R_m)$ such that $c_0, ..., c_k \in D$, each f_i is a function of many arguments on all of D, each R_j is a relation of many arguments on all of D, $0 \leqslant k, n, m$, $D \neq \emptyset$. The relational type of M is given by the sequence $((k), (a_0, ..., a_r), (b_0, ..., b_m))$, where each f_i is a_i-ary, each R_j is b_j-ary.

Definition 1.2. A procedural type is a sequence (n, e, p, q, a, b, c), where (a, b, c) is a relational type; $1 \leqslant n; 0 \leqslant e, p; 1 \leqslant q$.

Definition 1.3. If $\alpha = (a, b, c) = ((k), (a_0, ..., a_r), (b_0, ..., b_m))$ is a relational type and (n, e, p, q, a, b, c) is a procedural type then the α-symbols are the symbols $d_0, ..., d_k, F^0_{a_0}, ..., F^r_{a_r}, A^0_{b_0}, ..., A^m_{b_m}$. The (n, e, p, q, a, b, c)-atoms are the expressions $A^i_j(r_{e_1}, ..., r_{e_j})$ and $\sim A^i_j(r_{e_1}, ..., r_{e_j})$, where A^i_j is an α-symbol, and each e_t has $1 \leqslant e_t \leqslant n + 1 + e$.

The intention is that r_e refers to the (contents of the) eth register.

Each formalized algorithmic procedure (fap) will have a procedural type (n, e, p, q, a, b, c), interpreted as follows:

(a) n is the number of arguments on which the fap operates

(b) $r_1, ..., r_n$ are called the imput registers and r_{n+1} is called the output register

(c) $r_{n+2}, ..., r_{n+1+e}$ are called the auxiliary registers

(d) $r_{n+1+e+1}, ..., r_{n+1+e+p}$ are called the counting registers

(e) the fap has exactly q instructions

(f) the fap is intended for application to structures with relational type (a, b, c).

The q instructions must be numbered from 1 to q. Each instruction must be of the form "if E holds then do B; then go to instruction i", or of the form "if E holds then do B; then terminate", where E is either

(a) an (n, e, p, q, a, b, c)-atom or

(b) "r_i is empty" or "r_i is not empty", for some register r_i or

(c) "r_i has 0" or "r_i does not have 0", for some counting register r_i or

(d) $0 = 0$;

B is either

(a) "nothing" or

(b) "clear r_i" or

(c) "alter r_i so that its contents are the same as the contents or r_j", where r_i is a counting register if and only if r_j is a counting register or

(d) "add 1 to r_i" or "delete 1 from r_i if it contains something other than 0; otherwise do nothing", where r_i is a counting register

(e) "alter r_i so that its content is 0" or "alter r_j so that its content is d_j", where r_i is a counting register, r_j is not a counting register

(f) "replace r_i by the result of applying F^j_l to the contents of $(r_{t_1}, ..., r_{t_l})$ if all l registers are nonempty; otherwise do nothing", where r_i as well as each r_t are not counting registers and $l = a_j$.

Thus the fap will be given by (finite sets of) strings of symbols.

The action of a fap of type (n, e, p, q, a, b, c) applied to a structure M of relational type (a, b, c) at arguments $x_1, ..., x_n$ in the domain D of M is given as follows. Initially each x_i is placed in register r_i, and all other registers are empty. Then we go to instruction 1. Then one proceeds by checking whether the instruction one is at applies, where the (a, b, c) symbol A^i_j refers to the j-ary relation R_i of M. If one of the registers mentioned in the antecedent of an instruction which is an atom is empty, then that instruction does not apply. Each d_i refers to the constant c_i of M, each F_i^j to f_i^j of M. If one is at an instruction which does not apply then one passes to the next instruction in order; instruction 1 is next in order after the last instruction, q. Whenever one comes to "terminate", one terminates. If, after termination, the output register is empty, then we say that the flap applied to M at $(x_1, ..., x_n)$ yields $-$. If, after termination, the output register has y, then we say that the fap applied to M at $(x_1, ..., x_n)$ yields y. If termination never occurs then we say that the fap applied to M at $(x_1, ..., x_n)$ is undefined.

Thus every fap of type (n, e, p, q, a, b, c) applied to any structure of type (a, b, c) defines an n-ary partial function and an n-ary partial relation.

Definition 1.4. Let β be a formalized algorithmic procedure (fap) with procedural type (n, e, p, q, a, b, c), M a structure with relational type (a, b, c) and domain D. Then PFCN(β, M) is the n-ary partial function on D given by PFCN(β, M) $(x_1, ..., x_n) = y$ if and only if β applied to M at $(x_1, ..., x_n)$ yields y and $y \in D$. PREL(β, M) is the n-ary partial relation on D given by PREL(β, M) $(x_1, ..., x_n)$ is false if and only if β applied to M at $(x_1, ..., x_n)$ yields $-$; is true of and only if β applied to M at $(x_1, ..., x_n)$ yields some $y \in D$. P stands for "procedure".

Definition 1.5. Let β be a formalized algorithmic procedure. Then β is said to be a formalized algorithmic procedure without counting if and only if the procedural type of β is some $(n, e, 0, q, a, b, c)$.

We now introduce our generalized Turing algorithms.

We have a two-way infinite tape divided into squares of equal size equipped with a reading head which, at any given moment, is positioned at and reading one particular square. At any given moment, every square is either blank or has exactly one entity printed on it (either an element of D or an auxiliary symbol) and the machine is in one particular state. We let $a_0, a_1, ...$ be the auxiliary symbols, and let $q_0, q_1, ...$ be the states. This device, and its list of auxiliary symbols and states and a relational type $\alpha = ((k), (a_0, ..., a_r), (b_0, ..., b_m))$ will be fixed throughout the definition of the α-generalized Turing algorithms (the gTA of type α).

This device, when confronted with a particular structure M of type α and a gTA of type α (to be defined below) and an ordered list $(x_1, ..., x_n)$, $0 < n$, of elements of D, places $(x_1, ..., x_n)$ on the tape, places itself in state q_0, and starts to follow the gTa of type α. By "places $(x_1, ..., x_n)$ on the tape" we mean that the device chooses n adjacent squares of tape and prints $x_1, ..., x_n$

on the squares, in that order, and puts its reading head on the leftmost square of the adjacent ones chosen. Thus, after placing $(x_1, ..., x_n)$ on the tape, all but the adjacent squares chosen are blank.

Fix $M = (D, c_0, ..., c_k, f_0, ..., f_r, R_0, ..., R_m)$. At any given moment in the course of a computation we associate an $m + 3$-tuple $(q_i, x, e_0, ..., e_m)$, which describes the "local situation" is which the device finds itself at that given moment, given by

(1) the device is in state q_i

(2) x is the set of all auxiliary symbols printed on the square being read. Thus either $x = \emptyset$ and that square is blank or has an element of D on it, or else $x = \{a_k\}$ for some $0 \leqslant k$

(3) for each $1 \leqslant j \leqslant m$ we ahve $e_j = 0$ if the b_j squares to the right of the square being read all have element of D placed on them, and in order from left to right are $y_1, ..., y_{b_j}$ and $R_j(y_1, ..., y_{b_j})$; 1 otherwise. This $(q_i, x, e_0, ..., e_m)$ is called the α-situation.

An α-command is an imperative of one of the following sorts:

(1) "move the reading head one square to the left and become in state q_i", written Lq_i

(2) "move the reading head one square to the right and become in state q_i", written Rq_i

(3) "erase the entity (if any) printed on the square being read and become in state q_i", written Bq_i

(4) "erase the entity (if any) printed on the square being read and print the entity x and become in state q_i", written xq_i, where x is some d_j or is some a_j

(5) "switch the contents of the square being read with the contents of the square immediately to the right and become in state q_i", written SRq_i

(6) "switch the contents of the square being read with the contents of the square immediately to the left and become in state q_i", written SLq_i

(7) "erase the entity (if any) printed on the square being read and print the element of D, $F^j_{b_j}(y_1, ..., y_{b_j})$, and become in state q_i, where $y_1, ..., y_{b_j}$ is the list of elements of D that are placed on the b_j squares to the right of the square being read", written $F^j_{b_j}q_i$.

Note that all these α-commands can be followed in any α-situation $(q_i, x, e_0, ..., e_m)$ except command (5), which can be followed if and only if the b_j squares directly to the right of the square being read all have elements of D placed on them.

An α-instruction is a string of the form $(s \to t)$, where s is an α-situation and t is an α-command. This is interpreted as "if s holds then do t".

Finally, an α-algorithm is a finite set of α-instructions such that whenever both $(s \to t)$ and $(s \to t)$ are elements we have $t = r$.

We now describe the circumstances in which, when following an M-algorithm A using imput $(x_1, ..., x_n)$ the device discontinues computation. This occurs when, in the course of computation, an α-situation, s, arises such that either there is no α-command t with $(s \to t) \in A$ or some $(s \to t) \in A$ but t is an α-command of kind 5) above which cannot be obeyed. At the point of discontinuation, the device is said to yield as output whatever is on the square that is being read, which may either be a blank, an auxiliary symbol, or an element of D.

Definition 1.6. Let β be a generalized Turing algorithm of type $\alpha = ((k), (a_0, ..., a_r), (b_0, ..., b_m))$, M a structure with relational type α and domain D, $1 \leqslant n$. Then TFCN(n, β, M) is the n-ary partial function on D given by TFCN$(n, \beta, M)(x_1, ..., x_n) = y$ if and only if β applied to M at imput $(x_1, ..., x_n)$ yields y and $y \in D$. TREL(n, β, M) is the n-ary partial relation on D given by TREL$(n, \beta, M)(x_1, ..., x_n)$ is false if and only if β applied to M at imput $(x_1, ..., x_n)$ yields a blank; is true if and only if β applied to M at imput $(x_1, ..., x_n)$ yields something other than a blank.

We now introduce effective definitional schemes.

Definition 1.7. A definitional type is a sequence (n, Fcn, a, b, c) or (n, Rel, a, b, c), where $1 \leqslant n$, (a, b, c) is a relational type.

Definition 1.8. Let $\alpha = (n, x, (k), (a_0, ..., a_r), (b_0, ..., b_m))$ be a definitional type. Then the α-terms are given by

(1) each variable v_i, $1 \leqslant i \leqslant n$ is an α-term

(2) if $s_1, ..., s_{a_j}$ are α-terms then $F^j_{a_j}(s_1, ..., s_{a_j})$ is an α-term

(3) each d_i, $0 \leqslant i \leqslant k$, is an α-term.

The α-atomic formulae are strings of the form $A^i_{b_i}(t_1, ..., t_{b_i})$, or $\sim A^i_{b_i}(t_1, ..., t_{b_i})$, where $t_1, ..., t_{b_i}$ are α-terms.

Definition 1.9. Let α be a definitional type. Then the α-conditions are strings of the form $E_1 \& E_2 \& ... E_t$, $0 \leqslant t$, where each E_i is an α-atomic formula, and no E_i is $\sim E_j$, $i \neq j$. Two α-conditions are called incompatible just in case one of the conjuncts of one is the negation of one of the conjuncts of the other.

Definition 1.10. Let $\alpha = (n, \text{Fcn}, a, b, c)$ be a definitional type. Then the α-clauses are strings $(C \to t)$, where C is an α-condition and t is an α-term.

$(C \to t)$ is interpreted as "if C holds then it is t at arguments $(v_1, ..., v_n)$".

Definition 1.11. Let $\alpha = (n, \text{Rel}, a, b, c)$ be a definitional type. Then the α-clauses are strings $(C \to \text{true})$ or $(C \to \text{false})$, where C is an α-condition.

$(C \to \text{true})$, $(C \to \text{false})$ are interpreted as "if C holds then it is true at arguments $(v_1, ..., v_n)$", and "if C holds then it is false at arguments $(v_1, ..., v_n)$".

Definition 1.12. Let α be a definitional type. Then an α-definition is a set (possibly infinite) of α-clauses such that any two distinct elements have incompatible antecedents.

Definition 1.13. Let α be a definitional type. Then an effective definitional scheme of type α is an α-definition which is, as a set of strings, recursively enumerable in the sense of ordinary recursion theory.

Definition 1.14. Let β be an effectibe definitional scheme of type (n, Rel, a, b, c) and let $M = (D, c_0, ..., c_k, f_0, ..., f_r, R_0, ..., R_m)$ be a relational structure of relational type (a, b, c). Then we write DREL(β, M) for that n-ary partial relation on D given by DREL$(\beta, M)(x_1, ..., x_n)$ is true if and only if there is some element $(C \to \text{true})$ of β such that C holds when each v_i is replaced by x_i and each d_i by c_i and each F_i^j by f_j and each A_i^j by R_j; is false if and only if there is some element $(C \to \text{false})$ of β such that C holds when each v_i is replaced by x_i and each d_i by c_i and each F_i^j by f_j and each A_i^j by R_j.

Definition 1.15. Let β be an effective definitional scheme of type (n, Fcn, a, b, c) and let $M = (D, c_0, ..., c_k, f_0, ..., f_r, R_0, ..., R_m)$ be a relational structure of relational type (a, b, c). Then we write DFCN(β, M) for that n-ary partial function on D given by DFCN$(\beta, M)(x_1, ..., x_n) = y$ if and only if there is some element $(C \to t)$ of β such that C holds and t is y when each v_i is replaced by x_i and each d_i by c_i and each F_i^j by f_j and each A_i^j by R_j.

We will use the abbreviation eds for "effective definitional scheme".

Definition 1.16. Let β_1 be either a fap of type (n, e, p, q, a, b, c) or a gTa of type (a, b, c) or an eds of type (n, Fcn, a, b, c), and let β_1 be either a fap of type (n, e, p, q, a, b, c) or a gTa of type (a, b, c) or an eds of type (n, Fcn, a, b, c). Then β_1 is n-Fcn-equivalent to β_2 if and only if for all M of type (a, b, c) we have XFCN(n, β_1, M) = YFCN(n, β_2, M), where X, Y are appropriately either P or T or D, and n is mentioned whenever $X(Y)$ is T. One defines n-Rel-equivalence similarly.

We now wish to prove that any gTa of type (a, b, c) is n-Fcn-equivalent (n-Rel-equivalent) to some eds to type (n, Fcn, a, b, c) $((n, \text{Rel}, a, b, c))$, and vice versa. We first consider the first half.

We must first define the auxiliary notion of pseudo eds of type α.

Definition 1.17. Let α be a definitional type. Then a pseudo α-definition is a set of α-clauses such that any two elements whose antecedents are not incompatible have the same consequent.

Definition 1.18. Let α be a definitional type. Then a pseudo eds of type α is a pseudo α-definition which is, as a set os strings, recursively enumerable in the sense of ordinary recursion theory.

In addition we extend the notions DREL and DFCN and n-Fcn-equivalent and n-Rel-equivalent to pseudo eds's.

We also need the auxiliary notion of α-formula, for definitional types α.

Definition 1.19. Let (n, x, a, b, c) be a definitional type. The α-formulae are given by

(1) each α-atomic formula is an α-formula

(2) F_1 & ... & F_r, $0 \leqslant r$, is an α-formula if each F_i is

(3) $\sim F$ is an α-formula if F is.

It is clear what it means for an α-condition to imply an α-formula, in the sense of propositional logic.

Lemma 1.1.1. If β is a pseudo eds of type $\alpha = (n, x, a, b, c)$ then there is an eds γ of type α such that β is x-equivalent to γ.

Proof. Let f be a total function from ω onto β which is recursive in the sense of ordinary recursion theory. Define a total function g from ω into α-formulae by $g(0)$ = that α-condition C_0 such that for some x we have $f(0) = (C_0 \to x)$, $g(k+1) = C_1 \;\&\; \bigwedge_{0 \leqslant i \leqslant k} \sim g(i)$, where there is some y with $f(k+1) = (C_1 \to y)$. Then g is recursive in the sense of ordinary recursion theory. Let h be a total function ω into sets of α-conditions given by $h(k) \simeq$ $\{C : C$ is an α-condition and C implies $g(k)$ and C mentions every α-atomic formula or its negation that is mentioned in $g(k)\}$. Then h will be recursive in the sense of ordinary recursion theory. Finally set $\gamma = \{(C \to x)$: for some k we have $C \in h(k)$ and $(f(k) \to x) \in \beta\}$. Then γ is the desired eds of type α.

This Lemma 1.1.1 is essentially an effective version of the theorem of topology that every open set in the Cantor space can be written as the union of pairwise disjoint basic open sets.

Theorem 1.1. Let (a, b, c) be a relational type and let $0 < n$ and let β_1 be a fap of some type (n, e, p, q, a, b, c) and let β_2 be a gTa of type (a, b, c). Then there are eds γ_1, γ_2 of type (n, x, a, b, c) such that β_1 is n-x-equivalent to γ_1 and β_2 is n-x-equivalent to γ_2.

Proof. Let γ be any (n, x, a, b, c)-condition. We consider a modified action, called by y-action, of β_1 and β_2. For β_1, in the imput registers one places each variable v_i in register r_i. One proceeds with the instructions "if E holds then do $B; z$" with the following conventions. Each non-counting register will either be empty or else will contain an (n, x, a, b, c)-term. Unless E is an (n, e, p, q, a, b, c)-atom E is self-explanatory. If E is an (n, e, p, q, a, b, c)-atom and not all of the registers mentioned in E are nonempty, then the standard action is used; namely, E does not apply, and one passes to the next instruction. If they all are nonempty, then E is interpreted as an (n, x, a, b, c)-atomic formula. If when so interpreted, it is a conjunct of y, then it is considered to apply. If its negation is a conjunct of y then it is considered not to apply, and one passes to the next instruction. If neither it nor its negation is a conjunct of y then terminate *but do not yield any output whatsoever.*

If E is considered to apply, then do B. Doing B is self-explanatory unless B is "replace r_i by the result of applying F_l^j to the contents of $(r_{t_1}, ..., r_{t_l})$ if all l registers are nonempty; otherwise do nothing". In case they all are nonempty, they all contain (n, x, a, b, c)-terms, and so replace r_i by the (n, x, a, b, c)-term $F_l^i(s_1, ..., s_l)$, where s_i is the contents of r_{t_i}.

For the y-action of β_2, choose n adjacent squares and place the variables $v_1, ..., v_n$ on them in order and place the reader at the leftmost of the n squares chosen, and start in state q_0. Now for the subsequent y-action.

We now describe the subsequent action of β_2 at any stage: the device discontinues computation *and does not yield an output* if there is a $1 \leqslant j \leqslant m$ such that the b_j squares to the right of the square being read all have (n, x, a, b, c)-terms printed on them, $t_1, ..., t_{b_j}$, and both $A_{b_j}^j(t_1, ..., t_{b_j})$ and $\sim A_{b_j}^j(t_1, ..., t_{b_j})$ are not a conjunct of y. If there is so such j, then let $(q_i, x, e_0, ..., e_m)$ be such that the device is in state q_i, x is the set of auxiliary symbols printed on the square being read, and $e_j = 0$ if R_j, $t_1, ..., t_{b_j}$ are as above, and $A^j_{b_j}(t_1, ..., t_{b_j})$ is a conjunct of y; 1 otherwise. If there is no element $((q_i, x, e_0, ..., e_m) \to s)$ of β_2 then: the device discontinues computation and yields as output the contents of the square being read. If there is an element $((q_i, x, c_1, ..., c_m) \to s)$ of β_2, then follow the (a, b, c)-command s if it of kind (1)–(4). If s is of kind (5) then follow s after replacing "element of D, $F^j_{b_j}(y_1, ..., y_{b_j})$" by "$(n, x, a, b, c)$-term, $F^j_{b_j}(t_1, ..., t_{b_j})$" and replacing "$y_i$" by "$t_i$", and "elements of D" by "(n, x, a, b, c)-terms". If this modified s cannot be obeyed then terminate computation and yield as output the contents of the square being read.

Now take γ_1^* to be the pseudo eds of type (n, x, a, b, c) given by $\gamma_1^* \{(y \to t)$: the y-action of β_1 yields the (n, x, a, b, c)-term $t\}$ if $x = Fcn$; $\{(y \to \text{false}$: the y-action of β_1 yields $-\} \cup \{(y \to \text{true}$: the y-action of β_1 yields something other than $-\}$ if $x = \text{Rel}$. Take γ_2^* to be the pseudo eds of type (n, x, a, b, c) given by $\gamma_2^* = \{(y \to t)$: the y-action of β_2 yields the (n, x, a, b, c)-term $t\}$ if $x = \text{Fcn}$; $\{(y \to \text{false})$: the y-action of β_2 yields a blank$\} \cup \{(y \to \text{true}$: the y-action of β_2 yields something other than a blank$\}$. Finally choose γ_1, γ_2 to be eds's of type (n, x, a, b, c) which are, respectively, n-x-equivalent to γ_1^*, γ_2^*, by Lemma 1.1.1.

Theorem 1.2. Let (n, x, a, b, c) be a definitional type and let β be an eds of type (n, x, a, b, c). Then there is a gTa γ of type (a, b, c) such that β is n-x-equivalent to γ,

Proof. We only give an intuitive argument for the case $x = \text{Fcn}$, of the action when applied to a structure M of type (a, b, c) at imput $(x_1, ..., x_n)$.

Generate all the elements of β. Doing this much does not require *use* of the constant, function, and relation symbols, although the elements of β mention them, and are used as auxiliary symbols. Whenever an element $(C \to t)$ is generated, replace all occurrences of variables v_i by x_i. Next all (n, x, a, b, c)-terms occurring in this $(C \to t)$ are replaced by their values (elements of D) by first evaluating subterms of lowest complexity, and then of higher complexity. Then the (n, x, a, b, c)-atomic formuae are evaluated as true of false, and then finally C is evaluated as true or false. If C is evaluated as true, then the value of t previously computed is given as output. If it is evaluated as false, then the tape is cleared of $(C \to t)$ and the generation of β is continued.

We continue with our comparisons between *fap*, gTa, and *eds.*

Let $M(2)$ be the structure $(D, 0, 1, f, =)$ of type ((1), (2), (2)), where D^0

is given by

(a) $0, 1 \in D^0$

(b) $(s, t) \in D^0$ if $s, t \in D$.

Thus every element of $M(2)$ is given by a term; take $f(x, y) = (x, y)$.

Let D^+ be given by

(a) $0, 1, \infty \in D^+$

(b) $(s, t) \in D^+$ if $s, t \in D^+$.

Let $g : D^+ \to D^+$ be given by $g(x, y) = (x, y)$.

Definition 1.20. Let $0 < p$. Then $x \in D^0$ is called p-regular if and only if there is a finite sequence $(y_1, ..., y_n)$ such that

(1) each y_i is a sequence of elements of $D^+ \cup \{-\}$ of length p

(2) some entry of y_n is x

(3) each y_{i+1} differs from y_i in at most one entry, where some $z \in D^+$ was changed to some (z_1, z_2), or to 0, or to 1, or to $-$, or to z_1, or to ∞, where z_1, z_2 are both entries of y_i

(4) each entry of y_1 is either $0, 1, \infty$, or $-$. Such $(y_1, ..., y_n)$ are called p-good.

Lemma 1.3.1. For each $0 < p$ there is a $x_p \in D^0$ which is not p-regular; furthermore, no $y \in D^0$ is p-regular if x_p occurs as a subexpression of y.

Proof. Define the desired x_p inductively by $x_1 = (0, 1), x_{k+1} = ((y, z), (z, y))$, where $x_k = (y, z)$. Then one proves by induction on p that x_p is not p-regular. The basis case is trivial. If $(y_1, ..., y_n)$ is $k + 1$-good and x_{k+1} is an entry in y_n, then note that $x_{k+1} = (y, z)$, and by induction hypothesis both y and z are not k-good. But there has to be an entry in $(y_1, ..., y_n)$ where y occurs and z does not (or vice versa), unless $k = 1$, in which case the lemma is true by trial and error. As long as y stays, one is handicapped in one's efforts to obtain z together with y; in fact, one cannot make use of y for this purpose, and can only use k entries. Hence z cannot be obtained together with y in a $k + 1$-good sequence since z is not k-regular. (Of course, once y disappears, one has to obtain it all over again.)

Lemma 1.3.2. Let $\alpha = (1, e, p, q, (1), (2), (2))$ be a procedural type, and let β be any fap of type α. Then $\mathrm{PREL}(\beta, M(2))(x_{n+1+e}) \cong \mathrm{PREL}(\beta, M(2))(x_{n+2+e})$.

Proof. Note that there are n+1+e registers in which to place elements of D^0 in following β applied to $M(2)$. Using lemma 1.2.1 it is obvious that the action of β applied to $M(2)$ at x_{n+1+e} is isomorphic to the action of β applied to $M(2)$ at x_{n+2+e}.

Theorem 1.3. There is an eds of type (1, Rel, (1), (2), (2)) which is not 1-Rel-equivalent to any fap of type $(1, e, p, q, (1), (2), (2))$.

Proof: Consider the eds β given by $\beta = \{(A_2^0(v_1, t_k) \to \text{true} : k \text{ is even}, 0 < k\} \cup \{(A_2^0(v_1, t_k) \to \text{false} : k \text{ is odd}, 0 < k\}$, where $t_1 = F_2^0(d_0, d_1)$, $t_{k+1} = F_2^0(F_2^0(s_1, s_2), F_2^0(s_2, s_1))$, where $t_k = F_2^0(s_1, s_2)$. Then use lemma 1.2.2 to show that $\text{DREL}(\beta, M(2)) \neq \text{PREL}(\gamma, M(2))$ for all fap γ of type $(1, e, p, q, (1), (2), (2))$. Note that we have shown much more than the theorem. Namely, we have shown that there is a fixed structure with equality and such that every element is given by a term, on which there is a simple eds which cannot be simulated by any fap.

From the proof of Theorem 1.2 we easily obtain the following.

Corollary 1.3.1. There is a structure M with equality and with every element of the domain given by a term and eds's β_1, β_2 of type (1, Rel, (1), (2), (2)), (1, Fcn, (1), (2), (2)) such that for all fap's γ_1, γ_2 of type $(1, e_1, p_1, q_1, (1), (2), (2))$, $(1, e_2, p_2, q_2, (1), (2), (2))$ we have $\text{DREL}(\beta_1, M) \neq \text{PREL}(\gamma_1, M)$ and $DFCN(\beta_2, M) \neq PFCN(\gamma_2, M)$.

Proof: Take $M = M(2)$, β_1 as β above, $\beta_2 = \{(A_2^0(v_1, t_k) \to 0 : k \text{ is even}, 0 < k\} \cup \{(A_2^0(v_1, t_k) \to 1 : k \text{ is odd}, 0 < k\}$. If $\text{PFCN}(\gamma_2, M) = \text{DFCN}(\beta_2, M)$, then it would be easy to modify γ_2 to γ_3 such that $\text{PREL}(\gamma_3, M) = \text{DREL}(\beta_1, M)$, which is impossible.

It is clear from the proofs that Corollary 1.3.1 holds for any extension of $M(2)$.

Corollary 1.3.2. Suppose $M = (D \cup E, 0, 1, c_2, ..., c_k, f^*, f_1, ..., f_r, =, R_1, ..., R_m)$ is a relational structure with D = domain of $M(2)$, $D \cap E = \emptyset$, $f^*(x) = f(x)$ if $x \in D$; 0 otherwise, and for each i either f_i is not monadic and $\text{Rng}(f_i) \subset E$ of f_i is monadic and $\text{Rng}(f_i \upharpoonright E) \subset E$ and for all $x \in D$, $f_i(x) = f(x)$ or $f_i(x) \in E$. Then the conclusion of Corollary 1.2.1 holds (when the types are changes to suit M).

We now compare fap without counting to fap with counting.

Definition 1.21. A structure $M = (D, c_0, ..., c_k, f_1, ..., f_r, R_0, ..., R_m)$ is called ω-rich if and only if there is a 1–1 function g from all of ω into D such that

(1) $g(0)$ is some c_i.

(2) for some f_j, binary R_e, we have $g(n+1) = f_j(g(n), g(n), \ldots, g(n))$ and for all $n_1, n_2 \in \omega$ we have $R_e(g(n_1), g(n_2)) \equiv n_1 = n_2$.

Theorem 1.4. If M is of relational type (a, b, c) and β is a fap of type (n, e, p, q, a, b, c) and M is ω-rich, then there is a fap γ of type $(n, e_1, 0, q_1, a, b, c)$ such that β is both n-Rel and n-Fcn-equivalent to γ.

Proof. If the reader understands why the predecessor function $P(x) = x \dot{-} 1$ if $x \neq 0$; 0 if $x = 0$ is given by some fap without counting applied to $(\omega, 0, S, =)$, $S(x) = x + 1$, then he will understand the proof of this theorem.

So, given $k \in \omega$ we want to get $P(k)$. First produce 0 and check whether $k = 0$. If so, give out 0. If not, produce 0 together with $S(0)$. Check for $k = S(0)$. If so, give out 0. If not, change 0 to $S(0)$ and $S(0)$ to $S(S(0))$. Continue in this way, always having a pair m, $S(m)$. Whenever you find $S(m) = k$, give out m.

Definition 1.22. Let $0 \leqslant n$. Then M_n is $(\{0, 1, \ldots, n\}, f, =)$, where $f(i) = i + 1$ if $i < n$; 0 otherwise.

Lemma 1.5.1. Let $\alpha = (1, e, 0, q, (\ ), (1), (2))$ be a procedural type. Then any fap β, of type α, applied to M_k at 0 terminates if and only if it terminates in $\leqslant ((n+2)^{(e+1)}) \times q$ steps.

Lemma 1.5.2. Any partial recursive function is the result of some fap without counting applied to $(\omega, 0, S, P, R)$, where $R(x) \equiv x = 0$.

Proof. This is well known. One uses the characterization of the partial recursive functions as those obtainable from the sucessor, constant, and projection functions by successive applications by primitive recursion and μ-operator.

Lemma 1.5.3. There is a total recursive function $g : \omega \to \{0, 1\}$ such that for each fap, β, of type any $\alpha = (1, e, 0, q, (\ ), (1), (2))$ there is a k such that $\mathrm{PREL}(\beta, M_k)(0)$ is defined if and only if $g(k) = 0$.

Proof. Diagonalize over fap of type α, using Gödel numbers and Lemma 1.5.1.

Lemma 1.5.4. There is a fap, β, to type some $(1, e, p, q, (\ ), (1), (2))$ such that for every fap γ of type any $(1, e_1, 0, q_1, (\ ), (1), (2))$ there is a k such that $\mathrm{PFCN}(\beta, M_k) \neq \mathrm{PFCN}(\gamma, M_k)$ and $\mathrm{PREL}(\beta, M_k) \neq \mathrm{PREL}(\gamma, M_k)$.

Proof. Choose β with the following action applied to any structure $(D \cup \{0\}, f, R)$ at 0. Take 0. Apply f, obtaining $0, f(0)$. Test for $R(0, f(0))$. If negative then change $f(0)$ to $f(f(0))$ and leave 0 untouched. Test $R(0, f(f(0)))$, etcetera, until a positive answer comes. Meanwhile, record on a counting register how many times, say k, it took to get a positive answer (if one is ever obtained). If one is obtained take this k and compute $g(k)$. If the answer is 0 never terminate. If the answer is 1, terminate and yield 0.

Theorem 1.5. There is a structure M of type $((\ ), (1), (2))$ with equality, and a fap, β, of type some $\alpha = (1, e, p, q, (\ ), (1), (2))$ such that for any fap, γ, of type $(1, e_1, 0, q_1, (\ ), (1), (2))$ we have $\mathrm{PFCN}(\beta, M) \neq PFCN(\gamma, M)$ and $\mathrm{PREL}(\beta, M) \neq \mathrm{PREL}(\gamma, M)$.

Proof. Using Lemma 1.5.4. we take $M = (D, f, =)$ where $D = \{(n, m) : n, m \in \omega$ and $0 \leqslant m \leqslant n\}$, $f((n, m)) = (n, m+1)$ if $n \neq m$; 0 otherwise, and choose β exactly as in Lemma 1.5.4.

We now wish to find a restricted class of structures for which every eds is equivalent to some fap.

Definition 1.23. Let $\alpha = (n, x, a, b, c) = (n, x, (k), (a_0, ..., a_r), (b_0, ..., b_m))$ be a definitional type and let $0 < p$. Then the *p*-α-sequences are sequences $(y_1, ..., y_q)$ such that

(1) each y_i is a sequence of α-terms of length p

(2) every entry of y_1 is either a variable v_i, $1 \leqslant i \leqslant n$, or a constant symbol d_i, $0 \leqslant i \leqslant k$, or $-$

(3) every entry of y_{i+1} is either an entry of y_i or a variable or a constant or $-$ as in (2), or is $F^i_{a_i}(t_1, ..., t_{a_i})$ where each t_j is an entry of y_i

(4) y_{i+1} differs from y_i in at most one place. A *p*-α-term is an α-term which is an entry in some entry of some *p*-α-sequence.

We need some informally stated and informally proved lemmas.

Lemma 1.6.1. Fix $0 < p$, $\alpha = (n, x, a, b, c)$ a definitional type. Then using only p auxiliary registers one can go algorithmically from arguments $x_1, ..., x_n$ and a Gödel number e of a *p*-α-term to its value when each v_i is assigned x_i.

Proof. From the Gödel number e obtain an index of an appropriate *p*-α-sequence, and on the p auxiliary registers, follow it using x_i for v_i, emptiness for $-$, c_i for d_i, and f^j_i for F^j_i.

Lemma 1.6.2. Fix $0 < p$, $\alpha = (n, x, a, b, c)$ a definitional type. Let q be the maximum of the a_i, $0 \leqslant i \leqslant r$, where $b = (a_0, ..., a_r)$. Then, using the equality relation and only $(q + 1) \times p$ auxiliary registers, one can go algorithmically from arguments $x_1, ..., x_n$ and a Gödel number of an α-term to the value of that α-term if that value, as well as the values of its subterms are the same as the values of some p-α-terms (at $x_1, ..., x_n$); nontermination otherwise.

Proof. From the α-term t, the Gödel number of a sequence of α-terms, $s_1, ..., s_e$, $s_e = \alpha$, is obtained such that each s_i is either a variable $v_1, ..., v_n$, or a constant $d_0, ..., d_k$, where $(k) = a$, or $F^j_{a_j}(t_1, ..., t_{a_j})$ for some α-terms t_l which occur previously to s_i in the sequence $s_1, ..., s_e$. Then one wishes to find a Gödel number of a sequence of length e of Gödel numbers of p-α-terms $\beta_1, ..., \beta_e$ whose values at $x_1, ..., x_n$ are respectively the values of $s_1, ..., s_e$. So one enumerates Gödel numbers of sequences of length e of Gödel numbers of p-α-terms $\beta_1, ..., \beta_e$ on a counting register. Then one checks whether the value of each β_i is obtained from the values of the β_j, $j < i$ in the same way that the α-term s_i is obtained from the α-terms s_j, $j < i$. If it matches for all $1 \leqslant i \leqslant e$ then give out the value of β_e. To do this checking for each $1 \leqslant i \leqslant e$, one needs only to have, at one time, at most the values of q p-α-terms β_j, $j < i$. Each of this q are evaluated, the first being evaluated at the first p auxiliary registers, using Lemma 1.6.1, etcetera, so that we still have the last p registers of the $(q + 1) \times p$ available. On these last p, β_i is evaluated. It is then tested for equality with the appropriate function of the structures applied to these q values just obtained.

Definition 1.24. Let $\alpha = (n, x, a, b, c)$ be a definitional type and let $0 < p$ and let M be a structure of relational type (a, b, c). Then M is called p-structural if and only if the equality relation is a relation of M and for all $x_1, ..., x_n \in D$ and for all α-terms t there is a p-α-term s whose value in M at $x_1, ..., x_n$ is the same as the value of t in M at $x_1, ..., x_n$. M is called structural if M is p-structural for some $0 < p$.

Theorem 1.6. Let $\alpha = (n, x, a, b, c) = (n, x, (k), (a_0, ..., a_r), (b_0, ..., b_m))$, $b_i = 2$ for some fixed i, be a definitional type and let β be eds of type α. Then there is a fap γ of type some (n, e, p, q, a, b, c) such that for any structural M of type (a, b, c) with R_i as the equality relation we have DREL(β, M) = PREL(γ, M) or DFCN(β, M) = PFCN(γ, M) depending on whether x = Rel or x = Fcn.

Proof. The proof is straightforward using Lemma 1.6.2. See the proof of Theorem 1.2.

Corollary 1.6.1. Let K be a nonempty finite set, $K = \{x_1, ..., x_k\}$, $1 \leqslant k$, and K^* be the set of all nonempty strings of elements of K. Let $f_i : K^* \to K^*$ be given by $f_i(y) = yx_i$. Let $M = (K^*, x_1, ..., x_k, f_1, ..., f_k, =)$. Then for every partial recursive function on K^* of n arguments, g, and any partial recursively enumerable relation on K^* of n arguments, R, in the sense of ordinary recursion theory, there are fap's without counting β_1, β_2 of some types $(n, e, 0, q, (k-1), (1, 1, ..., 1), (2))$ and $(n, e_1, 0, q_1, (k-1), (1, 1, ..., 1), (2))$ such that $\text{PFCN}(\beta_1, M) = g$ and $\text{PREL}(\beta_2, M) = R$.

Proof. Use Theorem 1.6, and Theorem 1.4 to get counting registers. It is obvious that every partial recursive function and partial recursively enumerable relation on K^* is given by an eds applied to M.

Definition 1.25. Let M be a structure with equality of relational type $\alpha = (a, b, c) = ((k), (a_0, ..., a_r), (b_1, ..., b_m))$, $M = (D, c_0, ..., c_k, f_a, ..., f_r, R_1, ..., R_m)$, R_i as equality. Let D^* be the set of all nonempty strings of elements of D. Thus $D \cap D^* = \emptyset$. Let f be given by $f(x, y) = xy$ if $x \in D^*$, $y \in D$; x otherwise. Let g be given by $g(x) = y$ if $x \in D^*$ and y is the last entry of x; x if $x \in D$. Let h be given by $h(x) = y$ if $x \in D^*$, x has length more than 1, and y is the result of deleting the last entry from x; x if $x \in D$; y if $x = (y)$. Then $M^* = (D \cup D^*, c_0, ..., c_k, f_0^*, ..., f_r^*, f, g, h, R_0^*, ..., R_m^*)$, where $f_j^*(x) = f_j(x)$ if $x \in D$; x otherwise, $R_j^*(x_1, ..., x_{b_j}) \equiv R_j(x_1, ..., x_{b_j})$ if each $x_q \in D$; false otherwise, if $j \neq i$, $R_i^*(x, y) \equiv x = y$.

Theorem 1.7. Let $\alpha = (n, x, a, b, c) = (n, x, (k), (a_0, ..., a_r), (b_0, ..., b_m))$ be a definitional type with $b_i = 2$ for some fixed i, and let β be an eds of type $\alpha_0 = (n, x, (k), (a_0, ..., a_r, 2, 1, 1), (b_0, ..., b_m)) = (n, x, a_0, b_0, c_0)$. Then there is a fap γ of type some $(n, e, 0, q, a_0, b_0, c_0)$ and an eds β_1 of type α such that for all M of type α with R_i as the equality relation we have $\text{DREL}(\beta, M^*) = \text{PREL}(\gamma, M^*)$ or $\text{DFCN}(\beta, M^*) = \text{PFCN}(\gamma, M^*)$ according to whether $x =$ Rel or $x =$ Fcn, and the restriction of $\text{DREL}(\beta, M^*)$ to D($\text{DFCN}(\beta, M^*)$ to D) is the same as $\text{DREL}(\beta_1, M^*)$ ($\text{DFCN}(\beta_1, M^*)$).

Proof. Use Theorem 1.4 to obtain counting registers and Theorem 1.6. Note that M^* is structural.

Definition 1.26. Let D be a nonempty set. Then a weak pairing mechanism for D is a pair (P, f) such that P is a 1–1 total binary function on D and f is a partial monadic function on D such that for all $x, y \in D$ we have $f(P(x, y)) = y$.

Theorem 1.8. Let $\alpha = (n, x, a, b, c)$ be a definitional type and let β be an eds of type α. Furthermore let M be any structure, with domain D, of type (a, b, c) such that

(1) the equality relation on D is REL(γ, M) for some fap γ_0 of type some $(2, e_0, p_0, q_0, a, b, c)$

(2) there is a weak pairing mechanism (P, f) for D and fap's γ_1, γ_2 of types, respectively, some $(2, e_1, p_1, q_1, a, b, c)$ and $(1, e_2, p_2, q_2, a, b, c)$.

Then there is a fap γ of type some (n, e, p, q, a, b, c) such that DREL(β, M) = PREL(γ, M) or DFCN(β, M) = PFCN(γ, M) according to whether x = Rel or x = Fcn. Furthermore if p_0 and p_1 can be taken to be 0 and $P(x, y) \neq y$ for all $x, y \in D$, then p can be taken to be 0.

Proof. Although M by itself need not be structural, it is when P, f are adjoined. But P, f are obtained by fap's applied to M. So (the proof of) Theorem 1.6 applies. If $p_0, p_1 = 0$ then we can utilize the proof of Theorem 1.4 to insure that p can be 0 by noticing that if $P(x, y) \neq y$ for all $x, y \in D$ then given arguments $x_1, \ldots, x_n$, we may take the first, x_1, and have the infinite sequence without repetition, $x_1, P(x_1, x_1), P(P(x_1, x_1), x_1), \ldots$.

Section 2

We now apply effective definitional schemes, or equivalently, generalized Turing algorithms to generalize elementary recursion theory. In so doing, we will isolate the computational meaning of the proofs in elementary recursion theory.

Definition 2.1. Let M be a relational structure of type $((\ ), b, c)$, Then we let FCN(M) be the set of all TFCN(n, β, N) such that N is of type some (a, b, c) and β is a gTa of type (a, b, c) and $0 < n$ and M, N have the same domain and relations and functions. Let REL(M) be the set of all TREL(n, β, N) such that N is of type some (a, b, c) and β is a gTa of type (a, b, c) and $0 < n$ and M, N have the same domain and relations and functions. Throughout this section we will always assume that structures have infinite domains, and no constants.

Thus we allow arbitrary constants from the domain of M to, say, insure that all the constant functions are FCN(M).

FCN(M) is our generalization of partial recursive function and REL(M) is our generalization of partial recursively enumerable relation. The computational meaning of partial recursively enumerable relations is this: given arguments we can test them for the relation. We may eventually terminate testing and give yes, or may terminate testing and give no, or we may never terminate testing.

In many cases a theorem of elementary recursion theory may not lift to arbitrary structures, but does hold for a meaningfully restricted class of structures.

We consider four notions of recursively enumerable sets, which coincide in the case of ordinary recursion theory: domains of partial recursive functions, ranges of partial recursive functions, ranges of total recursive functions, and domains of partial recursively enumerable relations. We also consider two notions of recursive sets: recursively enumerable sets whose complements are recursively enumerable, and sets whose total characteristic relation is a recursively enumerable relation.

We consider these elementary theorems of ordinary recursion theory:

A. all four notions of recursively enumerable sets coincide

B. the two notions of recursive set coincide

C. the recursively enumerable sets are closed under intersection, union, and projection

D. the partial recursive functions satisfy enumeration

E. the partial recursive functions are closed under composition

F. the partial recursively enumerable relations satisfy enumeration

G. the recursively enumerable sets are not closed under complementation

H. every recursively enumerable set is the projection of some recursive set

I. the partial recursively enumerable relations are just those with a partial recursive partial characteristic function

J. the partial recursive functions are just those with a recursively enumerable partial characteristic relation

K. each partial recursively enumerable relation can be uniformized by a partial recursive function

L. there is a total recursive successor function which exhausts the domain

M. every infinite recursively enumerable set is the range of some one-one total recursive function.

We organize the elementary theorems into groups according to the assumptions needed on structures M in order for these theorems to lift.

1(a) domains of partial recursive functions are the same as domains of partial recursively enumerable relations

(b) every domain of a monadic partial recursive function is the range of some monadic partial recursive function

(c) sets whose total characteristic relation is recursively enumerable are just the domains of partial recursive functions whose complement is the domain of some partial recursive function

(d) the domains of partial recursive functions are closed under pairwise intersection and union

(e) the partial recursive functions are closed under generalized composition

(f) the partial characteristic function of a partial recursively enumerable relation is partial recursive.

2(a) the ranges of partial recursive functions are closed under pairwise intersection and union

(b) the ranges of total recursive functions are closed under pairwise intersections and unions

(c) the partial recursively enumerable relations are just those partial relations whose partial characteristic function is partial recursive

(d) the partial characteristic relation of a partial recursive function is partial recursively enumerable.

3(a) there is a partial recursive enumeration of the r-ary partial recursive functions

(b) there is a partial recursively enumerable enumeration of the k-ary partial recursively enumerable relations

(c) the domains of partial recursive functions are not closed under complementation.

4(a) the domains of partial recursive functions are closed under projection

(b) every partial recursively enumerable relation can be uniformized by a partial recursive function.

(c) every partial function with partial recursively enumerable partial characteristic relation is partial recursive

5. The ranges of partial recursive functions are not closed under complementation.

6(a) the range of every partial recursive function is the domain of some partial recursive function

(b) the ranges of total recursive functions are exactly the nonempty domains of monadic partial recursive functions

(c) the domain of every partial recursive function is the projection of some set with partial recursive total characteristic function

(d) there is a total recursive successor function which exhausts ω

(e) every infinite recursively enumerable set is the range of some one-one total recursive function.

We consider three conditions on structures M.

I. The equality relation is in REL(M)

II. FCN(M) has a pairing mechanism and REL(M) is not trivial; i.e., there is a (P, f, g) such that P is 2-ary, 1–1, and total, f, g are monadic $f(P(x, y)) = x$, $g(P(x, y)) = y$, and $P, f, g, \in$ FCN(M), and there is, for some k, a k-ary $S \in$ REL(M) and $a_1, ..., a_k, b_1, ..., b_k$ such that $S(a_1, ..., a_k)$ is true, $S(b_1, ..., b_k)$ is false.

III. M is computably finitely generalted; i.e., there is a sequence $x_1, ..., x_k$ $\in M$ and a sequence $f_1, ..., f_r \in$ FCN(M) such that each f_i is total and every element of D can be written as a term involving only the f_i and the x_j.

Definition 2.2. Let f be a k-ary partial function. Then the partial characteristic relation R of f is the partial $k + 1$-ary relation given by $R(x_1, ..., x_k, y) =$ true if $f(x_1, ..., x_k) = y$; false if $f(x_1, ..., x_k) = z$ and $z \neq y$; undefined if $f(x_1, ..., x_k)$ is undefined. Let S be a partial k-ary relation. Then g is called a partial characteristic D-function of S just in case for some $a \neq b, a, b, \in D$, we have $f(x_1, ..., x_k) = a$ if $S(x_1, ..., x_k)$ is true; b if $S(x_1, ..., x_k)$ is false, undefined if $S(x_1, ..., x_k)$ is undefined.

Definition 2.3. Let R be a $k + 1$-ary partial relation. Then f uniformizes R if and only if f is a k-ary partial function such that whenever $(\exists y)$ $(R(x_1, ..., x_k, y)$ is true) then $R(x_1, ..., x_k, f(x_1, ..., x_k))$ is true, and whenever $(\forall y)$ $(R(x_1, ..., x_k, y)$ is either false or undefined) then $f(x_1, ..., x_k)$ is undefined.

Definition 2.4. Let Y be a set of k-tuples. Then the total characteristic relation of f is the k-ary total relation given by $R(x_1, ..., x_k) \equiv (x_1, ..., x_k) \in Y$.

Theorem 2.1. All elementary theorems numbered 1 above lift to all structures.

Proof. (a) domains of elements of FCN(M) are the same as domains of elements of REL(M). To see this let f be a k-ary element of FCN(M). Compute at imput $(x_1, ..., x_k)$ as follows: compute $f(x_1, ..., x_k)$, and if and when an output is obtained, given any output. Conversely let R be a k-ary element of REL(M). Compute at imput $(x_1, ..., x_k)$ as follows: test for $R(x_1, ..., x_k)$.

If and when an output is obtained, give output x_1.

(b) every domain of a monadic element of FCN(M) is the range of some monadic element of FCN(M). Let f be monadic, $f \in$ FCN(M). Compute at imput x as follows: compute $f(x)$. If and when an output is obtained, give out output x.

(c) sets whose total characteristic relation is in REL(M) are just the domains of elements of FCN(M) whose complement relative to D is the domain of some element of FCN(M). Let Y be a set of k-tuples, R its total characteristic relation. Compute at imput $(x_1, ..., x_k)$ as follows: test for $R(x_1, ..., x_k)$. If and when a negative answer is obtained, give output x_1. Conversely, let Y be a set of n-tuples which is the domain of $f \in$ FCN(M), and $D-Y$ is the domain of $g \in$ FCN(M). Then compute at imput $(x_1, ..., x_k)$ as follows: compute, simultaneously, $g(x_1, ..., x_k)$ and $f(x_1, ..., x_k)$. If and when a value is obtained for the first, yield a blank; if and when a value is obtained for the second, yield an auxiliary symbol.

(d) the domains of elements of FCN(M) are closed under pairwise intersection and union. Left to the reader.

(e) if $f, g_1, ..., g_n \in$ FCN(M), f is n-ary, each g_i is k-ary, $h(x_1, ..., x_k) \simeq f(g_1(x_1, ..., x_k), ..., g_n(x_1, ..., x_k))$, then $h \in$ FCN(M). Compute at imput $(x_1, ..., x_k)$ as follows: compute, simultaneously, $g_1(x_1, ..., x_k), ...,$ $g_n(x_1, ..., x_k)$. If and when values $y_1, ..., y_n$ are obtained, respectively, then compute $f(y_1, ..., y_n)$. If and when a value is obtained, give it as output.

(f) any partial characteristic D-function of a partial relation in REL(M) lies in FCN(M). Let $R \in$ REL(M) be k-ary, let $a \neq b$, $a, b \in D$. Compute at imput $(x_1, ..., x_k)$ as follows: test $R(x_1, ..., x_k)$. If and when a positive answer is obtained, yield a. If and when a negative answer is obtained, yield b.

Theorem 2.2. The elementary theorems numbered 2 lift to all structures satisfying I above.

Proof. (a) the ranges of elements of FCN(M) are closed under intersection and union. Let $\alpha, \beta \in$ FCN(M), and k-ary. For intersection choose $a \in$ Rng$(\alpha) \cap$ Rng(β). If no such a exists then we are done. Compute at imput $(x_1, ..., x_{2k})$ as follows: compute $\alpha(x_1, ..., x_k)$ and $\beta(x_{k+1}, ..., x_{2k})$. If and when both values are obtained, test them for equality. If and when a positive answer is obtained, give that value as output. If and when a negative answer is obtained, give output a. For union, choose $a \in$ Rng(α). If no such a exists then we are done. Set $h(x_1, ..., x_k, y) \simeq \alpha(x_1, ..., x_k)$ if $y = a$; $\beta(x_1, ..., x_k)$ otherwise.

(b) the ranges of total elements of FCN(M) are closed under intersection and union. We have so arranged the proof of (a) that it also constitutes proof of (b).

(c) The elements of REL(M) are just those partial relations all of whose partial characteristic D-functions are in FCN(M). It is obvious that for any structure M, any partial characteristic D-function of an element of REL(M) is an element of FCN(M). Now let R be a k-ary partial relation, f a partial characteristic D-function of R, using a, b. Compute at imput $(x_1, ..., x_k)$ as follows: compute $f(x_1, ..., x_k)$. If and when a value, y, is given, test for $y = a$. If and when a positive answer is obtained, yield a blank; if and when a negative answer is obtained, yield an auxiliary symbol.

(d) the partial characteristic relations of elements of FCN(M) are elements of REL(M). Let $f \in$ FCN(M) be k-ary. Compute at imput $(x_1, ..., x_k, y)$ as follows: compute $f(x_1, ..., x_k)$. If and when a value z is obtained such that $z \neq y$ yield a blank. If and when a value z is obtained such that $z = y$ yield an auxiliary symbol.

Theorem 2.3. The elementary theorems numbered 3 lift to all structures satisfying II above.

Proof. (a) for all $0 < r$ there is an $r + 1$-ary $\phi_r \in$ FCN(M) such that for every r-ary $f \in$ FCN(M) there is an $x \in D$ with $\phi_r(x, x_1, ..., x_r) \simeq f(x_1, ..., x_r)$.

We give an intuitive argument. Fix $0 < r$, and fix (P, f, g) as a pairing mechanisms for M, and let S, $a_1, ..., a_k$, $b_1, ..., b_k$ be as in Condition II. We first show how to code finite sequences $(i, x_1, ..., x_p)$, $i \in \omega$, $x_1, ..., x_p \in D$. We give an intuitive algorithm which takes an arbitrary element $x \in D$ and either never terminates computation or else winds up with some $(i, x_1, ..., x_p)$, $i \in \omega$, $0 \leqslant p$, and such that any $(i, x_1, ..., x_p)$, $i \in \omega$, $0 \leqslant p$, $x_1, ..., x_p \in D$ has a corresponding x. We write $G(x) = (i, x_1, ..., x_p)$. G is partial.

Let $f^*(n, x)$ and $g^*(n, x)$ be the two partial functions given by $f^*(0, x) = g^*(0, x) = x$, $f^*(n + 1, x) = f(g^*(n, x))$, $g^*(n + 1, x) = g(f^*(n, x))$.

Take x. Compute, successively, $f^*(0, f(x))$, $f^*(1, f(x))$, ..., $f^*(k - 1, f(x))$. If and when this is accomplished, test for the k-ary partial relation S at these k arguments arranged in the order in which they were obtained.

If and when these truth values are computed, then compute k more values: $f^*(k, f(x))$, ..., $f^*(2k - 1, f(x))$, if that test is negative; if positive, then do not compute k more values, and count the number of times, α, that this procedure had to be applied in order to obtain a positive answer. Continue in this way. Clearly $1 \leqslant \alpha$. Then if α is not of the form $2^i 3^p$, then $G(x)$ is undefined. If $\alpha = 2^i 3^p$, then successively compute $f^*(0, g(x))$, $f^*(1, g(x))$, ..., $f^*(p - 1, g(x))$. If and when this is accomplished, and the values are

$x_1, ..., x_p$, then set $G(x) = (i, x_1, ..., x_p)$. If some part of the above is never accomplished, then $G(x)$ is undefined.

To show that G is onto, let $(i, x_1, ..., x_p)$ have $i \in \omega, x_1, ..., x_p \in D$, $0 \leqslant p$. It suffices to choose x such that for each $1 \leqslant i \leqslant k$ we have $f^*((k \cdot 2^i 3^p) + i, f(x)) = a_i$, and for all $j < 2^i 3^p$ we have $f^*((k \cdot j) + i, f(x)) = b_i$, and such that $f^*(i, g(x)) = x_i$ for all $1 \leqslant i \leqslant p$. Since P is total and 1–1, this is easily done and is left to the reader.

Now note that there is a total function ψ from ω onto the set of all eds's of type $(r, \mathrm{Fcn}, (\ ), b, c)$, where M has type $((\ ), b, c)$, which is recursively enumerable in the sense of ordinary recursion theory; this uses the Kleene enumeration theorem for ordinary recursion theory.

We now define $\phi_r(x, x_1, ..., x_r)$. Consider $G(x) = (i, y_1, ..., y_p)$. If $G(x)$ is undefined, then $\phi_r(x, x_1, ..., x_r)$ is undefined. Consider $\psi(i)$. Consider the new eds d obtained by replacing each occurrence of any v_{r+j}, $1 \leqslant j \leqslant p$, in $\psi(i)$, by y_j.

Set $\phi_r(x, x_1, ..., x_r) \simeq \mathrm{DFCN}(d, M)\,(x_1, ..., x_r)$. It is intuitively clear that $\phi_r \in \mathrm{FCN}(M)$.

(b) the proof that REL(M) satisfies enumeration is analogous.

(c) for each k, the domains of k-ary elements of MCF(M) are not closed under complementation.

Let ϕ_1 be as in (a) above. Then $\lambda x(\phi_1(x, x)) \in \mathrm{FCN}(M)$. Let $Y = \{x : \phi_1(x, x) \text{ is defined}\}$. Suppose $D - Y$ is $\mathrm{Dom}(\alpha)$, $\alpha \in \mathrm{FCN}(M)$, α monadic. Choose a such that $\lambda x(\phi_1(a, x)) = \alpha$. Then $a \in Y \equiv \alpha(a)$ is defined $\equiv a \in \mathrm{Dom}(\alpha) \equiv a \in D - Y$, which is a contradiction.

Theorem 2.4. The elementary theorems numbered 4 lift to all structures satisfying condition III.

Proof. (a) the domains of elements of FCN(M) are closed under projection. Let $\alpha \in \mathrm{FCN}(M)$ be $k+1$-ary. Let $Y = \{(x_1, ..., x_k) : (\exists y)\,(\alpha(x_1, ..., x_k, y)$ is defined$)\}$. Let $y_1, ..., y_n, f_1, ..., f_p$ computably generate M. Compute at imput $(x_1, ..., x_k)$ as follows: start listing the terms involving $y_1, ..., y_n$, $f_1, ..., f_p$ and evaluate them and compute $\alpha(x_1, ..., x_k, z)$ for evaluations z. Do not get hung up on one test. Spend time on all computations that arise, using the well known backtrack technique of ordinary recursion theory. If and when a computation terminates, yield x_1 as output. The proof of (b) below will be similar.

(b) every element of REL(M) can be uniformized by an element of FCN(M). Let R be a $k+1$-ary element of REL(M). Let $y_1, ..., y_n, f_1, ..., f_p$ computably generate M. Compute at imput $(x_1, ..., x_k)$ as follows: start listing the terms involving $y_1, \ldots, y_n, f_1, \ldots, f_p$, and evaluate them and test for $R(x_1, \ldots, x_k, z)$, for evaluations z. Do not get hung up on one test. Spend time on all tests, using the well known backtrack technique of ordinary recursion theory. If and when a positive answer is obtained for some evaluation z, set $f(y_1, \ldots, y_k, z)$.

(c) every partial function with partial characteristic relation in REL(M) is in FCN(M). Let f be k-ary, R be the partial characteristic relation, $R \in$ REL(M). Let $y_1, \ldots, y_n, f_1, \ldots, f_p$ computably generate M. Compute at arguments $(x_1, \ldots, x_k)$ as follows: list the terms involving $y_1, \ldots, y_n, f_1, \ldots, f_p$ and evaluate them. Using the backtrack technique, test for $R(x_1, \ldots, x_k, y)$, for evaluations y. If and when a positive answer is obtained for some evaluation y give out y.

Theorem 2.5. The elementary theorem numbered 5 lift to all structures M satisfying both I and II.

Proof. The ranges of elements of FCN(M) are not closed under complementation. Let ϕ_1 be as in 3(a), and note that α, given by $\alpha(x, y) = x$ if $\phi_1(x, y) = x$; undefined otherwise, lies in FCN(M). Then $\text{Rng}(\alpha) = \{x : x \in \text{Rng}(\lambda y(\phi_1(x, y)))\}$. Then by diagonalization, D-Rng(α) is not the range of any monadic element of FCN(α), and hence not of any element of FCN(α).

Theorem 2.6. The elementary theorems numbered 6 lift to all structures satisfying conditions I and III. Furthermore, condition II follows from conditions I and III.

Proof. Let $y_1, \ldots, y_n, f_1, \ldots, f_p$ computably generate M. (a) the range of every element of FCN(M) is the domain of some element of FCN(M). Let $\alpha \in$ FCN(M), α k-ary. Compute at argument x as follows: list the terms involving $y_1, \ldots, y_n, f_1, \ldots, f_p$ and evaluate them. Using the backtrack technique, compute $\alpha(z_1, \ldots, z_k)$ for evaluations $z_1, \ldots, z_k$. If and when such a computation of $\alpha(z_1, \ldots, z_k)$ yields x, give x as output. One uses the equality relation in testing for $\alpha(z_1, \ldots, z_k) = x$.

(b) the ranges of total elements of FCN(M) are exactly the nonempty domains of monadic elements of FCN(M). Let α be monadic, $\alpha \in$ FCN(M), let $Y = \text{Dom}(\alpha)$, $Y \neq \emptyset$, $a \in Y$. Compute at argument x as follows: list the terms involving $y_1, \ldots, y_n, f_1, \ldots, f_p$ and evaluate. Wait till x comes up (here we need equality). Count the number of steps needed to obtain x, say q steps. Then compute $\alpha(x)$ for q steps. If an output for $\alpha(x)$ is obtained, give output x. If not, give output a.

(c) the domain of every k-ary element of FCN(M) is the projection of some set of $k+1$-tuples whose total characteristic relation is in REL(M). Let $\alpha \in$ FCN(M), α k-ary. Compute at imput $(x_1, ..., x_k, y)$ as follows: list and evaluate the terms involving $y_1, ..., y_n, f_1, ..., f_p$, and wait till y comes. Here equality is used. Count how many steps it takes to arrive at y, say q steps. Compute $\alpha(x_1, ..., x_k)$ for q steps. If a value is not obtained for $\alpha(x_1, ..., x_k)$, yield the blank as output. If a value is obtained, yield an auxiliary symbol as output.

(d) there is a total one-one monadic element $\alpha \in$ FCN(M) and an $x \in D$ such that x, α generate M. Take x to be the evaluation of the first term involving $y_1, ..., y_n, f_1, ..., f_p$. Compute at imput z as follows: list the terms involving $y_1, ..., y_n, f_1, ..., f_p$ and evaluate them. Using equality, wait till z is obtained. Then compute the next term and give the answer as output.

(e) every infinite domain of a monadic element of FCN(M) is the range of some one-one total element of FCN(M). Let Y be infinite, $Y = \mathrm{Dom}(\alpha)$, $\alpha \in$ FCN(M), α monadic. Compute at imput x as follows: list the terms involving $y_1, ..., y_n, f_1, ..., f_p$ and evaluate them. Using equality, wait till x is obtained. Let x be the qth element obtained, $0 < q$. Then list the terms $y_1, ..., y_n, f_1, ..., f_p$ and evaluate them and compute $\alpha(z)$, for evaluations z, using the bracktrack technique. Continue in this way until α is computed at q distinct arguments z, and then give out the last (the qth) one as output.

Condition II follows from conditions I and III. Compute at arguments y, z as follows: list the terms involving $y_1, ..., y_n, f_1, ..., f_p$ and evaluate them. Wait till y, z are evaluated, respectively the qth and rth obtained. Then continue until the $2^q 3^r$th such is obtained, and give this as output. Then P is one-one total and binary. We leave it to the reader to complete the pairing mechanism.

Theorem 2.7. Suppose $M = (D, f_1, ..., f_n, R_1, ..., R_m)$ satisfy conditions I and III. Then there is an isomorphism φ from M onto $N = (\omega, g_1, ..., g_n, S_1, ..., S_m)$ such that FCN(N) is the class of partial functions on ω which are partial recursive in $(g_1, ..., g_n, S_1, ..., S_m)$ in the sense of ordinary recursion theory, and REL(M) is the class of partial relations on ω which are recursively enumerable in $(g_1, ..., g_n, S_1, ..., S_m)$ in the sense of ordinary recursion theory.

Proof. Left to the reader.

We conjecture that these theorems are best possible. In other words, no elementary theorems numbered 2 lift to all structures satisfying II and III; none numbered 3 lift to all structures satisfying III; none numbered 4 lift to all structures satisfying I and II; the one numbered 5 does not lift to all structures satisfying I; the one numbered 5 does not lift to all structures satisfying II and III; the ones numbered 6 do not lift to all structures satisfying I and II; the ones numbered 6 do not lift to all structures satisfying II and III.

Section 3

We have indicated throughout the Introduction that Turing's analysis could probably be significantly improved in relation to configurational computability; it seems, instead, to be an analysis of linear computability. The problem is to fill in the blank in: Turing's operations are to finite linear configurations as are ―――――――――― to arbitrary finite configurations. Perhaps one will first successfully deal with some restricted class of finite configurations.

Our effective definitional schemes provide informative equivalents to generalized Turing algorithms. It would be interesting to find similarly informative equivalents to formalized algorithmic procedures both with and without counting. Perhaps one can use suitably restricted effective definitional schemes for this purpose.

It should be possible to elegantly generalize the usual schemata for defining partial recursive functionals via primitive recursion and μ-operator to obtain schemata for defining the (equivalence type of) formalized algortihmic procedures with or without counting, and generalized Turing algorithms. This might bear on the relationship between our notions and the prime "computability" of Moschovakis [2]; there Moschovakis introduces prime "computability" not in terms of the action of devices, but in terms of schemata. We have not investigated the relation between our work and that of [2].

We conjectured in Section 2 that our theorems of Section 2 were best possible. Some parts of this conjecture are trivial. Others seem to require genuine combinatorial arguments.

It is clear from glancing at the arguments in Section 2 that only certain basic facts are used about generalized Turing computability. It would be interesting to isolate these facts, and perhaps, obtain decision procedures. Of course the basic facts are dependent on which of the conditions I, II, III are placed on the structures M beforehand.

Computer scientists tell me that our algorithmic procedures without counting are similar to their program schemata, and have referred me to Paterson [3]. It would be interesting to see whether more of the work of computer scientists fits into the framework presented here. Perhaps the way in which the basic definitions of Section 1 are set out lead to the best ways of expressing basic distinctions familiar to computer scientists.

References

[1] R. Gandy, Computable functionals of finite type I, in: Sets, models, and recursion theory, ed. J. Crossley (North-Holland, Publ. Co., 1967).
[2] Y. Moschovakis, Abstract first order computability I, Trans. Am. Math. Soc. 138 (April, 1969).
[3] M. Paterson, Program schemata, in: Machine intelligence 3, ed. Donald Michie (Edinburgh University Press, 1968).
[4] J. Shepherdson and H. Sturgis, Computability of recursive functions, J. for the Association for Computing Machinery (1963) 217.

SEMANTIC PROOF OF THE CRAIG INTERPOLATION THEOREM FOR INTUITIONISTIC LOGIC AND EXTENSIONS. PART I

Dov M. GABBAY

The Hebrew University of Jerusalem,
Mathematics Institute, Oxford University

Introduction

In this paper we shall give, using the methods of [1], a semantic proof of a form of Robinson's consistency theorem for the intuitionistic predicate calculus. This form is equivalent to the Craig interpolation theorem, for any extension of the intuitionistic predicate calculus.

The Craig theorem was proved syntactically by Schutte [4] and Nagashima [3], for the minimal calculus as well. Our semantic proof also works for the minimal calculus.

We shall treat some other extensions in Part II.

We would like to mention here that the general form of Robinson's theorem which is stronger than Craig's theorem will be refuted in Part II. Our plan is as follows: In Section 1 we give some definitions and state the results. In Section 2 and 3 we give the proofs and in Part II we turn to some other systems.

Before we turn to define the concepts and give the proofs let us say a few words about the motivation.

Our main concept is that of a theory as a pair of sets of sentences see Section 1. This concept is essential to the proof since Craig's theorem is equivalent to a version of Robinson's theorem. (See (4b).) Also this concept gives us a certain form of weak negation since we can in a theory demand that certain sentences will be false.

We cannot add a special connective $\sim_\omega$ to our intuitionistic language. This will give us too strong a system, with the connective $\sim\sim_\omega$ being a S4* modal necessity operator $\Box = \sim\sim_\omega$. This is so since in the translation of the intuitionistic logic into the modal logic S4* without the Barcan formula, is translated as $\Box \sim_\omega$. This concept of a theory as a pair is also essential in treating the extension of the intuitionistic logic with the axiom $\forall x(\phi \vee \psi(x) \to (\phi \vee \forall x \psi(x))$ where x is not free in ϕ.

Now the idea of the proof is the following. Let us take the classical case, and suppose that we have two classically consistent theories such that the intersection in the common language is complete. If these two theories are Henkin-saturated then the union is trivially consistent, since we can construct a model from the theories themselves. So to prove the theorem in the general case one can extend the given theories to Henkin-saturated ones that satisfy the same assumptions.

In the intuitionistic case a model is obtained not from a single theory but from a partially ordered set of saturated theories and so what we do in Section 1 is to construct two partially ordered sets of saturated theories that are isomorphic in a certain sense. This is done by extending our given theories and by transferring subtheories from one place to another.

1. Results

Definition 1. By an intuitionistic theory we mean a pair (Δ, Γ) of sets of sentences where Δ is the set of true sentences and Γ is the set of false sentences.

(Δ, Γ) is said to be complete iff for all ϕ either $\phi \in \Delta$ or $\phi \in \Gamma$. (Δ, Γ) is said to be consistent iff for no $\delta_1, ..., \delta_n \in \Delta, \gamma_1 ... \gamma_m \in \Gamma$ do we have $\vdash \delta_1 \wedge ... \wedge \delta_n \to \gamma_1 \vee ... \vee \gamma_m$. A Kripke structure is a model of (Δ, Γ) iff all sentences of Δ hold in the structure and all sentences of Γ fail to hold. In [2] we proved:

Theorem 2. (Δ, Γ) is consistent iff it has a model.

We now give the general form of Robinson's consistency statement.

(3) Let $(\Delta_1, \Gamma_1), (\Delta_2, \Gamma_2)$ be two consistent theories in the languages L_1 and L_2 respectively, with the property that the common theory (in the common language) $(\Delta_1 \cap \Delta_2, \Gamma_1 \cap \Gamma_2)$ is complete, then $(\Delta_1 \cup \Delta_2, \Gamma_1 \cup \Gamma_2)$ is consistent.

The following is true:

Theorem 4. (a) The general form of Robinson's consistency statement (3) fails to hold see Part II.

(b) The version of (3) for the case of $\Gamma_1 \subseteq \Gamma_2$ (and hence Γ_1 is in the common language) is equivalent to Craig interpolation theorem.

The proofs will be given in Section 3.

2. Robinson's theorem

In this section we give the proof of Robinson's theorem for the case (4b) (i.e. $\Gamma_1 \subseteq \Gamma_2$).

Let (Δ_0, Γ_0), (Θ_0, Γ_0^*) be two consistent theories in the languages L_0 and M_0 respectively such that

(5) Γ_0 is a set of formulae in the common language $L_0 \cap M_0$ and $\Gamma_0 \subseteq \Gamma_0^*$.
(6) $(\Delta_0 \cap \Theta_0, \Gamma_0)$ is a complete theory in the language $L_0 \cap M_0$.

We intend to construct a Kripke model of $(\Delta_0 \cup \Theta_0, \Gamma_0^*)$. We assume our languages to be countable and to contain no individual constants which are not in the common language.

Let C_1, C_2 ... be a sequence of pairwise disjoint countable sets of individual constants. Define:

$$L_1 = L_0 + C_1$$
$$M_1 = M_0 + C_1$$
$$\cdots\cdots\cdots$$
$$L_{n+1} = L_n + C_n$$
$$M_{n+1} = M_n + C_n$$
$$\cdots\cdots\cdots$$
$$L_\omega = \bigcup_n L_n \quad M_\omega = \bigcup_n M_n \,,$$

where $L + C$ is the language obtained from L by adding all the constants of C.

Definition 7. A theory (Δ, Γ) is said to be saturated iff

(a) $\phi, \psi \in \Delta \Rightarrow \phi \wedge \psi \in \Delta$.

(b) $\phi \vee \psi \in \Delta \Rightarrow \phi \in \Delta$ or $\psi \in \Delta$.

(c) $\exists x \phi(x) \in \Delta \Rightarrow$ for some u of the language $\phi(u) \in \Delta$.

(d) $\Delta \vdash \phi \Rightarrow \phi \in \Delta$.

In the definition of saturation Γ plays no role.

It is known, see [5], that any theory (Δ, Γ) can be extended to a saturated and complete theory (Δ', Γ') in a language with $\aleph_0$ more individual constants such that: $\Delta \subseteq \Delta'$ and $\Gamma \subseteq \Gamma'$.

We now turn to the construction of our model.

First note that (Δ_0, Γ_0) is not necessarily saturated. We may extend (Δ_0, Γ_0) to a complete and saturated theory (Δ_1, Γ_1') in the language L_1. We are not interested, however, in Γ_1' and so we will only regard (Δ_1, Γ_0). It is a saturated theory in the language L_1.

Note that $\Delta_0 \subseteq \Delta_1, \Gamma_0 \subseteq \Gamma_1'$. We now turn to prove some basic lemmas:

Lemma 8. The following theory is consistent: $(\Theta_0 \cup \{\psi \mid \psi \in \Delta_1 \cap M_1\}, \Gamma_0^*)$.

Proof. Otherwise for some $\theta_0 \in \Theta_0$, some $\psi_1 \in \Delta_1 \cap M_1$ and some $\gamma_0^* \in \Gamma_0^*$ we have: $\vdash \theta_0 \wedge \psi_1 \rightarrow \gamma_0^*$.

Note that Δ_1 and Θ_0 are closed under conjunctions and Γ_0^* under disjunctions.

So

$$\vdash \psi_1 \rightarrow (\theta_0 \rightarrow \gamma_0^*)$$

$$\vdash (\exists c_1 \ldots)\psi_1 \rightarrow (\theta_0 \rightarrow \gamma_0^*)$$

where $c_1 \ldots$ are the new constants that appear in ψ_1 and not in $\theta_0 \rightarrow \gamma_0^*$.

Now $(\exists c_1 \ldots)\psi_1$ is a sentence of language $L_0 \cap M_0$. Now since the $L_0 \cap M_0$-language-part of (Θ_0, Γ_0^*) is complete we get that $(\exists c_1 \ldots)\psi_1$ is either in Θ_0 or in Γ_0^*. If it is in Γ_0^* then it is in Γ_0 by (6). Now since $\Gamma_0 \subseteq \Gamma_1'$ we get that $(\exists c_1 \ldots)\psi_1 \notin \Delta_1$ which contradicts $\psi_1 \in \Delta_1$.

So we conclude that $(\exists c_1 \ldots)\psi_1 \in \Theta_0$, and therefore $(\theta_0 \rightarrow \gamma_0^*) \in \Theta_0$, but this contradicts the consistency of (Θ_0, Γ_0^*). Therefore we conclude that the above theory is consistent.

(9) Now we extend the theory of (8) to a saturated and complete theory in the language M_2. Let us denote this theory by (Θ_2, Γ_2^*) and let Γ_2 stand for $\Gamma_2^* \cap L_2$. We have that $\Gamma_0 \subseteq \Gamma_2$.

Lemma 10. The following is consistent: $(\Delta_1 \cup \{\psi_2 \mid \psi_2 \in \Theta_2 \cap L_2\}, \Gamma_2)$.

Proof. Otherwise for some $\delta_1 \in \Delta_1$, $\psi_2 \in \Theta_2 \cap L_2$ and $\gamma_2 \in \Gamma_2$ we have:

$$\vdash \delta_1 \wedge \psi_2 \to \gamma_2$$

$$\vdash \delta_1 \to (\psi_2 \to \gamma_2)$$

$$\vdash \delta_1 \to (\forall c_2 \ldots)(\psi_2 \to \gamma_2)$$

where c_2 ... are the new constants. Now since this is a sentence L_1 and $\delta_1 \in \Delta_1$ we get that $(\forall c_2 \ldots)(\psi_2 \to \gamma_2) \in \Delta_1 \cap M_1$ and so by our construction in (8) we get $(\forall c_2 \ldots)(\psi_2 \to \gamma_2) \in \Theta_2$. But this is a contradiction since $\psi_2 \in \Theta_2$ and $\gamma_2 \in \Gamma_2^*$; and (Θ_2, Γ_2^*) is consistent.

(11) Now extend the theory of (10) to a saturated and complete theory (Δ_3, Γ'_3) in the language L_3, with the properties that $\Gamma_2 \subseteq \Gamma'_3$ and $\Delta_2 \cup \{\psi_2\text{'s}\} \subseteq \Delta_3$. Again we are interested only in (Δ_3, Γ_2).

Lemma 12. The following is consistent: $(\Theta_2 \cup \{\psi_3 | \psi_3 \in \Delta_3 \cap M_3\}, \Gamma_2^*)$.

Proof. Otherwise for some $\theta_2 \in \Theta_2$, $\psi_3 \in \Delta_3 \cap M_3$ and $\gamma_2^* \in \Gamma_2^*$ we get $\vdash \theta_2 \wedge \psi_3 \to \gamma_2^*$, $\vdash \exists c_3 \psi_3 \to (\theta_2 \to \gamma_2^*)$.

Now if $\exists c_3 \psi_3 \in \Gamma_2^*$ then it is in Γ_2 and since $\Gamma_2 \subseteq \Gamma'_3$ ot cannot be in Δ_3 which contradicts $\psi_3 \in \Delta_3$ ((Δ_3, Γ'_3) is complete!).

So $\exists c_3 \psi_3 \in \Theta_2$ and so $(\theta_2 \to \gamma_2^*) \in \Theta_2$ which is a contradiction.

Now using the set of (12) we construct (Θ_4, Γ_4^*) and continue to construct (Δ_5, Γ'_5), (Δ_5, Γ_4) and so on. We obtain the following sequences.

$$(13) \qquad (\Delta_0, \Gamma_0), \quad (\Delta_1, \Gamma_0), \quad (\Delta_3, \Gamma_2) \ldots$$

$$(\Theta_0, \Gamma_0^*), \quad (\Theta_2, \Gamma_2^*), \quad (\Theta_4, \Gamma_4^*) \ldots$$

The following holds:

(a) $(\Theta_{2n}, \Gamma_{2n}^*)$ is a complete and saturated theory.

(b) $(\Delta_{2n+1}, \Gamma_{2n})$ is a saturated theory.

(c) $\Theta_{2n} \subseteq \Theta_{2n+2}$, $\Gamma_{2n}^* \subseteq \Gamma_{2n+2}^*$ and $\Gamma_{2n} = \Gamma_{2n}^* \cap L_{2n}$.

(d) $\Delta_0 \subseteq \Delta_1 \subseteq \Delta_3 \subseteq \ldots$.

Now let $\Delta_\omega = \bigcup_n \Delta_n$, $\Gamma_\omega = \bigcup_n \Gamma_n$, $\Gamma_\omega^* = \bigcup_n \Gamma_n^*$, $\Theta_\omega = \bigcup_n \Theta_n$. We now show that

(e) $(\Delta_\omega, \Gamma_\omega)$ $(\Theta_\omega, \Gamma^*_\omega)$ fulfill properties (5) and (6) and are saturated in the language L_ω, M_ω respectively.

Proof. Clearly these theories are saturated and $\Gamma_\omega \subseteq \Gamma^*_\omega$. Now to show that $(\Delta_\omega \cap \Theta_\omega, \Gamma_\omega)$ is a complete $L_\omega \cap M_\omega$ theory let α be a sentence in this language. So $\alpha \in L_{2n} \cap M_{2n}$ for some n. Now $(\Theta_{2n}, \Gamma^*_{2n})$ is complete and so either $\alpha \in \Theta_{2n}$ or $\alpha \in \Gamma^*_{2n}$. If $\alpha \in \Theta_{2n}$ it will be also in Δ_{2n+1} and if $\alpha \in \Gamma^*_{2n}$ it is certainly in Γ_{2n} and so either $\alpha \in \Delta_\omega \cap \Theta_\omega$ or $\alpha \in \Gamma_\omega$.

(14) Now we conclude that we may assume that our original pair (Δ_0, Γ_0) (Θ_0, Γ^*_0) is a pair of *saturated* theories and we shall prove the theorem for this case.

We now turn to another series of lemmas and constructions.

(15) Let $\phi = (\phi_1 \to \phi_2) \notin \Delta_0$. So clearly $(\Delta_0 \cup \{\phi_1\}, \{\phi_2\})$ is consistent. We now extend this theory to a complete and saturated theory $(\Delta_1(\phi), \Gamma_1(\phi)^*)$ in the language L_1. Let $\Gamma_1(\phi)$ be $\Gamma_1(\phi)^* \cap M_1$.

Lemma 16. The following is consistent: $(\Theta_0 \cup \{\psi \mid \psi \in \Delta_1(\phi) \cap M_1\}, \Gamma_1(\phi))$.

Proof. Otherwise for some $\theta_0 \in \Theta_0$, $\psi_1 \in \Delta_1(\phi) \cap M_1$ and $\gamma_1 \in \Gamma_1(\phi)$ we have

$$\vdash \theta_0 \wedge \psi_1 \to \gamma_1$$

$$\vdash \theta_0 \to (\psi_1 \to \gamma_1)$$

$$\vdash \theta_0 \to (\forall c_1 \ldots)(\psi_1 \to \gamma_1)\,.$$

Now since $\theta_0 \in \Theta_0$ we get that $(\forall c_1 \ldots)(\psi_1 \to \gamma_1) \in \Theta_0$ and so $(\forall c_1)(\psi_1 \to \gamma_1) \in \Delta_0 \subseteq \Delta_1(\phi)$ which is a contradiction.

Notice that we did not use the fact that we started with (Δ_0, Γ_0) and then transferred the ψ's to Θ_0. We could have started with $\phi_1 \to \phi_2 \notin \Theta_0$ and transferred to Δ_0. It does not matter whether we have Γ_0 or Γ^*_0 as second coordinate!

(17) Now extend the above pair of (16) to a saturated theory in the language M_2. Let us denote this theory by $(\Theta_2^\phi, \Gamma_2^{(\phi)})$. We are only interested in $(\Theta_2^\phi, \Gamma_1(\phi))$.

Lemma 18. The following is consistent: $(\Delta_1(\phi) \cup \{\psi_2 \mid \psi_2 \in \Theta_2^\phi \cap L_2\}, \Gamma_1(\phi)^*)$.

Proof. Otherwise for some $\delta_1 \in \Delta_2(\phi)$, $\psi_2 \in \Theta_2^\phi \cap L_2$ and $\gamma_1^* \in \Gamma_1(\phi)^*$ we have

$$\vdash \delta_1 \wedge \psi_2 \to \gamma_1^*$$

$$\vdash \exists c_2 \psi_2 \to (\delta_1 \to \gamma_1^*) .$$

Now if $\exists c_2 \psi_2$ (being in L_1) is not in $\Delta_1(\phi)$ then it is in $\Gamma_1(\phi)^*$, and since it is in M_1 as well, it must be in $\Gamma_1(\phi)$ but this contradicts $\psi_2 \in \Theta_2^\phi$.

So $\exists c_2 \psi_2 \in \Delta_1(\phi)$ and so $\delta_1 \to \gamma_1^* \in \Delta_1(\phi)$ but $\gamma_1^* \in \Gamma_1(\phi)^*$ and $\delta_1 \in \Delta_1(\phi)$, a contradiction.

(19) Now extend this pair to a saturated and complete theory $(\Delta_3(\phi), \Gamma_3(\phi)^*)$.

We can proceed now and show that the following is consistent: $(\Theta_2^\phi \cup \{\psi_3 \mid \psi_3 \in \Delta_3 \cap M_3\}, \Gamma_3(\phi))$ for otherwise for some $\theta_2, \psi_3, \gamma_3$

$$\vdash \theta_2 \wedge \psi_3 \to \gamma_3$$

$$\vdash \theta_2 \to (\forall c_3)(\psi_3 \to \gamma_3) .$$

So since $\theta_2 \in \Theta_2^\phi$ we get that $\forall c_3(\psi_3 \to \gamma_3) \in \Theta_2^\phi$ and so $\forall c_3(\psi_3 \to \gamma_3) \in \Delta_3(\phi)$ which is a contradiction.

We continue in this manner to obtain the following sequences

(20) $(\Delta_1(\phi), \Gamma_1(\phi)^*), (\Delta_3(\phi), \Gamma_3(\phi)^*) \ldots$

$(\Theta_2^\phi, \Gamma_1(\phi)), (\Theta_4^\phi, \Gamma_3(\phi)) \ldots .$

These sequences have properties (13) (a)–(e) stated for our case and in addition to this we have that

(21) $\phi_2 \in \Gamma_1(\phi)^*$.

We therefore have that in the limiting theories $(\Delta_\omega(\phi), \Gamma_\omega(\phi)^*), (\Theta_\omega^\phi, \Gamma_\omega(\phi))$ we have $\phi_1 \in \Delta_\omega(\phi), \phi_2 \in \Gamma_\omega(\phi)^*$.

(22) We may therefore conclude the following: Given $\Delta_0 \not\vdash \phi_1 \to \phi_2$ we can construct a saturated and complete theory $(\Delta_1(\phi), \Gamma_1(\phi)^*)$, which we may assume to be in the language L_1, such that $\phi_1 \in \Delta_1(\phi)$ and $\phi_2 \in \Gamma_1(\phi)^*$, and $\Delta_0 \subseteq \Delta_1(\phi)$. At the same time we can construct a saturated theory $(\theta_1^\phi, \Gamma_1(\phi))$ which is in the language M_1 with the properties

(a) $\Theta_0 \subseteq \Theta_1^\phi, \Gamma_1(\phi) = \Gamma_1(\phi)^* \cap M_1$

(b) $(\Delta_1(\phi) \cap \Theta_1^\phi, \Gamma_1(\phi))$ is a complete and saturated $M_1 \cap L_1$-theory.

A similar construction may be done for a $\phi = \phi_1 \to \phi_2 \notin \Theta_0$. We then denote the theories by $(\Theta_1(\phi), \Gamma_1(\phi)^*), (\Delta_1^\phi, \Gamma_1(\phi))$.

(23) Now assume that for some $\delta = \forall x\beta(x)$ we have that $\Delta_0 \not\vdash \forall x\beta(x)$ then for some u in L_1 we have $\Delta_0 \not\vdash \beta(u)$ and so $(\Delta_0, \{\beta(u)\})$ is a consistent theory. We can therefore repeat the process of (15)–(21) to obtain two pairs $(\Delta_1(\delta), \Gamma_1(\delta)^*), (\Theta_1^\delta, \Gamma_1(\delta))$ fulfilling (20) and also $\beta(u) \in \Gamma_1(\delta)^*$.

(24) Moreover, in the construction of $(\Delta_1(\phi), \Gamma_1(\phi)^*)$ we used no more than the fact that (Δ_0, Γ_0) and (Θ_0, Γ_0^*) fulfill (6) (used in (16)).

Now $(\Delta_1(\phi), \Gamma_1(\phi)^*)$ and $(\Theta_1^\phi, \Gamma_1(\phi))$ also fulfill (6) and therefore we can construct for example $(\Delta_2(\phi)(\epsilon), \Gamma_2(\phi)^*(\epsilon)^*)$ etc.

(25) For any pair $(\Delta, \Gamma), (\Theta, \Gamma^*)$ in the languages L_n, M_n, respectively, such that (6) holds and for any $\phi = \phi_1 \to \phi_2$ (or $\delta = \forall x\beta(x)$, such that $\Delta \not\vdash \phi$ (or $\Delta \not\vdash \delta$) we can construct theories $(\Delta_{n+1}(\phi), \Gamma_{n+1}(\phi)^*)$ and $\Theta_{n+1}(\phi), \Gamma_{n+1}(\phi))$ etc. as in (22) and (23).

Now before constructing our models let us recall that all our theories are saturated.

(26) We now construct two structures. One is associated with the Δ-theories and one is associated with the Θ-theories, we shall prove that these two structures are isomorphic in the common language $L_0 \cap M_0$.

We now define two sets S_Δ and S_Θ composed of ordered pairs whose first member is a theory and whose second member is a finite string of formulas. We shall also give a 1–1 map: $S_\Delta \to S_\Theta$.

(a) Let $((\Delta_0, \Gamma_0), \langle 0 \rangle) \in S_\Delta$ and $((\Theta_0, \Gamma_0^*), \langle 1 \rangle) \in S_\Theta$ and say that these two are *related*.

(b) For any $\phi = \phi_1 \to \phi_2(\forall x\beta(x))$ such that $\Delta_0 \not\vdash \phi_1 \to \phi_2(\Delta_0 \not\vdash \forall x\beta(x))$, put $((\Delta_1(\phi), \Gamma_1(\phi)^*), \langle 0\rangle^\frown\langle\phi\rangle) \in S_\Delta$ and $((\Theta_1^\phi, \Gamma_1(\phi)), \langle 1\rangle^\frown\langle\phi\rangle) \in S_\Theta$ and say that these two elements are related. Similarly for the case of $\forall x\beta(x)$. $\langle x\rangle$ denotes the sequences whose element is x and $^\frown$ denotes concatenation.

(c) For any $\phi = \phi_1 \to \phi_2 \notin \Theta_0$ put $((\Theta_1(\phi), \Gamma_1(\phi)^*), \langle 1, \phi\rangle) \in S_\Theta$ and $((\Delta_1^\phi, \Gamma_1(\phi)), \langle 0, \phi\rangle) \in S_\Delta$ and say that they are related. Similarly for $\forall x\beta(x) \notin \Theta_0$, we say that these elements were constructed in stage 1.

(d) Induction clause: Let $((\Delta, \Gamma_1, \langle 0\rangle^\frown t)$ and $((\Theta, \Gamma_2, \langle 1\rangle^\frown t)$ be any pair of related elements constructed in stage n. We assume the respective pair of theories to fulfill (6). Repeat the construction of (b) and (c) above for this pair of theories. The right-hand members of the new elements of S_Δ and S_Θ will be of the form $\langle 0\rangle^\frown t^\frown\langle\phi\rangle$ or $\langle 1\rangle^\frown t^\frown\langle\phi\rangle$ for ϕ as in (b) and (c).

Now S_Δ and S_Θ are our sets of possible worlds which we shall use to build our two models as is done in [5].

Note that from our construction it follows that if $(\Delta, \Gamma_1), (\Theta, \Gamma_2)$ are left-hand members of related elements then $(\Delta \cap \Theta, \Gamma_1 \cap \Gamma_2)$ is a complete and saturated theory of the common language.

Define on S_Δ: $((\Delta, \Gamma), t) \leqslant ((\Delta', \Gamma'), t')$ iff $\Delta \subseteq \Delta'$ and t is initial segment of t'.

Now let $[P(x_1 \dots x_n)]_{((\Delta,\Gamma),t)} = 1$ iff $P(x_1 \dots x_n) \in \Delta$ for atomic P. A similar definition would turn S_Θ into a model.

Claim. For any ϕ in L

$$(27) \qquad [\phi]_{((\Delta,\Gamma),t)} = 1 \quad \text{iff } \phi \in \Delta.$$

Proof. For atomic ϕ this holds by definition. $\wedge, \vee, \exists$ present no difficulties since the theories are saturated. For $\phi_1 \to \phi_2 \notin \Delta$ take $\Delta(\phi)$. We have that $\phi_1 \in \Delta(\phi)$ since $\phi_2 \notin \Delta(\phi)$ since $\phi_2 \in \Gamma(\phi)^*$. This yields the case of $\to$, and similarly for $\forall$. We regard $\sim\phi$ as $\phi \to 0$.

(28) A similar lemma holds for the Θ-model.

Now to show that these two models are isomorphic in the language $L_0 \cap M_0$ take the isomorphism as "*being related to*" and identity as the 1–1 map which preserves the common language part. In fact we have constructed the sister-theories so that we shall have isomorphism. Recall that all constants of M_0 or L_0 are in the common language.

By (27)–(28) we conclude that e.g. S_Δ is a model of $(\Delta_0 \cup \Theta_0, \Gamma_0^*)$.

3. Craig's theorem

Theorem 29. Robinson's statement (3) for theories such that $\Gamma_1 \subseteq \Gamma_2$ holds for the minimal and intuitionistic logics.

Proof. Our proof works for both cases. In fact it works for positive logic as well.

Theorem 30. Robinson's statement in (29) is equivalent to Craig interpolation theorem.

Proof. I: Assume Craig theorem. Now if $(\Delta_1 \cup \Delta_2, \Gamma_2)$ is not consistent then for $\phi_1 \in \Delta_1, \phi_2 \in \Delta_2, \gamma_2 \in \Gamma_2$ we have

$$\vdash \phi_1 \wedge \phi_2 \rightarrow \gamma_2$$

$$\vdash \phi_1 \rightarrow (\phi_2 \rightarrow \gamma_2)\,.$$

By Craig's theorem we have that for some $\phi \in L_1 \cap L_2$

$$\vdash \phi_1 \rightarrow \phi, \qquad \vdash \phi \rightarrow (\phi_2 \rightarrow \gamma_2)\,.$$

We conclude that $\phi \in \Delta_1$ and therefore $\phi \in \Delta_2$ and so $(\phi_2 \rightarrow \gamma_2) \in \Delta_2$ which is a contradiction.

II: Assume Robinson's theorem. Let $\vdash \phi_1 \rightarrow \phi_2$. Define

$$\Delta_0 = \{\phi \in L_1 \cap L_2 \mid \vdash \phi_1 \rightarrow \phi\}\,.$$

If for some $\phi \in \Delta$ we have $\vdash \phi \rightarrow \phi_2$ we are finished. Otherwise $(\Delta_0, \{\phi_2\})$ is consistent. Let us extend this pair to a complete L_2-theory (Δ_2, Γ_2). Let Δ, Γ be the common $L_1 \cap L_2$ part. We claim that $(\Delta \cup \{\phi_1\}, \Gamma)$ is a consistent theory. Otherwise for some $\phi \in \Delta, \gamma \in \Gamma$ we have

$$\vdash \phi \wedge \phi_1 \rightarrow \gamma$$

$$\vdash \phi_1 \rightarrow (\phi \rightarrow \gamma)$$

and so $(\phi \rightarrow \gamma) \in \Delta_0 \subseteq \Delta_2$ which is impossible.

Now $(\Delta \cup \{\phi_1\}, \Gamma), (\Delta_2, \Gamma_2)$ are consistent with $\Gamma \subseteq \Gamma_2$ and Δ, Γ complete in the common language. So $(\Delta_2 \cup \{\phi_1\}, \Gamma_2)$ is consistent. But this contradicts $\phi_2 \in \Gamma_2$.

Theorem 31. Craig's theorem holds for the minimal, intuitionistic and positive (without $\sim$) calculi.

To complete the readers' information, let us mention the following theorem:

Theorem 32. The following version # (Kreisel) of Robinson's theorem is equivalent to Craig's theorem for $\phi_1 \to \sim \phi_2$.

#: Let Δ_1, Δ_2, be two consistent theories in the languages L_1, L_2 respectively. Then if $(\Delta_1 \cap L_2) \cup (\Delta_2 \cap L_1)$ is consistent so is $\Delta_1 \cup \Delta_2$.

Remark 33. Theorems (30) and (32) are true for any extension of the minimal logic.

References

[1] D.M. Gabbay, Craig theorem for modal logics.
[2] D.M. Gabbay, Model theory for intuitionistic logic I.
[3] T. Nagashima, An extension of the Craig-Schütte interpolation theorem, Ann. Japan Assoc. Phil. Sci. 3 (1966) 12–18.
[4] K. Schütte, Der Interpolationssatz der intuitionistischen Prädikatenlogik, Math. Ann. 148 (1962) 192–200.
[5] R. Thomason, On the strong semantic completeness of the intuitionistic predicate calculus, J. Symbolic Logic 33 (1968) 1–7.

SEMANTIC PROOF OF CRAIG'S INTERPOLATION THEOREM FOR INTUITIONISTIC LOGIC AND EXTENSIONS, PART II [1)]

Dov M. GABBAY
Mathematics Institute, Oxford University

In this part we shall continue to apply the methods introduced in Part I and prove (in Section 1) the Craig's theorem for some important extensions of intuitionistic logic. We shall also refute (in Section 2) the strong form of Robinson's consistency statement for the intuitionistic logic (see (3) Part I). We shall rely heavily on Part I and the reader is urged to have it in front of him. Our numbering of theorems and definitions is consecutive with Part I.
We shall consider two axiom schema:

(34a) $\forall x \sim\sim \phi(x) \to \sim\sim \forall x \phi(x)$,

or equivalently

(34b) $\sim (\forall x \sim\sim \phi(x) \wedge \sim \forall x \phi(x))$,

(35) $\sim \phi \vee \sim\sim \phi$.

The system I + (34) appears in several contexts in the literature. It has property (36) below and it appears in connection with Spector's translation in [3].

(36) The following holds ([2]):
(a) $\vdash_C \phi$ iff $\vdash_{(34)} \sim\sim \phi$,

(b) $\vdash_C \phi$ iff $\vdash_{(34)} \phi$.

1) Supported by the Roral Society under the Royal Society – Israel Academy exchange programme.

Where in (b) ϕ contains only conjunctions, negations and universal quantifiers.

This is the generalisation of the translation of the classical propositional logic into the intuitionistic. (36) does not hold for I as was shown by Curry, and I + (34) is the smallest extension for which (36) holds (see [2]).

1. The results

Theorem 37. Craig's interpolation theorem holds for the following extensions of the intuitionistic predicate calculus:

(a) I + (34),

(b) I + (35),

(c) I + (34) + (35).

Before we embark on the proofs we need a completeness theorem for these systems:

Theorem 38.

(a) The system (37a) is complete for the class of all Kripke structures with partially ordered world systems fulfilling:

$$\forall x\, \exists y (xRy \wedge \forall z (yRz \to y = z)) .$$

(b) (37b) is complete for the class of all structures fulfilling:

$$\forall x \forall y \exists z (xRz \wedge yRz)$$

(c) (37c) is complete for the class of all structures fulfilling:

$$\exists x (\forall z (zRx) \wedge \forall y (xRy \to x = y)) .$$

Proof.

(38a) was proved in [2].

(38b) can be easily proved using a standard Henkin construction, and using Lemma (39) below.

(38c) follows from (a) and (b).

To proceed and prove (37) we shall modify constructions (25) and (26) of Part I. To be able to do that we need some lemmas:

Lemma 39. Let Δ be a saturated and consistent (35) – theory. Let $\Delta_1 \supseteq \Delta$, $\Delta_2 \supseteq \Delta$ be two consistent (35) – theories, then $\Delta_1 \cup \Delta_2$ is (35) – consistent.

Proof. otherwise for some $\phi_1 \in \Delta_1$ and $\phi_2 \in \Delta_2$ we have:

$$\vdash_{(35)} \phi_1 \rightarrow \sim \phi_2 .$$

(We assume that our theories are closed under conjunctions and provability.)

Let $c_1, \ldots$ and $c_2, \ldots$ be the constants and variables that appear in ϕ_1 and not in ϕ_2 (in ϕ_2 and not in ϕ_1, respectively). We have:

$$\vdash_{(35)} (\exists c_1, \ldots)\phi_1 \rightarrow (\forall c_2, \ldots) \sim \phi_2 .$$

Now since

$$\vdash_{(35)} \sim (\exists c_1, \ldots)\phi_1 \vee \sim\sim (\forall c_1, \ldots)\phi_1$$

and since $(\sim \exists c_1, \ldots)\phi_1$ is definitely not a member of Δ we conclude that $\sim\sim (\exists c_1, \ldots)\phi_1 \in \Delta$. Now since

$$\vdash (\phi \rightarrow \psi) \rightarrow (\sim\sim \phi \rightarrow \sim\sim \psi)$$

$$\vdash \sim\sim \forall x \sim \phi(x) \rightarrow \forall x \sim \phi(x) .$$

We get that $(\forall c_2, \ldots) \sim \phi_2 \in \Delta$ which is a contradiction.

Lemma 40. Let (Δ_0, Γ_0) and (θ_0, Γ_0^*) be two saturated (35) – theories fulfilling (5) and (6) of Part I and let $\Delta_1, \Delta_2, \Delta_3, \Theta_1$ and Θ_2 be saturated (35) – theories with the following properties:

(a) $\Delta_0 \subseteq \Delta_1, \Delta_0 \subseteq \Delta_2, \Theta_0 \subseteq \Theta_1, \Theta_0 \subseteq \Theta_2$

(b) $\Delta_1 \cup \Delta_2 \subseteq \Delta_3$,

(c) $\Delta_2 \cap M_1 = \Theta_1 \cap L_1 =$ a saturated theory

$\Delta_2 \cap M_2 = \Theta_2 \cap L_2 =$ a saturated theory.

Then $\Theta_1 \cup \Theta_2 \cup (\Delta_3 \cap M_3)$ is (35) – consistent.

Proof. Assume otherwise; then for some $\tau_1 \in \Theta_1, \tau_2 \in \Theta_2, \psi_3 \in \Delta_3 \cap M_3$:

$$\vdash_{(35)} \tau_1 \wedge \tau_2 \rightarrow \sim \psi_3.$$

Let $c_1, \ldots$ ($c_2, \ldots$, and $c_3, \ldots$ respectively) be the variables and constants that appear in τ_1 (in τ_2 and ψ_3 respectively) and not in the others, then:

$$\vdash_{(35)} (\exists c_1, \ldots)\tau_1 \rightarrow (\forall c_2, \ldots)(\tau_2 \rightarrow (\forall c_3, \ldots) \sim \psi_3);$$

one can easily see that $\sim\sim (\exists c_1, \ldots)\tau_1 \in \Theta$ and so

$$\sim\sim (\forall c_2, ...)(\tau_2 \to (\forall c_3, ...) \sim \psi_3) \in \Theta$$

and so

$$(\forall c_2, ...) \sim\sim (\tau_2 \to (\forall c_3, ...) \sim \psi_3) \in \Theta_2$$

and so

$$\sim\sim (\tau_2 \to (\forall c_3, ...) \sim \psi_3) \in \Theta_2$$

and so

$$\sim\sim \tau_2 \to \sim\sim (\forall c_3, ...) \sim \psi_3 \in \Theta_2$$

and so

$$(\forall c_3, ...) \sim \psi_3 \in \Theta_2 .$$

We therefore conclude, since $\psi_3 \in \Theta_2 \cap L_2$ that $(\forall c_3, ...) \sim \psi_3 \in \Delta_2 \subseteq \Delta_3$ which is a contradiction.

Lemma 41. (a) Let (Θ_0, Γ_0) be a (34) – consistent theory, then there exists a saturated classical extension $\Theta \supseteq \Theta_0$.

(b) Let (Δ_0, Γ_0) and (Θ_0, Γ_0^*) be two saturated (34) – theories fulfilling (5) and (6) of Part I and let $\Theta_0 \subseteq \Theta_1$ with Θ_1 a classical theory then $\Delta_0 \cup (\Theta_1 \cap L_1)$ can be extended to a classical theory.

Proof. (a) Since Θ_0 is (34) – consistent, it has a model. In this model (see 38a) there are endpoints. Let Θ be the theory of any endpoint.

(b) Follows from (a) and (8) of Part I.

(42) Let us now proceed to show that Robinson's consistency statement of (4b) of Part I, which is equivalent to Craig's interpolation theorem, holds for the system I + (34). Let $(\Delta_0, \Gamma_0), (\Theta_0, \Gamma_0^*)$ be two (34) – consistent and saturated theories in the languages L_0 and M_0 respectively, fulfilling (5) and (6). We want to show that $(\Delta_0 \cup \Theta_0, \Gamma_0^*)$ has a model. For this purpose let us perform all the constructions of (25) and (26) in Part I; of course we must remember that our theories are all (34) – saturated.

Now in the course of the construction, besides the construction of theories of the form $(\Delta(\phi_1 \to \phi_2), \Gamma(\phi_1 \to \phi_2))$ and $(\Delta(\forall x\, \beta(x)), \Gamma(\forall x\, \beta(x)))$

etc. ... We shall also construct a theory $(\Delta(\infty), 0)$ which is for any Δ the classical theory promised in (41). (41b) also assures us that a respective Θ^{∞} may be found. Of course the indices in (26) of Part I must be modified as to allow (∞) and $\langle\infty\rangle$ to be used.

We thus get two structures S_Δ and S_θ which are isomorphic and fulfill (38a). Thus the proof of (37a) is completed.

(43) Let us now try to prove (4b) of Part I for the system of (37b). Here the construction of the two models, which we shall denote by $S_\Delta(35)$ and $S_\theta(35)$, is more complicated; and we shall proceed by stages:

Stage 0: Construct the two isomorphic sets of theories S_Δ and S_θ as in (26) of Part I.

Stage 1: Let Ω be any diagram of the form shown below, where the theories are taken from S_Δ and S_θ and where the isomorphism and other relations can be deduced from the diagram

$$\begin{array}{ccccccc} (\Delta',\Gamma') & & (\Delta'',\Gamma'') & \qquad & (\Theta',\Gamma') & & (\Theta'',\Gamma'') \\ & \diagdown \quad \diagup & & & & \diagdown \quad \diagup & \\ & (\Delta_0,\Gamma_0) & & & & (\Theta_0,\Gamma_0) & \end{array}$$

Now by (39) and (40) and by (8) – (14) of Part I, we can find a pair of theories; $(\Delta(\Omega), 0)$ and $(\Theta(\Omega), 0)$ that fulfill (5) and (6) of Part I and such that $\Delta', \Delta'' \subseteq \Delta(\Omega)$ and $\Theta', \Theta'' \subseteq \Theta(\Omega)$.

For each Ω let us now regard the new pair $(\theta(\Omega), 0)$ and $(\Delta(\Omega), 0)$ as a new apir fulfilling (5) and (6) and so by (25) and (26) we can construct two isomorphic models of theories which we shall denote by $S_{\Delta(\Omega)}$ and $S_{\Theta(\Omega)}$, now let:

$$W_1 = S_\Delta \cup \bigcup_\Omega S_{\Delta(\Omega)}$$

$$V_1 = S_\Theta \cup \bigcup_\Omega S_{\Theta(\Omega)}$$

$$R_1 = \{(\Delta', \Delta(\Omega)), (\Delta'', \Delta(\Omega)) \mid \Delta', \Delta'' \in \Omega\}$$

and let $R(W_1)$ be the transitive closure of $R_1 \cup R(S_\Delta) \cup \bigcup_\Omega R(S_{\Delta(\Omega)})$.

Similarly we define $R(V_1)$. We now have that $(W_1, R(W_1))$ and $(V_2, R(V_1))$ are isomorphic in the sense of (26), (i.e. just like the way S_Δ and S_θ are!).

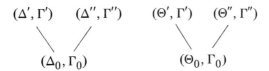

Stage n + 1: We regard $(W_n, R(W_n))$ and $(V_n, R(V_n))$ just like the S_Δ and S_θ of Stage 1 and thus construct $(W_{n+1}, R(W_{n+1}))$ and $(V_{n+1}, R(V_{n+1}))$. Let

$$S_\Delta(35) = \cup W_n$$

$$S_\theta(35) = \cup V_n$$

and define R on each of them in the proper manner. Clearly these sets fulfill (38b) and so give rise to the required model. Thus (37b) is proved.

(44). Let us now turn to prove (37c). For this case we proceed as follows:

First we regard our two theories (Δ_0, Γ_0) and (θ_0, Γ_0^*) as (37a) theories and so construct the two isomorphic worlds of theories S_Δ and S_θ as in the case of (37a) and so they have property (38a). But the theories concerned are also (37b) theories; we shall use this fact to prove that a theory $\Delta(\infty)$ may be added on top of S_Δ and an isomorphic theory $\Theta(\infty)$ may be added on top of S_θ. This will conclude the proof. To be able to do this we note the following three facts (we leave the details to the reader):

(a) From (35) it follows that in each model all the endpoints are elementarily equivalent.

(b) The union of the endpoints in each model is a (37c) – consistent theory. The proof is analogous to the proof of (39).

(c) The statement analogous to statement (40) holds.

Thus (37c) is proved.

2. Refutation of the strong Robinson's statement

The following example shows that statement (4a) of Part I is false for the intuitionistic predicate logic.

Example 45: Our structure A is in the language with one unary predicate $P(x)$. S equals ω. R is the natural order $\leqslant$. A_{2n} are all isomorphic. They have $\aleph_0$ members x with $P(x) = 1$ and $\aleph_0$ members x with $P(x) = 0$.

A_{2n+1} are all isomorphic. They have $\aleph_0$ members and every $x \in A_{2n+1}$ is such that $P(x) = 1$. We describe the imbedding of A_n in A_{n+1} (or the "identifying" of elements if we want $A_n \subseteq A_{n+1}$).

We write $A_n = \{\aleph_0^+; \aleph_0^-\}$ where $\aleph_0^+$ represents the $\aleph_0$ elements x such that $P(x) = 1$ and $\aleph_0^-$ represents the $\aleph_0$ elements such that $P(x) = 0$. Arrows denote the embedding.

$A_4 \quad \{\aleph_0^+ \parallel \aleph_0^-\}$

$\uparrow$

$A_3 \quad \{\aleph_0^+\}$

$\uparrow \ \nwarrow$

$A_2 \quad \{\aleph_0^+ \mid \aleph_0^-\}$

$\uparrow$

$A_1 \quad \{\aleph_0^+\}$

$\uparrow \ \nwarrow$

$A_0 \quad \{\aleph_0^+ \mid \aleph_0^-\}.$

We require of the embedding f shown by $\uparrow$ in the diagram that in the next (higher) model there are $\aleph_0$ elements not in the range of f such that $P(x) = 1$ (i.e. in $\aleph_0^+$).

A and its truncation at A^{2n} are obviously isomorphic. Also A and A^{2n+1} are elementarily equivalent, since if ϕ holds in A it holds also in A^{2n+1} and if ϕ holds in A^{2n+1} then it holds also in A^{2n+2} which is isomorphic to A. (Where A^m is the model defined by $S^m = \{n \mid n \geqslant m\}$.)

Theorem 46 (refutation of a strong Robinson consistency statement (4a)). There exist two consistent theories (Δ_1, Θ_1), (Δ_2, Θ_2) such that the common theory (Δ, Θ) is complete but $(\Delta_1 \cup \Delta_2, \Theta_1 \cup \Theta_2)$ is *not* consistent.

Proof. Take Example 45. Our language L has only one unary predicate $P(x)$. Let L_1 be L with a new unary predicate P_1 and let L_2 be L with a new unary predicate P_2. Let (Δ, Θ) be the complete theory of the model of the example (Θ is actually empty). Now let

$$\Delta_1 = \Delta \cup \{\forall x (P_1(x) \vee P(x)); \exists x P_1(x)\}$$

$$\Theta_1 = \Theta \cup \{\exists x (P_1(x) \wedge P(x))\}$$

$$\Delta_2 = \Delta \cup \{\forall x (P_2(x) \vee P(x))\}$$

$$\Theta_2 = \Theta \cup \{\exists x (P_2(x) \wedge P(x)), \exists x P_2(x)\}.$$

(Δ_i, Θ_i) both say that in the base model $[P(x)]_0 = 1 \leftrightarrow [P_i(x)]_0 = 0$.

(Δ_1, Θ_1) is consistent since it holds at A_0 of Example 35. When one takes $[P_1(x)]_0 = 1$ iff $[P(x)]_0 = 0$ and $[P_1(x)]_m = 1$, for $m > 0$.

(Δ_2, Θ_2) holds at A_1 where one takes $[P_2(x)]_1 = 1$ iff $[P(x)]_1 = 0$ and $[P_2(x)]_m = 1$ for $m > 1$.

The union $(\Delta_1 \cup \Delta_2, \Theta_1 \cup \Theta_2)$ is not consistent.

Open problems.

(1) Is Craig's theorem true for the logic of constant domains? i.e. for $I + \forall x(\phi \vee \psi(x)) \to (\phi \vee \forall x\, \psi(x))$. Where x is not free in ϕ.

(2) Is the strong Robinson's statement refutable for the logic in (1)? (by the way, it is refutable for (37a) – (37c)).

(3) In view of (36) we deduce that if $\vdash_{(34)} \phi_1 \to \phi_2$ and ϕ_1 and ϕ_2 are built up using $\sim, \wedge, \forall$ only then the interpolation formula ϕ may be taken to be in $\sim, \wedge, \forall$ only. Is it true for the intuitionistic logic?

References

[1] D.M. Gabbay, Semantic proof of Craig's interpolation theorem for intuitionistic logic and extensions Part I, Manchester Proc. (North-Holland Publ. Co., 1969).
[2] D.M. Gabbay, Applications of trees to intermediate logics, to appear.
[3] C. Spector, Provably recursive functionals, in: Recursive function theory (AMS, 1962).

KREISEL'S WORK ON THE PHILOSOPHY OF MATHEMATICS – I. REALISM

J.M.B. MOSS
University of Wales, Aberystwyth, Wales

1. Introduction

From the references in his writings, Wittgenstein and Gödel emerge as the two most important intellectual influences on Kreisel (hereafter K). He knew the former from 1942 to 1951, and attended his seminars; though they "spent a lot of time together talking about the foundations of mathematics" [W I, p. 157], I have not attempted to spell out this influence in detail, since K did not write on philosophical topics until after he had reacted against it. However, his autobiographical statement, c. 1960, should be noted: "I believe that early contact with Wittgenstein's outlook has hindered rather than helped me to establish a fruitful perspective on philosophy as a discipline in its own right, and not merely for example as methodology of highly developed sciences". He goes on to add that "... most philosophy of the day, including consciously anti-Wittgensteinian brands, does not seem much better. Exceptions are, in my opinion, the philosophical writings of Bernays and Gödel [1], ..., the usefulness of [whose] ideas ... is, at least in my opinion, not confined to the area of mathematical philosophy" [W II, p. 251].

Indications in his early writings, even though some appeared in philosophical journals and conference proceedings, confirm that at that time K regarded philosophy as nothing more than "methodology of highly developed sciences" Both the remark: "a philosophy should explain what makes arithmetic *transparent* and *certain*" [VHT, p. 121] and his work of that period on the constructive content of classical mathematics reflect, in different ways, a philosophical ideology more than a philosophy. So there is no need to discuss here *either* the 'no-counter-example' interpretation of (formal) arithmetic and

[1] In [W I] and [W II], K refers specifically to the four papers by Gödel and Bernays reprinted in [B & P], and also to Bernays (1950) and Gödel (1958), which have not yet appeared in translation.

parts of analysis, in which he shows how statements of these branches can be replaced (interpreted) by an ω-sequence of quantifier-free formulae, *or* his early discussion [NAM I] of the constructive content of Cantor's diagonal argument.

It is clear that contact with Gödel, dating from *circa* 1955, has considerably influenced K's recent view of philosophy and the philosophy of mathematics. However, though in certain comparatively obvious ways this indebtedness is both clear and frequently acknowledged, it may be more significant at the methodological level than for the future development of the philosophy of mathematics. For K raises once again – and in a particularly sharp form – the traditional philosophical problems about the nature and verification of philosophical propositions; and though his major claims are not all established, I believe that his work contains philosophical insights which are both distinctive and potentially exciting.

It may prove helpful to the reader who has noted the frequent references in K's papers to applied mathematics (or mathematical physics) to recall that K's first two published papers and, I am told, some wartime research were on topics in this field. Also, the (very English) attitude, occasionally displayed in his writings, which may be connected with this, namely that foundations of mathematics is like applied mathematics, and very unlike 'real', i.e. pure, mathematics (e.g. [Sch, p. 204, 18]) should perhaps be recalled before it is forgotten.

2. Preliminary bibliographical note

The first papers containing a contribution to philosophy are [W I] and [HPr_1], both of which date from about 1958. During the following years, apart from the explicitly philosophical [W II] and the review [RS], a number of technical papers include discussion of philosophical issues raised by various types of constructivism. But the most decisive philosophical contribution is in a group of papers published or written in about 1965, namely [Saa], [Lak], [Sch], and Appendix II of [Kaka]. The additions to [HPr_2], the revised version of [HPr_1], and the review [RW] also date from this period. Recent papers include [SPT], [Lux], [Har], [Br], and [Max], but in so far as these contribute to philosophical issues, they mainly elaborate points of view presented earlier. In addition, K has five very recent papers on foundational topics which are listed below in the bibliography, and of these special mention should be made of [FPD]. Though mainly a critique of formalism, its discussion of the bearing of conservative extension results upon the evidence for and truth-value of mathematical propositions is important in connection with strong realism; I am grateful to K for sending me a preprint.

Much of K's argument revolves around three (types of) position which are sometimes labelled Formalism, Idealism and Realism, of which the first is the "baddie" while the other two are taken seriously and, it is suggested, might turn out not to exclude each other. Before presenting these views, it might be helpful to mention that, in his recent writings, K holds that philosophical questions are frequently unambiguous, and have answers [2], which will sometimes emerge through a mixture of technical progress in the foundations of mathematics and conceptual clarification. Lest the reader should feel that this view is absurd, let him consider, in the spirit of a traditionally conceived logicism, that the aim ot much of philosophy – and of almost the whole of philosophy of mathematics – has often been taken to be the search for certain elusive necessary truths, and on this view philosophy and logic (including mathematics) aim to discover those necessary truths which do not depend for their expression upon either the empirical world or the theories of the physical sciences. If, therefore, mathematics is a branch of philosophy, it is natural to believe that philosophical questions are (frequently) unambiguous, answerable, and answerable as a result of mathematical, i.e. philosophical, progress. (*Addendum.* In a comment on this passage, K writes, in a letter of 17 March 1970, that he does not regard "necessary truths as particularly elusive" or the empirical world as contingent, and adds "From [Saa], at the end, onwards, I have always contrasted philosophical work with (empirical) *case studies*; not with empirical subject matter. What I am optimistic about is the use of *qualitative* features of experience as a means of understanding experience, in contrast to the case studies which the empiricist tradition emphasises." See p. 419 below.)

3. Definitions

I start by joining together some quotations to serve as definitions of the views to be discussed. Though the words are Kreisel's, I have not indicated omissions, so I leave out quotation marks.

[2] I.e. correct answers, not dependent on pragmatic considerations. In 1952, however, K had written [RM, p. 365], "It is not reasonable to ask without qualification whether a formal system is a *correct* formalisation of some given branch of mathematics, because the features which are to be considered characteristic of, or essential to, a branch are determined by the applications to which it happens to be put". See also [Sch, p. 211, top] and [SPT, p. 361].

The data of foundations consist of the mathematical experience of the working mathematician; the general problem of foundations is to analyse this experience as a *whole*.

Crude formalism holds that mathematics consists of assertions of the form: – a concretely given configuration has been constructed by means of a given mechanical rule. No general statements about such configurations belong to mathematics.

The most important point of the formalist doctrine is this: all questions which go beyond such elementary acts of recognition as to whether or not a sequence of symbolic expressions is formed according to some finite number of mechanical rules, are regarded as outside mathematics.

The positivist doctrine considers informal derivations either as unreliable or as irrelevant to mathematics. According to this doctrine, the traditional problems are ignored rather than resolved on the ground that they lack precision.

The idealist conception is that mathematics is a free mental activity; intuitionism, a solipsistic doctrine, seems to be the only version that has been developed, and here one confines oneself to thoughts which are about mathematical activity and not about external objects. The essence of the idealist conception consists precisely in its assumption that there are relatively few notions that are *basic* in our mathematical experience.

Realism is the assumption that there are basic elements (sets and the membership relation) with the properties assumed in the cumulative hierarchy; that is, the existential assumptions of set theory are valid. The realist position is here taken in its *strong* form, as involving the existence of sets of high ordinal; but not in its *strict* form – that *all* mathematical notions are built up from the notion of set by means of logical definitions in the language of pure set theory ([Saa, p. 95], [Kaka, p. 220], [Sch, pp. 224, 225], [Kaka, pp. 163, 173], [Saa, pp. 129, 187, 188], [Sch, pp. 191, 245], [Sch, pp. 213, 219].)

In what follows, I aim to give a critical analysis of some central features of the above three versions of realism. Formalism and idealism are mentioned only in so far as they bear upon the realist thesis [3]. There is a brief note at the end on K's conception of general philosophy, and also two brief passages in which I attempt an account of a version of realism which might meet some of the difficulties raised against the thesis; it will be clear that these attempts are highly tentative.

[3] The sequel will contain a detailed discussion of K's approach to constructive mathematics and to the various versions of formalism and idealism defined above.

4. Three versions of realism

The definitions cited above give rise to three distinct versions of the thesis, which can be called weak realism, strong realism, and strict realism, respectively. The weak form claims only "the existence of *some* mathematical objects" [Sch, p. 219]; though K finds it ambiguous, it is implied by the other two forms, and ambiguities in its meaning will help to determine that of the strong and the strict forms.

Strict realism holds in effect that set theory is the only foundation for mathematics, and is rejected for reasons which are of some general interest, which I proceed to discuss. Set theory is here understood, following Zermelo (1930), a paper which K describes as giving an "unsurpassingly clear foundation of set theory" [Har, p. 13], and Gödel (1944), as the theory of a cumulative rank structure (c.r.s.). A given set or 'collection' of not necessarily spatio-temporal objects constitutes the lowest level (rank zero) of the structure (or hierarchy); this might be, for example, $\emptyset$, N [4], a generic subset of N, or a collection of contingently existent 'atoms', though in this last case additional problems will arise concerning modalities. The sets of higher ranks – a rank is an ordinal number – are then built up through repetition of the sum-set and power-set operations applied to sets of lower rank. Specifically, if V_α is the set of objects of rank α, where $\alpha \neq 0$, then $V_\alpha = \bigcup_{\beta<\alpha} (V_\beta \cup \mathcal{P}(V_\beta))$ [5].

Note: it is sometimes held, though not by K, that the *only* coherent notion of a set is as an element of a c.r.s., or even that the intuitive notion of set is that of a collection formed at some stage in the construction of such a hierarchy (Shoenfield, 1967, p. 238). I have elsewhere characterised these views (in a paper: "Conventionalism in Logic") as "the Stanford heresy", but Myhill pointed out to me that this ascription presupposes a (different) orthodoxy. At any rate, it is clear that, without substantial further analysis, there is no objection to the existence of competing foundational theories, whether these are called set theories or theories of classes, structures, collections, categories, or extensions of properties.

By writing down sentences which characterise the rank structure built up from the null-set, the set-theoretic consequences of the first-order versions of these principles can be proved (Montague, Lévy) to include the theorems of Zermelo-Fraenkel (ZF) set theory, and the ordinal-theoretic consequences of

[4] $\emptyset$ = the null-set, N = the set of natural numbers. Also (below) CH = Cantor's continuum hypothesis.

[5] Sometimes, for metamathematical work, a narrower collecting operation is used.

the first-order axioms expressible in set theory can be proved using any of the standard definitions of ordinals in ZF. However, though certain partial reductions (both ways) between sets and ordinals are possible, both notions are primitive for the cumulative structure, and it cannot be proved either that ordinals are sets or that sets are ordinals. Despite the possibility of translatability results, it is evident that the members of the c.r.s. are not the only mathematical objects [6], and indeed it is not to be expected either that all propositions of mathematics, if expressible at all, are expressible within the language of set theory, or that the truth-values of those which are so expressible can be determined using set-theoretic reasoning. Indeed, on the analogy of extending the language of arithmetic by a truth predicate, or a hierarchy of them, K expresses his belief [Sch, p. 215], [Saa, p. 111], [Lak, p. 150] that resolution of e.g. CH will require notions not expressible in the language of set theory, and he mentions (e.g. [Kaka, p. 189], [Saa, pp. 117, 118]) the possibility of an alternative and perhaps more fundamental realistic foundation for mathematics in which set-theoretic notions would be definable but not vice-versa. These alternatives are not very fully explored, though theories based either on one of Quine's systems, or on category theory, might be felt to be interesting candidates – though K rejects both these suggestions (see [Mac]).

In the same way as Gödel (1944), K mentions three notions of set which have been conflated in the literature [Lak, p. 143], [Kaka, pp. 171, 172], [Saa, pp. 99, 188], [Sch, pp. 218, 219], namely: (1) an analogue of finite collections; (2) what is characterised by the c.r.s. described above; and (3) the extension of a property. (1) would clearly be inadequate as a foundation for mathematics, and K does not discuss it. In connection with (3), he notes [Kaka, pp. 173, 174, 189] that, though it has not been studied extensively, it can plausibly be supposed to be a more fundamental notion for a realistic foundation of mathematics than the better understood (2). However, K sometimes claims that on interpretation (3), the unrestricted comprehension axiom also holds, which implies that the principles of (2-valued) logic must fail [Saa, p. 188]. But since these principles hold for realistic views, (3) could then yield a (set-theoretic) foundation for idealistic mathematics *only*. The general approach which lies behind (3) might, however, prove fruitful, and it is interesting to note in this connection that work is in progress (by

[6] The translatability of (first-order) ordinal arithmetic into set theory [Sch, p. 272] fails to extend to second-order logic, which might also be felt to count against the thesis of strict realism. I am greatly indebted to Saul Kripke, Peter Aczel, and Mike Yates for discussions on this point.

Dana Scott) on the foundations for a theory in which, unlike the rank structure, self-application of mathematical objects (functions) can occur.

Inter alia, the above considerations show the falsity of strict realism. Also, a technical result in connection with axioms of infinity (not mentioned above) which K formulates as "any axiom system for set theory is inadequate or unjustified set-theoretically" [Sch, p. 217] should be noted. Assuming that an adequate system for set theory is capable of some measure of justification (see Section 8 below), this again implies that strict realism is false; I have not attempted to present this argument rigorously, and to do so would, as Yates has emphasised, require discussion of the technical question: what distortions arise through the use of approximations provable within the c.r.s. for sentences about the structure which cannot be exactly mirrored within the system?

Strong realism is committed to the existence of sets, some containing an infinite number of members, which exist independently of minds. But I postpone discussion of this thesis to Section 9, concentrating in the next four sections on the thesis of weak realism, *viz*. that there *are* mathematical objects.

5. Weak realism

I take one of K's most important philosophical remarks to be the following: "the notion of a mathematical object is defective because one has no clue for using it to provide satisfactory answers to the (philosophically significant) questions which it should answer, but no case has been made out that it cannot do so – rather like philosophy itself" [W I, p. 138]. Two years later, in 1960, K announced his reluctant conversion to the view that there must be abstract structures [W II, p. 245]; and central to his more recent refutations of formalist doctrines is his attack on the thesis that there are no mathematical objects. But it is not clear from his recent writings how far he has succeeded in advancing beyond the claim that there must be mathematical objects, and his latest comment, in footnote 10 of [Lux], is very much in the spirit of the 1958 remark quoted above. Assuming that a realistic view is possible, he asks, how can it be employed (see also [Sch, pp. 219–221])? But the prior question is: is a realist view possible? For fundamental to all versions of realism is a whole-hearted commitment to the view that mathematical necessity is not trivial; that there are, in an absolute sense of "possible", possible worlds in which mathematical objects do not exist. And it is far from clear what this possibility might amount to.

The problem just raised is essentially the more general one: is ontology possible? (Cf. the last four words of the last quotation above.) Consider the following argument: if there are propositions of ontology, they – or at least those relevant to a realist philosophy of mathematics – are (presumably) either necessarily true or necessarily false. But if this is so, no observations can count either for or against them, and no other evidence is immune to the objections of the sceptic. Hence, because it seems impossible to test ontological claims, their truth-value makes no difference. And even if the verifiability principle is rejected for empirical knowledge, a case can be made out for rejecting ontological claims of *this* sort as meaningless.

In addition, it is certainly false that abstract objects causally influence our organism; the alternative is magic, which is (empirically) absurd (see [Sch, p. 268]). This fundamental point is often overlooked, but it is surely correct; if I can be said to be apprehending a (particular) proposition, it is not because my mind or organism is causally affected by any external abstract entity, and if there are external mathematical objects, they have no explanatory rôle in connection with mathematical behaviour; in short, whatever we *do* want to say about mathematical knowledge, it is unreasonable to hold that mathematical and intellectual behaviour can be explained by assigning a causal activity to abstract entities. What interpretation, then, can be given to the thesis of weak realism? For though K claims [Saa, p. 186] that "the realist view has never been developed far enough to be put to a real test", the argument above suggests that it is in principle incapable of test, and that it is therefore obscure as to what *any* formulation of it might amount to. Until this objection is met (in Section 9), K's remarks: "what characterises the difference between e.g. the idealist and the realist view is what aspects of (crude) experience are regarded as significant and suitable for study" and "I do not know a formulation of the realist view for which experience establishes that there are infinite sets (in the sense of set theory)" [Saa, pp. 186, 190] are premature. And, by what is said above, if the aim of foundations is to analyse mathematical *experience*, realism is not directly relevant.

Obscure though the thesis is, could it, nevertheless, be true? K, at least, expresses slight unease with (one aspect of) Gödel's thoroughgoing Platonism: "it seems to be unusual in the natural sciences to have rival theories where one *knows* that they will not conflict on observable consequences" [Saa, p. 187 – see also p. 186], and [Sch, pp. 220, 221], which seems to acknowledge a conceptual (philosophical) difference between ontological assumptions in physics and in mathematics [7]. But because of this difference, no account has been given of the meaning of mathematical assertions. (*Addendum*. In his letter of 17 March 1970, K says that he has doubts about Gödel's *particular* tests by arithmetic consequences, but – implicitly – not about Gödel's general platonist position. This would agree with the doubts about the necessary-contingent distinction mentioned above (p. 413), but I cannot see that it throws light upon the meaning of sentences of e.g. mathematical analysis or set theory.)

A tentative attempt to make sense of a realistic theory is as follows – whether either Kreisel or Gödel would subscribe to any part of it, I do not know. Some true necessary propositions are vacuous, i.e. true in all possible worlds (of course, the notion of a possible world requires independent explanation). Some or all true mathematical propositions, however, are not true in all possible worlds but rather, as Descartes held, we are not able to make intelligible to ourselves any of the possible worlds in which they are false. From this point of view it might be illuminating to interpret a possible world as a Boolean-valued model of set theory [7a], so that though the axioms of ZF would be necessarily true, and the axiom of choice, if true, necessarily so, CH would be true in some worlds but false in others. Some mathematical truths would therefore be truths of logic [8], while the rest would be ontologically akin to general physical truths though, for some philosophically unimportant reason, immune to observational test.

It would be interesting if this idea could be developed, but even if this were done, its application to philosophical issues would require both a general account, (presumably) not dependent upon either linguistic or epistemological theses, of the range of the sentences of logic, and an indication as to how the truth-values of the propositions expressed by the distinctively mathematical sentences might be established, since observation can be of no help.

[7] Note however the following (puzzling) comment: "It would be agreed that the realist assumption of external mathematical objects is not more *dubious* than that of physical objects" [Saa, p. 186]. But it would not be agreed; no nominalist would accept the assumption that there are mathematical objects.

[7a] This interpretation of modal logic in terms of Boolean-valued models was suggested to me by Peter Aczel.

[8] I omit discussion of a (quite basic) point, viz. that what makes a truth a logical truth is not what it says but how it is expressed; thus "all bachelors are male", though analytic, is not a sentence of logic, unlike "all unmarried males are male", which expresses the same proposition. But if some mathematical propositions are true in all possible worlds, it can be argued (though I omit the argument here) that, if they are expressible, the sentences which express them are sentences of logic, and therefore that they are logical truths.

Very little work has been done on either of these, though in connection with the former there is some recent work by Hintikka in support of the view that many logical truths of first-order logic are synthetic a priori, and not "surface tautologies" like the logical truths of propositional and monadic quantificational logics (see Hintikka (1968), and references cited therein in footnotes 1 and 2; J.L. Mackie (1966) is a variation on a related theme). However, Church's theorem on the impossibility of a decision procedure for first-order logic plays a central rôle in Hintikka's argument, and a result which limits what can be known rather than what there is can hardly be important in connection with an *ontological* [9] thesis. Thus, if there is a distinction to be drawn between logic and 'substantial' mathematics, the whole of first-order logic and perhaps also formal arithmetic will surely be on the logic side.

On the second problem, K (and also Dummett) have explored, somewhat tentatively, the view that the various non-standard models of number theory presuppose the standard model. (See e.g. [Lak, pp. 165–169] and [Lux].) Though K gives a partial affirmative answer for one explanation of "presuppose" (in terms of definability), it is doubtful if his approach could be used to given even a 'moral' certainty for or against CH.

There is also the pragmatist position espoused by Gödel in: "There might exist axioms so abundant in their verifiable consequences, shedding so much light upon a whole field, and yielding such powerful methods for solving problems ... that, *no matter whether or not they are intrinsically necessary,* they would have to be accepted *at least in the same sense* as any well-established physical theory" (1947, in [B & P, p. 265; also pp. 264–269]; my italics). But if some mathematical propositions lack 'intrinsic necessity', it is hard to see how their truth-values can be established in a similar way to those of propositions of a physical theory – see two pages back for a reference to K's reservations about Gödel's suggestion.

Thus, if a realist view aims to distinguish between (substantial) mathematical propositions and (trivial) logical ones [10] on the grounds of some intrinsic distinction between them, no specification has been made of any such difference, and no method given by which the truth-value of the mathematical propositions could be established.

[9] A.J.T.D. Wisdom once wrote "Epistemology ... and ontology ... are one" (1948, p. 229). Fundamental to all versions of realism, and to some of idealism, is the rejection of Wisdom's thesis.

[10] I have not discussed the realist view of M.R. Cohen and Blanshard, which holds that all propositions of logic are substantial; there is, to my mind, a basic implausibility in the claim that, for example, "everything is identical with itself" is not a trivial truth, but a non-trivial general "principle of being".

To return, then, to the mathematical object: what role does it play? In his recent writings, K makes, as I see it, two main points. Firstly, that the various versions of formalism are false and that there must be mathematical objects, so that the key question becomes: are these objects mind-dependent (idealism) or do they exist externally to ourselves (realism)? Secondly, that it is through the use of higher-order logic that we may come to be able to justify the existential claims of a relatist view.

The following points now call for discussion:

(1) The relation between idealism and realism (Section 6).

(2) The falsity of formalism, and, more generally, the relationship between formalism and realism (Section 7).

(3) The relation between first-order logic, second-order logic and set theory (Section 8).

(4) The nature and possibility of justification of realist theses in general, and strong realism in particular (Section 9).

6. Realism and idealism

The present paper does not seek to examine K's substantial and important contributions to intuitionism, though the sequel contains a discussion of constructivist mathematics in general, and of K's work on finitist and predicativist foundations in particular. Here I confine myself, very briefly, to two general points that arise in connection with his versions of realism and idealism: firstly, their relationship with the traditional philosophical theses associated with the words "realism" and "idealism", and, secondly, their compatibility.

As mentioned above, realist doctrines for K entail, at least, the existence of *some* mathematical objects, i.e. abstract entities. However, the traditional philosophical doctrines whereby realism is opposed to idealism, rather than nominalism, have emphasised the existence of the external (so-called) material world, against its mental reinterpretation by e.g. Berkeley and his latter-day successors. Now first-year philosophy students are taught (as one says) that realism as an epistemological doctrine bears no relation to ontological realism, viz. the thesis that there are abstract entities (universals), and even those opposed to a distinction between epistemology and ontology have seemed disinclined to assimilate questions of justification to questions of existence. Though the received doctrine just mentioned is a gross over-simplification, ignoring as it does *both* epistemological questions about universals, deriving from Locke's conceptualism, *and* the ontological commitments, sometimes hidden, of idealists from Berkeley [11] to Bradley and of realistic views about laws of nature, it is nevertheless *only* an oversimplification. Thus if Wisdom's reductionist claim (see above, footnote 9) is rejected, then, although ontological problems arise in connection with epistemological realism, and vice versa, the two theses are quite different.

Now K holds both (1) that there are mathematical objects which must exist either in our minds or in the external world, and (2) that Idealism and Realism can be sharply formulated as mathematical theses, and when this has been achieved, it may turn out that they do not exclude one another. The apparent conflict between (1) and (2) is resolved by: "... a non-realist view of mathematical evidence is not directly *inconsistent* with the existence of of mathematical objects. But it ... has a *negative* consequence for the realist view; for if it is shown to 'account' for our mathematical experience, then ... mathematical experience does not *force* one to adopt a realist interpretation" ([Sch, p. 222]) [12]; but problems remain in connection with both the role of intellectual intuition and the status of such specific theses as e.g. the law of excluded middle. As to the former, K claims that idealism involves notions with which we are directly presented, and even though we should not expect to be able to decide *all* the properties of our mental constructions [Sch, p. 267], nevertheless the discovery of idealist or intuitionist principles, unlike realist ones [13], is apparently to be carried out by some process of linguistic phenomenology or conceptual analysis. The rejection of the verificationist and post-Kantian approaches to idealism is especially noteworthy here, as also in K's rhetorical question "why should an idealistic conception reject

11) Not for nothing did C.S. Peirce once (1871) characterise Berkeley's philosophy as a "strange union of nominalism with Platonism" which illustrates the British national character, and which was earlier exemplified by John of Salisbury – though also by Abelard [CP VIII, pp. 11, 31].

12) Thus, though K is here discussing (an aspect of) epistemological realism, his primary concern is with an ontological aspect of it. Consider the following analogy: there are chairs, therefore either they exist in our minds or in the external world. If, for example, a non-realist view about chairs were shown to account for our knowledge of them, this would no doubt have a negative consequence for the general thesis of epistemological realism. (The difference of course is that chairs, assuming that they exist, causally affect our organism, while mathematical objects don't; see Section 9 for a suggestion of a realist view designed to meet *this* objection).

13) "If something in mathematical experience is to be really evident, it must be wholly about such experience, and not, for instance, ultimately about objects external to ourselves or even about concepts of such objects, for instance of sets" [Sch, pp. 242–43]. Though K characterises this doctrine only as "problematic", it raises a vital issue. For if it were true, how could any realist principles be discovered, since not through introspection?

ideal elements?" [Saa, p. 138], and his support for an idealist version of set theory. But, on this approach, why should idealism reject such a realist thesis as excluded middle?

To this question, the obvious answers of tolerance – let the intuitionist do mathematics his way – and multifurcation (of the concepts of disjunction and negation) are clearly too formalist in intent. Perhaps therefore K sees intuitionism as a purely epistemological thesis, but, if so, there seems to be *no* objection to the (classical) law of excluded middle. The Brouwer-Heyting thesis that the law is an illegitimate because metaphysical idealisation is not open to K, who raises no objection to metaphysical idealisations. But this leaves him no reason to accept the fundamental intuitionist claim that "in the study of mental mathematical constructions 'to exist' must be synonymous with 'to be constructed' " (Heyting, 1956, p. 2, or [B & P, p. 56]) – the (usual) language-dependent and mind-dependent concepts of existence and constructibility are not identical. Perhaps K's remarks about Cohen's notion of forcing [Sch, p. 253], [Saa, p. 107ff], which he strikingly anticipated in [FCS] and [STP], confirm the suggestion here made, but however this may be, the conclusion to be drawn is that K's *new* idealism in no way leads to the rejection of realist laws. Thus, unlike Brouwer and Heyting, he rejects, not the *classical* law of excluded middle, but only the *quite different* law which is expressed by the same symbols but where '$\neg$' and '$\vee$' have the intuitionist interpretation.

7. Realism and formalism

In contrast to K's optimistic view that philosophical questions, including some of the traditional problems in the philosophy of mathematics, frequently permit a sharp formulation and resolution, stands an alternative doctrine, roughly corresponding to one held by Curry and Carnap [Sch, p. 228], which runs as follows:

(Version I) Formal systems embody no metaphysical commitment. Though interpretations of these systems may serve empirical or pragmatic ends, no interpretation is privileged; all that can be said is that the axioms of the system become acceptable (or, at least, are accepted) for some interpretation. There are no mathematical objects (*or* – first alternative – it is senseless to speak of "mathematical objects" *or* – second alternative – if there were mathematical objects, it would make no difference to our experience and there could be no reason to believe in them), so that no question of a standard or privileged interpretation arises.

Consequently there is no such abstraction as truth – we agree in accepting or rejecting certain sentences, including metalinguistic ones, and that is all [14].

A second version of the doctrine, perhaps more familiar to philosophers, is:

(Version II) Ordinary language embodies no metaphysical commitment. We accept certain claims and meta-claims, i.e. analyses, but there *are* no meanings or necessities, although words can correctly be said to have meanings, and statements to be true in virtue of them [15]. There are no abstract entities (*or* ... as in I), and no philosophical question therefore as to whether any conceptual framework is privileged; though some might be more adequate for e.g. science. Consequently, there is no metaphysical problem of truth; we can merely undertake a conceptual analysis of the uses of the word "true", and that is all. In short, ontology is impossible.

The above views I and II elaborate a position which K sometimes characterises as "positivist-pragmatist" (e.g. [Sch, p. 203]); its three main theses, each of which he rejects, are:

(1) A 'conventional' view of logic,

(2) A denial of the real existence of abstract entities, including semantic notions themselves, and

(3) The assumption that every philosophical problem can be dissolved by careful examination of the appropriate (language-dependent) concept.

A second formalist doctrine which K attacks is that of Hilbert, whose central thesis is claimed to be the "separation of the foundations of mathematics from philosophy" [SPT, p. 321], or more specifically that the truth-value of any mathematical proposition which can properly be said to have one can be discovered, though perhaps only with great difficulty, by the use of a small number of very simple and evidently correct principles of reasoning.

A third suspect thesis is mechanism, which holds that mathematical experience can be explained as the output of an organism (brain) of finite com-

[14] Though K presents Hilbert as holding that "the notion of mathematical truth can have no place in mathematics as we know it" [B & P, p. 174], Hilbert allowed that some propositions, both observational and arithmetical, have a truth-value, and his version of formalism differs from I in essential respects – see below. Indeed Curry [B & P, p. 156, top] emphasises the distinction between Hilbert's conception of mathematics and (his own version of) formalism.

[15] The word "statement" is indeed a testament to the "inherited experience and acumen of many generations of men". That something that has both a meaning *and* a truth-value is called by "one of our common stock of words (which) embodies all the distinctions men have found worth drawing, and the connexions they have found worth marking, in the lifetimes of many generations" (J.L. Austin, pp. 133, 130) is quite remarkable.

plexity, obeying recursive laws.

Hilbert's thesis is held to be refuted and mechanism rendered doubtful by Gödel's incompleteness theorem [Sch, pp. 236–267, 255 l–8 to l–5, and 266–269]. Full discussion of this is postponed to the second paper, but it is important to remark here that it is not the formal theorem itseld which does the work, but interpretations superimposed upon the theorem as to what it 'means'; hence the conclusion of K's argument will not be acceptable to a formalist. Note that K's remark about: "Gödel's well known interpretation of his incompleteness theorem: either there are mathematical objects external to ourselves or they are our own constructions and mind is not mechanical" [Sch, p. 267] alludes to an unpublished Gibbs lecture by Gödel at Brown University in the late forties [16].

The relation between the positivist-pragmatist doctrine presented above and that of Hilbert is not entirely clear; Hilbert would surely have rejected the first two theses of the former position (as summarised above) and my impression is that he would not have cared about the last. It does not follow, however, that different considerations are needed to establish the falsity of the two versions, since the existence of an ontologically correct interpretation of a (set of) sentences refutes the positivist-pragmatist doctrine directly and, with Gödel's (1931) incompleteness (formal) results, also establishes the falsity of Hibert's conjecture. My present concern, however, is only with K's positive thesis, and specifically with his claim that, by means of "informal rigour", sometimes called "philosophical analysis" [Sch, pp. 202, 205], we are frequently able to give precise answers both to foundational and philosophical problems.

I consider briefly three of his examples of the use of informal rigour: (1) postulates for the natural numbers; (2) the c.r.s. hierarchy of sets; and (3) the completeness of first-order logic.

(1) The Dedekind-Peano-Peirce analysis of the number sequence yields immediately a categorical set of second-order axioms for the natural numbers. This example shows that informal considerations and the use of 'pictures' of structures (see Section 8 below) are useful for the discovery of axioms, but does it therefore follow, as K seems to suggest, that there is a (unique) realm of natural numbers which we have discovered, either in our minds or in Plato's heaven? Or does the mathematical result, together with the empirical fact that the Peano postulates for number theory are generally accepted, merely estab-

[16] On pp. 228, 229 and 269 of [Sch], K makes the stronger claim that Gödel's theorem both refutes the formalist view and establishes the "non-mechanistic character of the laws satisfied by the natural numbers".

lish that there is *some* coherent and perhaps interesting or applicable interpretation of the postulates? In the latter case, even if Frege's derivation of number theory from (second-order) logic had not led to paradox, it would have been established neither that numbers are external mathematical objects nor that our concept of them is unambiguous.

A related example concerns K's own discovery of a postulate system for ordinal arithmetic, categorical with respect to set theory ([Sch, p. 272], [Lak, p. 168], [Kaka, pp. 169, 170], and note 6 above). The question can again be asked: what follows from this example about the ontological status of mathematical objects, properties or structures?

(2) The c.r.s. (see above, Section 4) yields a second example to illustrate how axioms may be obtained by an analysis of evident properties. Even if sets are (defined as) Stanford sets (i.e. sets of something), and satisfy the Zermelo-Fraenkel axioms, it is premature to suggest, as K does (in [Lak, pp. 97–103], a discussion of Mostowski's paper in the same volume), that it is reasonable to reject future bifurcation or multifurcation of the concept of (Stanford) set. Indeed, it would seem impossible to exclude a priori that there is a coherent notion of set satisfying both the Zermelo-Fraenkel axioms and AD [17)], assuming that the latter is not refutable in ZF. Pragmatically, we need to allow the possibility of giving a natural interpretation to any *interesting* consistent system, and the widely quoted remark of Dirac, an author whom K often cites, in connection with understanding a physical theory, applies equally to foundational studies: "One may, however, extend the meaning of the word 'picture' to include *any way of looking at the fundamental laws which makes their self-consistency obvious*" [QM, 4th ed., p. 10]. No doubt in time we will come to feel that we understand or can picture such systems as "ZF + all sets of reals are Lebesgue measurable + DC" [18)], proved relatively consistent with "ZF + there is an inaccessible cardinal" by Solovay, and Quine's systems ML and NF, assuming their consistency.

It is sometimes argued that systems such as those just mentioned 'lack a semantics', but this is unimportant because, e.g. the meaning of '&' is given in terms of the (presupposed) meaning of the word 'and'. Semantic explanations in foundational work are circular, though this constitutes no barrier to their employment.

(3) *Completeness of first-order logic* (see [Lak, pp. 152–157], [Kaka, pp. 189–192], [Saa, pp. 114–117], [Sch, pp. 253–255]).

K presents this as the paradigm example [SPT, p. 335], cf. [Saa, p. 178],

[17)] AD = the (full) axiom of determinateness (Mycielski and Steinhaus).
[18)] DC = the axiom of dependent choices.

[Sch, p. 240] of how use of informal considerations can yield significant formal and foundational results. Three notions are involved, namely *D, V*, and *Val*, where:

$Dp \underset{\text{df}}{=} p$ is derivable in (say) the first-order part of Frege's *Begriffsschrift*.

$Vp \underset{\text{df}}{=} p$ is true in all set-theoretic structures, i.e. for the c.r.s.

$Val\, p \underset{\text{df}}{=} p$ is intuitively a logical truth.

The following two principles are assumed:

(1) $(\forall p)(Val\, p \supset Vp)$ – this is claimed to be recognised as correct, and

(2) $(\forall p)(Dp \supset Val\, p)$ – this follows since the axioms of the first-order calculus express logical truths, and the rules of inference are logically valid. Whence, from (1) and (2),

$$(3) \qquad (\forall p)(Dp \supset Vp)\,.$$

In addition, K claims that

$$(4) \qquad (\forall p)(Vp \supset Dp)$$

is established by Gödel's completeness theorem, and it follows that *D, V*, and *Val* are co-extensional. The intuitive notion *Val* has therefore been shown to have a precise meaning (extension) though this "doesn't mean that *Val* was vague before" 1930, ([Lak, p. 154]). He notes that though the mathematical result behind (4) was obtained by Skolem and Herbrand, they failed to obtain a completeness theorem because in (slightly) different ways they lacked the notion of *Val* – see [Sch, p. 255] and [SPT, p. 359, footnote].

Since Gödel's own proof does not mention the cumulative hierarchy, it might be helpful to clarify the connection between the notion of validity in his (1930) and *V* above. Gödel proves (non-finitistically, using the law of excluded middle and König's lemma) that the first-order calculus of *Principia* is complete if its propositional subsystem is, referring in connection with the latter to Bernays (1926) and not to Post's original proof. Gödel's definition of completeness, for both first-order and propositional calculi, involves a notion of validity which he explains [vH, p. 584, note 3] as: true for all substitution instances. He also remarks that it is sufficient to consider validity in a denumerable domain (note 4, same page). There is no reference in Gödel's paper to validity in set-theoretic structures, and though Bernays' later arithmetised version of the completeness theorem is based upon an axiomatically specified *V*, it is misleading to suggest, as K seems to, that Gödel's result or its original proof involves the notion of validity for the c.r.s.;

indeed, as he notes ([Sch, p. 254]; see also [Saa, p. 116, l-14 to -12 and p. 117, l-9], and [Lak, p. 157]). Gödel's result holds for a variety of notions of set.

The conclusion to which this leads is that, strictly speaking, what Gödel had proved (by 1930) was not (4) but

(5) $(\forall p)(Val\, p \supset Dp)$

while (4) was established rigorously only later. This in no way affects the philosophical moral that K has drawn, i.e. that the informal notion which makes (a formalised version of) Gödel's result a *completeness* theorem was shown to be determinate in extent.

In connection with his work on finitist and predicativist provability, K expresses the hope that it will become possible to find semantic concepts V_F and V_P co-extensional with the D_F and D_P for finitist and predicativist provability respectively proposed by K, Feferman and Schütte. At present, no such semantic notions have been found, so there is as yet no possibility of a completeness proof for the provability concepts; indeed, Feferman (1968) says that "it may not be possible to obtain as definite results for predicativity as Gödel's completeness theorem for predicate logic with respect to intuitive logical validity" (p. 133). But is Gödel's completeness theorem, therefore, *really* a formal result?

One (methodological) conclusion to which these three examples lead is surely that it is always possible that future technical progress will show that different notions have, up to a certain time, been conflated, bringing about a multifurcation of concepts. To establish that only certain lines of development are legitimate, much more would need to be done. It might, for example, become possible to show that the definability results to which K appeals in connection with non-standard arithmetic do indeed establish that the non-standard concept presupposes the standard one, but even then the arguments used to establish such a presupposition could not be expected to defeat the sceptical follower of Wittgenstein or Skolem. Also, to suppose that different mathematical objects exist corresponding to each arbitrary consistent system leads to a Quinean slum, so that drastic measures are needed to reduce population overcrowding. But no way of doing this has been justified, and the various dogmatisms, e.g. of Kronecker, of strict realism, and of contemporary nominalism have not been shown to be satisfactory. In particular, therefore, the existence of sets is not established by the empirical fact that there is a (not necessarily unique) common coherent concept of set, so that we are no nearer to answering the original question: what is meant by the existence of sets, or indeed of any mathematical objects? And K's optimism which holds that this philosophical problem can be resolved by technical progress seems to be unfounded, even allowing that the various versions of formalism are false; I have suggested above that his arguments purporting to establish this latter claim are circular.

8. Second-order logic and mathematical structures

Rejection of formalism has led to the development of systems of logic which cannot be represented mechanically, as by a Turing machine. Ways in which the language of logic might be enriched include the use of (a) additional (infinitary) quantifiers (Mostowski), (b) infinite formulae, or (c) second-order logic. K refers only briefly to (a), but hopes that (b) and (c) together will lead to new foundational results [Lak, p. 191]. A specific hope is that new primitive notions, perhaps suggested by work on second-order logic, will enable us to determine the truth-value of (e.g.) CH [Lak, pp. 150–152]. It has, however, to date proved more rewarding to extend set theory by means of axioms of infinity, e.g. reflection principles, which can be expressed without extending the language [Lak, p. 152]. The difficulty with K's programme is that, though there are properties which can only be defined in either an enriched first-order or a second-order language [19], knowledge of what can be achieved in a logic based on such languages is not yet sufficiently clear. Thus no completeness result connecting V and Val for second-order logic is known, though K believes that one is to be expected. A related question, to which the answer is also unknown, is: what is the second-order analogue of ω in the Löwenheim-Skolem theorem, formulated as

$$(\forall p \subseteq \omega)(\forall \alpha > \omega)(p \in V_\alpha \equiv p \in V_{\omega+1}),$$

where V_α, for α an ordinal, is the set of sets of the c.r.s. of ranks $< \alpha$? [Kaka, p. 191].

[19] For example, the full strength of second-order logic is not needed for a categorical characterisation of N, for which an axiom in an infinitary language corresponding to the (semi-formal) ω-rule suffices. But the property WO of being a well-ordering, though second-order definable, is, even restricted to $V_{\omega+1}$, not first-order definable (Tarski, 1936). On the other hand, Lopez-Escobar (1966) proved that WO cannot be defined by any set of sentences of an infinitary language, if only finite strings of quantifiers occur; and languages which allow infinite strings of quantifiers are known to be of comparable strength to second-order languages. Thus, unlike N, WO is *essentially* a second-order notion, though preanalytically it seems a clear one.

Perhaps because of the influence of the doctrines mentioned in Section 7, the philosophical importance of second-order logic and of infinitary logics is far from clear ([Sch, pp. 260–262], [Kaka, pp. 192–194], cf. [Lak, pp. 99–102]). Essentially, K argues for their significance by opposition to the thesis of structuralism, a currently fashionable version of formalism espoused by Mostowski and Benacerraf, and in Bourbaki's historical note [19a]; though K's onslaught on Bourbaki is not an attack on the thesis about to be described.

The structuralist view allows the legitimacy of the attempt to characterise abstract structures, but denies that any of these are ontologically privileged; hence multifurcation of concepts is to be expected. Since it would be circular to attempt to characterise either the c.r.s., or the notion of second-order consequence in this way, K draws the important conclusion that the structuralist thesis is "incoherent" in allowing *some* infinite structures but in denying that any are privileged. See also his remarks on 'meaning and use' in [Sch, pp. 262–266].

But, in view of the possibility of foundational theories alternative to the c.r.s., can it be supposed to be a privileged structure? Two arguments sometimes used in support of this thesis are (a) that, unlike other foundational theories, it is an extremely rich structure for which, moreover, we possess a semantics and (b) that in particular, we recognise how it avoids the paradoxes. Though K flirts with both of these, I suspect that he rejects the possibility of a mathematical foundation for mathematics (though [W I, p. 143] is not clear on this point), holding rather that there need be no (consistent) mathematical theory in which every piece of mathematical work can be embedded – although there may be such a theory for work done up to a given time, in view of the reluctance of mathematicians to use the full power of the abstract assumptions that they allow (see [W II, p. 242] and [Sch, pp. 218, 220]).

What K says about (a) and (b) is of interest, and I summarise and comment briefly.

In 1958 and 1965 ([W I, p. 157] and [Saa, pp. 100, 101]) he uses the argument of Russell's paradox to show that the property (of a set) of not being self-membered lacks a "definite extension". Thus if a set A contains only non-self-membered sets, so does its proper superset $A \cup \{A\}$, and there can be therefore no set of all non-self-membered sets. (Menger (1928) makes a similar observation.) This line of argument fails, however, to yield an imme-

[19a] See also Bourbaki: L'architecture des mathématiques, in F. le Lionnais (ed.): *Les Grands Cowarts de la Pensée Mathématique* (1948 and 1962). An English translation of the article appears in Amer. Math. Monthly 57 (1950).

diately satisfactory resolution of the paradoxes, for it leads to the Poincaré vicious-circle principle, which, by allowing only sets which can be shown to have a definite extension by their manner of construction, disqualifies substantial parts of classical analysis as nonsense.

However, the c.r.s., in which there are sets definable only by quantifying over sets of which they are members, might be felt to result from a natural weakening of the vicious circle principle, since the absence of a predicative definition casts no doubt upon existence, if the fundamental structure is taken for granted. This is sometimes expressed by saying that, for this (minimal) fundamental structure, "all impredicativities are reduced to one special kind, namely the existence of certain large ordinal numbers ... and the validity of recursive reasoning for them" (Gödel, 1944, in [B & P, p. 227]). And for the c.r.s., since no set is a member of itself, the class of all non-self-membered sets is the universe, i.e. not a set.

Even the minimal c.r.s. is, therefore, a very rich structure, which we understand if we understand the notion of an (arbitrary) ordinal. But the fact that there is this common core concept of a set implies neither that there *are* sets, nor that any such structure is foundationally privileged.

The c.r.s. is, of course, a powerful structure, and because a great many mathematical structures can be embedded in it, an important one. But a foundation for mathematics needs to be, if not self-justifying, at least self-corroborating, and the technical result by K mentioned in Section 4 seems to preclude this. In addition, any such structure which is privileged only by being strong can hardly help to resolve philosophical worries about its existential commitments. The moral to be drawn from this is that second-order logic determines existential commitments, and therefore which structures are important, and not vice versa.

It is therefore necessary to turn to suggest how the lower levels of the c.r.s. are related both to the possibility of physical experience, and to logic – in short, to reconsider the programme of logicism. Since K believes [Lux, note 10] that the logicist criticisms of other approaches are valid, he may well have some sympathy with the attempt. I therefore return to the discussion left suspended at the end of Section 5 as to the role of the mathematical object.

9. Realism, strong realism, and the mathematical object

The position arrived at above was that if there are external mathematical objects, no method of establishing their existence or their properties is at hand, and that it is therefore fundamentally obscure as to what the realist thesis amounts to. Nevertheless, if formalism is false, as may be evident, although K has not, as I think, proved it, there are mathematical objects, which must either be our constructions or exist externally. Since external mathematical objects could not causally affect us, it remains therefore to give an account firstly of the relation between mathematical objects or properties and the physical world, and secondly of how human mathematical experience (and intellectual experience generally) can arise from the interconnection of the organism and the external world. It could be surmised that, as a first approximation, the two accounts will be related in much the same way as the primary and secondary properties of matter on the (Galileo-Descartes-) Locke account, though with an additional degree of complexity since intellectual rather than sensory experience is to be explained.

I consider only the first problem, and start from the basic logicist contention that cardinal numerosity is a property of (objectifications of) properties [20], and number an abstract object related to numerosity as the reference of "the concept *horse*" (which for Frege is not a concept but an object) to what "is a horse" designates. If this is correct, as I shall assume it to be, the thesis of weak realism is reduced to that of the existence of objectifications of external properties of objects. That there are external properties of objects follows from (1) the (non-contingent) existence of universals and (2) the (contingent) existence of the external world, and of primary qualities – this shows that the *two* realist theses (as opposed to nominalism and as opposed to idealism respectively) mentioned in Section 6 are involved. (1) is easily established, since on the assumptions made, it does not seem possible to deny that (say) having 2 instances is, genuinely, a common property. With regard to (2), though the existence of external objects is not seriously in doubt, that there are external properties might be; this would certainly be disputed by Wittgenstein, and other naive realists, and also by those who see contemporary physics (quantum theory) as justifying an idealistic philosophy. (If this last mentioned view were correct, and it has not, to my mind, been made plausible, the very tenuous link between idealism in physics and in mathematics just hinted at would nevertheless survive.)

It remains, therefore, to justify the transition from the existence of properties to that of their objectifications, a step sometimes called "abstraction", though the psychological overtones of the word load it with idealist associa-

[20] It is interesting that Hilbert agreed with this. See the 'logicist' analysis of 'There are two F's' on p. 137 of [HA], and also the remark of p. 136 that "it becomes possible to subsume the theory of numbers under logic".

tions, which are certainly misleading if Frege is wrong in holding to a sharp distinction between objects and properties, and dubious if he is correct. But regardless of whether Frege is right or wrong, the problem is: which properties of objects are (or correspond to) objects? By Cantor's theorem not all properties, indeed almost none, can, but by the basic logicist assumption, some must, since any property, including numerosity, is a property of *objects*.

Since we lack a general criterion for deciding which properties are objectifiable [21)], how can it be shown that cardinality properties are among them? However, a finite cardinality is necessarily a property of some property, hence an infinite number of abstract objects is needed if an empirical assumption about the number of atoms is to be avoided, and these objects can best be taken to be the natural numbers themselves. Weak realism is thus established. Strong realism normally requires in addition that the property of being a natural number be objectifiable. Here I shall only say, dogmatically, that I can see no argument against this; if Frege is wrong (see above), it would seem highly implausible that natural numbers and properties of atoms are objects, and not the property of being a natural number; if Frege is right, it seems only slightly less implausible that this property should not correspond to an object. Nevertheless, formally, an axiom of the sethood of N is needed for both Quine's ML [22)] and the c.r.s., which leads to regret that Quine is correct in his claim that the considerations central to the resolution of the paradoxes are not yet sufficiently well understood.

10. Philosophy and the philosophy of mathematics

At the beginning of [W I], K makes the following very clear distinction between "(i) the *foundations of mathematics* or *philosophy of mathematics* and (ii) *general philosophy* in the following way: the former applies to philosophically interesting differences between various parts or aspects of mathematics, the latter, in so far as it is concerned with mathematics, applies

21) If limitation of size is the criterion, we are led to the c.r.s. for which it is known (Montague, 1965) that the theorems of first-order ZF with atoms can be correlated with logical truths of a higher-order logic; though the rank conception is used in the definition of higher-order logical truth, which reduces the epistemological significance of the results. But logicism, as presented above, is incompatible with ZF and the c.r.s., since no positive integer or non-zero ordinal is an object if size (i.e. cardinality) determines which properties are objectifiable.

22) Though it is a theorem of ML that there are infinite sets, N cannot be one of them, if ML is consistent.

to philosophically interesting differences between mathematics and other intellectual activities. For instance, the questions 'what is a (correct) proof' or, generally, 'what is mathematics' belong to general philosophy, 'what is a constructive proof' or 'what is a predicative concept' belong to the philosophy of mathematics" [W I, p. 135]. Since therefore general and mathematical philosophy differ only in their degree of generality, little remains to be said at the ontological level about the former, mathematics being the field of intellectual activity with the most extensive abstract existential commitments.

Any such ontological conception of philosophy is sharply opposed to that implicit in the writings of Wittgenstein, which K characterises as semi-nominalist, semi-behaviourist, and as identifying metaphysics with grammar [W II, pp. 240, 244]. For K, since there are mathematical objects and abstract structures, and hence meanings to be expressed, ontology must be prior to grammar, and cannot be reduced to it. Wittgenstein overemphasised, he thinks, various empiricist theses relating to "the grammar of [elementary] mathematics" [W I, p. 144] and ignored the fact that the mathematical characterisation of concepts and structures can bring about an advance in philosophy of mathematics, and therefore, by the distinction drawn above, also in general philosophy [23].

However, in his critique of Wittgenstein, K also wrote "I do not believe that the foundations for [philosophy as a discipline in its own right] are yet laid" [W II, p. 238], though seven years later, this was changed to "Philosophical clarification of a concept does not usually consist in reduction to another set of concepts ... but in becoming aware of the properties of the concept, and not looking past in" [Lak, p. 102]. Taken literally, the latter viewpoint is that of conceptual analysis, but although this is the method to be used in the analysis of idealist notions (see Section 6), it is at variance with both K's ontological approach to realism and his rejection of the Wittgensteinian thesis just mentioned that metaphysics is grammar. The suspicion must be therefore that K's conception of philosophy is not yet sufficiently developed to answer the (quite basic) question: how can it be established that various sceptical and reductionist theses are false?

"I find myself involv'd" in such a labyrinth, that, I must confess, I neither know how to correct my former opinions, nor how to render them consistent." So Hume in the Appendix to *A Treatise of Human Nature*, on the epistemological problem of personal identity. K's work brings out a similar

[23] Historically, it is clear that the absurdity of various fashionable theses in the field of general philosophy is shown by even very simple results and theories in mathematics and the physical sciences; the pedagogical moral is obvious.

dilemma in the field of the philosophy of mathematics. On the one hand, formalism is false; there are abstract structures, almost all of which are not representable by a mechanism – indeed, the truth-set may be uncountable – and second-order notions are legitimate, so that set theory is not *just* another structure. On the other hand, mathematical propositions are necessarily true or necessarily false, hence ontologically different from contingent propositions, and (ultimately) incapable of being established or confirmed by observation or evidence. So that, if true, they are contentless, or might just as well be. Hence the existence of mathematical objects is at best explanatorily useless and at worst a meaningless hypothesis, and thus it can be helpful to think *as if* there are abstract structures, only as a guide to the choice of fruitful axiom systems.

If, as K believes, traditional philosophical theses can be sharpened and resolved, this can be done for the above two points of view. But how?

11. Postscript

Thus, what puzzles me is a version of the old empiricist slogan: *if nothing in experience (or determined by the conceptual framework that we take as given) can distinguish between two apparently different positions, then it is empirically (or conceptually) meaningless to suppose a difference between them.* I do not know whether or not this slogan is true – assuming of course that its truth-value is not determined by the given conceptual framework. I *think*, however, that K rejects it, although some of his remarks suggest that if we look hard enough we can always (in principle) discover *some* clearly meaningful difference between any two non-synonymous philosophical positions, so that the slogan holds because of the falsity of its protasis. This optimistic view, whether or not K really holds it, would appear to be refutable from cardinality considerations, though it might be interesting to find weak assumptions which refute it.

For contemporary first-order logic, though not for Frege's (intended) logical system, weak realism is not a logical truth. Hence, if Frege's framework can be made consistent, the question arises as to how a decision between the two frameworks might be justified. My conjecture is that this question is sufficiently precise for it to be shown that *no* non-circular justification for weak realism exists, and therefore that the thesis that there are mathematical objects remains obscure, in a fundamental way.

Again, in connection with strong realism, for which I have attempted to sketch the outlines of a partial defence in Section 9, there is a question which divided Keyser and Russell more than 65 years ago (in the Hibbert Journal 2 and 3) viz: is the axiom of infinity, as Keyser claimed and Russell (at that time) denied, a new 'presupposition' of thought? If so, as the c.r.s. holds, then the meaning of the axiom is obscure, in the above sense; that is, not only have we so far failed to discover what considerations determine the truth-value of strong realism, within a conceptual framework not committed to infinite sets, but (perhaps) within any such conceptual framework, there are no considerations to be discovered.

The realist view is, no doubt, "natural" [Sch, p. 220], and one reason for believing it is therefore independent of any critique of formalist positions, but the question remains: how can any version of a realist view be *shown* to be true, or formalist views false? If the two conjectures just mentioned can be established, we are faced with the empiricist's dilemma, as presented by the slogan above: *the truth-value of realism (weak or strong) makes no difference*. Hence the meaning of certain sentences of which the truth-value of what they say is not "dubious" has not been explained, and consequently, there is something that we know that we do not understand. Though this conclusion belongs to a well-established theological tradition, one might prefer to conclude that realism, weak and strong, remain doubtful.

Bibliography

I. Works by Kreisel (only those referred to in the text are listed)

[NAM I] Note on arithmetic models for consistent formulae of the predicate calculus I, Fund. Math. 37 (1950).

[RM] The diagonal method in formalised arithmetic (review article of Mostowski's Sentences undecidable ...) Brit. J. Phil. Sci. 3 (1952–53).

[VHT] A variant to Hilbert's theory of the foundations of arithmetic, Brit. J. Phil. Sci. 4 (1953–54).

[FCS] A remark on free choice sequences and the topological completeness proofs, J. Symbolic Logic 23 (1958).

[HPr_1] Hilbert's programme, Dialectica 12 (1958) and Logica (Festschrift for Bernays) (1959).

[HPr_2] The revised edition of [HPr_1], in [B & P] – see Part II of bibliography.

[W I] Wittgenstein's Remarks on the foundations of mathematics, Brit. J. Phil. Sci. 9 (1958–59).

[W II] Wittgenstein's theory and practice of philosophy, Brit. J. Phil. Sci. 11 (1960–61).

[STP] Set theoretic problems suggested by the notion of potential totality, in: Infinitistic methods (Warsaw, 1961).

[RS] Review of papers in Scholz: Mathesis universalis, J. Symbolic Logic 28 (1963).

[Saa] Mathematical logic, in: Lectures on modern mathematics, Vol. III, ed. T.L. Saaty (1965).
[RW] Review of Wang, A survey of mathematical logic, Phil. Rev. 75 (1966).
[Kaka] Elements of mathematical logic (with J.L. Krivine), (North-Holland Publ. Co., 1967). The original French version (Dunod, Paris) is also dated 1967.
[Lak] Informal rigour and completeness proofs, and Comments on Mostowski, Recent results in set theory, in: Problems in the philosophy of mathematics, ed. I. Lakatos (1967).
[Sch] Mathematical logic; what has it done for the philosphy of mathematics?, in: Bertrand Russell, Philosopher of the century, ed. R. Schoenman (1967)
[SPT] A survey of proof theory, J. Symbolic Logic 33 (1968).
[Har] Foundations of mathematics, 1900–1950, in: Scientific thought 1900–1960, ed. R. Harré (1969).
[Lux] Axiomatisation of non-standard analysis ..., in: Applications of model theory, ed. W.A.J. Luxemburg (1969).
[Br] Luitzen Egbertus Jan Brouwer 1881–1966, with M.H.A. Newman, in: Biographical memoirs of Fellows of the Royal Society 15 (1969).
[Mac] Appendix (by K) to paper by Feferman, in: Reports of the Midwest Category Seminar, III, ed. S. MacLane (1969).
[My] Two papers in: Intuitionism and proof theory, eds. J. Myhill, A. Kino, and R. Vesley (1970.
[Sc] Observations on popular discussions of foundations, in: Proc. symp. pure math. 13, ed. D.S. Scott (1970).
[TN] Two notes on Foundations, in Dialectica (1970).
[FPD] The formalist-positivist doctrine of mathematical precision in the light of experience, in: L'âge de la science 3 (1970).
[*] Paper in this volume.

Part II

Austin, J.L., Philosophical papers (1961).
[B & P] Benacerraf and Putnam (eds.): Philosophy of mathematics, selected readings (1964) – includes [HPr_2], Bernays (1935) and (1959), Gödel (1944) and (1947) (revised), Curry (1954) and Heyting (1956).
Bernays, P., Axiomatische Untersuchungen der Aussagen-Kalküls der Principia Mathematica, Math. Z. 25 (1926).
Bernays, P., On platonism in mathematics (1935), in [B & P], translated from the French.
Bernays, P., Mathematische Existenz und Widerspruchsfreiheit, in: Etudes de philosophie des sciences (Festschrift for Gonseth) (1950). An outline in French appears in Revue Phil. 143 (1953).
Bernays, P., Comments on Ludwig Wittgenstein's Remarks ... (1959), reprinted in [B & P].
Curry, H.B., Remarks on the definition and nature of mathematics (1954, written 1939) reprinted in [B & P].
[QM]: Dirac, P.A.M., Quantum mechanics, 1st. ed. (1930), 4th ed. (1958).
Feferman, S., Autonomous transfinite progressions and the extent of predicative mathematics, in: Logic, methodology and philosophy of Science, III, eds. Rootselaar and Staal (1968).

Gödel, K., The completeness of the axioms of the functional calculi of logic (1930), in [vH], translated from the German.
Gödel, K., On formally undecidable propositions of Principia Mathematica and related systems I (1931), in [vH], translated from the German.
Gödel, K., Russell's Mathematical logic (1944), reprinted in [B & P].
Gödel, K., What is Cantor's continuum problem? (1947), revised and reprinted in [B & P].
Gödel, K., Über eine bisher noch nicht benützte Erweiterung des finiten Standpunktes (1958), in: Dialectica 12 and Logica (Bernays Festschrift).
Heyting, A., Disputation, in: Intuitionism, an introduction (1956), reprinted in [B & P].
[HA]: Hilbert and Ackermann, Principles of mathematical logic (1950), translated from the 2nd German edition (1938).
Hintikka, K.J.J., Are mathematical truths synthetic a priori?, J. Phil. 65 (1968).
Lopez-Escobar, E.G.K., On defining well-orderings, Fund. Math. 59 (1966).
Mackie, J.L., Proof, Proc. Arist. Soc. Suppl., 40 (1966).
Menger, K., Bemerkungen zu Grundlagenfragen (esp. II), Jahrb. deut. math. Ver. 37 (1928).
Montague, R., Set theory and higher-order logic, in: Formal systems and recursive functions, eds. Crossley and Dummett (1965).
[CP VIII]: Peirce, C.S., Review of Fraser, The works of George Berkeley (1871), reprinted in: Collected papers, Vol. VIII, ed. A.W. Burks (1958).
Shoenfield, J., Mathematical Logic (1967).
Tarski, A., Foundations of the calculus of systems (1936), in: Logic, semantics and metamathematics (1956), translated from the German.
[vH] Van Heijenoort, J., ed., From Frege to Gödel (1967) – includes Gödel (1930 and 1931).
Wisdom, A.J.T.D., Note on the new edition of Professor Ayer's Language, truth and logic (1948), reprinted in: Philosophy and psychoanalysis (1953).
Zermelo, E., Über Grenzzahlen und Mengenbereiche, Fund. Math. 16 (1930).

RAMSEY'S THEOREM DOES NOT HOLD IN RECURSIVE SET THEORY

E. SPECKER
Zurich, Switzerland

1. Let R be a binary symmetric relation on the set N of natural numbers. A subset T of N is called "R-homogeneous" if either $R(x, y)$ holds for all x, y in T such that $x \neq y$ or $\neg R(x, y)$ holds for all x, y in T such that $x \neq y$. Ramsey's theorem states that there exist infinite R-homogeneous subsets for every binary symmetric relation R on N [3].

Theorem. There exists a recursive binary relation R on N such that no recursively enumerable infinit subset of N is R-homogeneous [1)].

The proof of the theorem is based on the existence of two recursively enumerable sets of incomparable degrees of unsolvability [1, 2].

2. Let f_1, f_2 be one-to-one functions from N into N. Define a function $r(f_1, f_2)$ from N into the set of rational numbers as follows:

$$r(f_1, f_2)(n) = \sum_{k=0}^{n} [2^{-f_1(k)} - 2^{-f_2(k)}] .$$

Clearly, $r(f_2, f_1) = -r(f_1, f_2)$. Define a binary relation $R(f_1, f_2)$ on N as follows: $R(f_1, f_2)(n_1, n_2)$ holds if and only if either

$$n_1 \leqslant n_2 \quad \text{and} \quad r(f_1, f_2)(n_1) \leqslant r(f_1, f_2)(n_2)$$

1) The result of this paper has been presented at the Universities of Bristol and Manchester in March 1966. The present publication is due to the suggestion of G. Kreisel who has also informed me that the same result (and extensions of it) have been proved in the meantime by C.G. Jockusch [4]. See also Yates [5] (this proceedings).

or

$$n_2 \leqslant n_1 \quad \text{and} \quad r(f_1, f_2)(n_2) \leqslant r(f_1, f_2)(n_1) .$$

$R(f_1, f_2)$ is symmetric: $R(f_1, f_2)(n_1, n_2)$ holds if and only if $R(f_1, f_2)(n_2, n_1)$ holds. $R(f_1, f_2)$ is recursive in the functions f_1, f_2 and is therefore recursive if the functions f_1 and f_2 are recursive.

Let S_i ($i = 1, 2$) be the image of N by f_i.

3. Lemma. Let f_1 be a one-to-one function from N onto S_1 and let T be an infinite subset of N. Then there exists a function g from N into N satisfying the following conditions:

(a) g is recursive in f_1, S_1, T.

(b) g is a function from N into T.

(c) For all m, k in N such that $g(m) \leqslant k$ we have $m + 1 \leqslant f_1(k)$.

Proof. Let f_1^{-1} be the inverse function of f_1; f_1^{-1} is recursive in f_1, S_1 and maps S_1 onto N. Let t be the one-to-one monotonic function from N onto T; t is recursive in T and for all n in T we have $n \leqslant t(n)$ and $t(n) \in T$. Define h, $h : N \to N$, as follows

$$h(m) = \operatorname*{Max}_j \, [(\exists x)(x = m \wedge f_1^{-1}(x) = j] + 1 .$$

h is recursive in f_1^{-1} and therefore recursive in f_1, S_1; for all k, m such that $h(m) \leqslant k$ we have $m + 1 \leqslant f_1(k)$. Define g, $g : N \to N$, as follows: $g(m) = t(h(m))$. The function g clearly satisfies conditions (a), (b) and (c).

4. Let f_i, $i = 1, 2$, be one-to-one functions from N into N, let the binary relation $R(f_1, f_2)$ be defined as in Section 2 and let T be an infinite R-homogeneous subset of N, i.e. a subset such that either

(1) for all n_1, n_2 in T such that $n_1 < n_2$ the inequality $r(f_1, f_2)(n_1) \leqslant r(f_1, f_2)(n_2)$ holds, or

(2) for all n_1, n_2 in T such that $n_1 < n_2$ the inequality $r(f_1, f_2)(n_2) \leqslant r(f_1, f_2)(n_1)$ holds.

We shall show in the next section: if (1) is satisfied, then S_2 is recursive in S_1, f_1, f_2, T.

Applying this result to the function $r(f_2, f_1)$ we obtain: if (2) is satisfied, then S_1 is recursive in S_2, f_1, f_2, T. This and the following remark will complete the proof of the theorem: Let S_1, S_2 be two recursively enumerable sets and let f_i ($i = 1, 2$) be recursive one-to-one functions from N onto S_i. Let T'

be an infinite recursively enumerable R-homogeneous subset of N; T' contains an infinite recursive subset T which is also R-homogeneous. f_1, f_2 and T being recursive, a set recursive in S_1, f_1, f_2, T is recursive in S_1. The existence of an infinite recursively enumerable R-homogeneous set T' implies therefore that either S_1 is recursive in S_2 or S_2 is recursive in S_1. Neither of this need be, as there exist recursively enumerable sets of incomparable degrees.

5. Let f_i, $i = 1, 2$, be one-to-one functions from N onto S_i, let r be the function defined by

$$r(n) = \sum_{k=0}^{n} [2^{-f_1(k)} - 2^{-f_2(k)}]$$

and let T be an infinite set such that for all n_1, n_2 in T such that $n_1 < n_2$ the inequality $r(n_1) \leqslant r(n_2)$ holds. Let g be a function as established in the lemma of Section 3: g is recursive in f_1, S_1, T; $g(m)$ is an element of T for all m in N and for all k, m in N such that $g(m) \leqslant k$ we have $m + 1 \leqslant f_1(k)$. Define the set S_2' as follows:

$$m \in S_2' \quad \text{if and only if } (\exists x)(x \leqslant g(m) \wedge m = f_2(x)) .$$

S_2 is clearly recursive in f_2, g and therefore recursive in S_1, f_1, f_2, T. We show $S_2' = S_2$.

(1) If $m \in S_2'$ then $(\exists x) m = f_2(x)$ and $m \in S_2$.

(2) Assume $m \in S_2$. In order to prove $(\exists x)(x \leqslant g(m) \wedge m = f_2(x))$ it suffices to show the following: For all j such that $g(m) + 1 \leqslant j$ we have $f_2(j) \neq m$. Assume therefore $g(m) + 1 \leqslant j$ and choose n in T such that $j \leqslant n$. $g(m)$ and n being elements of T such that $g(m) < n$ we have

$$\sum_{k=0}^{g(m)} [2^{-f_1(k)} - 2^{-f_2(k)}] \leqslant \sum_{k=0}^{n} [2^{-f_1(k)} - 2^{-f_2(k)}] \;;$$

therefore

$$\sum_{k=g(m)+1}^{n} 2^{-f_2(k)} \leqslant \sum_{k=g(m)+1}^{n} 2^{-f_1(k)} .$$

If $g(m) + 1 \leqslant k$, then $m + 1 \leqslant f_1(k)$; the function f_1 being one-to-one, we have

$$\sum_{k=g(m)+1}^{n} 2^{-f_2(k)} \leqslant \sum_{j=m+1}^{q} 2^{-j} < \sum_{j=m+1}^{\infty} 2^{-j} = 2^{-m}$$

(q being a sufficiently big number).

From

$$\sum_{k=g(m)+1}^{n} 2^{-f_2(k)} < 2^{-m}$$

we conclude $m \neq f_2(k)$ for all k such that $g(m) + 1 \leqslant k \leqslant n$; as n can be chosen arbitrarily large, we finally have $m_2 \neq f_2(k)$ for all k such that $g(m) + 1 \leqslant k$.

6. The known proofs of Ramsey's theorem show that there exist arithmetical R-homogeneous sets for recursive relations R. A somewhat careful analysis of one of these proofs establishes the fact that there exists such a set in the class $\forall\exists\forall \cap \exists\forall\exists$.

Problem. What is the first class C in the arithmetical hierarchy such that for every recursive (binary symmetric) relation R there is a R-homogeneous set in C? (Editors' note: this problem has been completely answered by Jockusch – see [4] and [5].)

References

[1] F.P. Ramsey, On a problem in formal logic, Proc. London Math. Soc. 30 (1930) 264–286.

[2] R.M. Friedberg, Two recursively enumerable sets of incomparable degrees of unsolvability, Proc. Natl. Acad. Sci. U.S. 43 (1957) 236–238.

[3] A.A. Mučnik, On the unsolvability of the problem of reducibility in the theory of algorithms, Dokl. Akad. Nauk. SSSR (N.S.) 108 (1956) 194–197.

[4] C.J. Jockusch, Ramsey's theorem on recursion theory (to appear).

[5] C.E.M. Yates, A note on sets of indiscernibles, (these proceedings).

A NOTE ON ARITHMETICAL SETS OF INDISCERNIBLES

C.E.M. YATES
Manchester University, England

Specker has proved that there is a recursive partition (of the set **N** of natural numbers) which possesses no infinite recursively enumerable (Σ_1^0) set of indiscernibles; the existence of some infinite set of indiscernibles is simply the familiar theorem of Ramsey [3]. Specker's proof depends on the solution of Post's Problem by making use of two Σ_1^0 sets of incomparable degrees of unsolvability. Our main purpose in this note is to present (in Section 1) an entirely different proof based on the theory of retraceable sets.[1]. A shorter proof has been obtained more recently by Jockusch [1], who in fact obtains a stronger theorem (replacing Σ_1^0 by Σ_2^0) discussed at the end of Section 1. Specker observes at the end of his paper that the usual proof of Ramsey's theorem guarantees the existence of an infinite Δ_3^0 set of indiscernibles (for a recursive partition) and asks to what extent this can be improved. In terms of arithmetical classification this question has been completely answered by Jockusch by means of the negative result just mentioned combined with the positive result that Δ_3^0 can be improved to Π_2^0. We present our own proof of this second result in Section 2 to serve as a focal point for the discussion there. Once again it depends on an elementary 'priority argument'.

As usual, a set is called Σ_n^0, Π_n^0, Δ_n^0 if it is definable in prenex normal form with n alternating quantifiers where the first quantifier can be chosen to be respectively existential, universal or both. We use $\mathbf{P}_n(S)$ for the set of all n-element subsets of a set $S \subseteq N$. A *k-partition* of $\mathbf{P}_n(S)$ is simply a sequence of k disjoint sets whose union is $\mathbf{P}_n(S)$. If $A = (A_0, ..., A_k)$ is a k-partition of $\mathbf{P}_n(S)$ and $I \subseteq S$ then I is a *set of indiscernibles for A* if $\mathbf{P}_n(I) \subseteq A_j$ for some $j \leqslant k$ [2]. For other background concepts the most useful

[1] I first became aquainted with Specker's theorem at his Manchester talk in March 1966 and soon afterwards discovered the different proof expounded in Section 1. I am grateful to Georg Kreisel for informing me of Jockusch's work and relivening my interest in this topic, and also to Jockusch for various informative and critical comments.

[2] Specker uses the term 'R-homogeneous set' in his contribution to this volume [5].

reference is Rogers' book [4] ; in particular, there is some lengthy discussion of retraceability in the exercises to Chapter 9 of [4] , and in combination with [2] , [6] and [7] this should be more than adequate. For convenience, however, we recall the basic definitions. A set X (with elements $x_0, x_1, \ldots,$ in their natural order) is *retraceable* if there is a partial recursive function F such that $F(x_{n+1}) = x_n$ for all n and $F(x_0) = x_0$; F is said to retrace X. A partial recursive function is a *retracing function* if it retraces at least one infinite set. There is no loss of generality for most purposes in assuming a retracing function F to be *normalised* (or *special* as it has previously been called): i.e. $\mathrm{im}(F) \subseteq \mathrm{dom}(F)$ and $F(x) \leqslant x$ for all x.

Our notation will differ somewhat from [4] . We use R_e to denote the eth Σ_1^0 set in some standard recursive enumeration of these sets, and R_e^s to denote the finite subset of R_e enumerated at stage s in some standard recursive approximation to R. Small Greek letters such as σ and τ will be used to denote *strings*, i.e. finite sequences of integers. For our present purposes all strings may be supposed to be strictly monotonic. The length $\mathrm{lh}(\sigma)$ of a string σ is the number of elements in it, and for any two strings σ and τ we use $\sigma * \tau$ to represent the concatenation of σ and τ, i.e. the string obtained by listing the elements of τ after those of σ. We use $\emptyset$ to denote the null string (this should cause no confusion with the null set). Finally, for any function F we use $F[n]$ to denote the string $(F(0), \ldots, F(n))$; this notation is also used for the corresponding substrings of strings.

1. The proof we give of Specker's theorem depends on the existence of a retracing function F which retraces exactly one infinite set X and which is such that both X and $\overline{X}$ are immune. The existence of such a function F was first proved by McLaughlin [2] , and its construction involves an interesting elementary 'priority argument' which is roughly of the same level as that used to solve Post's Problem. In order to make this note self-contained and because this theorem is probably not well-known, we begin this section with a short proof of it which uses the e-state method of presentation.

Recall that a set is *immune* if it is infinite but has no infinite Σ_1^0 subset. It is easy to prove that any retraceable set is either recursive or immune and so it will be sufficient to arrange that $\overline{X}$ is immune. Also, it is easy to see that if F retraces an infinite set X such that $\overline{X}$ is immune then F can not retrace any other infinite set; for, if r_0 belongs to such another infinite set, then $\{y : (\exists z)(F^{(z)}(y) = r_0)\}$ is an infinite Σ_1^0 subset of $\overline{X}$. Hence it is sufficient to prove:

Theorem 1.1 (McLaughlin). There is a (finite-one) retracing function F which retraces a set X such that $\overline{X}$ is immune.

Proof. F will be defined as the limit of a recursive sequence of finite functions $\{F^s\}$, the construction of F^s proceeding stage-by-stage alongside the approximation $\{R_e^s\}$ to $\{R_e\}$. Before describing the construction of F^s we shall need some preliminary definitions.

First, an *e-state* μ is a string $(e_0, ..., e_k)$ with $e_0 < ... < e_k \leqslant e$. An e-state μ is *higher* than an e-state μ' if the least number in one but not the other belongs to μ. Notice of course that there are only finitely many e-states for any particular e.

Secondly, a string σ *satisfies e at stage s* if $\sigma(x) \in R_e^s$ for some $x \in \mathrm{dom}(\sigma)$. More generally, σ *satisfies an e-state* $\mu = (e_0, ..., e_k)$ *at stage s* if for each $j \leqslant k$ there is an x_j such that $\sigma(x_j) \in R_{e_j}^s$.

Lastly, a string σ is *eligible* if it is strictly monotonic and $\sigma(x) \geqslant 2x$ for all x.

Now, we can turn to the definition of F. The essential part of this definition is in fact the definition of an eligible string σ^s. Let σ^0 and F^0 both be empty. For $s > 0$ there are two cases.

Case 1: there is a number e and an eligible string σ which satisfies a higher e-state than σ^{s-1} at stage s, and *either* (a) $\sigma = \sigma^r$ for some $r < s$ *or* (b) $\sigma = \sigma^{s-1} * `k'$ for some $k \notin \mathrm{dom}(F^{s-1})$.

In this case we let e^s be the least such e and σ^s the first string (in some standard enumeration of the strings) corresponding to e^s.

Case 2: otherwise. We leave $\sigma^s = \sigma^{s-1}$.

Having defined σ^s it is easy to define F^s: it is simply the smallest function extending F^{s-1} which retraces the elements of σ^s. The existence of F^s is made possible by the condition $k \notin \mathrm{dom}(F^{s-1})$ in Case 1(b) above (which is the only case in which $F^s \neq F^{s-1}$). Hence it is easy to see by induction on s that $F^{s-1} \subseteq F^s$ for all $s > 0$. The required function F is the union of the sequence $\{F^s\}$.

The set X which we claim has the properties stated in the theorem is in a sense the limit of the sequence $\{\sigma^s\}$. Before this can be made clear, we need the simple but important lemma that follows. Notice that among the e-states which are satisfied at various stages, there must be a highest one μ_e. Let $s(e)$ be the first stage s at which σ^s satisfies μ_e.

Lemma A. $\sigma^s \supseteq \sigma^{s(e)}$ for all $s \geqslant s(e)$.

Proof. Suppose there is, for *reductio ad absurdum*, a least stage $s_0 \geqslant s(e)$ such that $\sigma^s \not\supseteq \sigma^{s(e)}$. Then there exists an f such that σ^{s_0} satisfies a higher f-state than σ^{s_0-1} at stage s_0. Moreover, $f > e$ because $\sigma^{s_0-1} \supseteq \sigma^{s(e)}$. Now, there is no k such that $\sigma^{s_0} = \sigma^{s_0-1} * `k'$, and so $\sigma^{s_0} = \sigma^r$ for some $r < s_0$. But r can not be $\geqslant s(e)$ because then $\sigma^{s_0} \supseteq \sigma^{s(e)}$, and also r can not be $< s(e)$ because each σ^r for $r < s(e)$ satisfies a lower e-state and hence a lower f-state (since $f > e$) than $\sigma^{s(e)}$ and σ^{s_0-1}. This is the required contradiction.

Let $\sigma_e = \sigma^{s(e)}$. It follows from Lemma A that $\sigma_0 \subseteq \sigma_1 \subseteq \ldots$; let X be the range of $\lim_e \sigma_e$. Then X is infinite because σ_e is eligible for each e. It is not, however, immediately clear that X is infinite. This follows from the immunity of $\overline{X}$ which in turn follows from the next lemma.

Lemma B. For each e, if R_e is infinite then σ_e satisfies e.

Proof. Suppose, for *reductio ad absurdum*, that e is not an element of μ_e. Let $s_0 \geqslant s(e)$ be such that there exist a number k_0 in $R_e^{s_0}$ which exceeds $2\mathrm{lh}(\sigma_e)$ and all the elements of σ^s for $s < s(e)$. We claim that σ^{s_0} satisfies a higher e-state than μ_e, which is the required contradiction. For, either σ^r satisfies a higher e-state than σ_e, for some $r < s$, or σ^{s_0} is inevitably defined to be $\sigma_e * `k_0'$. To see this, we have only to verify that $k_0 \notin \mathrm{dom}(F^{s_0-1})$. But if $k_0 \in \mathrm{dom}(F^{s_0-1})$ then k_0 belongs to σ^r for some r such that $s(e) \leqslant r < s_0$; in which case σ^r satisfies a higher e-state than μ_e.

It follows immediately from Lemma B that $\overline{X}$ is immune, and in particular that X is infinite. It is clear that X is retraced by F since σ_e is retraced by $F^{s(e)}$ for each e. This concludes the proof of Theorem 1.1.

As we observed before Theorem 1.1, we can immediately deduce the stronger result we need:

Corollary 1.2. There is a (finite-one) retracing function which retraces exactly one infinite set X and is such that both X and $\overline{X}$ are immune.

Now, we turn to deducing Specker's theorem.

Definition 1.3. Let F be a retracing function. The *partition of* $\mathbf{P}_2$ *(im(F)) associated with F* is $\{A_0, A_1\}$ where

$$\{x, y\} \in A_0 \leftrightarrow (\exists z)_{z \leqslant \max\{x,y\}}(F^{(z)}(x) = y \vee F^{(z)}(y) = x)\,,$$

$$\{x, y\} \in A_1 \leftrightarrow \{x, y\} \notin A_0\,.$$

Notice that any subset of a set retraced by F is a set of indiscernibles for the associated partition.

Theorem 1.4. Let F be a retracing function which retraces exactly one infinite set X and is such that both X and $\overline{X}$ are immune. Then the partition of $\mathbf{P}_2(\mathrm{im}(F))$ associated with F has no infinite Σ_1^0 set of indiscernibles.

Proof. Let $A = \{A_0, A_1\}$ be the partition of $\mathbf{P}_2(\mathrm{im}(F))$ associated with F. If I is an infinite set of indiscernibles for A then either $I \subseteq X$ or I differs from a subset of X by one element; for, if $\mathbf{P}_2(I) \subseteq A_0$ then $J = \{F^{(z)}(x) : x \in I\}$ is an infinite set retraced by F, and if $\mathbf{P}_2(I) \subseteq \overline{A}$, then I cannot contain more than one element of X. Since X and $\overline{X}$ are both immune, the theorem follows.

Corollary 1.5 (Specker). There exists a recursive partition $\{B_0, B_1\}$ of $\mathbf{P}_2(\mathbf{N})$ which possesses no infinite Σ_1^0 set of indiscernibles.

Proof. Let F be a retracing function having the properties required by Theorem 1.4; its existence was proved in Corollary 1.2. Let G be a one-one recursive function whose range is $\mathrm{im}(F)$. Define

$$\{x, y\} \in B_i \leftrightarrow \{G(x), G(y)\} \in A_i \qquad i = 0, 1$$

where $A = \{A_0, A_1\}$ is the partition associated with F. Then $B = \{B_0, B_1\}$ is a recursive partition of $\mathbf{P}_2(\mathbf{N})$ because G is one-one. Let I be an infinite set of indiscernibles for B. It follows that if

$$I' = \{G(x) : x \in I\}$$

the I' is an infinite set of indiscernibles for A. Clearly, if I is Σ_1^0, then I' is Σ_1^0; we deduce that I is not Σ_1^0.

Jockusch has announced [1] a stronger theorem, namely that there is a recursive partition of $\mathbf{P}_2(\mathbf{N})$ which has no infinite Σ_2^0 set of indiscernibles. Unfortunately, the present method can not be used to deduce this stronger theorem, because if F retraces X and F is *not* finite-one then for some x, y such that $F(y) = x$ the set $\{u : F(u) = x \;\&\; u \neq y\}$ is an infinite Σ_1^0 subset of $\overline{X}$. Also, by Theorem 5.2 of our paper [7], if a finite-one retracing function retraces a unique infinite set then that set must be of degree $\leqslant \mathbf{0}^{(1)}$. These two observations make it unlikely that any significant extension of this particular method will be found.

2. As Specker observes in [5], an analysis of the usual proof of Ramsey's Theorem reveals that any partition A of $\mathbf{P}_2(\mathbf{N})$ possesses a set of indiscer-

nibles which is Δ_3^0 in A. In the present section we give a proof of Jockusch's recent observation that Δ_3^0 can be replaced by Π_2^0. The proof is a nice example of a special procedure for reducing the quantifier-form of a set having certain prescribed properties. This general procedure is usually referred to as 'the priority method', although this now familiar nomenclature is rather inappropriate and could be more accurately replaced by 'the approximation method' or 'the trial-and-error method'.

It will probably be helpful to first sketch the proof for Δ_3^0. So let $A = \{A^0, A^1\}$ be a partition of $\mathbf{P}_2(\mathbf{N})$. We make an approximation $\{I_e\}$ to the required set I of indiscernibles as follows. I_e is defined by induction on e, simultaneously with a partition $A_e = \{A_e^0, A_e^1\}$ of I_e. Letting $I_0 = \mathbf{N}$ we define

$$A_e^j = \{x : x \in I_e \; \& \; \{I_e(0), x\} \in A^j\} \qquad j = 0, 1,$$

where $I_e(0), I_e(1), ...,$ is a listing of the elements of I_e in their natural order. Then we define:

$$J(e) = \begin{cases} 0 & \text{if } A_e^0 \text{ is infinite} , \\ 1 & \text{otherwise} , \end{cases}$$

and let $I_{e+1} = A_e^{J(e)}$. It is clear that I_e is uniformly of degree $\leqslant \mathbf{a}^{(2)}$ where $\mathbf{a}$ is the degree of A, but we can say no more than this (equivalently, I_e is uniformly Δ_3^0 in A). Let $E_0(0), E_0(1), ...,$ and $E_1(0), E_1(1), ...,$ be listings of those e such that $J(e) = 0$ and $J(e) = 1$ respectively. If for each $j \in \{0, 1\}$ we set

$$I^j(x) = I_{E_j(x)}(0)$$

for all x, and let $I^j = \{I^j(0), I^j(1), ...\}$, then it is easy to see that I^0, I^1 are sets of indiscernibles for A which are of degree $\leqslant \mathbf{a}^{(2)}$ (equivalently, belong to $\Delta_3(\mathbf{a})$). At least one of I^0, I^1 must be infinite.

Using the extension of the Kreisel–Shoenfield basis theorem which we mentioned in [7], it is possible to obtain a set of indiscernibles of degree $< \mathbf{a}^{(2)}$.

Theorem 2 (Jockusch). Let A be any 2-partition of $\mathbf{P}_2(\mathbf{N})$. Then A possesses an infinite set of indiscernibles which is Π_2^0 in A.

Proof. Let $A = \{A^0, A^1\}$ be a 2-partition of $\mathbf{P}_2(\mathbf{N})$. We make a *double* approximation $\{I_{ef}\}$ to the required set I of indiscernibles; the sequence $\{I_{ef}\}$ will be uniformly of degree $\leqslant \mathbf{a}^{(1)}$ where $\mathbf{a}$ is the degree of A. It is this extra approximation which enables us to reduce the quantifier-form of I. For each e, $\lim_f I_{ef}$ will exist, and letting $I_e = \lim_f I_{ef}$ we then form I^0, I^1 from the sequence $\{I_e\}$ in the usual way as outlined before this theorem. I^0 will $\in \Pi_2^0(\mathbf{a})$ and if I^0 is finite then I^1 will be infinite and $\in \Pi_2^0(\mathbf{a})$, so one of these two sets satisfies our demands.

We define I_{ef} for each f by induction on e, simultaneously defining a 2-partition $A_{ef} = \{A_{ef}^0, A_{ef}^1\}$ of I_{ef}. Let $I_{0f} = \mathbf{N}$ for all f and define

$$A_{ef}^j = \{x : x \in I_{ef} \,\&\, \{I_{ef}(0), x\} \in A^j\}$$

for each $j \in \{0, 1\}$, where $I_{ef}(0), I_{ef}(1), ...,$ is a listing of the elements of I_{ef} in their natural order. We next define

$$J(e,f) = \begin{cases} 0 & \text{if } (\exists y)_{y \geqslant f} \quad (y \in A_{ef}^0)\,, \\ 1 & \text{otherwise}\,. \end{cases}$$

Note that if $e \leqslant e'$ then $I_{e'f} \subseteq I_{ef}$ for all f.

Clearly, for each e if $\lim_f I_{ef}$ exists then $J(e, f)$ and $I_{e+1,f}$ can change at most once for $f \geqslant F(e)$, where $F(e)$ is the least f such that $I_{ef} = \lim_f I_{ef}$. It is important to note that F is monotonic, as can be verified by careful examination of the construction. Since $\lim_f I_{0f}$ exists by definition, we conclude that $J(e)$ (= $\lim_f J(e, f)$) and I_e (= $\lim_f I_{ef}$) are both defined. Similarly, A_e^j (= $\lim_f A_{ef}^j$) exists for all e and $j \in \{0, 1\}$. Another important observation is that if $I_{ef} \neq I_{e,f-1}$ then $I_{ef}(0) \geqslant f$; this ensures that $I_e(0) \geqslant F(e)$.

Next, note that for each e, if I_{ef} is infinite then one or other of A_{ef}^0, A_{ef}^1 must be infinite, so that $I_{e+1,f}$ is infinite. Hence we conclude that I_e is infinite for all e.

Finally, it is not difficult to verify that J, $\{I_{ef}\}$ and $\{A_{ef}^j\}$ are uniformly of degree $\leqslant \mathbf{a}^{(1)}$.

We now define I^0, I^1 from $\{I_e\}$ as before. Let $E_0(0), E_0(1), ...,$ and $E_1(0), E_1(1), ...,$ be listings of those e such that $J(e) = 0$ and $J(e) = 1$ respectively (both taken in their natural order). For each $j \in \{0, 1\}$ we set

$$I^j(x) = I_{E_j(x)}(0)$$

for all x, and let $I^j = \{I^j(0), I^j(1), ...\}$. This concludes the construction.

Our first claim is that both I^0 and I^1 are sets of indiscernibles for A. We claim in fact that

$$\mathbf{P}_2(I^j) \subseteq A^j$$

for each $j \in \{0, 1\}$. To prove this, let $\{I^j(x_1), I^j(x_2)\}$ be a typical element of $\mathbf{P}_2(I^j)$ with $x_1 < x_2$. Then

$$I^j(x_2) = I_{E_j(x_2), F(E_j(x_2))}(0) \in I_{E_j(x_1)+1, F(E_j(x_2))}$$

because $I_{e', f} \subseteq I_{e, f}$ for all f and e, e' such that $e' \geqslant e$. But then it follows that $I^j(x_2) \in I_{E_j(x_1)+1}$ because $F(E_j(x_2)) \geqslant F(E_j(x_1)+1)$ and so $I_{E_j(x_1)+1} = I_{E_j(x_1)+1, F(E_j(x_2))}$. Hence $I^j(x_2) \in A^j_{E_j(x_1)}$ and so $\{I^j(x_1), I^j(x_2)\} = \{I_{E_j(x_1)}(0), I^j(x_2)\} \in A^j$, which is the desired result. Our second claim is that $I^0 \in \Pi^0_2(\mathbf{a})$ and that if I^0 is finite then $I^1 \in \Pi^0_2(\mathbf{a})$. The latter part of the claim follows from the membership in $\Pi^0_2(\mathbf{a})$ of I, where $I = I^0 \cup I^1 = \{I_e(0) : e \geqslant 0\}$. To see first that $I \in \Pi^0_2(\mathbf{a})$, observe that

$$y \in I \leftrightarrow (\exists e)_{e \leqslant y} (\exists f)_{f \leqslant y} (I_{ef}(0) = y$$
$$\& \, (\forall d)_{d \leqslant e} (\forall g)_{g > f} (J(d, g) = J(d, f))) \, .$$

Then to see that $I^0 \in \Pi^0_2(\mathbf{a})$, a slightly more complicated expression is necessary:

$$y \in I^0 \leftrightarrow (\exists e)_{e \leqslant y} (\exists f)_{f \leqslant y} (I_{ef}(0) = y$$
$$\& \, (\forall d)_{d \leqslant e} (\forall g)_{g > f} (J(d, g) = J(d, f))$$
$$\& \, (\forall h)_{h > f} (J(e, h) = 0)) \, .$$

(The final clause holds because, for $h > F(e)$, $J(e, h)$ can only change from 0 to 1, not from 1 to 0.) Either I^0 is infinite in which case it is the required set, or I^1 differs from I by a finite set and so $\in \Pi^0_2(\mathbf{a})$.

This is in a sense the best possible result because of Jockusch's other theorem mentioned in Section 1: there is a recursive partition with no infinite Σ^0_2 set of indiscernibles. There are, however, other ways in which one might seek to strengthen it. For example, one might hope to prove that if $\mathbf{b}$ is r.e. in $\mathbf{0}^{(1)}$ and $> \mathbf{0}^{(1)}$ then every recursive partition has an infinite set of indiscernibles which is of degree $\leqslant \mathbf{b}$. We can not see, however, how to adapt the

the proof above to the usual techniques that are appropriate for this sort of problem. A much fuller investigation of the various possible extensions of Theorem 2 can be found in [1].

References

[1] C.G. Jockusch, Ramsey's Theorem and Recursion Theory (to appear).
[2] T.G. McLaughlin, Splitting and decomposition by regressive sets, 11, Can. J. Math. 19 (1967) 291–311.
[3] F.P. Ramsey, On a problem in formal logic, Proc. Lond. Math. Soc. 30 (1930) 264–286.
[4] H. Rogers, Theory of recursive functions and effective computability (McGraw Hill, 1967).
[5] E. Specker, Ramsey's theorem does not hold in recursive set theory, this proceedings.
[6] C.E.M. Yates, Recursively enumerable sets and retracing functions, Z. Math. Logik und Grund. Math. 8 (1962) 331–345.
[7] C.E.M. Yates, Arithmetical sets and retracing functions, Z. Math. Logik und Grundlagen der Math. 13 (1967) 193–204.